Cell
BIOLOGY

Molecular and Cell
BIOCHEMISTRY

Cell

BIOLOGY

SMITH AND WOOD

CHAPMAN & HALL

University and Professional Division

London · Glasgow · New York · Tokyo · Melbourne · Madras

UK	Chapman & Hall, 2–6 Boundary Row, London SE1 8HN
USA	Chapman & Hall, 29 West 35th Street, New York NY10001
JAPAN	Chapman & Hall Japan, Thomson Publishing Japan, Hirakawacho Nemoto Building, 7F, 1-7-11 Hirakawa-cho, Chiyoda-ku, Tokyo 102
AUSTRALIA	Chapman & Hall Australia, Thomas Nelson Australia, 102 Dodds Street, South Melbourne, Victoria 3205
INDIA	Chapman & Hall India, R. Seshadri, 32 Second Main Road, CIT East, Madras 600 035

First edition 1992

Reprinted 1993 (with corrections)

© 1992 Chapman & Hall

Typeset in 10/11½pt Palatino by EJS Chemical Composition,
Midsomer Norton, Bath, Avon
Printed in Hong Kong

ISBN 0 412 40740 X

British Library Cataloguing in Publication Data

Cell biology.–(Molecular and cell biochemistry)
I. Smith, C.A. II. Wood, E.J. III. Series
574.8

ISBN 0–412–40740–X

Library of Congress Cataloging-in-Publication Data

Smith, C.A. (Chris A.)
Cell biology/C.A. Smith and E.J. Wood
p. cm.–(Molecules and cell biochemistry)
Includes bibliographical references and index.
ISBN 0–412–40740–X (pb.)
1. Cytology. I. Wood, Edward J., 1941– . II. Title.
III. Series.
QH581.2.S55 1991
574.87—dc20   91–12368
                                                                          CIP

Copy Editors: Sara Firman and Judith Ockenden
Sub-editor: Simon Armstrong
Production Controller: Marian Saville
Layout Designer: Geoffrey Wadsley (after an original design by Julia Denny)
Illustrators: Ian Foulis & Associates
Cover design: Amanda Barragry

Contents

This book is one of a series of brief fundamental texts for junior under-graduates and diploma students in biological science. The series, Molecular and Cell Biochemistry, covers the whole of modern biochemistry, integrating animal, plant and microbial topics. The intention is to give the series special appeal to the many students who read biochemistry for only part of their course and who are looking for an all-encompassing and stimulating approach. Although all books in the series bear a distinct family likeness, each stands on its own as an independent text.

Many students, particularly those with less numerate backgrounds, find elements of their biochemistry courses daunting, and one of our principal concerns is to offer books which present the facts in a palatable style. Each chapter is prefaced by a list of learning objectives, with short summaries and revision aids at the ends of chapters. The text itself is informal, and the incorporation of marginal notes and information boxes to accompany the main text give a tutorial flavour, complementing and supporting the main narrative. The marginal notes and boxes relate facts in the text to applicable examples in everyday life, in industry, in other life sciences and in medicine, and provide a variety of other educational devices to assist, support, and reinforce learning. References are annotated to guide students towards effective and relevant additional reading.

Although students must start by learning the basic vocabulary of a subject, it is more important subsequently to promote understanding and the ability to solve problems than to present the facts alone. The provision of imaginative problems, examples, short-answer questions and other exercises is designed to encourage such a problem-solving attitude.

A major challenge to both teacher and student is the pace at which biochemistry and molecular biology are advancing at the present time. For the teacher and textbook writer the challenge is to select, distil, highlight and exemplify, tasks which require a broad base of knowledge and indefatigable reading of the literature. For the student the challenge is not to be over-whelmed, to understand and ultimately to pass the examination! It is hoped that the present series will help by offering major aspects of biochemistry in digestible portions.

This vast corpus of accumulated knowledge is essentially valueless unless it can be used. Thus these texts have frequent, simple exercises and problems. It is expected that students will be able to test their acquisition of knowledge but also be able to use this knowledge to solve problems. We believe that only in this way can students become familiar and comfortable with their knowledge. The fact that it is useful to them will mean that it is retained, beyond the last examination, into their future careers.

The present series was written by lecturers in universities and former polytechnics who have many years of experience in teaching, and who are also familiar with current developments through their research interests. They are, in addition, familiar with the difficulties and pressures faced by present-day students in the biological sciences area. The editors are grateful for the co-operation of all their authors in undergoing criticism and in meeting requests to re-write (and sometimes re-write again), shorten or extend what they originally wrote. They are also happy to record their grateful thanks to those many individuals who very willingly supplied illustrative material promptly and generously. These include many colleagues as well as total strangers whose response was positive and unstinting. Special thanks must go to the assessors who very carefully read the chapters and made valuable suggestions which gave rise to a more readable text. Grateful thanks are also due to the team at Chapman & Hall who saw the project through with good grace in spite, sometimes, of everything. These include Dominic Recaldin, Commissioning Editor, Jacqueline Curthoys, formerly Development Editor, Simon Armstrong, Sub-editor, and Marian Saville, Production Controller.

Finally, though, it is the editors themselves who must take the responsibility for errors and omissions, and for areas where the text is still not as clear as students deserve.

Contributors

DR R.G. BARDSLEY   *Department of Applied Biochemistry and Food Sciences, University of Nottingham, Nottingham, UK.   Chapter 10.*

DR M.M. DAWSON   *Department of Biological Sciences, the Manchester Metropolitan University, Manchester, UK.   Chapter 11.*

PROFESSOR F.M. GOÑI   *Universidad del Pais Vasco, Bilbao, Spain. Chapters 2 and 9.*

PROFESSOR M. GRIFFIN   *Department of Life Sciences, Nottingham Trent University, Nottingham, UK.   Chapter 4.*

DR P.J. HANSON   *Aston University, Birmingham, UK.   Chapter 8.*

DR P. KUMAR   *Department of Biological Sciences, the Manchester Metropolitan University, Manchester, UK.   Chapter 6.*

PROFESSOR J.M. MACARULLA   *Universidad del Pais Vasco, Bilbao, Spain. Chapters 2 and 9.*

DR N.K. PACKHAM   *Department of Biochemistry and Genetics, Newcastle University, Newcastle, UK.   Chapter 5.*

DR C.A. SMITH   *Department of Biological Sciences, the Manchester Metropolitan University, Manchester, UK.   Chapters 1 and 3.*

DR I. THOMSON   *Department of Biological Sciences, Napier University, Edinburgh, UK.   Chapter 12.*

DR D. WEST   *Department of Immunology, University of Liverpool, UK. Chapter 7.*

DR E.J. WOOD   *Department of Biochemistry and Molecular Biology, University of Leeds, Leeds, UK.   Chapter 3.*

Preface

The book offers students an introduction to the vast body of knowledge that constitutes cellular biology. It differs from typical biology texts in that, although it is largely descriptive, it is written from the point of view from biochemistry and molecular biology. This is, indeed, a trend in the whole of biology today, which is endeavouring to understand biological phenomena in chemical terms. Such a trend implies a familiarity with the chemical structure making up the fabric of cells and with the ways in which cell chemicals interact to produce the phenomenon called life. Although, of course, we would like to know a great deal more, it must be said that scientists have been reasonably successful in understanding many of the fundamental processes that, only a generation ago, had seemed wholly intractable.

The first two chapters offer a view of eukaryotic cells, prokaryotic cells and viruses to set the scene. The inclusion of viruses, which are not cells, may seem perverse. However, aside from any interest in their pathology, viruses (and bacteriophages) have been of inestimable importance in the development of biochemistry and molecular biology. These relatively simple systems have helped biologists to understand the workings of more complex organisms and have also provided the biological tools to engineer cellular activities for human purposes.

The following four chapters are concerned with the inner workings of cells. Chapter 3 describes what is known about the prodigious feat of packing the vast length of the DNA double helix into a relatively tiny cell nucleus, while still allowing the genes to be accessible for transcription and control. Chapters 4 and 5 look at membranes and at the two main energy-producing organelles, the mitochondria and the chloroplasts, which depend on the properties of their membranes to carry out their functions. Chapter 6 deals with the largely proteinaceous structures that form the cytoskeleton, in which there have been important advances in our knowledge in recent years. In contrast with the internal skeleton of cells, Chapter 7 is concerned with the extracellular matrix. Until a few years ago the extracellular matrix was thought to consist of inert supporting materials, but modern studies have revealed it to be a highly dynamic structure crucial to the activities of the cell.

The remaining five chapters considers cells at a more specialized level. Chapter 8 addresses the question of how cells communicate with each other to achieve unity of purpose, and the subject of Chapter 9, nerve cells, is used to illustrate an extreme of cell specialization as an adaptation to function. Chapter 10 on muscle contraction continues the theme of specialization and considers the structure and activities of various types of muscle cells. Chapter 11, on the immune response, examines the concepts of biological defence and recognition of self and non-self through the activities of the specialized cells of the immune system. The final chapter on differentiation and development could, today, form the topic of a whole volume on its own. It asks, and tries to answer, the question of how cells differentiate into the wide range of structures we see in multicellular organisms with their division of labour and their exquisite control of cellular activities.

A principal aim of this book is to convey the idea of unity of structure and purpose at cellular level. Despite the enormous range, and beauty, of the multifarious cell types, each is but a variation of the same basic plan. A nerve cell and a muscle cell, for example, despite their obvious differences in structure and purpose, both contain similar sets of organelles, need to be bound within membranes, and depend on the same biochemical reactions for their survival.

Abbreviations

A	adenine (alanine)
ACP	acyl carrier protein
ACTH	adrencorticotrophic hormone
ADP	adenosine diphosphate
Ala, A	alanine
AMP	adenosine monophosphate
cAMP	adenosine 3′,5′-cyclic monophosphate
Arg, R	arginine
Asn, N	asparagine
Asp, D	aspartic acid
ATP	adenosine triphosphate
ATPase	adenosine triphosphatase
C	cytosine (cysteine)
CDP	cytidine diphosphate
CMP	cytidine monophosphate
CTP	cytidine triphosphate
CoA, CoASH	coenzyme A
CoQ, Q	coenzyme Q, ubiquinone
Cys, C	cysteine
d-	2-deoxy-
D	aspartic acid
d-Rib	2-deoxyribose
DNA	deoxyribonucleic acid
cDNA	complementary DNA
e^-	electron
E	glutamic acid
E	oxidation–reduction potential
F	phenylalanine
F	the Faraday (9.648×10^4 coulomb mol^{-1})
FAD	flavin adenine dinucleotide
Fd	ferredoxin
fMet	N-formyl methionine
FMN	flavin mononucleotide
Fru	fructose
g	gram
g	acceleration due to gravity
G	guanine (glycine)
G	free energy
Gal	galactose
Glc	glucose

Gln, Q	glutamine
Glu, E	glutamic acid
Gly, G	glycine
GDP	guanosine diphosphate
GMP	guanosine monophosphate
GTP	guanosine triphosphate
H	histidine
H	enthalpy
Hb	haemoglobin
His, H	histidine
Hyp	hydroxyproline (HOPro)
I	isoleucine
Ig G	immunoglobulin G
Ig M	immunoglobulin M
Ile, I	isoleucine
ITP	inosine triphosphate
J	Joule
K	degrees absolute (Kelvin)
K	lysine
L	leucine
Leu, L	leucine
ln x	natural logarithm of x $= 2.303 \log_{10} x$
Lys, K	lysine
M	methionine
M_r	relative molecular mass, molecular weight
Man	mannose
Mb	myoglobin
Met, M	methionine
N	asparagine
N	Avogadro's number (6.022×10^{23})
N	any nucleotide base (e.g. in NTP for nucleotide triphosphate)
NAD^+	nicotinamide adenine dinucleotide
$NADP^+$	nicotinamide adenine dinucleotide phosphate
P	proline
Pi	inorganic phosphate
PPi	inorganic pyrophosphate
Phe, F	phenylalanine
Pro, P	proline
Q	coenzyme Q, ubiquinone
Q	glutamine
R	arginine
R	the gas constant ($8.314 \, J \, K^{-1} \, mol^{-1}$)
Rib	ribose
RNA	ribonucleic acid

mRNA	messenger RNA
rRNA	ribosomal RNA
tRNA	transfer RNA
s	second
s	sedimentation coefficient
S	Svedberg unit (10^{-13} second)
S	serine
SDS	sodium dodecylsulphate
Ser, S	serine
T	thymine
T, Thr	threonine
TPP	thiamine pyrophosphate
Trp, W	tryptophan
TTP	thymidine triphosphate (dTTP)
Tyr, Y	tyrosine
U	uracil
UDP	uridine diphosphate
UDP-Glc	uridine diphosphoglucose
UMP	uridine monophosphate
UTP	uridine triphosphate
V	valine
V	volt
Val, V	valine
W	tryptophan
Y	tyrosine

Greek alphabet

A	α	alpha	N	ν	nu
B	β	beta	Ξ	ξ	xi
Γ	γ	gamma	O	o	omicron
Δ	δ	delta	Π	π	pi
E	ε	epsilon	P	ϱ	rho
Z	ζ	zeta	Σ	σ	sigma
H	η	eta	T	τ	tau
Θ	θ	theta	Y	υ	upsilon
I	ι	iota	Φ	ϕ	phi
K	κ	kappa	X	χ	chi
Λ	λ	lambda	Ψ	ψ	psi
M	μ	mu	Ω	ω	omega

1

Cells:
an introduction

Objectives

After reading this chapter you should be able to:

☐ describe the advances made over the last 300 years in studying cell structure and functions;

☐ outline the theory and uses of a variety of types of light microscopes;

☐ evaluate the advantages and disadvantages of using electron microscopy in biology;

☐ discriminate between organisms on the basis of their cellular structure;

☐ describe the structure and functions of organelles from eukaryotic cells;

☐ summarize the major methods of disrupting and fractionating cells and tissues.

1.1 Introduction

The first use of the word *cell* is attributed to Hooke in 1665. Using a primitive optical microscope Hooke noticed that cork was composed of many regular structures, which reminded him of little rooms (Fig. 1.1). The application of microscopy to the study of biological materials over the next 200 years revealed that all observable organisms were composed of cells. However, the idea that the cell is the *basic structural unit* of all organisms was not formally proposed until 1839 by Schleiden and Schwann. That the cell is the basic unit of life was restated by Virchow (1855): *omnis cellula e cellula* (every cell comes from a cell).

Organisms composed of more than one cell are described as **multicellular**; an organism composed of a single cell would be called **unicellular** or **acellular**. Unicellular organisms are necessarily complex biochemically. An *Amoeba* (Fig. 1.2a) needs to be able to carry out *all* the biochemical operations involved in the maintenance and reproduction of the organism. In multicellular organisms, individual cells may be specialized for a restricted number of roles (Fig. 1.2b). This specialization pays dividends in efficiency. Individual cells of multicellular organisms are biochemically simpler than free-living single-celled organisms.

Tremendous advances have been made in the last 30–40 years in understanding the detailed structure and complex functioning of the cell. Many of these advances have been made possible by the use of techniques and instrumentation that enable cells to be disrupted and the functions of their individual components analysed, besides the advances made in microscopy which have allowed the fine structure of cells to be examined in

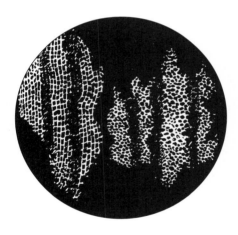

Fig. 1.1 A specimen of cork as drawn by Hooke.

---

cell: from the Latin for small room.

Reference Alberts, B., Bray, D., Lewis, J., Raff, M., Roberts, K. and Watson, J.D. (1989) *Molecular Biology of the Cell*, 2nd edn, Garland, New York, USA. Full of information and a wealth of figures and photographs.

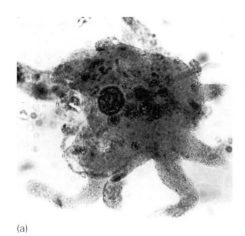
(a)

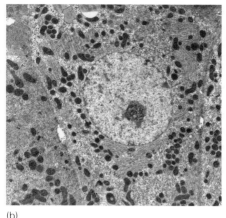
(b)

Fig. 1.2 (a) Unicellular organism: *Amoeba* sp. (×240). Courtesy of M.J. Hoult, Department of Biological Sciences, Manchester Polytechnic, UK. (b) Multicellular organism: rat liver cells (×2560). Courtesy of P.L. Carter, Department of Biological Sciences, The Manchester Metropolitan University, UK.

□ A simple microscope contains only a single lens; a compound microscope employs a system of lenses to produce an image.

□ Chromatic aberration is the failure of a lens bring light rays of different wavelengths to the same focus producing a spectrum. The lens acts partly as a prism.

detail. Because of the importance of microscopy this chapter commences with a basic description of this technique.

1.2 Microscopy

Early microscopists, such as van Leeuwenhoek (1632–1723), used a simple microscope (Fig. 1.3) while others such as Hooke employed primitive compound microscopes (Fig. 1.4). The latter suffered from **chromatic aberration** because only simple lenses were employed. Following the introduction of improved achromatic microscopes by Lister in 1827, light microscopy has developed to extremely high standards.

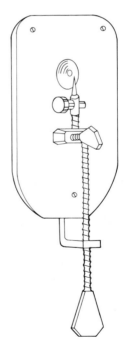

Fig. 1.3 van Leeuwenhoek simple microscope. The specimen was mounted on the needle and its position adjusted by means of the screw threads. Redrawn from Bradbury, S. (1984) *The Optical Microscope in Biology*, Edward Arnold, London, UK.

Fig. 1.4 Compound microscope of the type used by Hooke.

Modern light microscopy

Modern light microscopes can produce images of specimens to about ×2000 magnification. Greater magnification is limited by the **resolution** achievable with visible light. The resolving power of a lens is the closeness two objects can be in proximity to one another and still be perceived as separate structures. The resolution (r) of a microscope, using light of wavelength λ, is defined as:

$$r = \frac{0.61\lambda}{NA}$$

Reference Darnell, J., Lodish, H. and Baltimore, D. (1990) *Molecular Cell Biology*, 2nd edn, Scientific American Books, New York, USA. Another good book on cell biology.
Reference de Duve, C. (1984) *A Guided Tour of the Living Cell*, Vols 1 & 2, Scientific American Books, New York, USA. An idiosyncratic approach to describing the structure and functions of the cell. Many varied and beautiful illustrations, a considerable number designed to show that art imitates nature.

where NA is the numerical aperture of the lens. NA is the product of the sine of the angular aperture, u (Fig. 1.5), and the refractive index, n, of the medium through which the rays pass: usually air, but oil may be used. In practice, the maximum NA available for light microscopes is about 1.4 and thus the limit of resolution for visible light of wavelength 560 nm is about 0.24 μm. Two objects closer than this will be observed as a single structure.

CONVENTIONAL LIGHT MICROSCOPES are systems of lenses designed to magnify the specimen in two stages, and to give an even and adequate illumination of the specimen. Figure 1.6 shows how a magnified image of the specimen is produced. The objective lens forms a real, enlarged image within the barrel of the microscope. This image is the object of the eyepiece lens, which forms a second, virtual image of the specimen. This final image is magnified according to the product of the magnifying power of the two lenses. The eyepiece lens can only magnify the image formed by the objectives lens: it cannot improve upon its resolution.

Objective lens

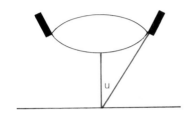

Fig. 1.5 Definition of the angle u.

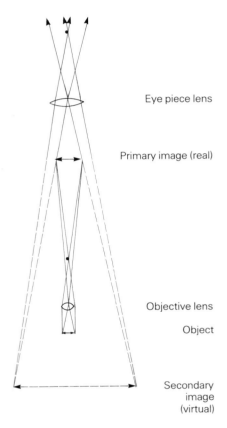

Eye piece lens

Primary image (real)

Objective lens

Object

Secondary image (virtual)

Fig. 1.6 Image formation in a compound microscope. Redrawn from Hopkins, C.R. (1978) *Structure and Functions of Cells*. W.B. Saunders.

The specimen in an optical microscope is usually illuminated by a light beam that is modulated by a series of lenses and diaphragms. The substage condenser lens provides an angular cone of light which just fills the angular aperture of the objective lens (Fig. 1.7) and so the NA of both objective and condenser lenses must be closely matched. Beneath the condenser lens is the iris diaphragm which is adjustable and is used to limit the angle of the cone of light leaving the condenser. The iris diaphragm must be adjusted correctly for each objective lens if that lens is to achieve optimum resolution. If the

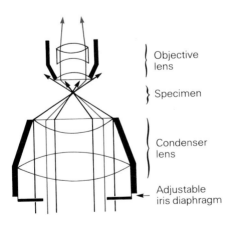

Objective lens

Specimen

Condenser lens

Adjustable iris diaphragm

Fig. 1.7 Use of iris diaphragm to illuminate objective lens. The coloured rays show the normal adjustment. Redrawn from Hopkins, C.R. (1978) *Structure and Functions of Cells*. W.B. Saunders.

Exercise 2

What method(s) should be used to control light intensity with a light microscope?

aperture is opened to its maximum extent, giving the maximum NA for the objective lens, excessive glare will result. However, if the aperture is not opened sufficiently, the full NA and hence resolution, will not be achieved. Thus a balance between optimal illumination and maximum resolution must be selected. Typically the condenser is usually set to illuminate two-thirds to three-quarters of the aperture of the objective lens in use (Fig. 1.7).

Conventional bright-field light microscopy is by far the most widely used type of light microscopy employed. However, many other light microscopy techniques are available.

DARK-FIELD MICROSCOPY, in its simplest form, uses a central patch stop in the condenser to prevent illumination of the specimen by direct light (Fig. 1.8a). The image appears bright against a dark background and unstained specimens viewed against a dark field show high contrast (Fig. 1.8b). However, the resolution is only as good, and, indeed, sometimes worse than with conventional light microscopy.

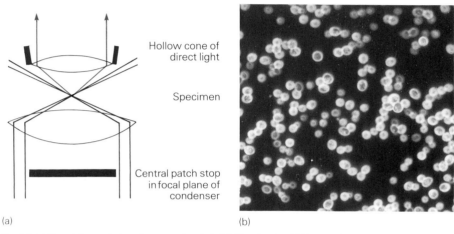

(a)   Hollow cone of direct light   Specimen   Central patch stop in focal plane of condenser   (b)

Fig. 1.8 (a) Use of the central patch stop to give dark field illumination. (b) Yeast cells viewed by dark field microscopy (×535).

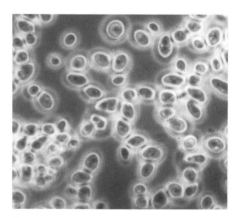

Fig. 1.9 Yeast cells viewed by phase contrast microscopy (×1260).

PHASE-CONTRAST MICROSCOPY allows **phase changes** in the light on passing through a specimen to be viewed. Normally the variations in the properties of visible light which can be perceived are changes in intensity (changes in wave amplitude) and change in colour (change in wavelength). The thin transparent sections obtainable from cells and tissues do not affect either of these very much unless the specimen is stained to heighten the contrast or colour of different regions. However, changes in the phase of the light passing through the specimen can be induced by variations in the thickness, or by refractive index differences within the specimen. The phase contrast microscope artificially increases the phase difference between rays from the specimen and those from the background by 90°. The resulting change in amplitude produces a visible image of the specimen. The contrast of this 'phase image' is very high, allowing living cells and tissues to be studied (Fig. 1.9).

FLUORESCENCE MICROSCOPY is based upon the ability of certain substances to absorb radiant energy (often ultraviolet radiation), and then re-emit energy as radiation of a longer, visible, wavelength.

The fluorescence microscope is essentially a conventional microscope with the inclusion of appropriate filters (Fig. 1.10). The illuminating beam is passed

Reference Bradbury, S. (1976) *The Optical Microscope in Biology*, Edward Arnold, London, UK. A concise introduction to how optical microscopes are used in biological sciences. Also includes a section on how to analyse optical images.

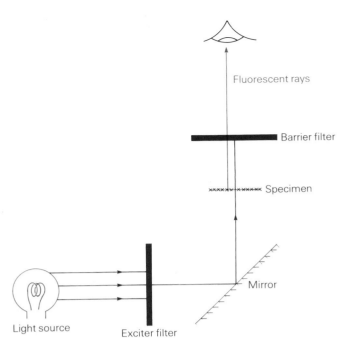

Fig. 1.10 Arrangement of a fluorescence microscope.

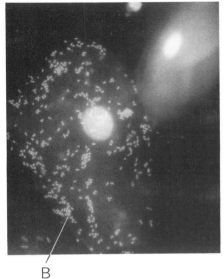

B

Fig. 1.11 Human cheek cells stained with acridine orange and viewed by fluorescence microscopy (×700). Note the bacteria (B) adhering to the cell surface. Courtesy of A. Curry, Withington Hospital, Manchester, UK.

through a filter so that the specimen receives radiation only of the appropriate exciting wavelength. A second filter removes the exciting radiation, but allows the emitted radiation to enter the eyepiece lens so that the image can be seen (Fig. 1.11).

Studying the internal organization of cells is restricted by the limited resolution of the light microscope which is at best about 0.24 μm. A typical bacterial cell is about 1 μm in diameter, while the largest cell in the human body, the ovum, has a diameter of 0.1 mm. Typical animal cells are 20–50 μm in diameter. Plant cells are generally larger. However, the resolution of the electron microscope allows cells to be studied in much greater detail.

Electron microscopy

All matter has a dual nature, both particulate and wave-like. Thus elections have wave-like properties and their wavelength (λ) depends upon the speed at which they are moving. The approximate wavelength of electrons, in nanometres (nm), over a potential difference of V volts, can be calculated from:

$$\lambda = \sqrt{1.5/V} \ \text{nm}$$

Thus electrons at 60 kV have an apparent wavelength of 0.005 nm, which should give a huge increase in resolution over the light microscope.

The phenomenally high resolving power theoretically available with electron microscopy unfortunately is not realizable. The resolving power is limited by specimen contrast and the highly imperfect electromagnetic lenses. In practice, the best resolution available with TEM is about 0.2 nm but this is only achieved when studying crystal lattices. The resolution observed when studying biological materials, including the internal structures of cells, is about 1 nm. This is of course a huge improvement over the light microscope and it has led to an enormous advance in knowledge of the ultrastructure of cells.

□ High-voltage electron microscopes accelerate electrons at 500 000 to 1 000 000 V. Such microscopes give very high resolution, and can be used to examine specimens up to 5 μm thick.

Reference Krstic, R.V. (1979) *Ultrastructure of the Mammalian Cell: An Atlas*, Springer-Verlag, Heidelberg, Germany. A series of superb line diagrams drawn from electron micrographs of cell organelles.

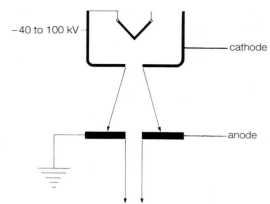

Fig. 1.12 The electron gun of an electron microscope.

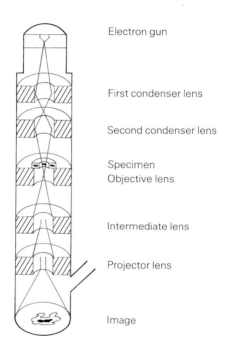

Fig. 1.13 Arrangement of electromagnetic lenses, and image formation in an electron microscope.

Electrons can be focused using **electromagnetic lenses** and therefore transmission electron microscopes (TEM) can be designed using the same basic plan as the conventional light microscope. The source of electrons is a small V-shaped piece of tungsten wire (Fig. 1.12) or **cathode**. When heated by a current at high voltage it becomes incandescent and emits electrons which are attracted to the positively charged **anode**, and pass, as a narrow beam, through a hole in its centre. The voltage operating between the cathode and anode is usually 40–100 000 V.

The arrangement of electromagnetic lenses used in focussing the electron beam and generating an enlarged image of the specimen is shown in Figure 1.13. Two condenser lenses are used making it possible to form a narrow beam of electrons, limiting the area of specimen irradiated and so minimizing damage to the specimen. The image of the specimen is formed by at least three lenses: an objective lens; a variable number of intermediate lenses; and a projector (equivalent to eyepiece) lens. The use of these additional lenses allows the magnification to be varied over a wide range, typically 1000–200 000-fold. The final image formed by the projector lens is visualized by allowing the electrons to impinge upon a fluorescent screen. This screen can be replaced by a camera to allow a photographic record, or **electron micrograph**, to be made of the specimen.

SCANNING ELECTRON MICROSCOPY (SEM) uses a fine beam of electrons to examine the *surface* of an object. The beam is focussed to a tiny spot on the surface of the specimen and systematically scanned across its surface. Low-energy, secondary electrons generated by this process are focussed on to a scintillation detector which converts each secondary electron impact to a flash of light. Each flash is amplified electronically and the resulting signal used to construct an image on a cathode-ray tube screen. The beam used to form this image is synchronized with the electron beam scanning the specimen and thus the image formed represents the surface of the specimen.

The magnification of the image is the ratio of the surface area of the cathode-ray screen divided by the area of the scanning electron beam. The resolving power of the SEM is much less than that of the TEM, usually of the order of 1–10 nm. However, the instrument has a great depth of focus, giving the two-dimensional image of the specimen an impressive three-dimensional appearance (Fig. 1.14).

Limitations of electron microscopy

Because electrons are absorbed and scattered by air, the inside of the electron microscope must be evacuated to a very high vacuum, and since biological

Reference Grimstone, A.V. (1986) *The Electron Microscope in Biology*, 2nd edn, Edward Arnold, London, UK. A little old, but still a concise, inexpensive introduction to electron microscopy.

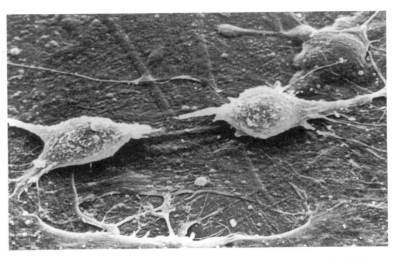

Fig. 1.14 Two fibroblasts (possibly daughter cells from a recent division) viewed by SEM (×750). Courtesy of A.J. Erroi, Department of Biological Sciences, The Manchester Metropolitan University, UK.

materials also absorb electrons, sections must be cut very thin indeed. The cutting of these sections, typically less than 100 nm, requires expensive microtomes using glass, sapphire or diamond knives. Specimens must be fixed with agents such as glutaraldehyde or osmium tetroxide that preserve their structure (Fig. 1.15). They are then dehydrated and embedded in a resin to support the tissue whilst sections of the required thickness are cut.

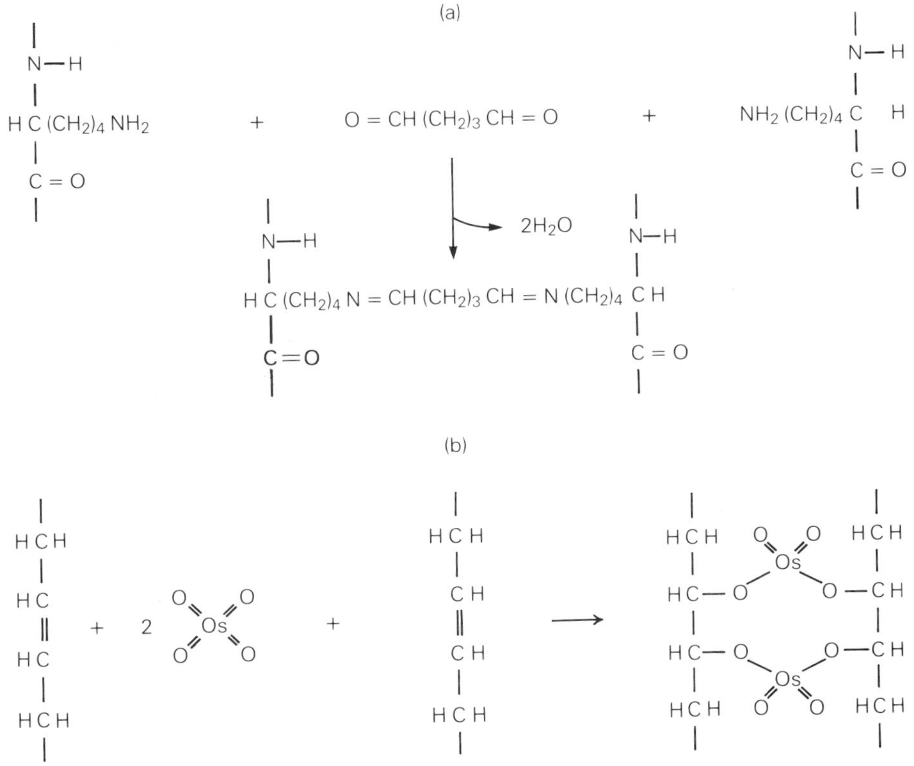

(a)
N—H
H C $(CH_2)_4$ NH_2
C = O
+
O = CH $(CH_2)_3$ CH = O
+
N—H
NH_2 $(CH_2)_4$ C H
C = O
$2H_2O$
N—H
H C $(CH_2)_4$ N = CH $(CH_2)_3$ CH = N $(CH_2)_4$ C H
C=O
N—H
C = O
(b)
H C H
H C
H C
H C H
+ 2
Os
O
O
O
O
+
H C H
C H
C H
H C H
H C H
H C—O
H C—O
H C H
Os
Os
H C H
O—C H
O—C H
H C H

Fig. 1.15 Mechanism of cross-linking by (a) glutaraldehyde, and (b) osmium tetroxide during the fixing of specimens for electron microscopy.

☐ A crude estimate of the thickness of a section can be made from the colour it appears due to interference patterns.

Colour	Thickness (nm)
Grey	30–60
Silver	60–90
Gold	90–150
Purple	150–190

Biological materials show little contrast unless stained with heavy metals such as lead, gold or osmium which are effective in scattering electrons. Heavy metal salts are deposited on the section and, being electron-dense, appear as dark areas on the image, contrasting with lighter unstained regions of the specimen. As a consequence of fixing, dehydration, embedding and sectioning procedures, followed by staining with heavy metals and viewing in a high vacuum, only dead biological material can be viewed by electron microscopy. Indeed, the harsh treatment of specimens before their examination by electron microscopy has often led to the charge that the structures viewed by this technique are likely to be artefacts. However, the majority of the evidence supports the current views of cell structure as correct. It must be borne in mind, nevertheless, that the beautifully structured, two-dimensional electron micrographs of cells are static visualizations of dynamic three-dimensional systems. The interior of the cell is a busy place! Structures move and change shape, and molecules are constantly being built up and degraded.

1.3 Structure of cells

Light microscopy of plant and animal cells shows little of the internal structural differentiation of the *protoplasm* (Fig. 1.16). The major structural feature visible within the cell is the **nucleus**, first observed by Brown in 1831, surrounded by a **nuclear envelope** and containing a densely staining region, the **nucleolus**. Electron microscopic studies (Fig. 1.17) show that the nuclear membrane is a double-membrane system punctured by nuclear pores.

A wealth of evidence has firmly established the nucleus as the control centre of the cell. It is here that the genome of an organism, contained in its DNA, is located. All somatic cells of multicellular organisms are **totipotent**; that is, they contain a complete set of genetic instructions. This was

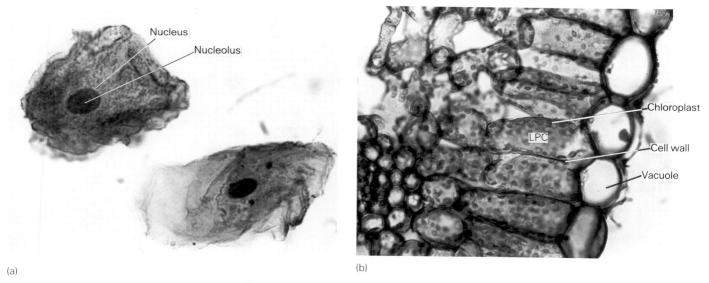

(a) (b)

Fig. 1.16 Photomicrographs of (a) human cheek cell (×1100), (b) leaf palisade cells (LPC) (×440). Courtesy of R.H. Bowling, Department of Biological Sciences, The Manchester Metropolitan University, UK.

protoplasm: *the material of the cell. It is composed of nucleoplasm, the material of nucleus; and cytoplasm, the rest of the cell. From the Greek protos, first; kytos, hollow vessel; plasma, form; and the Latin nucis, nut.*

Reference Hinshaw, J.E., Carragher, B.O. and Milligan, R.A. (1992) Architecture and design of the nuclear pore complex. *Cell*, **69**, 1133–1141. Detailed information on nuclear pore: good diagrams.

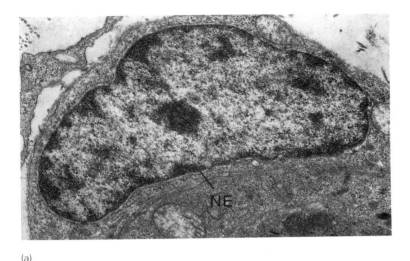

(a)

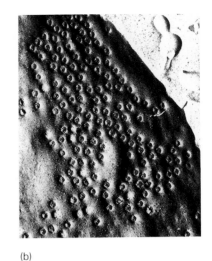

(b)

Fig. 1.17 (a) TEM of nucleus of an endothelial cell. Note the double membrane of the nuclear envelope (NE) (×13 500). Courtesy of P.L. Carter, Department of Biological Sciences, The Manchester Metropolitan University, UK. (b) Freeze-fracture preparation of nucleus of rodent spermatocyte. Plane of fracture through outer membrane of the nuclear envelope showing surface view of nuclear pores. Photograph kindly supplied by Emeritus Professor D.W. Fawcett, Harvard Medical School, USA.

spectacularly demonstrated by Gurdon in 1966 using the African clawed toad, *Xenopus laevis* (Fig. 1.18). The nucleus of a normal unfertilized egg was destroyed by ultraviolet irradiation. This nucleus was then replaced by one taken from a differentiated intestinal cell. The 'transplanted' nucleus came from a mutant strain of *X. laevis* having only one nucleolus per diploid nucleus. Some of these reconstituted eggs developed, via tadpoles, to adult frogs and these always contained mononucleolar nuclei. This showed that the complete genome of an organism resides in the nucleus. The structure and organization of nuclei and nucleoli are explained more fully in Chapter 3.

Outside the nucleus both animal and plant cells possess a granular **cytoplasm** which contains minute thread-like structures called **mitochondria** (Fig. 1.19). However, at the resolution of the light microscope, the fine structural details of these grains and threads cannot be observed.

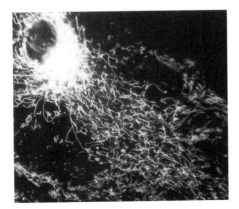

Fig. 1.19 Photomicrograph of cell stained with rhodamine 123, a dye specific for mitochondria. Note the filamentous appearance of the mitochondria. Courtesy of Dr L.B. Chen, Dana Farber Cancer Institute, Boston, USA.

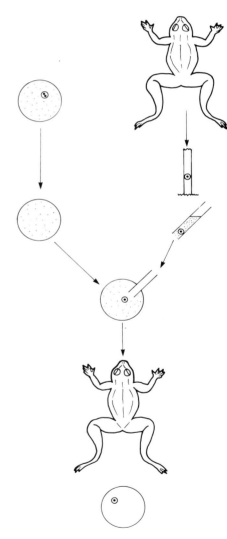

Fig. 1.18 Outline of Gurdon's experiment to demonstrate totipotency.

The general plan of a typical animal cell is shown in Figure 1.16 (a). Its boundary is indistinct since the **cell membrane** cannot be resolved. In contrast, the plant cell is sharply delimited by a **cell wall** external to the membrane. Plant cells are also characterized by the presence of a large *vacuole* and of **chloroplasts**, the sites of photosynthesis.

Cells of bacteria, including Cyanobacteria (Fig. 1.20) also possess granular cytoplasm and a cell wall but do not have a membrane-bound nucleus. Cells which possess a nucleus are called *eukaryotic* cells, while those lacking a nucleus are called *prokaryotic*. Prokaryotic cells are described more fully in Chapter 2.

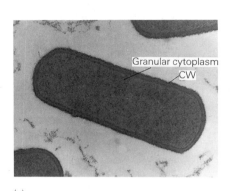

(a)

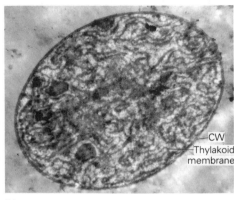

(b)

Fig. 1.20 Prokaryotic cells. (a) The bacterium *Bacillus subtilis* (×13 200). Courtesy of Dr I.D.J. Burdett, National Institute of Medical Research, Mill Hill, London, UK. (b) The cyanobacterium, *Anabaena azollae* (×27 900). Note the presence of cell walls (CW) but absence of nucleus in both cell types.

1.4 Classification of organisms by cell structure

Organisms can be classified into acellular (**viruses**), and those with cellular organization. This latter group contains all the entities 'intuitively' recognized as species. In 1937 Chatton proposed the terms *procariotique* and *eucariotique* for non-nucleated and nucleated cellular organisms respectively. The difference between the groups is much more fundamental than the simple presence or absence of a nucleus and some of the major features distinguishing the two groups are listed in Table 1.1. Eukaryotic organisms

Fig. 1.21 Photomicrograph of the fungus *Cladisporium resina* (×1080). Courtesy of Dr G. Hobbs, UMIST, Manchester, UK. This fungus is a serious pest and is able to grow on aircraft fuel.

Table 1.1 *Major differences between prokaryotes and eukaryotes*

Prokaryotes	Eukaryotes
Cell size: 1–10 μm diameter	Cell size: 10–100 μm diameter
DNA not in nucleus	DNA in nucleus
Circular chromosomes	Linear chromosomes of DNA. RNA and proteins are also present
Reproduction by simple binary division	Reproduction by mitosis
Cytoplasm largely undifferentiated	Cytoplasm highly differentiated into membrane-limited components, e.g. mitochondria, lysosomes
Simple bacterial flagella	Complex 9 + 2 arrangement to flagella and cilia
Many are strict anaerobes	Most are aerobic; anaerobic types are secondarily modified
Various photosynthetic forms	All photosynthetic types produce O_2
Multicellular types rare, no development of tissues	Metazoans with extensive development of tissues and organ systems common

vacuole: a large membrane-limited, fluid-filled cavity within the cell; from the Latin vacuus, empty space. Also called the tonoplast; from the Greek tonos, stretch, and plastos, formed.

eukaryotes and prokaryotes: organisms with and without a nucleus. From the Greek pro, before; eu, good, well; karyon, nut.

are further divided into four major groups: **Protoctista**, Fungi (Fig. 1.21). Animalia and Plantae. The first group contains all those organisms which by default are not clearly recognizable as being fungi, animals, or plants.

The application of electron microscopy to cytological studies since about 1950 has revealed a wealth of structural detail (Fig. 1.22) about cells.

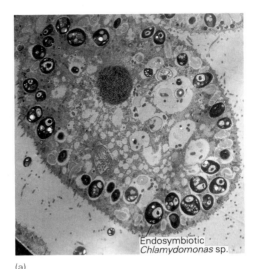

Endosymbiotic
Chlamydomonas sp.

(a)

(b)

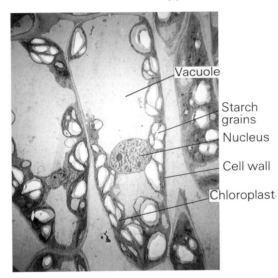

Vacuole

Starch grains

Nucleus

Cell wall

Chloroplast

(c)

Fig. 1.22 Cells viewed by TEM. (a) A unicellular animal, *Paramecium bursaria*. This cell contains the endosymbiotic unicellular green alga *Chlamydomonas* sp. arrowed (×1260). (b) Electron micrograph of nasal ciliated epithelium (×5000). Courtesy of Dr A. Curry, Public Health Laboratory, Withington Hospital, Manchester, UK. (c) Tobacco (*Nicotiana tabacum*) leaf cell (×1880). Courtesy of Dr E. Sheffield, Department of Cell and Structural Biology, University of Manchester, UK.

1.5 The cell membrane

In electron micrographs, the outline of animal cells is clearly seen to be a distinct membrane demarcating the cell from its surroundings. Plant cells have a similar membrane but this is normally tightly pressed against the inner face of cell wall (Fig. 1.22). This **plasma membrane** or *plasmalemma*, like all cell membranes, is composed of a lipid bilayer that contains a variety of protein molecules. In animal cells the outer surface of the plasmalemma is covered with a **glycocalyx**, which consists of oligosaccharide units projecting from the glycolipids and glycoproteins embedded in the outer layer of the membrane (Fig. 1.23).

See Chapter 4

---

protoctista: from the Greek protos, *first;* ktistos, *to establish. The term has been used since the nineteenth century to denote a single-celled organism; however, the kingdom Protoctista includes some multicellular types, such as certain algae.*
plasmalemma: from the Greek words, plasma, *form, a thing formed, and* lemma, *a husk or rind.*

Reference Maragulis, L. and Schwartz, K.V. (1987) *Five Kingdoms: An Illustrated Guide to the Phyla of Life on Earth*, 2nd edn, W.H. Freeman, New York, USA. Pictorial introduction to classifying organisms, illustrated throughout with many fine electron micrographs and line drawings.

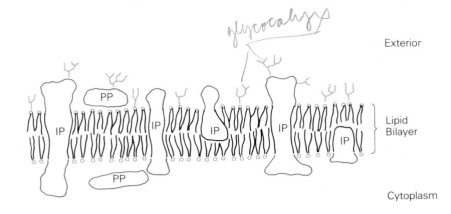

glycocalyx

Exterior

Lipid
Bilayer

Cytoplasm

Fig. 1.23 Cross-section of plasma membrane. IP, integral proteins; PP, peripheral proteins. The glycocalyx is tinted in grey.

The plasma membrane acts as an interface between the cell and its external environment. It is **selectively permeable** and regulates the movement of materials into and out of the cell. Multicellular organisms regulate the activities of their component cells by sending signals which are usually received by receptor molecules on the outer surface of the cell membrane. This signal is then relayed, often after being amplified, to the interior of the cell.

In mobile cells the plasmalemma is involved in cell locomotion, while in immobile types it mediates cell adhesion. The plasma membrane also plays a vital role in cell–cell interactions such as contact inhibition.

The plasmalemma is only one of a number of membrane systems found in eukaryotic cells. The interior of such cells is subdivided into a series of membrane-limited compartments which form distinct structural and functional units. The advantages conferred by such compartmentalization are explained in the following sections.

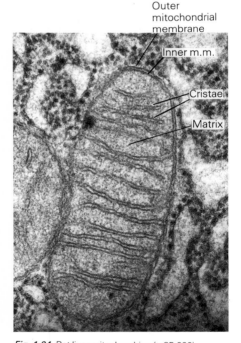

Outer mitochondrial membrane

Inner m.m.

Cristae

Matrix

Fig. 1.24 Rat liver mitochondrion (×35 000). Courtesy of Dr E.J. Wood, Department of Biochemistry and Molecular Biology, Leeds University, UK.

1.6 Membrane compartments

The interior of the cell has several types of compartments which differ in structure. Some are membrane-bounded *vesicles*, concerned with shuttling materials between different regions of the cell, while others, including **lysosomes** and **microbodies**, have more specific biochemical roles. Several membrane systems form extensive *cisternae*, while some organelles are discrete, complex structures. Examples of these are the nucleus, mitochondria and chloroplasts.

Mitochondria and chloroplasts

Mitochondria and chloroplasts are the sites of the production of metabolic energy. Virtually all eukaryotic cells contain mitochondria, while chloroplasts are restricted to plants.

MITOCHONDRIA are membrane-bounded organelles which lie free in the cytoplasm. In electron micrographs they appear as spherical or rod-shaped structures of diameter 0.5–1.0 μm and variable length (Fig. 1.24). Mitochondria are plastic and undergo rapid changes in shape in living

Reference Bretscher, M.S. (1985) The molecules of the cell membrane. *Scientific American*, **253(4)**, 86–90. A typical *Scientific American* overview: clear, interesting and beautifully illustrated.

cells (Fig. 1.25). Serial sections through some cell types show they contain a single mitochondrion forming an extensive reticular network (Fig. 1.26). Mitochondria are always bounded by a smooth, permeable outer membrane, but have a highly infolded, selectively permeable, inner membrane (Fig. 1.24), the two membranes being separated by an intermembrane space. The folds of the inner membrane, called **cristae**, project into an inner **matrix**. Cristae vary considerably in number and shape between the mitochondria of different cells, although in all cases they greatly increase the surface area of the inner membrane.

Mitochondria are concerned with the oxidation of fuel molecules and the concomitant production of ATP. For example, pyruvate formed during glycolysis in the cytosol is oxidatively decarboxylated in the mitochondrial matrix forming acetyl CoA.

$$\text{pyruvate} + NAD^+ + CoASH \rightarrow \text{acetyl CoA} + CO_2 + NADH + H^+$$

Acetyl CoA is degraded via the tricarboxylic acid (TCA) cycle to yield CO_2, NADH and $FADH_2$. The reduced coenzymes are reoxidized by feeding electrons into the electron transport chain of the inner mitochondrial membrane (Fig. 1.27), which eventually reduces oxygen.

The current of electrons, perhaps as many as 10^5 electrons flowing at once, is used to increase the concentration of protons in the intermembrane space relative to that in the matrix. As the inner mitochondrial membrane is impermeable to protons, this proton gradient represents a store of chemical potential energy and forms the link between electron transport and ATP generation. Protons may be returned to the matrix only via molecules of a **proton-dependent ATP synthetase** in the inner membrane. In the electron microscope the ATP synthetase is visible on the inner surface of cristae as knobs (Fig. 1.28).

CHLOROPLASTS are the sites of photosynthesis in the photosynthetic cells of green plants. While the chloroplasts of 'lower plants' exhibit a variety of structural forms (Fig. 1.29a–c), those of 'higher plants' normally appear as oval structures 1–10 μm long and up to 5 μm in diameter (Fig. 1.29d). Chloroplasts are surrounded by a double membrane enclosing a **stroma**-filled space. Within the stroma is a third system of highly folded membranes called **thylakoids**.

See Chapter 5

□ Serial sections are consecutive sections cut from a specimen, the two-dimensional images obtained from the series of photographs can then be used to construct the three-dimensional structure of the specimen as each successive image is superimposed upon the next.

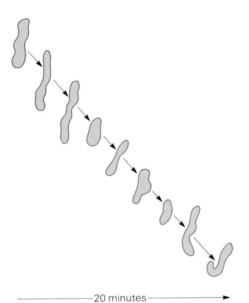

————20 minutes————→

Fig. 1.25 Representations of the rapid, observable changes in shape of a mitochondrion. Redrawn from Alberts, B. *et al.* (1983) *Molecular Biology of the Cell.* Garland, New York, USA.

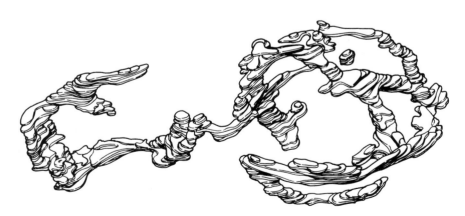

Fig. 1.26 Three-dimensional model of a single, large mitochondrion from a yeast cell. Redrawn from Hoffman, H.P. and Avers C.J. (1973) *Science*, **181**, 749–51.

vesicle: *any small fluid-filled, membrane-limited space within the cytoplasm. From the Latin* vescula, *small bladder.*
cisternae: *expanded, flattened spaces within the cytoplasm surrounded by a membrane. From the Latin* cisterna, *water reservoir.*

Reference Ernster, L. and Schatz, G. (1981) Mitochondria: an historical review. *Journal of Cell Biology*, **91**, 227s–235s. A wide ranging, interesting article. Good starting place for reading about the mitochondrion.

☐ Many of the functional proteins of mitochondria and chloroplasts are synthesized in the cytosol. These are transferred into the organelles using amino acid signal sequences which allow the proteins to be transferred into the organelles.

Exercise 5

Using Figure 1.24 estimate the relative increase in surface area achieved by folding the inner mitochondrial membrane into cristae.

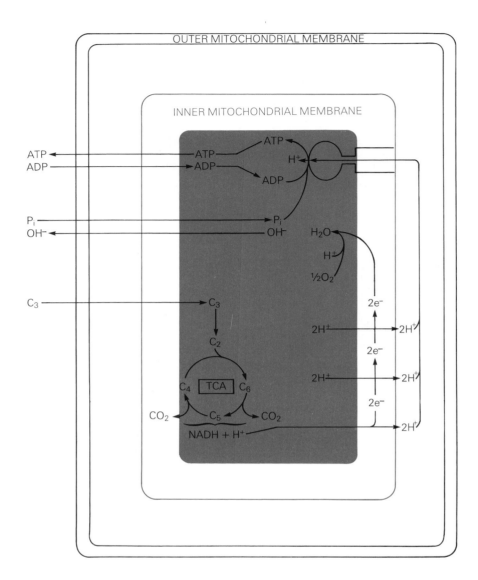

Fig. 1.27 The oxidative reactions of mitochondria.

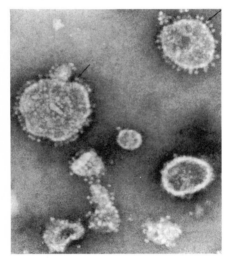

Fig. 1.28 Particles prepared from the inner mitochondrial membrane showing the distinctive ATP-synthetase complex (arrowed) (×125 000).

☐ Mitochondria and chloroplasts resemble bacteria and cyanobacteria in several respects, including size, possession of circular DNA and types of ribosomes. This has led to the suggestion that these eukaryotic organelles are evolutionary descendants of endosymbiotic bacteria and cyanobacteria.

The thylakoids are organized into stacks called **grana** which resemble piles of coins and the grana are interconnected by stromal thylakoids. The stroma has a granular appearance, but is essentially unstructured. It contains starch grains and lipid droplets, although their size and number depend upon the metabolic state and type of cell. Like the mitochondrial matrix, the stroma contains ribosomes and DNA allowing the chloroplast to have a semi-independent existence. In spite of their somewhat different appearances, mitochondria and chloroplasts have essentially similar structures (Fig. 1.30).

The light reactions of photosynthesis occur on the thylakoid membrane. Absorption of light by pigment molecules such as chlorophylls raises electrons to an energy level at which they can reduce $NADP^+$. The electron deficiency of the chlorophylls is made good by removing electrons from water in a process which supplies H^+ for the formation of NADPH and releases

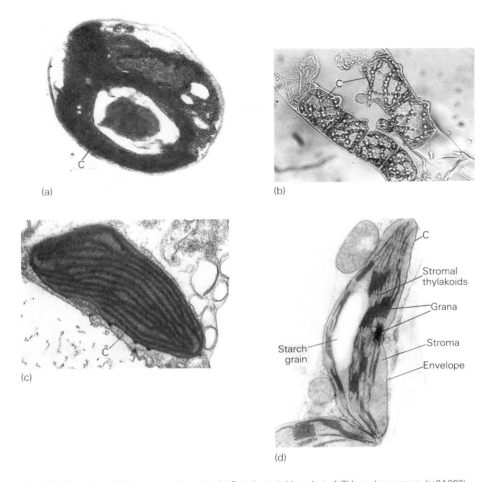

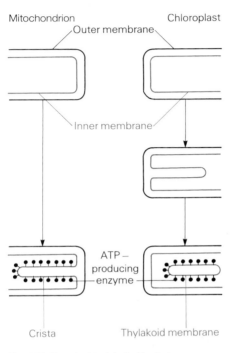

Fig. 1.29 Chloroplasts (C) from several species. (a) Cup-shaped chloroplast of *Chlamydomonas* sp. (×24 000). (b) Spiral chloroplast of *Spirogyra* sp. (×120). Courtesy of M.J. Hoult, Department of Biological Sciences, The Manchester Metropolitan University, UK. (c) Chloroplast of the brown alga *Fucus vesiculosus* (×20 100). (d) Chloroplast of the higher plant *Nicotiana tabacum* (×10 560). Courtesy of Dr E. Sheffield, Department of Cell and Structural Biology, University of Manchester, UK.

Fig. 1.30 Organizational similarities between mitochondria and chloroplasts.

oxygen (Fig. 1.31). As electrons flow through carriers in the thylakoid membranes, proton gradients are generated which can be used to drive ATP production in a very similar way to the process occurring in mitochondria. The ATP and NADPH-dependent biosynthetic reactions of the dark processes of photosynthesis occur in the stroma which contains the enzymes necessary to catalyse these processes (Fig. 1.31).

PLASTIDS, OTHER THAN CHLOROPLASTS. Chloroplasts are only one type of membrane-bounded structure in plant cells which are given the collective name of **plastids**. All higher plant cells contain some form of plastid although chloroplasts are both structurally and functionally the most complex and the most widespread. All plastids have a limiting double membrane and are characterized by the presence of lipid droplets (plastoglobuli) and the possession of an internal genome. Most plastids are discoid or spherical, although a variety of types are known which have a range of functions.

Exercise 6

Why is *light-independent* a better term than *dark reactions* for the ATP- and NADPH-producing reactions of photosynthesis?

Reference Bogorad, L. (1981) Chloroplasts. *Journal of Cell Biology*, **91**, 256s–270s. A good starting place for reading about the chloroplast.

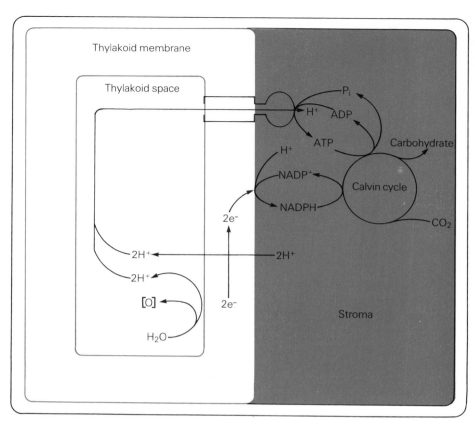

Fig. 1.31 The light and dark reactions of the chloroplast.

Proplastids often contain starch grains and crystalline materials. They form a heterogeneous group, and their biochemistry is on the whole poorly understood, although the proplastids of young leaf cells are known to differentiate to form chloroplasts. **Leucoplasts** are enlarged, colourless proplastids which do not contain starch or other deposits. **Amyloplasts** are proplastids which are completely filled with starch grains. They are mainly found in plant storage tissues, but it seems that the perception of gravity in plant roots is due to the differential settling of large amyloplasts found in cells near the root cap. **Chromoplasts** are plastids which contain coloured, yellow, orange and red pigments and which are responsible for imparting colour to floral structures and fruits. They develop from chloroplasts, the chlorophyll being degraded and replaced by carotenoids.

Unlike the membranes of plastids and mitochondria, the cisternal membranous systems of eukaryotic cells are limited by *single* membranes. These are generally thinner than the plasmalemma and have their own individual compositions (Table 1.2). Examples of such cisternal membrane systems are those forming the ***endoplasmic reticulum*** and the **Golgi apparatus** which together form the **endomembrane** system of the cells.

The endoplasmic reticulum

The endoplasmic reticulum (ER) forms the most extensive membrane system in most cells and particularly in those that are actively synthesizing materials

endoplasmic reticulum: *from the Greek* endon, *within;* plasma, *form, and the Latin* rete, *net (as used by gladiators).*

Table 1.2 *Composition of several biological membranes expressed as percentage by weight*

Membrane	Protein	Lipid	Carbohydrate
Plasma membrane			
human red blood cell	49	43	8
mouse liver	46	54	2–4
Myelin	18	79	3
Mitochondria			
outer	52	48	2–4
inner	76	24	1–2
Chloroplast			
spinach lamellae	70	30	6
Endoplasmic reticulum	50	50	0
Golgi apparatus	65	35	0
Nucleus (rat liver)	59	35	3

Data from several sources.

for secretion. The ER exists in two distinct morphologically different forms: the rough ER and the smooth ER. These are clearly distinguishable in electron micrographs by the presence of numerous ribosomes attached to the rough endoplasmic reticulum (RER) which give it its 'rough' appearance.

ROUGH ENDOPLASMIC RETICULUM normally consists of a series of interconnected flattened sacs, while the smooth endoplasmic reticulum (SER) is generally more tubular in appearance; in fact the two are interconnected to form a single membrane system (Fig. 1.32). Together they form the boundary of a single space, called the lumen, which anastomoses throughout the cytosol (Fig. 1.33). The functions of the cytosol are described more fully later. The ER is involved in the biosynthesis, modification and transport of materials throughout the cell.

SMOOTH ENDOPLASMIC RETICULUM is concerned with lipid biosyntheses and the detoxification of drugs and other potentially harmful **xenobiotics** such as pesticides and herbicides. These compounds are rendered less harmful by hydroxylations catalysed by mixed-function oxidases located in the ER membrane.

$$R\text{–}H + O_2 \xrightarrow{\quad NADPH + H^+ \quad NADP^+ \quad} R\text{–}OH + H_2O$$

The newly formed hydroxyl group is then usually modified by conjugation with a charged unit such as a sulphate or a glucuronate. The net effect of these reactions is to increase the water solubility of the original compound, allowing it to be transported to the kidney and excreted in urine.

Box 1.1
Detoxification of
xenobiotics

The detoxification of xenobiotics can sometimes lead to potentially disastrous results. In some cases the cell's 'detoxification' procedures can produce metabolites with *greater* toxicity than the parent compound. This increase in toxicity can occur at either the oxidative stage or the conjugation step (see main text). For example, the monooxygenases of endoplasmic reticulum membrane convert paracetamol and benzo(*a*)pyrene into biochemically active products which can cause liver necrosis and cancer respectively.

Reference Mulder, G. (1979) Detoxification or toxification? Modification of the toxicity of foreign compounds by conjugation in the liver. *Trends in Biochemical Sciences*, **4**, 86–90. A rather old, but still useful account of the functions of the endoplasmic reticulum in the detoxification of xenobiotics.

xenobiotic: *foreign to a living organism, usually applied to compounds such as drugs, pesticides, and the like.*

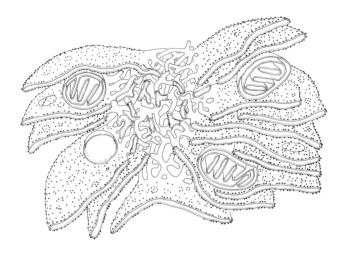

Fig. 1.32 Three-dimensional representation of a portion of the rough and smooth endoplasmic reticulum of a rat liver cell. Redrawn from Krstic, R.V. (1979) *Ultrastructure of the Mammalian Cell.* Springer-Verlag, Heidelberg, Germany.

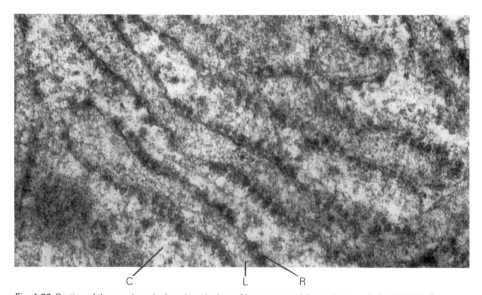

Fig. 1.33 Portion of the rough endoplasmic reticulum of human synovial membrane cells (×127 500). R, ribosome; C, cytosol, L, lumen. Courtesy of Dr A.R. Sattar, St Mary's Hospital, Manchester, UK.

The SER is a major site of lipid biosynthesis and is extensive in cells forming large amounts of lipids such as liver, active mammary gland and intestinal cells.

Triacylglycerols are synthesized and are often stored as energy-rich lipid droplets in the lumen. Phospholipids involved in membrane structure are also synthesized in the SER. The acyl chains anchor the incomplete lipid in the membrane, while a variety of polar head groups can be added to complete the phospholipid (Fig. 1.34). Enzymes involved in the biosynthesis of steroids from cholesterol are also found in the membranes of the SER and cells that secrete large amounts of steroid hormones such as those of testis and adrenal cortex have extensive SER systems.

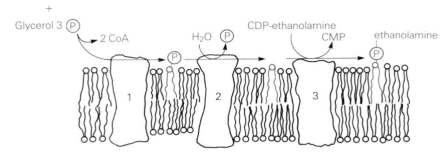

Fig. 1.34 Biosynthesis of the phospholipid, phosphatidylethanolamine. (1) Acyl transferase, (2) phosphatidate phosphatase, (3) CDP-ethanolamine diacylglycerol phosphoethanolamine transferase.

The SER of muscle cells is a highly specialized membrane system called the **sarcoplasmic reticulum**. This system is able to accumulate Ca^{2+} ions from the cytosol (sarcosol), a property crucial in the contraction of muscle fibres. RER is mainly associated with the transport and processing of proteins formed in the ribosomes attached to its cytosolic face. These proteins, for example zymogens and protein hormones, are largely destined for secretion from the cell or for integration into the cellular endomembranes and plasmalemma.

PROTEIN SECRETION. The steps involved in the biosynthesis and export of secretory proteins were first traced by Palade and coworkers using the pancreatic acinar cell (Fig. 1.35) as a model system. Palade was able to

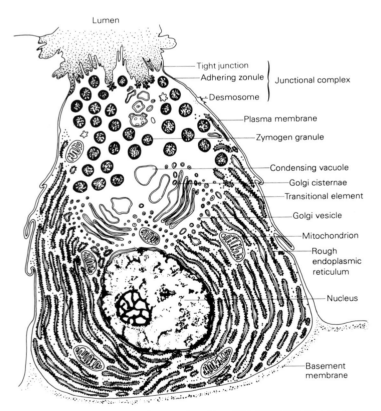

Fig. 1.35 Diagram of a pancreatic acinar cell. Redrawn from Case, R.M. (1978) *Biol. Rev.*, **53**, 211

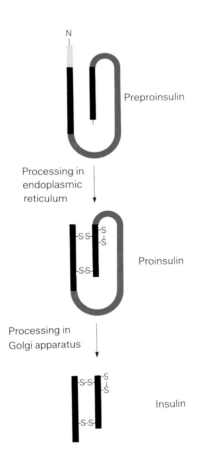

Fig. 1.36 Conversion of preproinsulin to insulin.

demonstrate that secretory proteins were synthesized on ribosomes of the rough endoplasmic reticulum before being translocated into the lumen of the reticulum system. From here the proteins were transported to the Golgi complex (see later), packaged into secretory vesicles and extruded from the cell into branches of the pancreatic duct. The products of the cells (zymogens) could then be conveyed to the digestive tract.

Subsequent to this work it has been shown that in general, secretory proteins are produced as **preproteins** which are converted to mature proteins during their secretory pathways (Fig. 1.36). Sabatini and Blobel suggested that proteins destined for secretion from the cell contain **signals** directing their export from the cell and the existence of these signals, called ER translocation signals, is now well established.

ER TRANSLOCATION SIGNALS are contained in the first 25–30 amino acid residues of the nascent protein and such signals have generally similar but not identical structures in most secreted proteins (Fig. 1.37). The extreme amino-terminus usually carries one or more basic charges, while the central core of the signal contains continuous stretches of hydrophobic residues. However, signal sequences differ considerably between different proteins and do not show any sequence homology (Table 1.3).

Translocation of proteins destined for the ER lumen begins in the usual fashion but when some 70–80 codons have been translated, approximately 30 amino-terminal amino acid residues of the signal sequence protrude from the ribosome. This signal is recognized and binds to a **signal recognition particle** (**SRP**). Upon binding of the SRP, translation of the remaining mRNA is arrested. Since signal sequences are not homologous, the SRP must recognize some secondary or tertiary structural feature of the signal sequence in order to allow specific binding and arrest of translation. The ribosome–SRP complex then binds to the ER membrane via a receptor or docking protein (Fig. 1.38). Once the ribosome–SRP complex has docked with the SRP receptor, the ribosome can interact directly with the RER membrane.

RIBOSOME–RER MEMBRANE INTERACTIONS are modulated by direct contact between the large subunit of the ribosome and proteins of the RER membrane. This is necessary since the SRP receptor lacks affinity for ribosomes. Two integral proteins of the RER membrane which have the necessary affinity have been identified and designated **ribophorins I** and **II**. Following transfer of the ribosome to its specific RER membrane binding

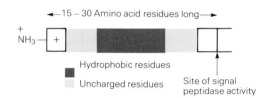

Fig. 1.37 Generalized amino-terminal hydrophobic signal.

See *Molecular Biology and Biotechnology*, Chapter 4

Table 1.3 *Sequences of some eukaryotic signal peptides (hydrophobic residues are coloured); the site of cleavage is shown by the arrow*

Protein	Sequence of signal peptide
Secretory	
Preproalbumin	Met.Lys.Trp.Val.Thr.Phe.Leu.Leu.Leu.Leu.Phe.Ile.Ser.Gly.Ser.Ala.Phe.Ser.↓Arg.
Preprolysozyme	Met.Arg.Ser.Leu.Leu.Ile.Leu.Val.Leu.Cys.Phe.Leu.Pro.Leu.Ala.Ala.Leu.Gly.↓Lys.
Preprolactin	Met.Asn.Ser.Gln.Val.Ser.Ala.Arg.Lys.Ala.Gly.Thr.Leu.Leu.Leu.Met.Met.Ser-
	Asn.Leu.Leu.Phe.Cys.Gln.Asn.Val.Gln.Thr.↓Leu.
Membrane	
Pre-vesicular stomatitis virus glycoprotein	Met.Ser.Ile.Gln.His.Phe.Arg.Val.Ala.Leu.Ile.Pro.Phe.Phe.Ala.Ala.Phe.Cys-
	-Leu.Pro.Val.Phe.Ala.↓His.
γ-subunit acetycholine receptor (*Torpedo*)	Met.Val.Leu.Thr.Leu.Leu.Leu.Ile.Ile.Cys.Leu.Ala.Leu.Glu.Val.↓Arg.Ser.Glu.

Data from several sources.

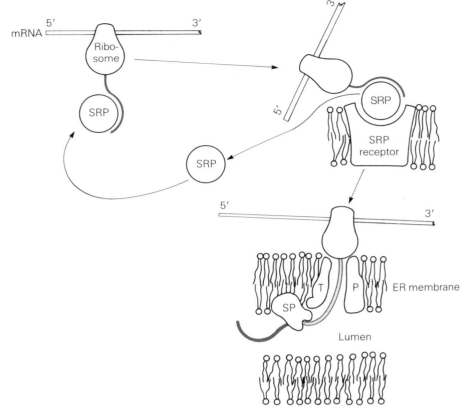

Fig. 1.38 The translocation of proteins across endoplasmic reticulum (ER) membrane. SRP, signal receptor particle; T and P are ribosome-binding proteins; SP is signal peptide peptidase.

sites, the SRP is released allowing it to be recycled and to deliver other ribosomes to the RER. The release of the SRP allows the elongation of the nascent polypeptide chain to resume. The growing polypeptide is now extruded through the RER membrane and passes into the lumen of the ER system. Not all proteins are completely translocated across the RER membrane and major exceptions are proteins which form integral components of the ER, Golgi, lysosomal and plasma membranes.

Membrane proteins

The translocation of membrane proteins is similar to that of secreted proteins but translocation is only partial and the protein is *inserted* into the membrane. Nascent proteins of membranes generally have typical amino-terminal signal sequences (Table 1.3), which are removed during integration of the protein. However, membrane proteins appear to have a 'stop' signal which becomes fixed in the lipid bilayer, causing the protein to span the membrane forming an integral membrane protein. The simplest modes of membrane integration are seen with proteins such as glycophorin or the vesicular stomatitis virus glycoprotein. Integration is achieved by a stop signal comprising a sequence of strongly hydrophobic residues (Table 1.4) sufficiently long to form an α-helix which can span the lipid bilayer (Fig. 1.39).

The cytoplasmic regions of the proteins are rich in basic residues which could not cross the hydrophobic interior of the membrane easily but can

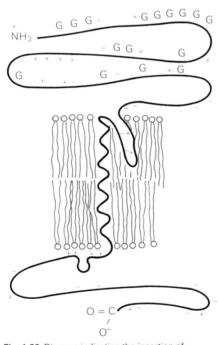

Fig. 1.39 Diagram indicating the insertion of glycophorin into the plasma membrane. G, the position of attached carbohydrates; + and −, the presence of charged amino acid residues.

Table 1.4 *Largely hydrophobic sequences responsible for anchoring some proteins in the membrane bilayer (coloured). Adjacent basic residues are shown.*

Protein	Sequence
Vesicular stomatitis virus glycoprotein	...Lys Ser.Ser.Ile.Ala.Ser.Phe.Phe.Phe.Ile.Ile.Gly.Leu.Ile.Ile.Gly.Leu.Phe.Leu- -Val.Leu Arg...
Glycophorin	...Lys.Lys.Ile.Leu.Arg.Arg Ile.Gly.Tyr.Ser.Ile.Leu.Leu.Ile.Thr.Gly.Ile.Val.Gly.Ala- -Met.Val.Gly.Phe.Ile.Ile.Leu.Thr.Ile.Glu.Pro.Glu.Ser.Phe.His.His.Ala.Leu- -Gln.Val Arg.Glu...
First domain of bacteriorhodopsin	...Thr.Gly Arg.Pro.Glu.Trp.Ile.Trp.Leu.Ala.Leu.Gly.Thr.Ala.Leu.Met.Gly.Leu- -Gly.Thr.Leu.Tyr.Phe.Leu.Val.Lys.Gly Met...

Data from several sources.

interact favourably with the polar groups of the membrane phospholipids. Following translocation, both membrane-bound and secreted proteins are transported through the membrane system.

Endomembrane transport

See *Molecular Biology and Biotechnology*

Once translocation is completed, proproteins fold into their native conformation. Most secretory proteins are modified by a further sequence of post-translational modifications all presumably essential if the proteins are to fulfil this role in the economy of the organism. Thus during their passage through the endomembrane system, proteins may be subjected to **disulphide bond formation**, **glycosylation**, **sulphation**, **phosphorylation** or possibly **acylation**.

Proproteins are transported from the ER to the Golgi apparatus in transition vesicles which bud off from the ER membrane and fuse to form elements of the Golgi apparatus.

The Golgi apparatus

The Golgi apparatus was observed by Golgi in 1898 following heating of sections of barn owl cerebellum in the presence of Os^{2+} and Ru^{2+} salts. Golgi observed a subcellular structure which he called the *apparel reticulaire interne*. This structure, and similar structures in other cells, were eventually called the Golgi apparatus or complex. Many cytologists regarded the Golgi apparatus as an artefact of the preparative procedures and it was only the application of electron microscopy in the mid-twentieth century that firmly established the existence of this organelle. The structure of the Golgi apparatus can be understood by considering three levels of organization: the **cisterna** (or **sacculus**), the **dictyosomes** or Golgi bodies and the **Golgi complex**.

CISTERNAE of the Golgi apparatus are limited by smooth membrane. They are typically flattened, usually with a central plate-like region which is often continuous with peripheral tubules and vesicles. The plate region is normally 0.5–1.0 μm in diameter, and the tubules are usually 30–50 nm in diameter and may be several μm long. These tubules often form a complex network and may link different dictyosomes.

DICTYOSOMES are stacks of cisternae and are characteristic of the Golgi apparatus. While there are usually 5–8 cisternae per dictyosome, it is not uncommon to find 30 or more in lower organisms. The number of dictyosomes in a cell is variable, ranging from zero in some fungi to several thousand in certain algal rhizoids.

Reference Rothman, J.E. (1985) The compartmental organisation of the Golgi apparatus. *Scientific American*, **253(3)**, 84–95. An overview, clear, interesting and very well illustrated.

Reference Singer, S.J. and Yaffe, M.P. (1990) Embedded or not? Hydrophobic sequences and membranes. *Trends in Biochemical Sciences*, **15**, 253–7. Excellent, readable overview of latest ideas on the integration of membrane proteins.

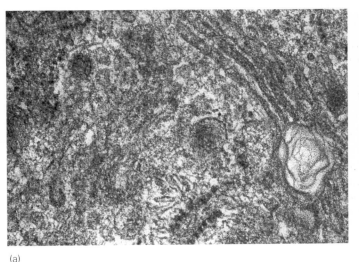
(a)

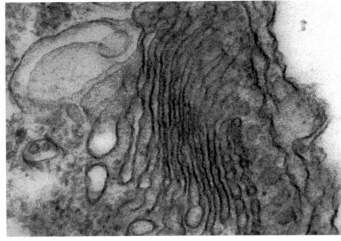
(b)

Fig. 1.40 (a) Golgi apparatus of human synovial membrane cell. Courtesy of Dr A.R. Sattar, St Mary's Hospital, Manchester, UK. (b) Golgi apparatus of the brown alga *Fucus vesiculosus* (×97 500).

THE GOLGI APPARATUS is composed of an association of dictyosomes. In mammalian cells the dictyosomes are arranged in compact zones, while in some plants and invertebrates they are dispersed throughout the cytoplasm (Fig. 1.40a,b). When all the dictyosomes in a cell are structurally interassociated, only one Golgi apparatus is considered to be present (Fig. 1.41), but it is possible for a cell to have several Golgi apparatuses, each composed of one or more dictyosomes. The Golgi apparatus, therefore, varies in form and extent depending upon the type of cell and on its metabolic state.

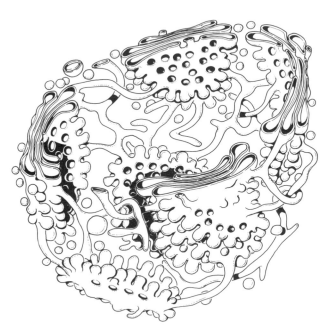

Fig. 1.41 Three-dimensional representation showing the interconnections between the dictyosomes of the Golgi apparatus. Redrawn from Krstic, R.V. (1975) *Ultrastructure of the Mammalian Cell*. Springer-Verlag, Heidelberg, Germany.

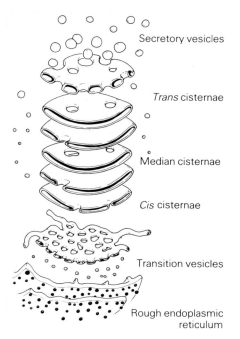

Fig. 1.42 Diagram to show the polarized nature of a dictyosome.

See *Molecular Biology and Biotechnology*, Chapter 5

DICTYOSOMES ARE POLARIZED, that is they have two opposing faces. These faces are now normally called the *cis* and *trans* faces (Fig. 1.42). Cisternae between the faces are called *median* cisternae.

Transition vesicles from the ER fuse to form the *cis* face of the Golgi apparatus (Fig. 1.42). Material for secretion from the cell such as zymogens and hormones is transferred to the Golgi apparatus in these vesicles. The secretory material is then transferred along the dictyosome in vesicles which bud from cisternae and migrate to, and fuse with, successive cisternae. **Secretory vesicles** are released from the distended rims of *trans* cisternae. Secretory vesicles contain concentrated amounts of material, which is released from the cell when the vesicles fuse with the plasma membrane (Fig. 1.42). This process is called **reverse endocytosis**.

In plant cells, materials required for cell wall deposition are transported out of the cell in vesicles derived from dictyosomes. Proteins destined for inclusion into the plasma membrane remain fixed in the ER membrane as described earlier. These are transported to the plasma membrane as part of the Golgi, then secretory vesicle membranes (Fig. 1.43). This emphasizes a peculiar feature of the endomembranes: the lumen of the system is topologically equivalent to the exterior of the cell.

BIOCHEMICAL ACTIVITIES OF THE GOLGI APPARATUS. Early views of the Golgi apparatus regarded it as a warehouse, where material for export from the cell was merely stored and concentrated, but it is now known that the Golgi apparatus is the site of many biosynthetic activities. For example, the degree of glycosylation of glycoproteins is modified and extended by **glycosidases** and **glycosyltransferases** present in Golgi membranes. Mannose residues are removed and *N*-acetylglucosamine residues added to glycoproteins in *medial* cisternae, while galactose, sialic acid and fucose residues extend the glycosylation in the terminal *trans* cisternae. Glycosylation of glycolipids also occurs in the Golgi apparatus.

The conversion of proproteins to mature, secretable forms of the proteins is catalysed by peptidases in Golgi cisternae and secretory vesicles. An example

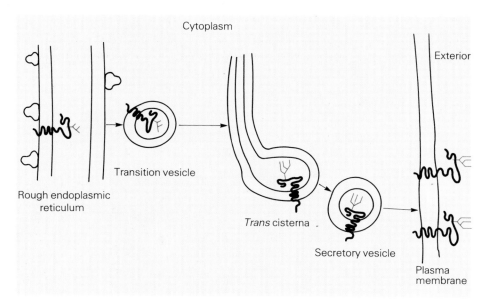

Fig. 1.43 Illustration of how integral proteins of the plasma membrane are inserted into the membrane.

is the conversion of proinsulin to insulin (Fig. 1.36). Sulphation and phosphorylation of lysosomal enzymes also occurs in the Golgi apparatus.

The fusion of secretory vesicles with the plasmalemma constantly increases the surface area of the cell and in active cells this could double the area of the plasma membrane in less than an hour. This is prevented by recycling of the vesicle membrane back to the Golgi apparatus.

The plasma membrane can also invaginate and pinch off vesicles for transport to other parts of the interior of the cell, a process called endocytosis.

Endocytosis

The most widely known forms of endocytosis are **pinocytosis** and **phagocytosis** which are responsible for the uptake of some soluble materials

Exercise 7

A cell of 20 μm diameter produces sufficient secretory vesicles in 30 min to replace completely its surface membrane. If the average diameter of a secretory vesicle is 250 nm, calculate the rate of vesicle production per hour by the Golgi apparatus. Assume both cell and vesicles are spherical.

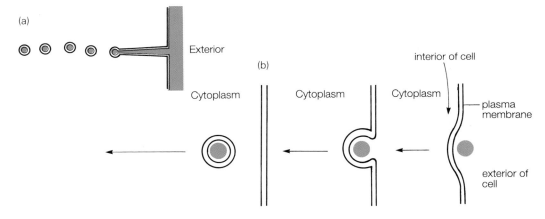

Fig. 1.44 Uptake of materials by the cell via (a) pinocytosis, (b) phagocytosis.

See *Biosynthesis*, Chapter 9

□ Transferrin is a serum protein which transports iron around the body. Any free iron would precipitate, as the solubility product of Fe (OH)$_3$ is about 4×10^{-36} mol dm^{-1}.

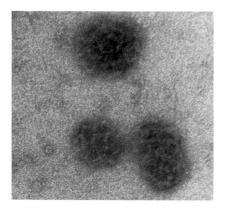

Fig. 1.45 Coated vesicle (×127 000). Courtesy of H.M. James, Department of Biological Sciences, The Manchester Metropolitan University, UK.

and suspended particles by cells (Fig. 1.44a,b). Receptor-mediated endocytosis takes place at specific sites on the plasma membrane called **coated pits**. Coated pits are lined with the protein **clathrin** on the cytoplasmic side of the membrane. Many ligands which must be taken up by the cell bind to receptors on the exterior of the coated pit. Such ligands include low-density lipoproteins rich in cholesterol, transport proteins such as transferrin and some peptide hormones. Binding to the receptors allows the ligands to be concentrated on to a small area of membrane which is then internalized by the formation of a **coated vesicle** (Fig. 1.45).

COATED VESICLES differ from 'normal' vesicles because they are surrounded by a cage-like coat of clathrin on their cytoplasmic surface. Other proteins also form part of the 'cage'. These additional proteins may act as recognition factors allowing coated vesicles to be directed to different intracellular sites. However, not all coated vesicles arise at coated pits on the cell surface. Cells appear to contain several populations of coated vesicles which transport materials between the Golgi apparatus and ER and plasma membranes. Coated vesicles are also important in the transfer of materials between the Golgi complex and lysosomes.

Lysosomes

Lysosomes are vesicular structures, limited by a single smooth membrane, from 50 nm to over 1 μm in diameter and containing hydrolases active at acid pH values (Fig. 1.46). Lysosomes arise initially as *primary* lysosomes, which appear to be derived from coated vesicles released from *trans* cisternae of the Golgi apparatus. Lysosomes contain about 60 hydrolytic enzymes whose concerted action will degrade most biological materials (Table 1.5). They were discovered by de Duve following the observation that the activity of acid hydrolases in homogenates of animal tissues increased with time. This led to the proposal that enzyme activity increased because of leakage from the organelle following lysis of its membrane; hence the name lysosome.

Coated vesicles arising from receptor-mediated endocytosis deliver their contents to a vesicle called an **endosome**. Fusion of an endosome with a

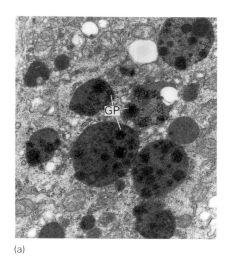

(a)

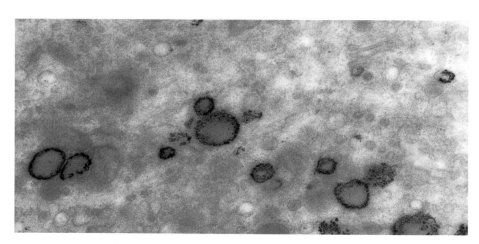

(b)

Fig. 1.46 (a) Lysosomes in a macrophage cell from a sufferer of rheumatoid arthritis who has been treated with gold injections. The gold particles (GP) present in some lysosomes are clearly visible (×14 110). (b) Lysosomes in a synovial cell visualized by aryl sulphatase activity (×19 380). Both photographs courtesy of Dr C.J.P. Jones, Department of Pathology, University of Manchester, UK.

Reference Castle, A.G. (1984) Phagocytosis. *Biologist*, **31**, 9–14. An extremely clear account of phagocytic activities of a variety of cells. Emphasizes the widespread importance of the process.

Reference Dautry-Varsat, A. and Lodish, H.F. (1984) How receptors bring proteins and particles into cells. *Scientific American*, **250(5)**, 48–54. A lucid, well-illustrated review article.

Table 1.5 *Examples of some lysosomal hydrolytic enzymes*

Enzyme	Biological substrate
Nucleic acid degradation	
Acid RNase	RNA
Acid DNase	DNA
Carbohydrate degradation	
β-Galactosidase	Galactosides
α-Glucosidase	Glycogen
Lysozyme	Bacterial cell walls
Mucopolysaccharide degradation	
β-Glucuronidase	Mucopolysaccharides
Lysozyme	Bacterial cell walls
Hyaluronidase	Hyaluronate, chondroitin sulphate
Protein degradation	
Cathepsins	Proteins
Collagenase	Collagen
Phosphatases	
Acid phosphatase	Most phosphomonoesters
Acid phosphodiesterase	Oligonucleotides
Sulphatase	
Arylsulphates	Organic sulphates

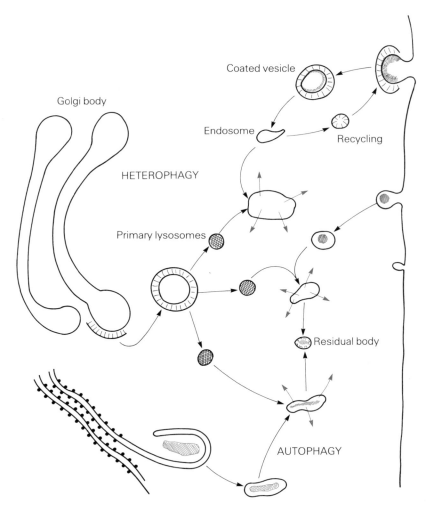

Fig. 1.47 Lysosomal activities.

Reference Farquar, M.G. (1983) Multiple pathways of exocytosis, endocytosis, and membrane recycling: validation of a Golgi route. *Federation Proceedings*, **42**, 2407–13.

A wide-ranging description of the pathways for membrane movements within the cell, emphasizing the importance of the Golgi apparatus and coated vesicles.

primary lysosome forms a *secondary* lysosome. The activity of a H⁺-ATPase in the lysosome membrane pumps protons into the intralysosomal space producing a pH of about 5 which activates the lysosomal enzymes. The fate of the ligand and receptor depends upon their specific nature. For example, peptide hormones are degraded while their receptors are recycled back to the plasma membrane (Fig. 1.47). However, either or both the receptor and ligand may be recycled, or degraded, or transported to other intracellular sites.

Phagocytic vesicles (**phagosomes**) also fuse with primary lysosomes to give secondary lysosomes. This allows digestion of the ingested material (digestive vesicle), and subsequent absorption of the products into the cytosol. Following digestive activity, a **residual body** containing non-degradable material may remain. Residual bodies are retained within the cell and may accumulate.

The digestion of *extra*cellular material is called **heterophagy**. However, apparently lysosomes can also degrade material of *intra*cellular origin, such as mitochondria and ribosomes. This process is called **autophagy** (Fig. 1.47).

It is apparent that the lysosomal membrane is unusual. Not only is it resistant to digestion by the hydrolases of the lysosome, but also, under normal circumstances, it is impermeable to both the enzymes and their substrates (Fig. 1.48). Substrates are delivered to the lysosome by fusion of the primary lysosome with other vesicles. Despite this, the lysosomal membrane is freely permeable to the low-molecular-weight products of hydrolysis.

☐ The Golgi apparatus is important in transferring membrane-limited vesicles back and forth between the endoplasmic reticulum, the plasmalemma and lysosomal systems. The Golgi apparatus is now recognized as occupying a central role in the intracellular traffic of membrane-limited vesicles.

LYSOSOMAL ENZYMES are synthesized on the RER. These are inserted into the ER membrane and glycosylated in the ER lumen (see earlier). The bound enzymes are then transported to the Golgi membrane. Lysosomal enzymes are distinguished from the other proteins in the Golgi complex because they are tagged with a **mannose 6-phosphate residue** in the *cis* cisternae (Fig. 1.49). These phosphorylated mannose groups allow the enzymes to be transported to domains of the Golgi membrane which will bud off to give primary

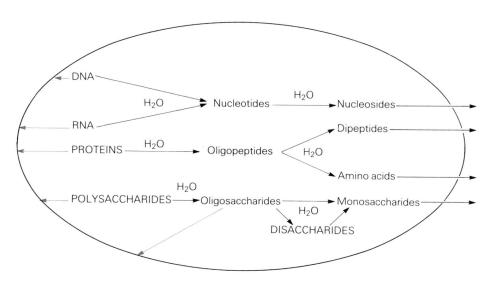

Fig. 1.48 Hydrolytic activities of lysosomes. Red arrows indicate materials to which the lysosomal membrane is impermeable.

Reference Dice, J.F. (1990) Peptide sequences that target cytosolic proteins for lysosomal proteolysis. *Trends in Biochemical Sciences*, **15**, 305–9. Typically readable *TIBS* article.

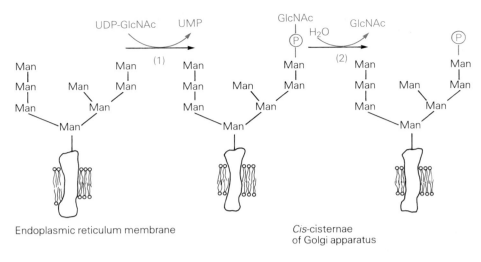

Man Man
Man Man Man
Man Man
Man

Endoplasmic reticulum membrane

Man Man
Man Man
Man

Cis-cisternae
of Golgi apparatus

Fig. 1.49 Phosphorylation of lysosomal enzymes. Step 1 is catalysed by UDP-*N*-acetylglucosaminyl-phosphotransferase, step 2 by GlcNAc phosphoglycosidase.

lysosomes. Intriguingly, the mannose-6-phosphate residues are removed once the enzymes reach the lysosome. Presumably this fixes the enzyme *in situ*, and preventing loss during membrane recycling activities.

Box 1.3
Clinical importance of lysosomes

Lysosomes are important in many clinical and medical aspects of biochemistry. For example, phagocytic cells in tissues of lungs and liver contain large lysosomes which are important in digesting foreign materials. Silicosis is a condition resulting from the inhalation of silica particles into the lungs which are then taken up by phagocytes. Reactions between the silica and the lysosomal membranes lead to the rupture of the membrane, the release of lysosomal enzyme and eventually the death of the phagocyte. The death of large numbers of phagocytes stimulates fibroblasts to deposit collagen fibres which decrease lung elasticity, impair breathing and contribute to the pathology of disease. Silicosis is similar to asbestosis and black lung disease, conditions caused by breathing in asbestos fibres and coal dust respectively.

The absence of specific hydrolase enzymes from lysosomes leads to the accumulation of substrate for that enzyme within the lysosomes. This usually has severe medical consequences. Many lysosomal storage diseases are known, each characterized by a specific enzyme defect. (See Biological Molecules, Chapter 7.)

In **inclusion cell** (I-cell) **disease**, hydrolases are secreted from the cell and this may have several underlying causes. One is a deficiency in *N*-acetylglucosaminyl-phosphotransferase activity. Thus, lysosomal enzymes are not recognized as such by the Golgi-sorting machinery, leading to their secretion.

Absence of mannose 6-phosphate receptors from the Golgi or plasma membranes are further causes. Lack of receptors in the Golgi membrane will mean the enzymes are not recognized and directed to the lysosomal system. Absence of plasmalemma receptor means that secreted enzyme would not be recognized at the cell surface, endocytosed and delivered to the lysosomal system. Receptors on the cell surface normally function as a salvage system to capture lysosomal enzymes that escape the usual Golgi-sorting apparatus.

In inflammation and diseases such as arthritis and in autoimmune diseases, lysosomal hydrolytic enzymes are released from phagocytes and other cells causing damage. Release may be due to cell death. Alternatively, it may be from living cells stimulated by abnormal conditions.

Reference Bainton, D.L. (1981) The discovery of lysosomes. *Journal of Cell Biology*, **91**, 665–765. A clear, interesting and informative account of the discovery of lysosomes and peroxisomes.

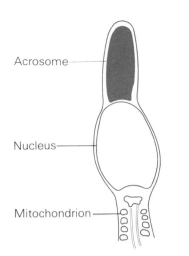

Acrosome

Nucleus

Mitochondrion

Fig. 1.50 Diagrammatic representation of head of a spermatozoon.

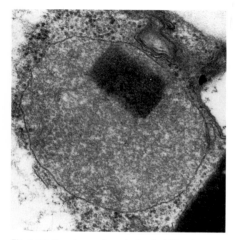

Fig. 1.51 Peroxisome from leaf cell of *Nicotiana tabacum* (×36 000). Note the large crystal, probably of catalase. Courtesy of Dr E. Sheffield, Department of Cell and Structural Biology, University of Manchester, UK.

See *Energy in Biological Systems*

☐ Peroxisomes are thought by many to be the relics of organelles having a role in energy metabolism but now superseded by the mitochondria.

ACROSOMES are specialized lysosomes which cover the head of animal spermatozoa (Fig. 1.50) and arise from the fusion of Golgi vesicles. Fertilization occurs when a spermatozoon fuses with an ovum. The acrosomal and plasma membranes fuse, allowing the acrosomal hydrolytic enzymes to be released into the surrounding medium. The outer coat of the ovum is then digested by the acrosomal enzymes, allowing the spermatozoon to contact the plasmalemma of the ovum. The plasma membranes of both then fuse, the cytoplasm of the ovum engulfing the nucleus of the spermatozoon. Fusion of the nuclei of ovum and spermatozoon completes fertilization.

DEGRADATIVE PROCESSES are often dependent upon lysosomal activities. For example, lysosomal enzymes are mainly responsible for the dissolution of larval tissue, such as during the metamorphosis of tadpoles into frogs. Extracellular degradations of tissues can be dependent upon lysosomal enzymes released from the cell. **Fibroblasts** release hydrolases to break down connective tissue before ossification can proceed. Bone is not a static tissue, but is constantly being degraded and rebuilt by **osteoclasts** and **osteoblasts** respectively. The extracellular release of lysosomal hydrolytic enzymes by osteoclasts plays an important role in bone reabsorption. Osteoclasts also secrete lactate to create a microenvironment of suitably low pH for optimal enzyme activity.

Microbodies

Microbodies are a heterogeneous group of small vesicle-like organelles, concerned largely with oxidation. They are usually oval or spherical with a diameter of 0.2–1.7 μm and bounded by a single smooth membrane. Microbodies are found in the liver and kidneys of vertebrates, in the leaves and seeds of plants as well as in protozoa, yeasts and other fungi. Microbodies consist of **peroxisomes**, **glyoxisomes**, **hydrogenosomes** and **glycosomes**. Peroxisomes and glyoxisomes are the best characterized of these.

PEROXISOMES were discovered by de Duve (Fig. 1.51) and they contain the enzymes **urate oxidase**, D-**amino acid oxidase** and **catalase**. Urate oxidase and catalase are synthesized in the cytosol, and transferred to the peroxisome by means of a **transit signal** which recognizes the peroxisomal membrane. This is similar to the method by which some mitochondrial and chloroplast proteins are transferred from the cytosol to their sites of activity (see Section 5.5).

Urate oxidase is a purine-catabolizing enzyme, converting urate into allantoin, carbon dioxide and hydrogen peroxide. Thus peroxisomes play a role in the breakdown of nitrogenous bases derived from nucleic acids but they are also involved in the degradation of L-α-hydroxyacids to oxoacids (and hydrogen peroxide) and of oxoacids to smaller products including acetyl CoA. The enzymes involved in this oxidation are different from those of the β-oxidation pathway of mitochondria.

D-Amino acid oxidase and catalase have protective functions. D-Amino acids may be absorbed from the gut following their release by the breakdown of the cell walls of gut bacteria. These 'unnatural' amino acids are degraded to give oxoacids and hydrogen peroxide (Fig. 1.52). Hydrogen peroxide is a powerful oxidizing agent and is potentially toxic. It is produced by the peroxisome-based reactions described above, as well as other cellular activities. Catalase catalyses the rapid degradation of H_2O_2 to water and oxygen. Alternatively, by a peroxidate process, it may be used to oxidize other substrates, such as phenols, aldehydes and ethanol (Fig. 1.52). Ethanol is produced by intestinal microorganisms, in addition to any taken with the diet, and is readily absorbed. The potentially dangerous effects of hydrogen peroxide are, therefore, eliminated by the activity of catalase.

References de Duve, C. (1983) Microbodies in the living cell. *Scientific American*, **248(5)**, 52–62. A wide-ranging review of the four known types of microbodies, by the discoverer of the peroxisome.

Reference Connock, M. (1986) New functions for peroxisomes. *Biologist*, **33**, 279–83. Clear, informative description of activities of peroxisomes and glycosomes.

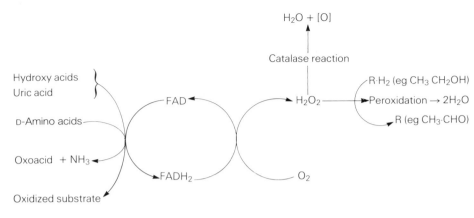

Fig. 1.52 Oxidative reactions of peroxisomes.

GLYOXYSOMES are microbodies of plant cells, in which the enzymes of the glyoxylate cycle are functionally more important than those involved in oxidative mechanisms. The glyoxylate cycle allows the relatively immobile fatty reserves of, for example, seeds, to be converted to sugars, and therefore more easily transported to growing tissues. The fats are degraded to acetyl CoA which feeds into the glyoxylate cycle to give succinate as a net product (Fig. 1.53). Succinate is then transported to the mitochondria where it becomes a substrate for gluconeogenesis.

See *Biosynthesis*, Chapter 3

HYDROGENOSOMES are a type of microbody found only in a group of protozoa called the **trichomonatids**. Although these flagellates lack mitochondria their hydrogenosomes contain enzymes that are able to catalyse the oxidation of pyruvate to acetate and CO_2 generating ATP in the process

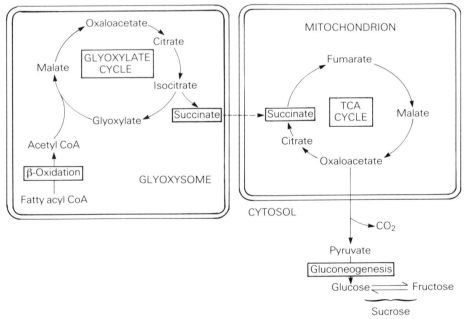

Fig. 1.53 Functional interrelationship of glyoxysomes and mitochondria.

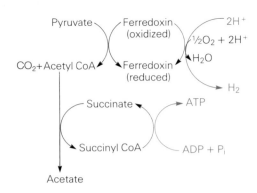

Pyruvate → CO$_2$+Acetyl CoA

Ferredoxin (oxidized) ⇄ Ferredoxin (reduced)

2H$^+$

½O$_2$ + 2H$^+$ → H$_2$O

H$_2$

Succinate ⇄ Succinyl CoA

ATP

ADP + P$_i$

Acetate

Fig. 1.54 Oxidative reactions and formation of ATP in hydrogenosomes.

(Fig. 1.54). If oxygen is available, the electrons released by this oxidation are transferred to oxygen. In anaerobic conditions, however, H$^+$ acts as the electron acceptor resulting in the production of molecular hydrogen. A further feature of trichomonatids is that their flavin–oxidase–catalase system is located in the cytosol and hence there is controversy over the classification of the hydrogenosome as a microbody.

Box 1.4
Clinical importance of microbodies

Trichomonatids have clinical importance as parasites of the genital tract of mammals, including humans. Since hydrogenosomes contain ferredoxin, they have the ability to reduce nitroimidazole derivatives to compounds that are highly toxic to the parasite. Treatment with nitroimidazole drugs is therefore standard for trichomonal infections. It is interesting that pathogenic strains of these organisms that have become resistant to this therapy have been isolated.

Trypanosomes are the causative agents of a number of debilitating diseases, including the sleeping sickness spread by tsetse flies in Africa, Chagas' disease of Latin America and leishmaniasis-type diseases which occur in various parts of the world. The glycolytic enzymes of the glycosome present in trypanosomes differ molecularly from the glycolytic enzymes found in the cytosol of the host cells. This fact is the basis of the search for drugs which will selectively poison the glycolytic enzymes of the trypanosome, but leave those of the host cells unaffected.

Brazilian 10 000 cruzado banknote featuring Professor Chagas.

Reference Fulton, A.B. (1982) How crowded is the cytoplasm? *Cell*, **30**, 345–7. An interesting minireview of the organization of water and proteins within the cytosol.

GLYCOSOMES are found in **trypanosomes**, a different group of flagellated protozoans. The major function of glycosomes seems to be the catabolism of glucose to smaller carbon compounds via some of the reactions of the glycolytic pathway. In all other organisms these reactions occur in the **cytosol** rather than in organelles. In aerobic conditions, glycosomes cofunction with mitochondria in the production of ATP; in anaerobic conditions the glycosome appears to be able to meet all the ATP requirements of the organism. The classification of glycosomes is controversial. Both are biochemically different, and probably unrelated, to peroxisomes.

1.7 The cytosol

The cytosol contributes up to 55% of the cell volume. It is about 20% by weight protein and has a gel-like consistency. Many of the proteins of the cytosol are enzymes concerned with intermediary metabolism. Thus the reactions of glycolysis and gluconeogenesis occur there, as do those concerned with the biosynthesis of other sugars, fatty acids, nucleotides and amino acids.

See Energy in Biological Systems and Biosynthesis

The cytosol of particular cell types also contains a variety of inclusions. For example, liver and skeletal muscle cells contain numerous glycogen granules, as energy stores. White adipose tissue cells contain droplets of triacylglycerols which often coalesce to give a single droplet occupying most of the cell's volume. Prominent granules, present in the cytosol of all cells, are the **ribosomes**. The subunits of ribosomes are highly complex aggregates of RNA and protein and are formed in the nucleolus from rRNA molecules and ribosomal proteins which are synthesized on pre-existing ribosomes. The role of ribosomes is to mediate protein biosynthesis.

The cytoskeleton

Amongst the most prominent structural proteins of the cell are those that make up the **cytoskeleton**.

The cytoskeleton is a complex, three-dimensional system of fibres ramifying throughout the cytosol. The cytoskeleton has important functions in cell mobility during, for example, embryonic development; the movement of organelles within the cell (for example, in secretion, phagocytosis), and in the separation of chromosomes during cell division. It also influences cell shape and is, therefore, important in cytodifferentiation. Three types of fibre are present in the cytoskeleton: **microfilaments**, **intermediate filaments** and **microtubules**.

MICROFILAMENTS have a diameter of 7 nm. They are composed of the protein **actin**, which can constitute up to 15% of the protein in actively mobile cells. It is, therefore, not surprising that microfilaments are the major *dynamic* constituent of the skeleton, allowing cells to move and change shape. The contractile role of actin has received most attention in skeletal muscle cells. The activity of actin within the cytoskeleton is controlled by a large number of **actin-associated proteins**.

INTERMEDIATE FILAMENTS have diameters of 8–11 nm, and are so-called simply because they are 'intermediate' in diameter between actin filaments and microtubules. They are found in virtually all types of vertebrate cells, and have an essentially similar, characteristic appearance when examined by electron microscopy.

Reference Goodsell D.S. (1991) Inside a living cell. *Trends in Biochemical Sciences*, **16**, 203–6. Interesting account of the crowded nature of cytoplasm. Fascinating diagrams.

MICROTUBULES are normally present as single fibres of about 25 nm diameter. They are composed of the protein **tubulin** (M_r 55 000). Microtubules form a scaffold maintaining the position of organelles, and stabilizing the shape of the cell, while giving the gel of the cytosol a more organized structure. Microtubules are able to dissociate and radically reorganize. For example, in dividing cells the microtubules are reassociated to form the mitotic spindle. Microtubules can also occur as *doublets* in **cilia** and **flagella**, and as *triplets* in their **basal bodies** and in **centrioles**. Basal bodies and centrioles are microtubule *nucleating centres*, from which the fibres grow, and are often called **microtubule organizing centres** (MTOCs).

CENTRIOLES consist of two hollow cylindrical bodies, usually lying at right angles to one another (Fig. 1.55). Each cylinder is about 0.4 μm long by 0.15 μm diameter. The walls of the cylinder consist of nine sets of microtubules, each set being made up of a triplet of fibres. The centrioles are embedded in a dense granular substance called **pericentriolar material** (PCM) (Fig. 1.55). Centrioles plus the PCM are normally called the **centrosome**. Centrioles are restricted to animals, protozoa and some fungi.

During the cell cycle, centrioles replicate producing daughter centrioles at right angles to the parental centrioles. The migration of the centrosomes to opposite poles of the cell allows the microtubules to be reorganized into a spindle apparatus, facilitating separation of the chromosomes at anaphase.

CILIA AND FLAGELLA of eukaryotic cells are built up from a microtubular framework enclosed by extensions of the cell membrane. Both have diameters of about 0.5 μm, but while cilia are 2–10 μm long, flagella are about 100–200 μm in length (Fig. 1.56a, b). In both cilia and flagella the microtubules are arranged in nine pairs running the length of the appendage, with two single

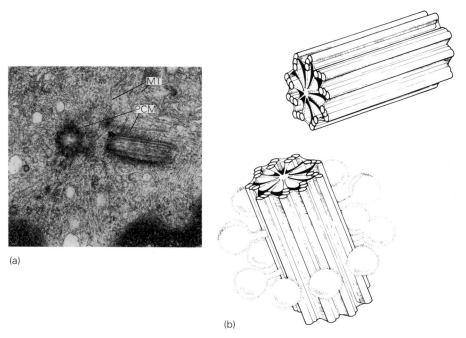

(a)

(b)

Fig. 1.55 (a) Composite TEM of a centrosome (×14 100). MT, microtubules; PCM, pericentriolar material. (b) Diagram showing centrosome. Redrawn from Krstic, R.V. (1979) *Ultrastructure of the Mammalian Cell.* Springer-Verlag, Heidelberg, Germany.

Reference Gibbons, I.R. (1981) Cilia and flagella of eukaryotes. *Journal of Cell Biology*, **91**, 107s–124s. A good starting place for learning about these appendages.

Reference Karsenti, E. and Maro, N. (1986) Centrosomes and the spatial distribution of microtubules in animal cells. *Trends in Biochemical Sciences*, **11**, 460–3. A short informative account of the centrosome and its influence upon microtubule structures.

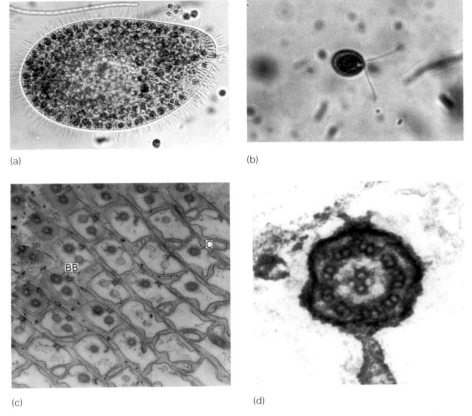

(a)

(b)

(c)

(d)

Fig. 1.56 (a) Photomicrograph of *Paramecium* (×380) showing surrounding cilia. (b) Photomicrograph of a *Chlamydomonas* showing two flagella (×610). Courtesy of M.J. Hoult, Department of Biological Sciences, The Manchester Metropolitan University, UK. (c) TEM through pellicle of *Paramecium* in surface plane showing cross-sections of cilia (C) and basal bodies (BB) (×10 050). Courtesy of P.L. Carter, Department of Biological Sciences, The Manchester Metropolitan University, UK. (d) TEM of cross-section of flagellum of *Pandorina* (×130 000).

tubules arranged down the centre (Fig. 1.56c, d). This so-called *9+2* arrangement is distinct from the *9 + 0* motif described for centrioles which is also found in basal bodies (Fig. 1.56c).

Basal bodies, or kinetosomes, of cilia and flagella have an identical structure to centrioles. They also act as microtubule organizing centres organizing the fibres of the cilia or flagella which emanate from these structures. Like centrioles, basal bodies replicate independently of the cell. New cilia and flagella can be generated from the new daughter basal bodies.

Cilia and flagella are mobile appendages. The hydrolysis of ATP is used to power the sliding of the microtubules past another, allowing the appendage to beat. Cells lining the air passage of vertebrate lungs have cilia whose co-ordinated beats sweep mucus and trapped particles from the inside of the lungs. Some unicellular organisms, such as the protozoan *Paramecium*, use cilia for locomotion while others use flagella as do spermatozoa.

1.8 Compartmentation of eukaryotic cells

Eukaryotic cells show a high level of structural complexity and their interior is divided into many membrane-limited compartments. Prokaryotes are structurally less complex, only Cyanobacteria showing any degree of internal membrane organization (Fig. 1.20), but are nevertheless a highly diverse,

Reference The whole issue of *BioEssays* (1987) **7(4)**. Has several articles devoted to different aspects of the cytoskeleton.

Box 1.5
Enveloped virus and cell compartments

Certain viruses have evolved to exploit the compartments of eukaryotic cells to a remarkable degree. One well-studied example is the Semliki Forest Virus (SFV).

The SFV infects a variety of animals. It is a togavirus; that is, the nucleocapsid or protein and nucleic acid is surrounded by a membrane (Latin: *toga*). The SFV has an RNA genome surrounded by only one type of capsid or C protein. The membrane is derived from the host cell's plasma membrane but contains three types of viral glycoproteins.

The viral glycoproteins recognize and bind to coated pits of host cells, and the virion is internalized in a coated vesicle. This vesicle fuses with an endosome which delivers the virion to a lysosome. However, the virion is not digested but instead fusion of the viral and lysosomal membranes releases the nucleocapsid into the cytosol. The viral RNA is released, and translated by the host's protein-producing machinery. This produces an RNA-dependent RNA polymerase, which codes for the production of more viral genomes. Translation of approximately the last third of the viral RNA gives the four viral proteins. The C protein is produced first, and on its completion is cleaved from the growing polypeptide. This exposes a signal sequence allowing the ribosome to dock with the rough endoplasmic reticulum. Further translation produces the viral glycoproteins which are inserted into the endoplasmic reticulum membrane, the Golgi and finally the plasma membrane. During their passage through the endomembrane system the proteins are glycosylated.

In the cytosol nucleocapsids are assembled from the C proteins and RNA genomes and migrate to the plasmalemma. Here a section of membrane, containing the viral glycoproteins wraps around each nucleocapsid, and the complete viral particle buds off from the cell, completing a marvellously adapted life cycle.

successful form of life. However, it has to be remembered that eukaryotic cells are larger than prokaryotic cells. The plasma membrane of prokaryotes carries out many functions associated with the internal membranes of eukaryotes, including electron transport and ATP production. It is apparent, therefore, that the increased size of eukaryotic cells must be associated with convolutions of their membranes, if the ratio of cell volume to membrane surface is to be maintained. Dividing the cytoplasm into separate compartments offers other advantages too. It provides enclosed intracellular compartments within which newly synthesized products can be transported. Further, the division of the cytoplasm into separate compartments means that distinct subenvironments, adapted to different functions, can be maintained within the cell. This value is, perhaps, best demonstrated in active lysosomes where a pH of 5 necessary for optimal lysosomal enzyme activity is normal but is considerably lower than the pH of about 7.4 typical of the rest of the cytoplasm.

The increased structural complexity of eukaryotic cells allows an elaborate division of labour which has enabled eukaryotic cells to reach high orders of functional complexity, and is, presumably, the basis of metazoan organization.

1.9 Cell fractionation

The functions of different parts of the cell may be investigated by disrupting the cell and separating it into its individual compartments. The process is called cell fractionation and consists of:

1. disruption or homogenization of cells, usually in a supportive iso-osmotic medium, to give a homogenate, and
2. separation of the particles and organelles in the homogenate for individual study.

Reference de Duve, C. and Beaufay, H. (1981) A short history of tissue fractionation. *Journal of Cell Biology*, **91**, 293s–299s. The title is self explanatory. C de Duve is one of the pioneers in studying the structure and activities of the cell.

HOMOGENIZATION is the disruption of the cells, generally in a carefully formulated aqueous medium. Osmotic pressure (usually provided by $0.25 \, mol \, dm^{-3}$ sucrose) and pH (normally buffered at about 7.0) are maintained at constant values, while the temperature is usually kept at 0–4°C. Mg^{2+} to stabilize membranes and β-mercaptoethanol ($2–5 \, mmol \, dm^{-3}$) to preserve protein structure and enzyme activities are often included.

Cells are disrupted in a variety of ways. The simplest is a **pestle and mortar** in which cells are ground up, often with sand or glass microbeads. This is the most commonly used procedure for disrupting plant and bacterial cells, where the presence of a tough cell wall makes homogenization difficult. **Ultrasonic vibrations** are also often used in disrupting bacterial cells. Animal cells are much less tough and are most often disrupted using **Waring blenders**, which can also be used with plant tissues, or **Potter homogenizers**. Waring blenders are like food-processors and have high-speed rotating blades (Fig. 1.57). Potter homogenizers consist of a cylindrical pestle of Teflon or glass which is rapidly rotated while being moved up and down in a stout glass tube of diameter slightly greater than the pestle (Fig. 1.58). The shearing forces generated depend upon the speed of rotation, the clearance between the pestle and container (0.1–0.15 mm), and the ratio of tissue to homogenizing fluid.

Fig. 1.57 A Waring blender.

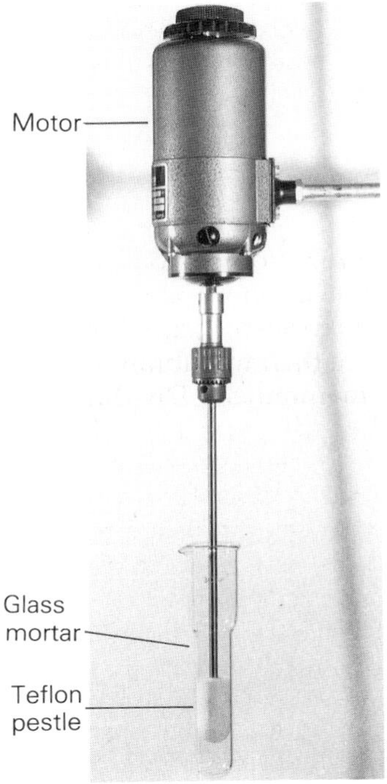

Motor

Glass mortar

Teflon pestle

Fig. 1.58 A Potter homogenizer.

The effectiveness of homogenization depends upon the amount of connective and vascular tissue present. The method works best for soft tissues, such as liver and kidney, but with many tissues, such as heart muscle or mammary gland, it is necessary to force the tissue through a fine steel mesh before homogenization to remove tough fibres.

SEPARATION OF THE HOMOGENATE into subcellular fractions is nearly always done by **centrifugation**. Organelles or particles which differ in size, density and shape sediment at different rates when subjected to an applied centrifugal field (G) and may therefore be separated from one another.

☐ Centrifugation of an homogenate will separate it into two parts: a 'pellet' at the bottom of the centrifuge tube, and a liquid 'supernatant' containing the unpelleted material.

homogenate: *of the same composition throughout. From the Greek* homos, *same; and* genos, *kind. Despite the term, homogenates are not of consistent composition throughout, and contain connective tissue fibres, unbroken cells, and organelles at various stages of disruption.*

References de Duve, C. (1975) Exploring cells with a centrifuge. *Science*, **189**, 186–94. The Nobel lecture by one of the founders of cell biology.

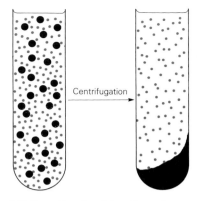

Fig. 1.59 Separation of particles of large size difference by differential centrifugation.

Differential centrifugation allows mixtures of particles to be separated on the basis of differences in their sizes. Homogenates are separated into a number of fractions by a stepwise increase in the relative centrifugal force (RCF). The values of **g** are chosen to separate organelles as shown, for example, in Figure 1.59. Differential centrifugation is mainly used to separate particles which differ in size by at least an order of magnitude. The main disadvantage of differential centrifugation is that the pellets are always contaminated, because of the initial uniform distribution of material in the centrifuge tube (Fig. 1.60).

Rate-zonal or **density-gradient centrifugation** exploits small differences in densities between particles. A small volume of homogenate is layered as a zone over a density gradient, often of sucrose. The density continuously increases down the tube (Fig. 1.61). The sample is centrifuged until the particles form discrete bands at intervals down the gradient. This technique allows organelles or membrane fractions of similar sizes, but different densities, to be separated.

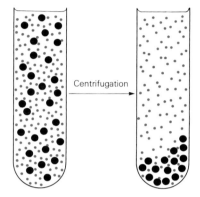

Fig. 1.60 Contamination of pellets of large particles with smaller particles, following differential centrifugation.

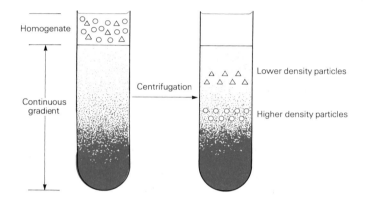

Fig. 1.61 Separation of particles of similar sizes but differing densities by rate-zonal centrifugation.

Table 1.6 *Approximate densities and diameters of some cellular organelles and membranes*

Organelle/membrane	Density (g cm^{-3})	Diameter (μm)
Golgi apparatus	1.06–1.10	–
Plasma membrane	1.16	–
Smooth endoplasmic reticulum membrane	1.16	–
Mitochondria	1.19	0.5–1.0
Lysosomes	1.21	0.5–2.0
Peroxisomes	1.23	0.2–1.5
Ribosomes	1.60–1.75	0.025

Exercise 8

A cell homogenate contains only three organelles which have diameters and densities of 10 μm, 1.30 g cm^{-3}; 1.3 μm, 1.20 g cm^{-3}; and 1.0 μm; 1.24 g cm^{-3}, respectively. Describe briefly how centrifugal techniques could be applied to separate these organelles. Details of speeds and times of centrifugation are not required.

Table 1.7 *Some marker enzymes which can be used to identify particular subcellular fractions*

Fraction	Marker enzyme
Nuclei	DNA polymerase
Mitochondria	Succinate dehydrogenase
Chloroplasts	Ribulose bisphosphate carboxylaseoxygenase
Endoplasmic reticulum	Glucose-6-phosphatase
Golgi apparatus	Galactosyltransferase
Lysosomes	Acid phosphatase
Microbodies	Catalase
Plasma membrane	Na$^+$,K$^+$-ATPase
Cytosol	Lactate dehydrogenase

A *combination* of differential and rate-zonal centrifugation applied to cellular homogenates (Table 1.6) allows cells to be fractionated into their various compartments. Cell organelles are identified not only by their appearance in electron microscopy, but by the activities of **diagnostic enzyme markers** (Table 1.7).

1.10 *Overview*

Advances in optical microscopy and the application of electron microscopy has allowed the structures of cells to be examined in fine detail. These studies have shown that eukaryotic cells are surrounded by a limiting membrane, with an interior divided into a system of membrane-bounded compartments. The disruption of cells followed by fractionation of the resulting homogenate has shown that the different cellular compartments are specialized for specific biochemical activities.

The nucleus of the cell contains the genome and is responsible for controlling cellular activities. Communications between the nucleus and cytoplasm appear to occur via pores in the nuclear envelope. Mitochondria and chloroplasts are complex organelles and are the main energy-transforming organelles of the cell. The endomembrane system of the cell consists of a series of membrane-limited compartments (endoplasmic reticulum, Golgi apparatus) important in biosynthetic activities, and the transfer of materials to the exterior of the cell. Several large vesicular organelles have important specialized functions. Lysosomes have roles in intracellular digestive activities, while microbodies generally perform specialized oxidative functions. Materials are transported between sub-cellular organelles by means of vesicles. Coated vesicles show the greatest structural organization, and probably occur in several different forms.

The shape and mobility of the cell is allowed by a system of fibrous proteins ramifying throughout the cytoplasm. The integrity of this cytoskeleton is organized to some extent by microtubule organizing centres, of which centrosomes are the most widespread.

Answers to Exercises

1. $r = \dfrac{0.61}{NA} = \dfrac{0.61 \times 260}{1.4} = 113$ nm

It is usually impracticable to use UV because human eyes cannot detect it, and glass absorbs it; silica lenses would have to be used which are either unavailable or extremely expensive.

2. Use dimmer switch on light source or insert absorbent filters into light path. Adjustments to iris diaphragm should not be made since this will affect u and resolution.

3. $\sqrt{\dfrac{1.5}{V}} = \sqrt{\dfrac{1.5}{1\,000\,000}} = 1.2 \times 10^{-3}$ nm

4. Area of image
   $= 0.18 \times 10^9 \times 0.15 \times 10^9$ nm^2

   Area of beam
   $= \dfrac{0.18 \times 10^9 \times 0.15 \times 10^9}{20\,000}$ nm^2
   $= 1.35 \times 10^{12}$ nm^2

5. Length of mitochondrion
   $= \sim 140$ units, width $= 5$ units

   circumference
   $= \sim (140 \times 2) + (55 \times 2)$ units

   Total length crista $= \sim 750$ units

   Area crista $= \sim 750 \times 2$ units

   % increase in area $=$

   $\sim \dfrac{1500 - ((140 \times 2) + (55 \times 2))}{(140 \times 2) + (55 \times 2)} \times 100$

   $= \sim 300\%$

6. These reactions are dependent upon the availability of NADPH and ATP, and so can occur in either light or dark conditions.

7. Area of cell surface
   $= 4\pi r^2 = 4\pi\ 10\,000^2$ nm^2

   Area of vesicles
   $= 4\pi r^2 = 4\pi\ 125^2$ nm^2

   No. of vesicles in 60 min
   $= \dfrac{10\,000^2}{125^2} \times 2 = 12\,800$

8. Use differential centrifugation to sediment particles of 10 μm diameter. Load supernatant on to sucrose gradient to separate particles of density 1.20 and 1.24 g cm^{-3} respectively.

FILL IN THE BLANKS

1. The _____ consists of _____ , intermediate _____ and _____ . The first consists largely of the protein _____ , the last _____ . Intermediate _____ are composed of a variety of proteins which are _____ . _____ form a scaffold in the cell. They are organized by _____ _____ _____ _____ . Centrosomes consist of two _____ and _____ _____ _____ . The _____ is formed during cell division by a _____ of the _____ .

Choose from: actin, cell-specific, centrioles, cytoskeleton, filaments (2 occurrences), microfilaments, microtubule organizing centres (MOCs), microtubules (3 occurrences), pericentriolar material (PCM), reorganization, spindle, tubulin.

MULTIPLE-CHOICE QUESTIONS

2. Indicate which item in the following groups is the odd one out:
A. nucleus, chloroplast, lysosome, mitochondrion
B. hydrogenosome, lysosome, peroxisome, glycosome
C. smooth endoplasmic reticulum, rough endoplasmic reticulum, Golgi apparatus, centrosome
D. centrosome, microtubules, microfilaments, centrioles

3. Arrange the following two columns in appropriate pairs linking their roles:

ATP synthesis	nucleus
protein synthesis	lysosomes
genome	hydrogenosomes
glycosylation	ribosomes
hydrolytic enzymes	mitochondria
H_2-production	Golgi apparatus

4 Indicate whether the following statements are true or false:
A. Signal sequences allowing translocations across the endoplasmic reticulum membrane are rich in basic amino acid residues.
B. Cilia and flagella always have a function in locomotion.
C. The Golgi apparatus is the only route for protein secretion.
D. All microtubule organizing centres show a common structure.

SHORT-ANSWER QUESTIONS

5. Arrange the following terms in an appropriate sequence to describe endocytosis: residual body, coated vesicle, secondary lysosome, coated pit, endosome, primary lysosome.

6. Calculate the weight of protons (H^+) which must be pumped into a lysosome of diameter 1.0 μm to reduce its pH from 7.0 to 5.0. If the transfer of two protons requires the hydrolysis of a molecule of ATP calculate the number of ATP molecules required to effect this reduction [Avogadro's number = 6.02×10^{23}].

7. The following table lists partial data associated with the isolation of an organelle specifically enriched in the enzyme **markerase**.

Step	Volume (cm^3)	Total protein (mg)	Markerase activity (U)	Specific activity (U mg^{-1})	Purification (fold)
Homogenate	120	1800	5.4×10^{-3}		1
Differential centrifugation	20*	180		2.83×10^{-5}	9.4
Rate-zonal centrifugation	6	40	4.9×10^{-3}	1.23×10^{-4}	

* Following resuspension of pellet.

(a) Complete the table.
(b) What is the average yield of the organelle?

ESSAY QUESTION

8. Write an essay explaining how the uptake of ^{14}C-leucine introduced into the cytosol of a cell can eventually become localized in primary lysosomes.

2

Bacteria and viruses

Objectives

After reading this chapter you should be able to:

☐ describe the biology and chemistry of bacteria, stressing the similarities with and differences from eukaryotic cells;

☐ relate the structure of viruses to their biological activities through the study of selected examples of bacterial, plant and animal viruses;

☐ discuss the importance of bacteria and viruses in the study of biochemistry and molecular biology, in health and in disease.

2.1 Introduction

Microorganisms or microbes are those organisms, most of which are unicellular, that are invisible to the naked eye and may only be seen using a microscope. Conventionally, microorganisms have been divided into four groups: fungi, protozoa, bacteria and viruses (Fig. 2.1). From a structural point of view, this classification may be extended as follows: both fungi and protozoans have differentiated nuclei with nuclear membranes, and are **eukaryotes**. Bacterial cells lack a differentiated nucleus and are **prokaryotes**. Viruses are not cellular in structure and cannot be said to be living.

Eukaryotic microorganisms are, for the biochemist, very much like higher organisms (Fig. 2.2) and will not be considered further in this chapter. Rather, attention will be focussed on prokaryotic organisms and viruses.

Two groups of bacteria may be distinguished: *eubacteria* and *archaebacteria*. Although both are morphologically similar, they are so different in phylogeny and molecular architecture that, from the biochemical point of view, they can be considered as different from each other as they are from eukaryotic cells (Fig. 2.3). Consequently, in this section the molecular structure of eubacteria and archaebacteria will be considered separately.

2.2 Eubacteria

The main structures that can be distinguished in eubacteria are surface structures, vacuoles and storage granules, and the nucleoid region (Fig. 2.4; Table 2.1). The major surface structures are the plasma membrane (the only surface structure common to all cells), cell wall, capsule, flagella and pili.

☐ An organism is a system capable of self-replication and self-repair that keeps itself at a minimum entropy level while increasing the entropy of its environment. Viruses are incapable of actively 'pumping out' entropy to the environment.

Exercise 1

Assume that the average diameter is 12 μm for an animal cell, 1.2 μm for a bacterium and 0.12 μm for a virus, and that all three are spheroid. Estimate the ratio of volumes: animal cell/bacterium/virus.

---

eubacteria and **archaebacteria:** *formed with the Greek roots* eu *and* archae *meaning good (or true in the sense of 'normal') and ancient respectively;* bacterium *is Greek for little staff.*

References Stanier, R.Y. *et al.* (1987) *General Microbiology*, 5th edn, Macmillan, London, UK. A textbook in basic microbiology, describing in great detail the phylogeny and structure of bacteria.

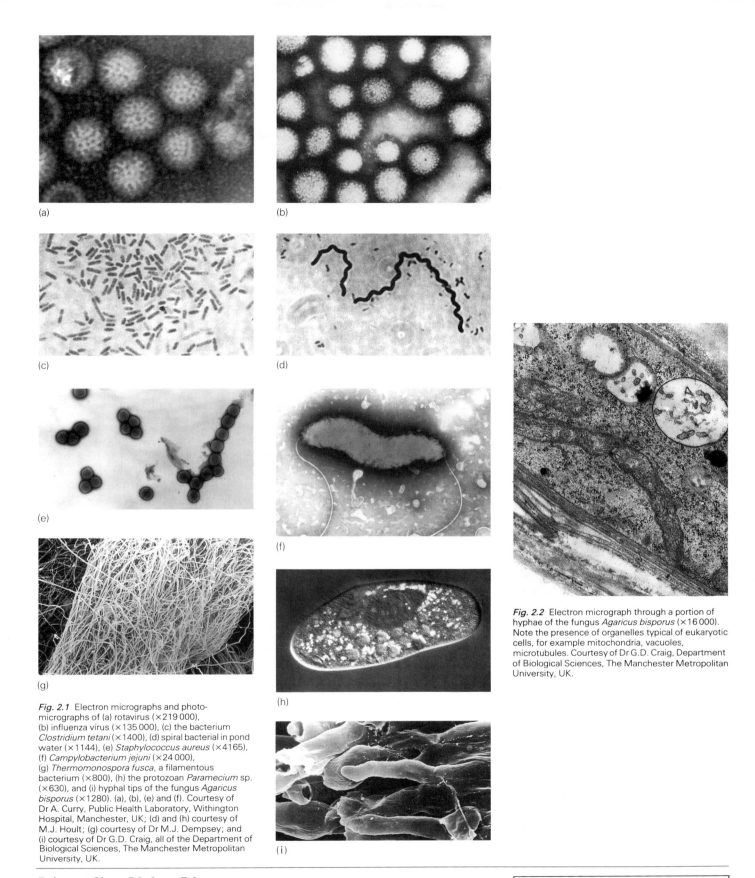

(a)

(b)

(c)

(d)

(e)

(f)

(g)

(h)

(i)

Fig. 2.1 Electron micrographs and photo-micrographs of (a) rotavirus (×219 000), (b) influenza virus (×135 000), (c) the bacterium *Clostridium tetani* (×1400), (d) spiral bacterial in pond water (×1144), (e) *Staphylococcus aureus* (×4165), (f) *Campylobacterium jejuni* (×24 000), (g) *Thermomonospora fusca*, a filamentous bacterium (×800), (h) the protozoan *Paramecium* sp. (×630), and (i) hyphal tips of the fungus *Agaricus bisporus* (×1280). (a), (b), (e) and (f). Courtesy of Dr A. Curry, Public Health Laboratory, Withington Hospital, Manchester, UK; (d) and (h) courtesy of M.J. Hoult; (g) courtesy of Dr M.J. Dempsey; and (i) courtesy of Dr G.D. Craig, all of the Department of Biological Sciences, The Manchester Metropolitan University, UK.

Fig. 2.2 Electron micrograph through a portion of hyphae of the fungus *Agaricus bisporus* (×16 000). Note the presence of organelles typical of eukaryotic cells, for example mitochondria, vacuoles, microtubules. Courtesy of Dr G.D. Craig, Department of Biological Sciences, The Manchester Metropolitan University, UK.

Reference Olsen, G.J., Lane, D.J., Giovannoni, S.J., Pace, N.R. and Stahl, D.A. (1986) Microbial ecology and evolution: a ribosomal approach. *Annual Review of Microbiology*, **40**, 337–66. Describes how the tools of molecular biology are applied to the study of microbial evolution.

Table 2.1 The structure of prokaryotic cells

Surface structures
    plasma membrane
    cell wall
    capsule
    flagella and pili
Vacuoles and storage granules
    gas vacuoles
    storage granules
    sulphur inclusions
Nucleoid
Cytoplasm
    ribosomes

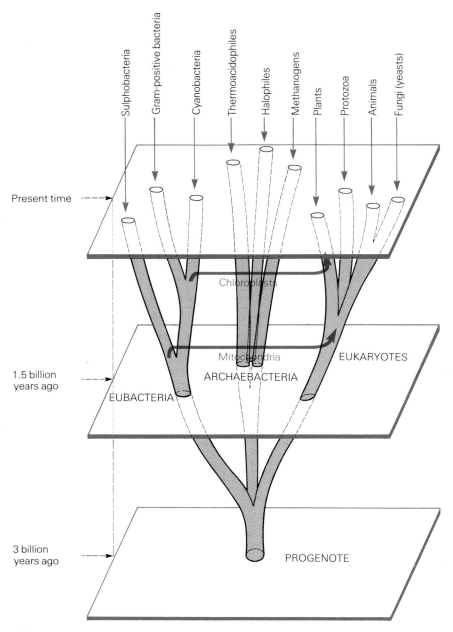

Fig. 2.3 An outline of the probable phylogenetic evolution of organisms.

Bacterial plasma membranes

These have a basic structure similar to that of all biological membranes. They consists of phospholipid bilayers with which specific proteins are associated, either through polar forces (extrinsic or peripheral proteins) or through hydrophobic bonds (intrinsic or integral proteins). From the chemical point of view, eubacterial membranes possess some peculiarities: (a) they do not contain sterols; (b) the phospholipid composition varies strikingly between species, and phospholipid classes not found in eukaryotes or archaebacteria are often present, and (c) membrane fluidity is regulated mainly through changes in fatty acid branching, instead of by varying the degree of fatty acid unsaturation, as occurs in eukaryote membranes.

See Chapter 4

Box 2.1
Cyanobacteria

See *Biosynthesis*, Chapters 2 and 5

Cyanobacteria, or blue-green bacteria, are Gram-negative photosynthetic eubacteria. They can perform oxygenic photosynthesis as well as nitrogen fixation. Consequently, cyanobacteria can grow in the light in a mineral medium exposed to air. Air provides them with CO_2 (carbon source) and N_2 (nitrogen source). Since nitrogen fixation is an anaerobic process, it does not coexist with oxygenic photosynthesis. Cyanobacteria resolve this problem by performing nitrogen fixation in a specialized type of cell called the heterocyst.

Cyanobacteria frequently form symbiotic associations with eukaryotic organisms, either providing fixed carbon to a heterotrophic partner, or fixed nitrogen to another photoautotropic organism. An example of agricultural importance is the symbiosis between the aquatic fern *Azolla* and the cyanobacterium *Anabaena azollae*. This symbiotic union provides a nitrogen source for rice cultivation in the paddy fields of southeast Asia, where *Azolla* is grown on the surface of paddy water among the rice plants.

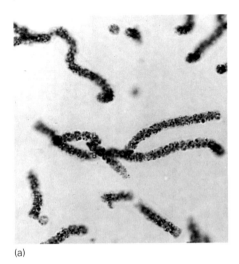
(a)

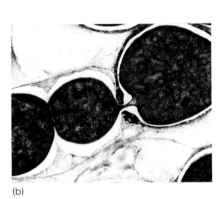
(b)

(a) Photomicrograph of free-living *Nostoc* sp. (cyanobacterium) (×900). Courtesy of M.J. Hoult, Department of Biological Sciences, The Manchester Metropolitan University, UK.
(b) Electron micrograph of symbiotic *Nostoc* sp. in a *Cycas*, showing vegetative cells and a thick-walled heterocyst (×8250). Courtesy of Professor G.A. Codd and Ms G. Alexandre, Department of Biological Sciences, University of Dundee, UK.

The bacterial plasma membrane is complex since, in addition to the functions carried out by eukaryotic plasma membranes, it also has to perform those of mitochondria, chloroplasts, and other membranous organelles found in eukaryotes. Thus, the bacterial plasma membrane contains permeases necessary for the uptake of various nutrients, together with electron transport chain systems for respiration and/or photosynthesis; ATP synthase molecules; enzymes for the biosynthesis of membrane and cell wall components, and binding sites for the bacterial chromosome.

This functional complexity may be the result of 'attempts' at morphological differentiation that occur in several eubacteria resulting in structures such as chromatophores or mesosomes. **Mesosomes** are membrane invaginations close to the cell division zone, that might be related to the formation of the transverse septum. **Chromatophores** contain the photosynthetic reaction centres of purple bacteria (Fig. 2.5).

☐ Mesosomes are now considered by many to be probable artefacts of the fixation of the bacterial cells for electron microscopy. However, they often appear at regions of the cell membrane associated with cell division. This supports the contention that the bacterial cell surface membrane is, indeed, differentiated into specialized regions.

☐ Chromatophores are readily purified in high yield. This is why most of our knowledge of bacterial photosynthesis relies on experimental data from photosynthetic bacteria, e.g. *Rhodopseudomonas*.

Reference Shuman, H.A. (1987) The genetics of active transport in bacteria. *Annual Review of Genetics*, **21**, 155–78. A clear and updated presentation, illustrating the application of molecular genetics to elucidate the structure and function of bacterial permeases.

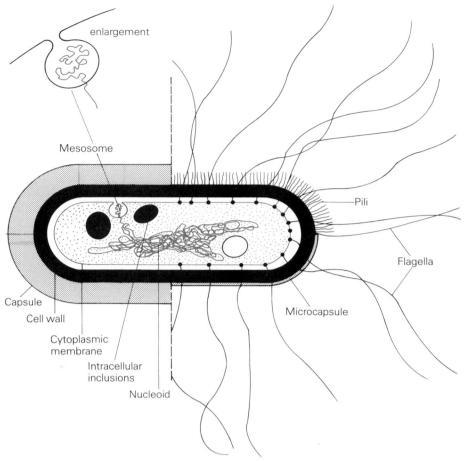

enlargement

Mesosome

Pili

Capsule

Cell wall

Flagella

Cytoplasmic
membrane

Intracellular
inclusions

Microcapsule

Nucleoid

Fig. 2.4 Principal structures of a generalized bacterial cell. To the left of the dotted line is shown a capsulated, non-flagellate rod and to the right a flagellate rod. Redrawn from Hawker, L.E. and Linton, A.H. (1979) *Microrganisms*, 2nd edn, Edward Arnold, London, UK.

Fig. 2.5 Electron micrograph showing (a) chromatophores in a photosynthetic bacterium (×17 400), and (b) a mesosome (×174 000). In both cases the structures appear separate from the cell membrane because the intervening membrane has been cut during specimen preparation. (a) Courtesy of Biophoto Associates. (b) Courtesy of Dr A. Curry, Public Health Laboratory, Withington Hospital, Manchester, UK.

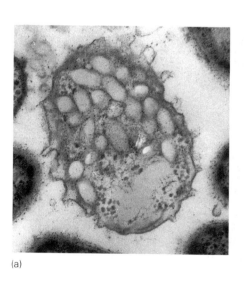

(a)

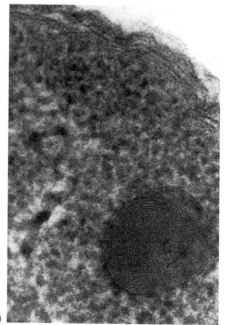

(b)

Reference Neidhardt, F.C., Ingram, J.L. and Schaechter, M. (1990) *Physiology of the Bacterial Cell*, Sinauer Associates, Inc., Sunderland, Massachusetts, USA. Splendid up-to-the-minute text. Sticks to its title.

Cell walls

Bacterial cell walls are required for an independent life since they offer protection against mechanical and osmotic shocks. Cell walls are present in all bacteria except **mollicutes**, formerly called **mycoplasams** (Fig. 2.6). For more than 100 years, bacteria have been divided into Gram-positive and Gram-negative, depending upon their retention of the crystal violet–iodine (complex) stain, the so-called Gram stain. It is known today that this difference is based on the chemical and structural characteristics of the cell wall (Fig. 2.7). In general, only Gram-negative bacteria contain lipids in their cell walls (Table 2.2).

□ *Acholeplasma laidlawii* is a mollicute that is extensively used in membrane research. The absence of cell wall makes the plasma membrane easy to isolate. In addition it is a fatty acid *auxotroph*, i.e. the bacterium incorporates virtually any kind of fatty acid in the medium into its lipids. This allows many studies on the roles of fatty acids in membrane physiology.

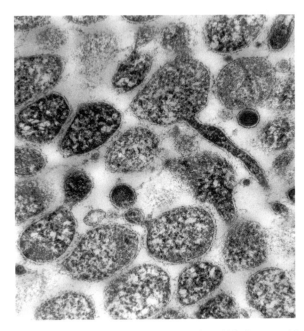

Fig. 2.6 Electron micrograph of the mollicute, *Mycoplasma iowae* (×35 800). Courtesy of Dr. J. Shareef, Department of Biological Sciences, The Manchester Metropolitan University, UK.

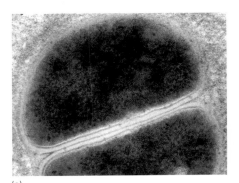

(a)

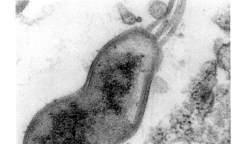

(b)

Fig. 2.7 Electron micrograph showing (a) two adjacent Gram-positive bacterial cells (*Streptococcus* sp.) (×60 000), and (b) a portion of the wall of a Gram-negative bacterium (*Gastospirillium* sp. from baboon stomach) (×40 000). Note the uniform appearance of the Gram-positive cell wall and the complex, multilayered structure of the Gram-negative. Courtesy of Dr A. Curry, Public Health Laboratory, Withington Hospital, Manchester, UK.

Table 2.2 *Surface structures of the prokaryotic cell with typical dimensions and chemical composition*

Structure	Location	Shape and size	Chemical composition
Membrane	Bounding layer of protoplast	Bilayer 8 nm wide	25% phospholipid 75% protein (by weight)
Cell wall	Layer immediately external to membrane	Gram-negative eubacteria: inner layer 2 nm, outer layer 7 nm	Peptidoglycan (murein) Phospholipids, proteins
		Gram-positive eubacteria: homogeneous layer 40 nm wide	Peptidoglycan (murein); teichoic acids; polysaccharides
		Archaebacteria: variable	Variable
Capsule	Diffuse layer external to wall	Homogeneous low-density structure	Diverse; usually a polysaccharide
Flagella	Anchors in protoplast traversing envelopes	Helical threads, 12 nm wide	Protein
Pili	Anchors in protoplast traversing envelopes	Straight threads, 20 nm wide	Protein

From Stanier, R.Y. *et al.* (1987) *General Microbiology*, 5th edn, Macmillan, London, UK, p. 146.

Box 2.2
Gram-positive and
Gram-negative bacteria

Bacteria are usually divided into Gram-positive and Gram-negative according to their behaviour towards a differential staining method introduced by the Danish bacteriologist Gram (1853–1938). Bacterial cells, heat-fixed on a microscope slide, are stained blue with crystal violet (Step 1). The so-called 'Lugol solution' (iodine and potassium iodide) is then added (Step 2). Iodine forms a complex with the blue stain that is water-insoluble but soluble in ethanol or acetone. Alcohol cannot penetrate the 'Gram-positive' bacterial cell walls because of their thick, hydrophilic, lipid-free composition. It can, however, diffuse across the lipid-rich cell wall of 'Gram-negative' bacteria. Consequently, ethanol treatment (Step 3) dissolves and removes the iodine-stain complexes of Gram-negative, but not of Gram-positive bacteria. Gram-negative bacteria are usually visualized after staining with a 'contrast stain' such as saffranin or basic fuchsin (Step 4). At the end of the process, Gram-positive bacteria appear stained blue–black while Gram-negative species are red.

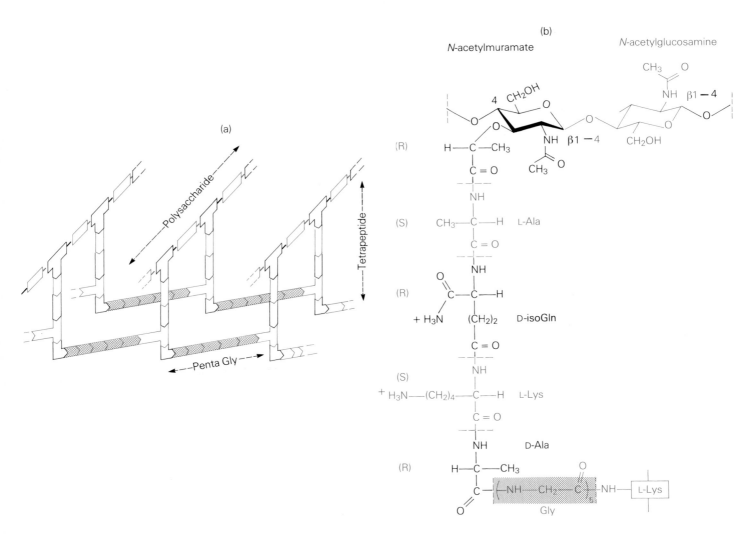

Fig. 2.8 (a) Schematic representation and (b) chemical structure of the repeating subunit peptidoglycan from *Staphylococcus aureus*.

An important type of macromolecule present in the cell walls of both kinds of eubacteria is **peptidoglycan** or **murein**. The peptidoglycan cell wall is a single molecule consisting of a bag-shaped, two-dimensional net. Sometimes as many as 20 of these network layers are superimposed and cross-linked to form the core of the cell wall. The fundamental structure of peptidoglycan consists of parallel chains of an acetylated polysaccharide, covalently linked through a branched oligopeptide (Fig. 2.8).

The peptidoglycan of *Staphylococcus aureus* has been studied in great detail. Its polysaccharide contains two sugar residues *N*-acetylglucosamine and *N*-acetylmuramate (glucosamine plus *ether*-bound propionate), in alternating sequence. The residues are connected by β1–4 glycosidic linkages, which impose a 180° rotation to each residue with respect to its neighbours. Note the structural resemblance to cellulose and chitin. The free carboxyl group of *N*-acetylmuramate is linked through an amide bond to the free amino end of an oligopeptide. In *S. aureus* this peptide contains L-alanine, D-isoglutamine, L-lysine and D-alanine. Isoglutamine receives its name because the carboxyl group involved in the peptide bond is the one in the side-chain, while the α-carboxyl is aminated. A cross-bridge consisting of five glycine residues links the carboxyl group of a terminal D-alanine to the ε-amino group of a lysine from a parallel chain. The regular repetition of this structure allows the parallel polysaccharide chains to be assembled through a series of transverse peptide chains giving rise to the two-dimensional network that was mentioned above.

In other bacterial species, isoglutamate substitutes for isoglutamine and, instead of L-lysine, L-hydroxylysine, L-ornithine or *meso*-diaminopimelate may be found. The glycine pentapeptide can also be substituted by other amino acid sequences, or, as in the case of *Escherichia coli*, by the direct link of the terminal D-alanine carboxyl group to the side amino group of a *meso*-diaminopimelate residue from a different chain.

Gram-positive bacterial cell walls contain, in addition to peptidoglycan, **polysaccharides** (differing widely according to species) and **teichoic acids** (Fig. 2.9). These are polymers of ribitol or glycerol linked through phosphodiester bonds; sometimes the polyalcohol is also linked to D-alanine, sugars or other biomolecules. The majority of teichoic acid molecules are covalently linked to peptidoglycan (wall teichoic acid) while a smaller fraction is attached to membrane glycolipids (membrane teichoic acid). The physiological role of teichoic acids is unknown, although they provide the cell wall with a network of negative charges. In addition, teichoic acids constitute major surface antigens in Gram-positive bacteria.

Gram-negative bacteria have only a small proportion of peptidoglycan in their cell wall. The wall also contains phospholipids, proteins and lipopolysaccharides. Lipopolysaccharides are characteristic of these cell walls.

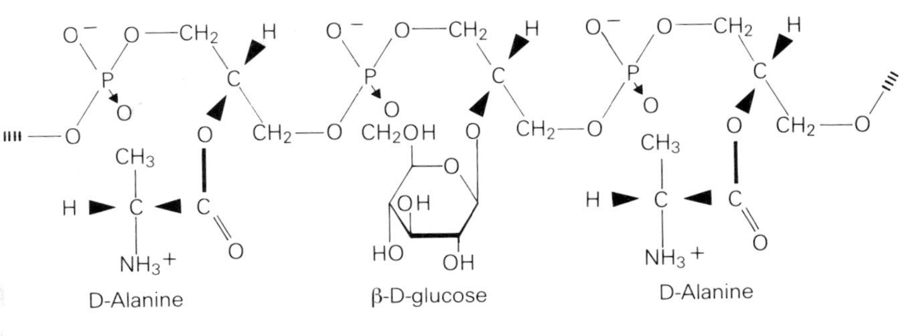

Fig. 2.9 Structure of the glyceroteichoate from *Lactobacillus arabinosus*.

Reference Baddiley, J. (1989) Bacterial cell walls and membranes. Discovery of the techoic acids. *Bioessays*, **10**, 207–10. A short essay which describes the discovery and probable functions of the techoic acids of Gram-positive cells.

Exercise 2

The peptidoglycan from the *Escherichia coli* cell wall differs from that found in *Staphylococcus aureus* in that the peptide chain is formed by L-alanine, D-glutamate, mesodiaminopimelate (mDAP, or 2R, 6S-diaminoheptane-dioate), and D-alanine. Interchain bonds exist between the terminal D-alanine and an mDAP from a different chain. Outline a possible structure for the two-dimensional lattice of *E. coli* peptidoglycan.

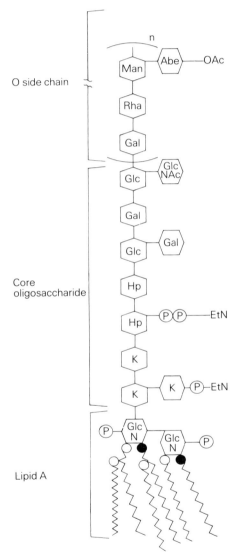

Fig. 2.10 Representation of the *Salmonella* lipopolysaccharide. Man, mannose; Abe, abequose; Rha, rhamnose; Gal, galactose; Glc, glucose; GlcNAc, *N*-acetylglucosamine; Hp, heptose; EtN, ethanolamine; K, 2-keto-3-deoxyoctonate; GlcN, glucosamine; ⓟ phosphate.

Salmonella are enterobacteria, i.e. bacteria usually found in the digestive tract of animals. *S. typhimurium* often lives in the intestine of birds, including chickens, without causing disease. The surfaces of eggs may carry residues of chicken faeces, and, with them, the bacterium. If this gains access to foods, rapid multiplication occurs, and ingestion of such dishes may lead to food-poisoning.*

Some egg-based emulsions, such as custard or mayonnaise, provide an ideal medium for the growth of *Salmonella*. Thus, it is strongly recommended that these should be consumed immediately after preparation. Industrial foods have been treated by pasteurization or otherwise, so that no active bacteria remain. Thus, at least from this point of view, they are safer than the home-made equivalents.

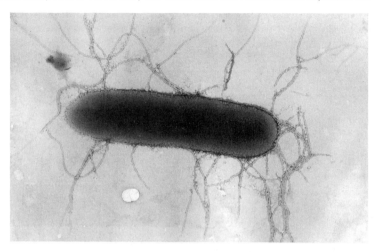

Electron micrograph of *Salmonella typhimurium* (×11 760). Courtesy of Dr A. Curry, Public Health Laboratories, Withington Hospital, Manchester, UK.

* Recently, an increasing number of egg-associated food poisoning cases have been caused by *S. entiridis*.

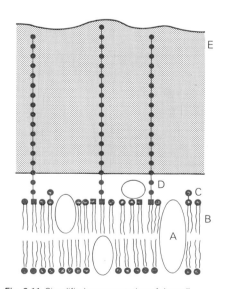

Fig. 2.11 Simplified representation of the cell wall structure of a Gram-positive bacterium. A, membrane protein; B, phospholipid; C, phosphatidyl glycolipid; D, glycolipid; E, cell wall.

These have an M_r in excess of 10 000 and show great chemical variability, according to species.

In the lipopolysaccharide of *Salmonella typhimurium*, the best studied example, three regions can be distinguished: the so-called lipid A, the core region R and the O side-chain (Fig. 2.10). **Lipid A** consists of a disaccharide to which two phosphate groups and six fatty acid residues are attached. One of the sugar residues is also covalently linked to a branched oligosaccharide (core oligosaccharide) formed by otherwise uncommon sugars. The core oligosaccharide, together with lipid A, forms the so-called **R core**. In turn, one of the sugar residues of the core oligosaccharide is attached to a branched polysaccharide, also consisting of uncommon sugars, the **O side chain**. Despite its chemical complexity, the lipopolysaccharide resembles membrane phospholipids in that it is *amphipathic*. In fact, both kinds of molecules share a number of physical properties.

The cell walls of Gram-positive bacteria appear homogeous when examined by electron microscopy (Fig. 2.7). They have a variable thickness of 10–80 nm, according to species. The fundamental structure is a single peptidoglycan macromolecule which surrounds the cell in a bag-like manner. The peptidoglycan framework is reinforced by cross-linking to chains of wall teichoic acids, while the membrane teichoic acids anchor the cell wall to the plasma membrane (Fig. 2.11).

The cell walls of Gram-negative bacteria are complex. A series of lipids and

amphipathic: *from the Greek for 'both feelings'. Molecules containing two regions or domains, one is lipophilic and one is hydrophilic, hence their peculiar physical properties.*

proteins form what is often called the **outer membrane** which covers a thin layer of peptidoglycan. The presence of two membranes defines a **periplasmic space**, containing some characteristic proteins (digestive enzymes). In the outer membrane, the inner lipid monolayer is formed of phospholipids, and the outer layer of lipopolysaccharide (Fig. 2.12).

The outer membrane contains proteins, mainly porins, which form unspecific channels, while other proteins function as selective permeases. The outer membrane is linked to the peptidoglycan layer through the **murein lipoprotein**. The carboxyl-terminus of this protein (M_r 7000) is covalently bound to peptidoglycan. Three fatty acid residues are covalently bound to its amino-terminus. These, in turn, are integrated into the inner monolayer of the outer membrane ensuring the peptidoglycan and inner monolayer are securely linked together. In addition, the cytoplasmic and outer membranes have contact regions called Bayer's junctions, which are of structural and

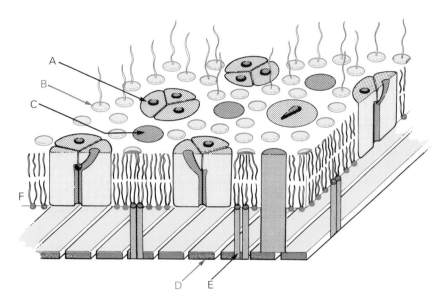

Fig. 2.12 A simplified representation of the cell wall structure of a Gram-negative bacterium (*E. coli*). Compare with the structure of a Gram-positive bacterium (Fig. 2.11). A, trimers of porin protein; B, lipopolysaccharide; C, outer membrane protein (OmpA); D, peptidoglycan (murein); E, murein lipoprotein; F, phospholipid.

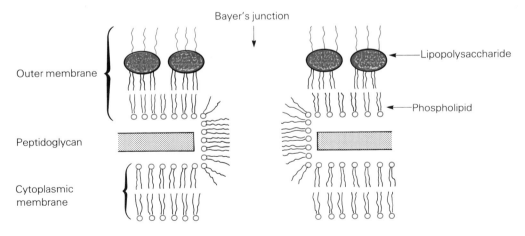

Fig. 2.13 Molecular structure of a Bayer's junction.

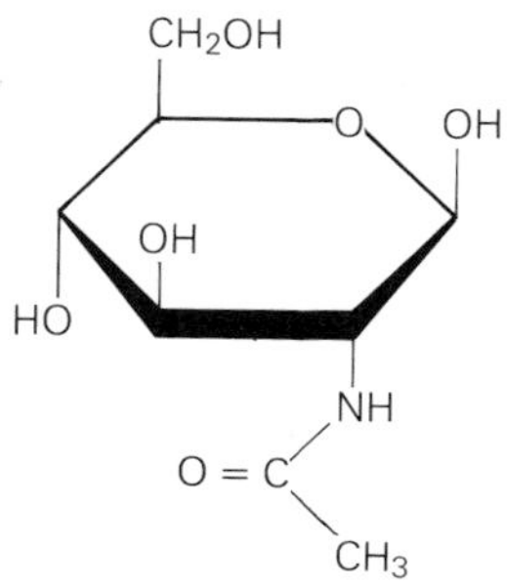

N-Acetylglucosamine

N-Acetylmuramate

Fig. 2.14 Structures of N-acetylglucosamine and N-acetylmuramate.

functional interest (Fig. 2.13). In these regions the outer layer of the plasma membrane is continuous with the inner layer of the outer membrane. The protein and lipid components of the outer membrane, synthesized on the inner side of the plasma membrane, are translocated through the Bayer's junctions to their final destination.

Biosynthesis of peptidoglycan

Peptidoglycan monomers are assembled in the cytosol, linked to a lipidic carrier (either bactoprenol or undecaprenol) and are transported across the plasma membrane. The monomers (N-acetylglucosamine and N-acetyl-muramate, Fig. 2.14) are polymerized on the outer face of the membrane, giving rise to intermediate structural forms about 30 disaccharide residues long. The latter diffuse in the periplasmic space until they reach the free ends of nascent peptidoglycan, where a **transpeptidase** binds the new unit while cleaving a terminal D-alanine (Fig. 2.15). β-Lactam antibiotics, such as penicillin, are bacteriostatic because they inhibit transpeptidase activity. As cells divide, the newly formed cells are highly sensitive to mechanical and osmotic shocks because they lack a cell wall.

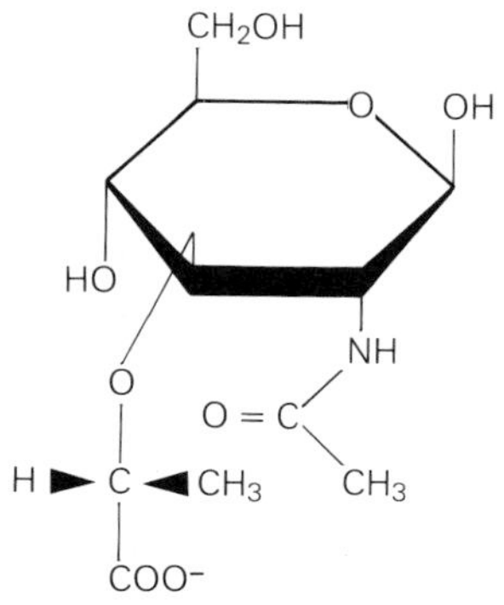

Capsules

Many bacteria possess an amorphous envelope of organic polymers called a capsule or slime layer which is found *outside* the cell wall (Fig. 2.16). In general, the capsule is formed by polysaccharides. Examples are the glucan in *Agrobacterium* or the typical pneumococcal polysaccharides of *Streptococcus pneumoniae* (Fig. 2.17). In the bacterial life cycle, the capsule performs only

---

Box 2.4
Antibacterial therapeutic agents

Chemicals which act as antimicrobial agents may, for convenience, be divided into three groups.

- *Disinfectants and antiseptics.* Disinfection kills disease-producing organisms and destroys their (toxic) products. Antiseptic agents act on microorganisms preventing their multiplication but do not necessarily kill them. They are applied externally to the body of invalids or to wounds or excretions. Common disinfectants and antiseptics include soaps and detergents, heavy metal compounds, such as mercurichrome, ethanol, phenols and cresol, and halogen compounds (e.g. iodine).
- *Chemotherapeutic agents.* These are drugs which, administered orally or parenterally to an animal or human, *selectively* destroy microbes without injuring the host. Examples include sulphonamides, which are competitive inhibitors of folic acid biosynthesis in many bacteria, and tuberculostatic drugs, such as isoniazid, which is a potent bacteriostatic agent against *Mycobacterium tuberculosis*.
- *Antibiotics.* These are substances, generally produced by microorganisms, which have the capacity to inhibit bacterial growth or to kill bacteria or other microorganisms. An enormous variety of antibiotics is known. Antibiotics may act through one or more of the following modes of action: (i) inhibition of cell wall synthesis (penicillin, bacitracin); (ii) interference with nucleic acid synthesis (actinomycin D, griseofulvin); (iii) inhibition of protein synthesis (chloramphenicol, streptomycin, erythromycin, tetracycline), and (iv) impairment of plasma membrane function (colistin, polymyxins).

Bacterial diseases can be classified according to the route of entry of the pathogenic agent:

- *Via the digestive tract.* Examples include typhoid fever (*Salmonella typhi*), enteric fever (*Salmonella typhimurium* among others), cholera (*Vibrio cholerae*), and bacterial dysentery (*Shigella dysenteriae*).
- *Via wounds.* This is the case with tetanus (*Clostridium tetani*), gangrene (*Clostridium perfringens*), brucellosis (*Brucella melitensis*), gonorrhoea (*Neisseria gonorrhoeae*), and syphilis (*Treponema pallidum*). Typically the clostridia are anaerobic and only thrive in deep, dirty wounds that lack a blood supply. Other organisms in this group may be transmitted via small lesions on the genitals.
- *Via exhaled droplets.* This is the infection route for diphtheria (*Corynebacterium diphtheriae*), tuberculosis (*Mycobacterium*), meningococcal meningitis (*Neisseria meningitidis*), or pneumococcic pneumonia (*Streptococcus pneumoniae*).
- *Via animal bites.* The most typical example is plague produced by *Yersinia* (formerly *Pasteurella*) *pestis*.

ancillary functions. It appears to be involved in cell–cell interactions and may provide some protection against antibacterial agents.

Some bacteria are capable of producing enormous amounts of extracellular material (Fig. 2.18).

☐ Two strains of the bacterium *Streptococcus pneumoniae*, only one of which was pathogenic, were used in the famous experiments of Griffiths and Avery and coworkers to demonstrate that the genetic material was nucleic acid.

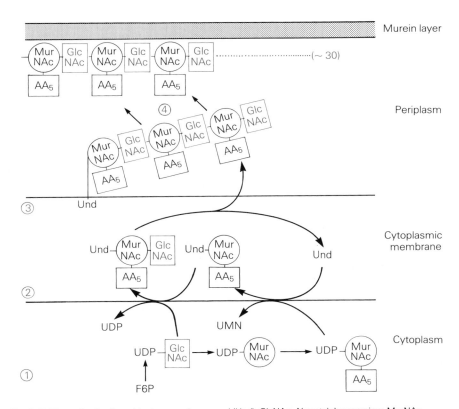

Fig. 2.15 Biosynthesis of peptidoglycan undecaprenol (Und). GlcNAc, *N*-acetylglucosamine; MurNAc, *N*-acetylmuraminate.

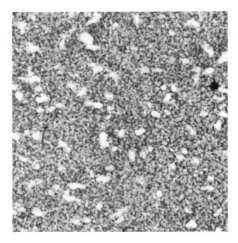

Fig. 2.16 The presence of capsules surrounding cells of *Klebsiella* sp. is indicated by the exclusion of Indian ink forming clear zones (×1140). Courtesy of Ms A.A. Leahy-Gilmartin, Department of Biological Sciences, The Manchester Metropolitan University, UK.

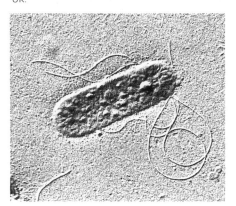

Fig. 2.19 Electron micrograph of *E. coli* with flagella (×13 000). Courtesy of P.L. Carter, Department of Biological Sciences, The Manchester Metropolitan University, UK.

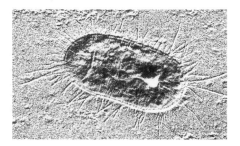

Fig. 2.20 Electron micrograph of *E. coli* with numerous pili (×16 200). Courtesy of P.L. Carter, Department of Biological Sciences, The Manchester Metropolitan University, UK.

☐ Sex pili are coded by a gene present in the F plasmid of *Escherichia coli*. Bacterial contact through sex pili may lead to transfer of genetic material (conjugation) between donor and recipient bacterial cells.

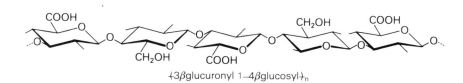

$+3\beta$glucuronyl 1—4βglucosyl$+_n$

Fig. 2.17 Chemical structure of the Type III capsule polysaccharide of *S. pneumoniae*.

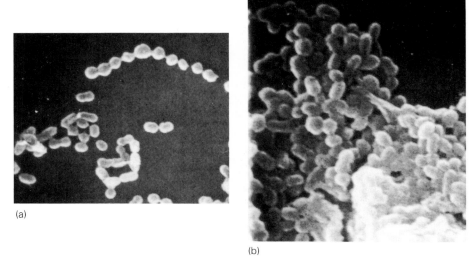

(a)

(b)

Fig. 2.18 Electron micrograph of *Streptococcus mutans*, an organism associated with dental plaque. (a) Growth in the absence of sugar (×2850), and (b) growth in the presence of sucrose. Note the presence of large amounts of extracellular polysaccharide in (b) allowing the bacterial cells to adhere to one another (×3660). Courtesy of Dr J. Verran, Department of Biological Sciences, The Manchester Metropolitan University, UK.

Flagella and pili

Flagella are extended protein structures, protruding from the cell, that are found in many kinds of bacteria, including vibrios, and bacilli to which they confer mobility (Fig. 2.19). They typically have a diameter of around 15 nm and may be up to 20 μm in length. Flagella are formed by the supramolecular association of subunits of flagellin, a protein of M_r 40 000 in *Salmonella*, which forms helical chains around an empty core. Flagellar motion depends directly on the electrochemical potential gradient across the cell membrane, since motion is coupled to the entry of protons into the cell through the basal end of the flagellum. Structures known as flagella and cilia can also be found in certain eukaryotic cells, but, despite their identical names, are of very different size, composition and structure.

Pili are similar in appearance to flagella, but are shorter, thinner, and are not responsible for bacterial motility (Fig. 2.20). **Pilin**, the protein subunit of pili from *E. coli*, has M_r 17 000 and forms helical assemblies about 7 nm in diameter. Pili allow the attachment of bacteria to the substratum, to eukaryotic cells, or, in some cases (**sex pili**) to other bacteria (Fig. 2.21).

Box 2.6
Microbial plant diseases

Plant diseases may be caused by viruses, fungi and bacteria. About 80 species of bacteria are known to cause plant diseases, although numerous pathovars, that is, strains which differ only in plant species they infect, are known.

Plant pathogenic bacteria cause the development of a wide range of differing symptoms in the plants they infect. Any given symptom may often result from bacteria of different genera. Conversely, each genus also contains some pathogens capable of causing different types of diseases:

Pea seedlings showing the effects of foot rot disease caused by the fungus *Fusarium solani* f. sp. *pisi*. The plant on the right is not affected; those on the left are severely diseased. Courtesy of Drs W.M. Whalley and M. Mabey, Department of Biological Sciences, The Manchester Metropolitan University, UK.

Blight in peas caused by the bacterium *Mycosphaerella* sp. Courtesy of Processors and Growers Research Organization, Peterborough, UK.

Galls on stem and roots of plant. Redrawn from G.N. Agrios (1988) *Plant Pathology*, 3rd edn, Academic Press.

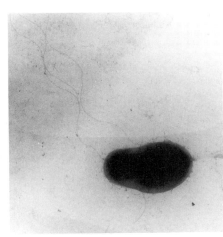

Electron micrograph of *Agrobacterium tumefaciens*. Courtesy Dr C.H. Shaw, Department of Biological Sciences, University of Durham.

The most common types of bacterial diseases of plants are those that cause spots on leaves, stems, blossoms and fruits. This can develop into a blight which appears as a rapidly advancing necrosis of the organs. The most common bacteria to cause spots and blights are *Pseudomonas* and *Xanthomonas* species and these affect a wide range of plant species.

Bacterial soft rots cause fleshy plant tissues to rot in the field or during storage. They are most often associated with fruits, vegetables and some ornamental plants. The associated foul smell is caused by the release of volatile compounds during bacterial breakdown of the tissues. In many cases, the bacteria involved are saprophytic rather than pathogenic, that is, they grow on tissues killed by true pathogens. Soft rots are caused largely by *Erwinia* and *Pseudomonas* species.

Galls are produced on stems and roots of plants, principally following infections by *Agrobacterium*, *Corynebacterium* and *Pseudomonas* species. The first group cause crown gall tumours because they contain the T_i plasmid whose products induce an hormonal imbalance in the plant (see Section 8.9). Some strains of *Agrobacterium* contain the plasmid R_1 which induces the condition called hairy root.

Relatively few canker diseases are caused by bacteria. However, some are so widespread and devastating that they result in great plant losses. Two of the most economically important are the bacterial cankers of stone and pome fruit caused by several *Pseudomonas* species, and bacterial canker of citrus fruit produced by infections of *Xanthomonas campestris* pv. *citri*.

Box 2.6 cont'd

Bacterial scabs are a group of diseases that largely affect the underground parts of plants. The symptoms are typically localized lesions affecting outer tissues. Economically important scab diseases are the common scab of potato and the soil pox of sweet potato caused by *Streptomyces scabies* and *S. ipomomoeae* respectively.

Bacterial diseases of plants are often difficult to control and frequently a combination of control measures is required. Planting healthy seed or seedlings, the use of resistant varieties and effective sanitation procedures, for example burning diseased plant material, are obvious starting points. Sterilization is only practical in small areas but may be done by steam or formaldehyde for greenhouse soil. Generally, the use of chemicals to control bacterial plant pathogens has been less successful than their use in controlling fungal infections. Foliar sprays containing Cu^{2+} have generally given the best results. In recent years, antibiotic sprays (usually streptomycin or oxytetracycline) have been used.

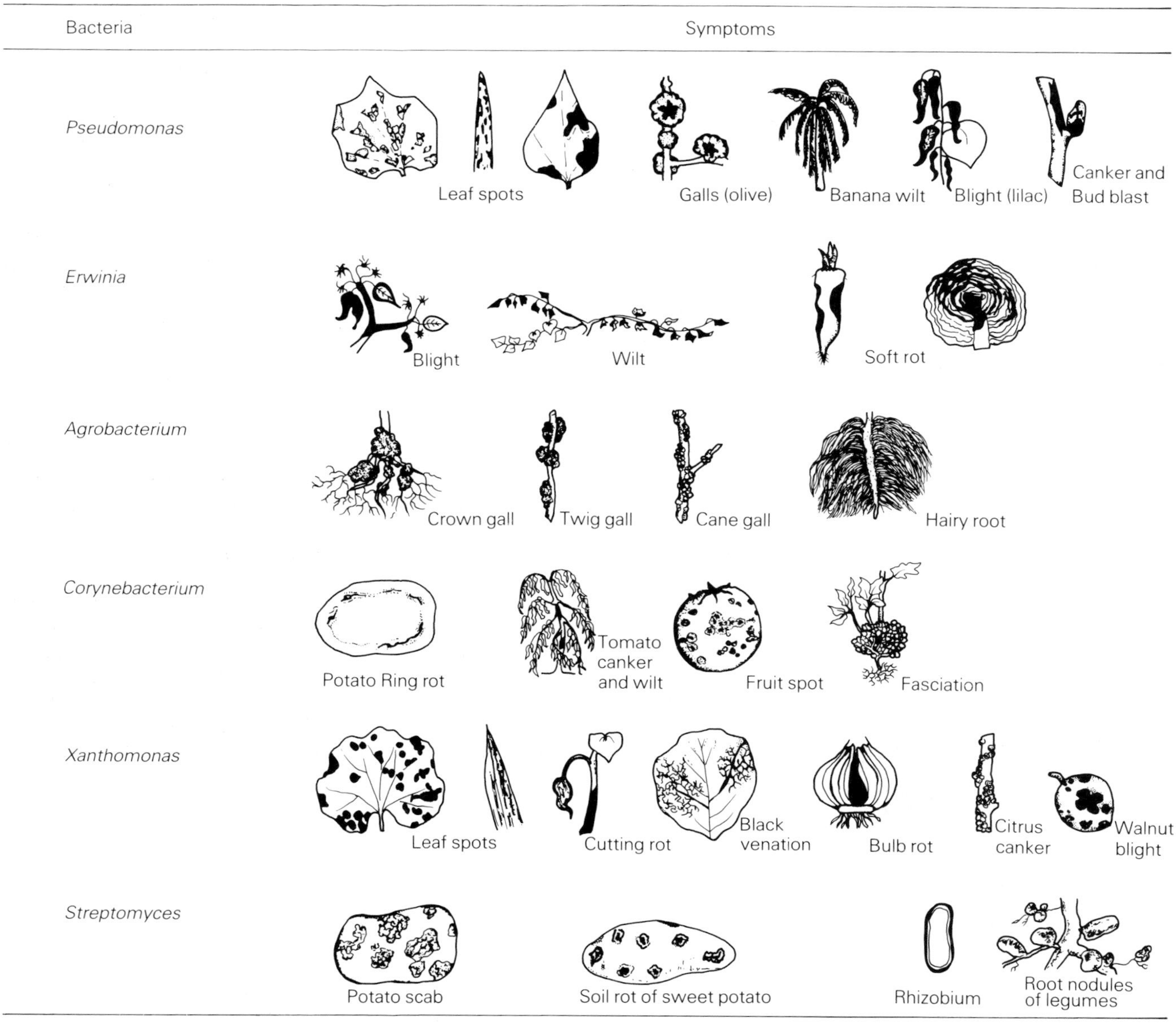

Bacteria	Symptoms

Pseudomonas

Leaf spots   Galls (olive)   Banana wilt   Blight (lilac)   Canker and Bud blast

Erwinia

Blight   Wilt   Soft rot

Agrobacterium

Crown gall   Twig gall   Cane gall   Hairy root

Corynebacterium

Potato Ring rot   Tomato canker and wilt   Fruit spot   Fasciation

Xanthomonas

Leaf spots   Cutting rot   Black venation   Bulb rot   Citrus canker   Walnut blight

Streptomyces

Potato scab   Soil rot of sweet potato   Rhizobium   Root nodules of legumes

Vacuoles and storage granules

The cytoplasm of some bacteria contains differentiated structures of various kinds, including **gas *vacuoles***, which are surrounded by a protein envelope; **storage granules**, consisting of nitrogen stores (cyanophycin), or carbon reserves (glucans, poly-β-hydroxybutyrate); **polyphosphate granules**, **sulphur inclusions**, and others (Table 2.3). The granular appearance of the cytoplasm is largely due to the presence of 70S ribosomes. Plasmids are also present.

Table 2.3 *Selected examples of reserve materials in prokaryotes*

Reserve materials	Bacteria
Carbon and energy sources	
glucans	Bacilli, Clostridia, many cyanobacteria and enteric bacteria
poly-β-hydroxybutyrate	*Pseudomonas, Azotobacter, Rhizobium, Vibrio*
Nitrogen sources	
cyanophycin (a copolymer of Arg and Asp)	Many cyanobacteria
Inorganic materials	
polyphosphates	Many bacteria

Nuclear region or nucleoid

Although the genetic material of prokaryotic cells is not surrounded by a nuclear membrane, special techniques allow the observation of a cytoplasmic zone particularly rich in DNA, called the nuclear region or nucleoid. Some authors even speak of 'bacterial nucleus'. The total length of unfolded bacterial DNA is about 1 mm, that is, about 500-fold greater than the average diameter of a bacterial cell (2 μm). Consequently, bacterial DNA must be

Exercise 3

A bacterium (F⁻) receives some genetic material, via conjugation, from another bacterial cell (F⁺). Suggest some biological advantages that the former bacterium may draw from this event.

See *Molecular Biology and Biotechnology*

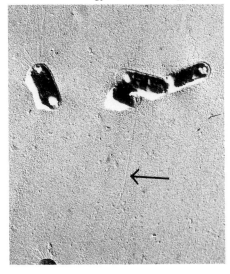

Fig. 2.21 Electron micrograph of *E. coli* with an extended sex pilus (arrowed) (×8 400). Courtesy of P.L. Carter, Department of Biological Sciences, The Manchester Metropolitan University, UK.

Box 2.7
Magnetosomes: a bacterial structure that detects the earth's magnetic field

Some flagellated bacteria, for example *Aquaspirillum magnetotacticum*, are capable of orientating their direction of motion along the lines of force of the earth's magnetic field. It is thought that this effect enables the cells to move towards the bottom of their watery habitats, the anaerobic sediments of bogs and marshes being their optimal environment.

These bacteria detect the earth's magnetic field using subcellular structures called **magnetosomes**. Magnetosomes are chains of particles of the iron oxide magnetite, Fe_3O_4, which are synthesized from soluble iron absorbed from the surroundings. (Magnetite is also a constituent of 'lodestone', a substance used in early forerunners of the compass.) The iron oxide particles are clearly visible in electron micrographs of the bacteria, and appear to be surrounded by a membrane. Magnetic fields have the effect of orientating the chain of particles by means of which the bacteria detect the field.

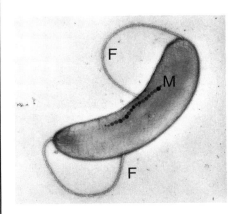

Electron micrograph of the bacterium *Aquaspirillum magnetotacticum* (×15 300). Magnetosomes (M) are prominent as a chain of dark particles within the cytoplasm. Two flagella (F) are present. Courtesy of Dr. R.P. Blakemore, Department of Microbiology, University of New Hampshire, USA.

---

vacuole: *from the Latin for 'empty'.*

elaborately folded and densely packed. This is made difficult, among other things, by the electrostatic repulsions of the negatively charged phosphate groups of the DNA. Charge neutralization is achieved through complexation to polyamines (spermine, spermidine) and basic proteins, that play a similar role to histones in eukaryotic cells as well as to mono- and divalent cations.

The isolated bacterial chromosome consists of a single large molecule of double-stranded circular DNA. It is supercoiled in the central regions of the cell, with transcription loops extending into the cytoplasm. The molecule is tightly folded, together with the electropositive molecules just mentioned, as well as considerable amounts of RNA polymerase and RNA. There is experimental evidence that at least part of this RNA plays some structural role in the chromosome. During its replication, the bacterial DNA molecule appears to be attached to the plasma membrane.

2.3 Archaebacteria

Archaebacteria form a heterogeneous group of bacteria and, judging from the base sequences of 16S ribosomal RNAs are, in evolutionary terms, very distant both from eubacteria and from eukaryotes (Fig. 2.22). This phylogenetic distance may explain the striking structural and functional peculiarities of archaebacteria: similar problems have been solved in very different ways. For instance, the **archaebacterial cell wall** presents a tremendous structural variety in the different archaebacterial groups (methanogens, halophiles, thermoacidophiles), but never contains murein. The **membrane** is characteristic because its lipids do not contain fatty acid esters, but rather, polyprenyl ethers of D-glycerol.

Energy metabolism in archaebacteria also has distinctive features. For example, methanogens lack cytochromes and quinones, suggesting that electron transfer reactions, if present at all, are very different from those

☐ Archaebacteria are often found in extreme environments (for example, high salt, high temperature or highly reducing conditions) and are the subject of much active research since their enzymes may be used for industrial applications under similar extreme conditions, for example, temperatures near 100°C.

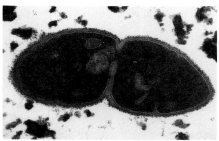

Electron micrograph of the Archaebacterium, *Methanobacterium ruminantium* isolated from a cow rumen. Courtesy Dr J.G. Zeikus, Michigan Biotechnology Institute and Dr K.A. Schwartz, University of Massachusetts. Reprinted from Zeikus, J.G. and Bowen, V.G. (1975) Comparative Ultrastructure of Methogenic Bacteria. *Canadian Journal of Microbiology*, **21**, 121–30.

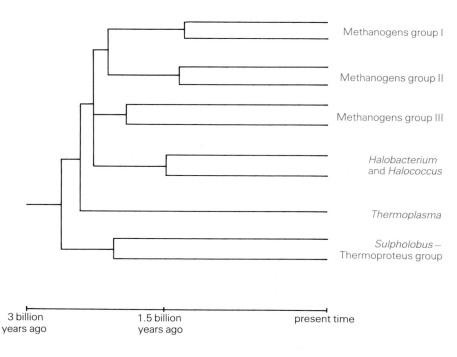

Methanogens group I

Methanogens group II

Methanogens group III

Halobacterium and *Halococcus*

Thermoplasma

Sulpholobus – Thermoproteus group

| 3 billion years ago | 1.5 billion years ago | present time |

Fig. 2.22 An outline of the probable phylogenetic evolution of Archaebacteria (compare with Fig. 2.3).

Reference Fewson, C.A. (1986) Archaebacteria. *Biochemical Education*, **14**, 103–15. A comprehensive student-orientated review.

occurring in other bacteria and eukaryotes. **Protein synthesis** is characterized by the absence of deoxyribothymidine in tRNA and the use of Met-tRNA (instead of fMet-tRNA) in the initiation complex. **Nucleic acid metabolism** also presents substantial differences. For example, the RNA polymerase of archaebacteria is very different from those found in eubacteria.

See *Molecular Biology and Biotechnology*, Chapter 4

2.4 Viruses

Viruses are supramolecular structures, consisting of a nucleic acid (either DNA or RNA) and one or a few specific proteins, sometimes surrounded by a lipid membrane (Fig. 2.23). Viruses are necessarily parasitic and are only capable of self-replication using the genetic machinery of a host cell. Animal, plant and bacterial cells are susceptible to attack by specific viruses. In the last case, they are called ***bacteriophages*** or, in short, phages (Table 2.4).

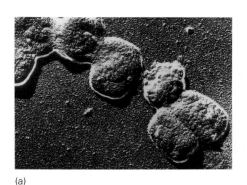
(a)

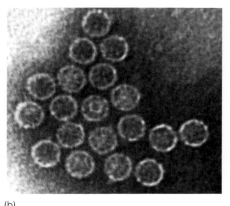
(b)

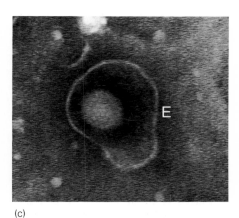
(c)

Fig. 2.23 Electron micrograph of the non-enveloped viruses (a) *Vaccinia* (×48 600), (b) φX174 (×257 500), and (c) the enveloped (E) virus of the Herpes group. Courtesy of North West Regional Virus Laboratory, Booth Hall Hospital, Manchester, UK.

Table 2.4 *Examples of viruses*

Name	Nucleic acid	No. of strands of nucleic acid	Approximate no. of genes	Host
φX174	DNA	1	5	bacteria
T4	DNA	2	160	bacteria
Vaccinia poxvirus	DNA	2	240	animals
Qβ	RNA	1	3	bacteria
Tobacco mosaic virus	RNA	1	6	plants
Poliovirus	RNA	1	8	animals
Influenza virus	RNA	1	12	animals
Reovirus	RNA	2	22	animals

Viruses were discovered at the beginning of the twentieth century. Around 1910, the French microbiologist d'Hérelle observed the presence of some substance in bacterial cultures that could pass through filters which were so fine that they retained bacteria. Furthermore, the filtrate so obtained was able to induce bacterial lysis in other cultures (d'Hérelle was actually describing a bacteriophage). Agents of this kind became known as 'filterable viruses': today are simply called viruses.

In the Introduction to this chapter it was stated that viruses are not cellular.

virus*: from the Latin for 'slimy liquid or poison'.*
bacteriophage*: 'bacteria eater'; compare macrophage, phagocyte.*

Reference Freifelder, D. (1987) *Molecular Biology*, 2nd edn, Jones and Bartlett, Boston, USA. Contains modern, clear presentation of the molecular biology of eukaryotic and bacterial viruses.

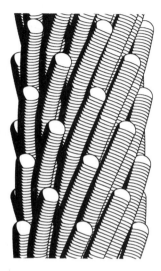

Fig. 2.24 Molecular model of portion of the filamentous bacteriophage M13. Several thousand identical protein subunits form a hollow shell which surrounds the DNA genome. The phage particle is about 6 nm in diameter and approximately 900 nm long (about 100 times longer than the segment in the figure). The subunits are 50 amino acid residue α-helices which run roughly parallel to the long axis of the phage, overlapping and interdigitating like fish scales. Courtesy of Dr D.A. Marvin, Department of Biochemistry, University of Cambridge, UK.

Their status as organisms is also debatable since organisms are considered to be cells or to be composed of cells. Thus although viruses show the properties of self-replication and mutation, they are not capable of self-repair, nor have they an energy transduction system, the last two being essential characteristics of organisms. This does not mean that viruses are incapable of performing or regulating complex metabolic processes, as will be seen.

Viral structure

The structures of different viruses are very varied but may be grouped into three basic models: the rod or filament, the spheroidal polyhedron, and the tailed spheroidal polyhedron. In **filamentous viruses**, the nucleic acid is arranged helically, the protein subunits protecting and stabilizing the nucleic acid genome (Fig. 2.24). A typical example is *tobacco mosaic virus* (TMV). In **spheroidal viruses**, the nucleic acid is condensed inside a protein envelope which is often organized into a multisided geometric figure. This is the case in adenoviruses (Fig. 2.25), a group of animal viruses that are capable of producing a range of diseases; some produce tumours in animals. The **tailed spheroid** shape is typical of bacteriophages, for example the T4 and lambda phages of the bacterium *E. coli* (Fig. 2.26).

Each complete viral particle is called a **virion**. The protein coat surrounding the **viral nucleic acid** is called a *capsid*. It is formed from a number of protein subunits, the **capsomers**. Each capsomer may be formed by one or more polypeptide chains which may, in turn, be different or identical. Nucleic acid and capsid together are called the **nucleocapsid**. This term is used mainly when the overall viral structure is complex: for example the phages consisting

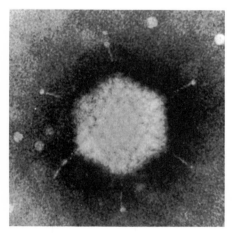

Fig. 2.25 Electron micrograph of adenovirus (×250 000). Courtesy of Dr G.E. Blair, Department of Biochemistry and Molecular Biology, University of Leeds, UK.

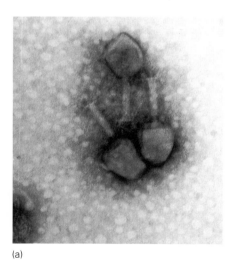

(a)

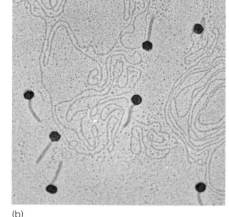

(b)

Fig. 2.26 Electron micrographs of (a) T4 (×141 000) and (b) lambda bacteriophages. Courtesy of Dr V. Virrankoski-Castrodenza.

mosaic: a group of plant diseases, of viral origin, characterized by mottling of the foliage.
capsid: from the Latin capsa, box.

of a nucleocapsid and a tail. In many cases, however, and particularly with animal viruses, the nucleocapsid is contained inside a lipoprotein membrane and such structures are called **enveloped viruses** (Figs 2.23c, 2.27). When the nucleocapsid is not surrounded by an envelope, the virus is said to be *naked*.

The principles that govern viral organization are easy to understand, as might be expected from their comparatively simple biochemical structure. Firstly, viral particles contain many copies of one or a few proteins. This allows a considerable economy of genes (a small genome is required). In turn, the repetitive structure demands a degree of symmetry and confers a simple geometry (rod, sphere, multisided) to the overall shape. Secondly, the more complex viruses are organized from **subassemblies** which are independently assembled and remain separate until the final stages of morphogenesis. For instance, in the tailed spheroidal phages, 'heads' and 'tails' make up independent subassemblies. Thirdly, the incorporation of the nucleic acid molecule is essential for the assembly of the viral particle. In *structural* assembly, the chain length of the nucleic acid is more important than its nucleotide sequence. This principle is frequently used in recombinant DNA studies: it means that lengths of DNA foreign to the viral genome can be incorporated into the viral nucleic acid and delivered to cells at infection.

General data on the structure of a given virus may be obtained using electron microscopy and chemical analysis. However, the fine structural details have been, and still are being, unravelled mainly by X-ray and electron diffraction studies. One example of each type of virus will now be described in some detail.

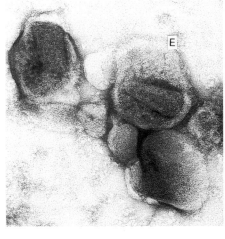

Fig. 2.27 Electron micrograph of Simian AIDS virus (SIV) (×102 000). The envelope (E) surrounding the nucleocapsid is clearly visible. Courtesy of Dr C. Grief, National Institute for Biological Standards and Controls, Hertfordshire, UK.

TOBACCO MOSAIC VIRUS (TMV) particles are rod-like, each consisting of about 2130 identical polypeptides (of 158 amino acid residues each) organized into a hollow helix 300 nm long and 18 nm diameter (Fig. 2.28). Each turn of the helix corresponds to 16.3 protein subunits. RNA winds coaxially with the protein, with three nucleotides bound to each subunit. The phosphate groups of the RNA molecule are neutralized by basic amino acyl residues of the protein (Fig. 2.29).

Exercise 5

The double-stranded DNA of *E. coli* bacteriophage T7 contains 39 936 base pairs. (a) Estimate the maximum number of proteins (assuming an average M_r of 24 000) that can be encoded in the T7 genome. The average molecular weight of a free amino acid is 138. (b) It is known that the T7 genome contains about 55 genes; comment on this result.

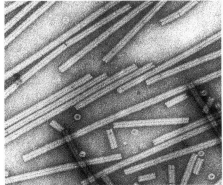

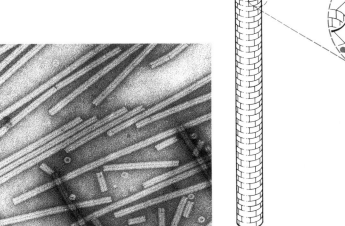

Fig. 2.28 Electron micrograph of tobacco mosaic virus (×100 000). Courtesy of C.M. Clay, Institute for Horticultural Research, Wellesbourne, UK.

Fig. 2.29 Structure of the tobacco mosaic virus (TMV) rod.

Reference Fields, B.N. *et al.* (eds) (1985) *Virology*, Raven Press, New York, USA. In particular, see Chapter 3 (by S. Harrison) for a very good description of viral structure.

Reference Zaitlin, M. and Hull, R. (1987) Plant virus–host interactions. *Annual Review of Plant Physiology*, **38**, 291–315. A review with an emphasis on novel aspects of virally induced plant diseases.

Particle assembly is believed to begin with the interaction of a 'starting loop', which exists in the RNA molecule near the 3' end (about 1 kb away), with the inner hollow part of a disk formed by 17 protein subunits. Binding of the polynucleotide causes, or is followed by, a conformational change in the disk, that now adopts a helical structure. From this initial assembly, growth proceeds through addition of protein subunits along both sides of the RNA strand, but extension is faster towards the 5' end (Fig. 2.30).

ADENOVIRUS particles are an example of a spheroidal virus. Their structure is far more complex than that of TMV. The capsid is icosahedral (Fig. 2.25) and consists of a highly ordered arrangement of **hexons** (in the faces) and **pentons** (in the vertices). Hexons are trimers of a protein of M_r 110 000, while pentons consist of a **base protein** (M_r 85 000) and a **projecting fibre** protein (M_r 62 000) (Fig. 2.31). The base protein is in contact with five hexons, the so-called peripentonal hexons, hence the name penton.

The adenovirus genome consists of double-stranded DNA (about 15 kb long). Two arginine-rich, basic proteins appear to play a role similar to that of histones in the eukaryotic chromosome. Adenovirus DNA is probably packed with the aid of these basic proteins inside preformed capsids, during viral assembly. In turn, capsid assembly requires the contribution of some viral proteins that do not become incorporated into the final mature structure.

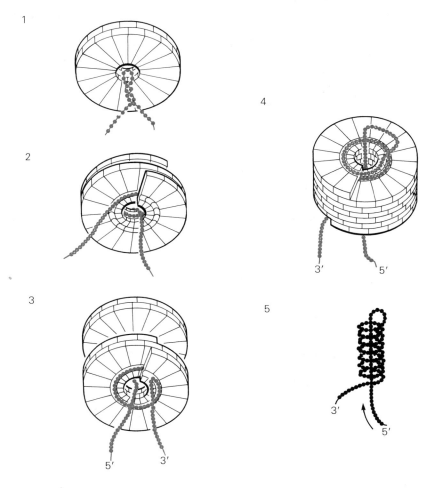

Fig. 2.30 Assembly of a TMV particle.

adenovirus: *from the Greek* aden, *gland.*

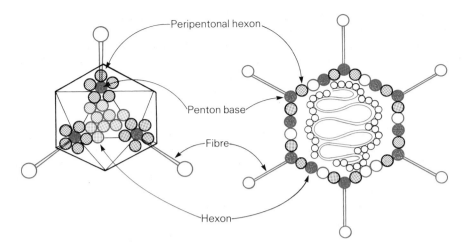

Fig. 2.31 Schematic structure of an adenovirus particle.

—————— *Exercise 7* ——————

MS-2, an RNA phage, is simple both in structure and life cycle. Its genome contains about 3600 nucleotides; the complete viral particle consists of the RNA molecule, plus 180 copies of the coat protein (CP) and a single copy of the attachment protein (A). Draw a possible viral geometry compatible with the above data.

BACTERIOPHAGE T4 has one of the most complex viral structures known. It is formed by a 'head' and a 'tail' (Fig. 2.26a) the tail being in turn composed of many different types of subunit. The icosahedral head measures about 112 nm in diameter, and the tail has a similar length. The tail consists of an inner **core**, a contractile **sheath** and a number of **fibres** (Fig. 2.32). In addition, a hexagonal **plate** at the free end of the tail appears to interact with a specific receptor (the Omp C protein, a porin) on the *E. coli* surface. Immediately after adsorption on to the host cell, the phage injects its DNA through the tail core and this is probably helped by contraction of the sheath.

Viral membranes

Some animal viruses are surrounded by a lipoprotein membrane (Figs 2.23c, 2.27), similar to and derived from the host cell plasma membrane. Viruses become surrounded by this portion of membrane while leaving the cell by a budding mechanism (Fig. 2.33). Viral membrane lipids are those found in the host cell membranes, but the proteins, often glycosylated, are encoded by

☐ During 1939–1941 Demerec isolated a series of phages from *E. coli* which were named T1, T2 … T7. They have been extensively used in biochemical studies. The nomenclature is somewhat misleading, since not all the T phages share the same structural or functional features. They are usually divided into four groups: T1; the T-even phages (T2, T4, T6); T3, T7; and T5.

☐ The complementation assay is one of the main techniques used to study viral assembly. In the case of T4, if two extracts of cells infected with phage mutants, one lacking heads and the other lacking tails, are mixed, phage particles form *in vitro*.
    The 'headless' extract can be fractionated and a component isolated which when added to the 'tailless' preparation allows tail formation.

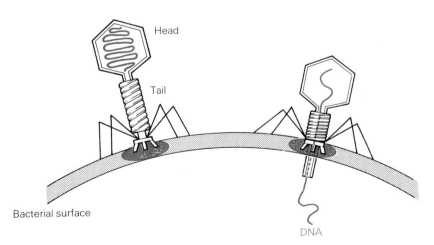

Fig. 2.32 Schematic of a T₄ bacteriophage before (left) and during (right) DNA injection.

Reference Stephens, E.B. and Compans, R.W. (1988) Assembly of animal viruses at cellular membranes. *Annual Review of Microbiology*, **42**, 489–516. An up-to-date description of this interesting aspect of viral physiology.

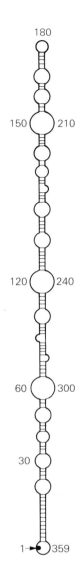

Fig. 2.34 Structures of a viroid. The figures correspond to the nucleotide sequence.

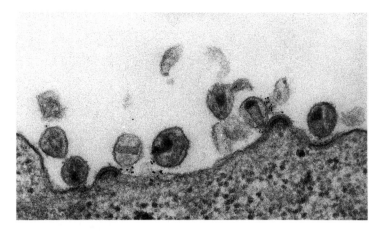

Fig. 2.33 Electron micrograph showing human immunodeficiency virus type 1 (HIV-1) particles budding from CCRF–CEM human leukaemia cells that are used to grow HIV-1 (×75 000). Courtesy of Dr D. Robertson and Professor R.A. Weiss, Institute for Cancer Research, Royal Cancer Hospital, UK.

viral genes, expressed within the host cell and subsequently inserted in the host plasma membrane.

The viral membrane is functionally important in at least two ways: (a) a receptor in the host cell surface is usually recognized by a viral membrane protein, and (b) the nucleocapsid enters the cell through fusion of the host and viral membranes. The viral membrane glycoproteins constitute major antigens which elicit the immune response after viral infection. New vaccines are being developed consisting of purified viral glycoproteins, often obtained through recombinant DNA technology, that are devoid of potential infection risks. These membrane glycoproteins are often water-insoluble; in this case they can be administered after reconstitution into semisynthetic phospholipid bilayers, in the form of **proteoliposomes**.

2.5 Viroids

Viroids consist of a single-stranded circular RNA molecule, about 400 nucleotides long, without specifically bound proteins (Fig. 2.34). Viroids are capable of self-replication in the nucleus of susceptible cells. They are much smaller than any known virus, but share with them the properties of intracellular parasitism and the possibility of extracellular survival. Viroids, like viruses, can be pathogenic, at least in plants. The mechanisms of both replication and *pathogenesis* of viroids remain a mystery.

| Box 2.9 Interferons | Interferons are a group of proteins of M_r 20–40 000 synthesized and secreted by vertebrate cells following a viral infection. Secreted interferons bind to the plasma membrane of other cells and induce an 'antiviral state' in them. In contrast with the specific immunity produced by an antibody, interferon confers resistance of the cells to a broad spectrum of viruses (see Chapter 11).

Interferons appear to lead to an antiviral state by stimulating the production of two enzymes: (a) a **protein kinase** acting on the eIF2 initiation factor, and thus inhibiting the formation of new polypeptide chains, and (b) a **2′,5′-oligoadenylate synthetase**, whose products activate a specific endonuclease that, in turn, hydrolyses the mRNA and rRNA molecules. The effects of interferon therefore include cessation of cell growth and proliferation (see *Molecular Biology and Biotechnology*). |

---

pathogenesis: from the Greek for 'the origin' of a disease.

Reference Smith, A.L. (1985) *Principles of Microbiology*, 10th edn, Times Mirror/Mosby, St Louis, USA. A very readable textbook emphasizing the biomedical applications of microbiology.

2.6 Prions

Prions are infectious macromolecules, protein in nature (M_r about 30 000), that appear to be responsible for some diseases in humans and animals, although ideas about their infectious nature are extremely controversial. Again, the mechanisms of replication and pathogenesis are unknown.

2.7 Bacteria and viruses in biochemical research

The importance of bacteria and viruses in the history of biochemistry can hardly be over-emphasized. It would be difficult to find a single chapter in this series in which studies using bacteria and viruses have not contributed essential knowledge and insights. Nowadays, microorganisms are as indispensable as ever in the biochemical laboratory. This is due to a number of properties shared by bacteria and virsues:

- structural simplicity and lack of compartmentation which facilitates structural and functional studies;
- easy growth in inexpensive growth media and simple storage requirements;
- the possibility of obtaining at will, and in a relatively short time, homogeneous populations of organisms;
- short replication times allowing genetic studies over many generations within a reasonable time scale;
- considerable variety, along with an essentially similar behaviour pattern, which allows a large choice of biochemical or biophysical properties within a given group of microorganisms;
- the ease of obtaining and propagating mutant forms of microorganisms.

These properties have allowed bacteria and viruses to play an essential role in biochemical research, and they still do.

Molecular genetics, which is so important to genetic engineering, was born of studies on bacteria and their viruses. Our knowledge of the genetic control of metabolism started with bacterial operons (for example, the *lac* operon from *E. coli*). The first membrane transport proteins to be described were bacterial permeases. Bacteria and viruses have made important contributions to genetic engineering and biotechnology. These biological systems have been used in the field of preventive medicine, particularly in mutagenicity and carcinogenicity tests.

□ The so-called integral (or intrinsic) membrane proteins cannot be isolated from their non-polar, lipidic environment, without loss of their structural and functional features. A methodology has been developed that allows the study of these proteins in purified preparations, called membrane protein reconstitution. Essentially, the intrinsic membrane protein is purified in the presence of a detergent (such as Triton X-100, octylglucoside or a bile acid). Detergents disrupt the membrane bilayer while providing a hydrophobic environment for the protein. The purified protein (now in detergent suspension) is mixed with membrane lipids (also in the form of lipid–detergent mixed micelles). Detergent removal through dialysis leads to the spontaneous formation of proteoliposomes, semisynthetic vesicles surrounded by one or more lipid bilayers in which the purified intrinsic protein is embedded.

See *Molecular Biology and Biotechnology*, Chapter 9.

□ Phage M13 is a filamentous *E. coli* bacteriophage containing a single-stranded, circular DNA. It does not kill the host cell or cause lysis. Instead, the viral particles leave the cell through budding. M13 has found extensive application in genetic engineering. It is the most frequently used vector for cloning single-stranded DNA; for example in DNA sequencing or for site-directed mutagenesis.

Liberation of M13 phage particles from an injected bacterium through budding. Note the single-stranded circular DNA and the incorporation of coat proteins from the bacterial membrane.

prion: *from proteinaceous infectious material.*

Reference Gabizon, R. and Prusiner, S.B. (1990) Prion liposomes. *Biochemical Journal*, **266**, 1–14. A starting point for reading about prions, Creutzfeldt-Jakob disease, and Bovine Spongiform Encephalopathy (BSE).

Reference Darnell, J. *et al.* (1990) *Molecular Cell Biology*, 2nd edn, Scientific American Books, New York, USA. Chapter 6 of this magnificent textbook is devoted to the use of cells and organisms in biological research.

2.8 Overview

Microorganisms are those organisms that can only be seen using a microscope. From the biochemical point of view, the most important microorganisms are bacteria and viruses.

Bacteria are prokaryotic microorganisms; that is, they lack a differentiated nucleus. The main differentiated structures in bacteria are the so-called 'surface structures', the vacuoles and the nucleoid region. Surface structures include the plasma membrane, cell wall, capsule, flagella and pili. Cell walls from Gram-positive bacteria contain mainly peptidoglycan, a bag-shaped molecule of parallel chains of sugars cross-linked with oligopeptides, plus polysaccharides and teichoic acids. In Gram-negative bacteria, peptidoglycan is found together with phospholipids, proteins and lipopolysaccharides.

Two groups of bacteria may be distinguished: eubacteria and archaebacteria, which probably separated early in evolution. Archaebacteria differ strikingly from eubacteria both structurally and functionally.

Viruses are structures consisting of a nucleic acid molecule (DNA or RNA) surrounded by a protein capsid, and, occasionally, by an outer lipoprotein membrane. Viruses may be filamentous, spheroidal, or tailed–spheroidal in shape. Viral assembly requires the presence of a nucleic acid molecule of the appropriate length. Viruses are intracellular parasites found in animal, plant or microbial cells. Some viruses become surrounded by a membrane while leaving the parasitized cell. In general, viral membrane proteins, but not lipids, are encoded by viral genes. Other infectious, but even simpler, structures are viroids and prions.

Bacteria and viruses constitute biochemical tools of great importance. Given their relatively simple structures and easy growth conditions, they were essential tools in the development of modern biochemistry.

Answers to Exercises

1. Since the volume of a sphere is related to the *cube* of the radius, the ratio of volumes will be 1000 000/1000/1.

2.

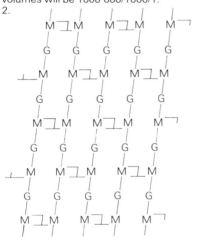

(G) *N*-acetylglucosamine
(M) *N*-acetylmuramate

3. In general, all the advantage associated with sexual reproduction: the acceptor may receive genes not previously possessed (e.g. antibiotic resistance genes), or native copies of genes previously present in non-functional form (e.g. mutated genes), etc. The possibility of receiving second copies of genes already present is much higher if the donor belongs to an Hfr strain; in this case, the acceptor will become temporarily diploid for part of its genome.

4. All the statements are compatible with the endosymbiotic theory.

5. (a) The average molecular weight of an amino acyl residue will be $138 - 18$ (M_r of water) $= 120$. Thus an average protein will contain $24 000/120 = 200$ amino acid residues. Since each amino acid is coded by three bases (mRNA) or three base pairs (DNA), each protein will require about $200 \times 3 = 600$ base pairs. Thus the maximum number of proteins will be 39 936 divided by 600; or about 67.

(b) The estimated number of proteins is similar to the actual number of genes. However, in practice not all genes code for proteins (some code for rRNA, or tRNA, or have regulatory functions) and the length of T7 proteins is known to vary widely from 29 to 883 amino acids.

6. (a) M_r is approximately 6×10^6;
(b) DNA; (c) double-stranded monomers;
(d) 251 monomers.

7.

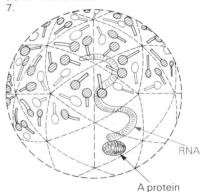

RNA

A protein

FILL IN THE BLANKS

1. Bacteria are cells _____ a differentiated nucleus. Viruses are _____ cellular. Bacterial plasma membranes do not contain _____ but possess _____ fatty acyl chains. Only Gram-negative bacteria contain _____ in their cell walls.

   In archaebacteria, the initiation complex for protein synthesis contains _____ , while in eubacteria, it contains fMet-tRNA, and, in eukaryotes _____ .

   The protein envelope surrounding the viral genome is termed _____ ; this is formed by a number of _____ _____ , the capsomers.

   The genome of tobacco mosaic virus contains single-stranded _____ , ϕX174 contains _____ DNA, while poxvirus contains _____ _____ .

Choose from: branched, capsid, double-stranded DNA, without, lipids, Met-tRNA (2 occurrences), not, protein subunits, RNA, single-stranded, sterols.

MULTIPLE-CHOICE QUESTIONS

2. A structure in the bacterial cell rich in enzymatic activity is the:
A. plasma membrane
B. cell wall
C. mesosome
D. capsule
E. flagellum

3. Bacterial cell walls:
A. contain antigens.
B. are synthesized by pathways sensitive to the action of certain antimicrobial compounds.
C. may be removed without destroying cell viability.
D. all of the above (A, B and C) are true.
E. none of the above (A, B or C) is true.

4. Viruses:
A. cannot be seen with a light microscope.
B. only grow within cells.
C. cannot be separated from media by filtration.
D. all of the above (A, B and C) are true.
E. none of the above (A, B or C) is true.

5. All bacteriophages:
A. cause bacteriolysis.
B. contain at least one molecule of nucleic acid.
C. produce genetic changes in their bacterial hosts.
D. all of the above (A, B and C) are true.
E. none of the above (A, B or C) is true.

SHORT-ANSWER QUESTIONS

6. (a) List three differences between eukaryotic and prokaryotic cells.
   (b) List three differences between eubacteria and archaebacteria.

7. Healthy humans carry in their bodies, on average, three bacteria for each of their own eukaryotic cells. These are mainly saprophytic bacteria living in the respiratory and digestive tracts, and on the skin. Estimate the percentage total volume occupied by bacteria in the human body. Assume the ratio of bacterial/eukaryotic cell diameters is 10 : 1.

8. DNA polymerase I from the thermophilic bacterium *Thermus aquaticus* has an optimal termperature for growth of 70–75°C. Could this have any practical implications in biotechnology?

9. In general, viruses of animal but not bacterial origin have genes consisting of exons and introns. Why is this?

10. Some viruses inject their nucleic acid into the host cell, while others deliver the DNA or RNA molecule on to the cell surface, the nucleic acid being internalized later. Design a simple experiment to find out whether a given virus follows one or the other of these methods of cell infection.

11. The M_r of the tobacco mosaic virus is 60 000 000. Estimate the weight (in grams) of a single virion. Assume a value of 6×10^{23} for Avogadro's number.

3
DNA packaging: the nucleus

Objectives

After reading this chapter you should be able to:

☐ cite evidence for nucleic acids as the carriers of genetic information;

☐ describe the structure of the bacterial nucleoid;

☐ describe the structure of the nucleus and associated structures;

☐ explain how genomic DNA molecules are folded in chromatin;

☐ outline the possible evolutionary origin of the nucleus.

3.1 Introduction

In the early part of this century, scientists searching for a chemical explanation of how the hereditary information is stored and used, were faced with a problem: was it the DNA or the protein of chromosomes that carried the genetic information? DNA was known to contain four bases but it was believed that its structure was far too simple to contain all the information to construct something as complicated, say, as a human being. DNA was held to have a role as a sort of scaffolding in the nucleus, and the genetic information in the genes was believed to be stored in the proteins of the nucleus. These proteins seemed sufficiently complex to hold the genetic information. It was not appreciated at that time that the major group of proteins of the nucleus, the histones, are relatively simple in structure and vary little between organisms. Now, of course, it is well established that exactly the opposite is true, namely that it is the DNA that carries the genetic information in all organisms except for a few viruses where RNA fills this role.

DNA consists of enormously long molecules which have to be precisely folded and packaged to fit into the nucleus in such a way as to allow their messages to be read and acted upon at appropriate times. Packaging the DNA with the basic histones forms 'chromatin' and this in turn is anchored to a fibrous proteinaceous lattice which gives the nucleus its morphology.

Despite the knowledge that the nucleus of eukaryotic cells was the genetic information store, all the early studies that provided evidence that DNA was the genetic material were carried out using bacteria and viruses. Prokaryotic cells do not have a nucleus and, indeed, lack histones. Nevertheless, they do contain DNA, complexed with proteins, although this is not easily visualized by the normal methods of histology.

Some of the early experiments that were seminal in confirming the idea that DNA was the genetic material will now be described. They led to the 'molecular biology revolution' that started in 1953 with the publication by Watson and Crick of their classic paper (Fig. 3.1) which related the *structure* of

☐ RNA and DNA are similar in information carrying capacity but the molecules do differ in a number of respects. DNA is generally much larger in size, RNA contains the sugar ribose rather than deoxyribose, and DNA has a greater propensity for forming double helices. Nevertheless, each of these types of molecule has a sequence of bases as its most characteristic feature. Therefore, it is perhaps not so surprising that certain viruses have RNA rather than DNA as their information store. RNA is used to carry information in *all* organisms — from DNA to protein — as messenger RNA.

MOLECULAR STRUCTURE OF NUCLEIC ACIDS

A Structure for Deoxyribose Nucleic Acid

WE wish to suggest a structure for the salt of deoxyribose nucleic acid (D.N.A.). This structure has novel features which are of considerable biological interest.

A structure for nucleic acid has already been proposed by Pauling and Corey[1]. They kindly made their manuscript available to us in advance of publication. Their model consists of three intertwined chains, with the phosphates near the fibre axis, and the bases on the outside. In our opinion, this structure is unsatisfactory for two reasons: (1) We believe that the material which gives the X-ray diagrams is the salt, not the free acid. Without the acidic hydrogen atoms it is not clear what forces would hold the structure together, especially as the negatively charged phosphates near the axis will repel each other. (2) Some of the van der Waals distances appear to be too small.

Another three-chain structure has also been suggested by Fraser (in the press). In his model the phosphates are on the outside and the bases on the inside, linked together by hydrogen bonds. This structure as described is rather ill-defined, and for this reason we shall not comment on it.

We wish to put forward a radically different structure for the salt of deoxyribose nucleic acid. This structure has two helical chains each coiled round the same axis (see diagram). We have made the usual chemical assumptions, namely, that each chain consists of phosphate diester groups joining β-D-deoxyribofuranose residues with 3′,5′ linkages. The two chains (but not their bases) are related by a dyad perpendicular to the fibre axis. Both chains follow right-handed helices, but owing to the dyad the sequences of the atoms in the two chains run in opposite directions. Each chain loosely resembles Furberg's[2] model No. 1; that is, the bases are on the inside of the helix and the phosphates on the outside. The configuration of the sugar and the atoms near it is close to Furberg's 'standard configuration', the sugar being roughly perpendicular to the attached base. There is a residue on each chain every 3·4 A. in the z-direction. We have assumed an angle of 36° between adjacent residues in the same chain, so that the structure repeats after 10 residues on each chain, that is, after 34 A. The distance of a phosphorus atom from the fibre axis is 10 A. As the phosphates are on the outside, cations have easy access to them.

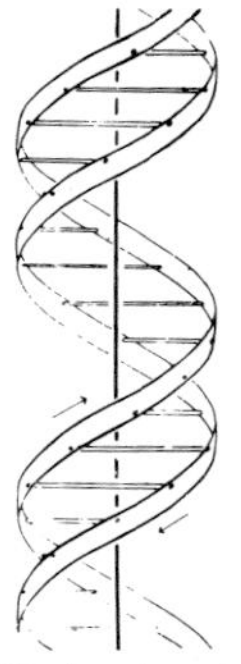

The structure is an open one, and its water content is rather high. At lower water contents we would expect the bases to tilt so that the structure could become more compact.

The novel feature of the structure is the manner in which the two chains are held together by the purine and pyrimidine bases. The planes of the bases are perpendicular to the fibre axis. They are joined together in pairs, a single base from one chain being hydrogen-bonded to a single base from the other chain, so that the two lie side by side with identical z-co-ordinates. One of the pair must be a purine and the other a pyrimidine for bonding to occur. The hydrogen bonds are made as follows: purine position 1 to pyrimidine position 1; purine position 6 to pyrimidine position 6.

If it is assumed that the bases only occur in the structure in the most plausible tautomeric forms (that is, with the keto rather than the enol configurations) it is found that only specific pairs of bases can bond together. These pairs are: adenine (purine) with thymine (pyrimidine), and guanine (purine) with cytosine (pyrimidine).

In other words, if an adenine forms one member of a pair, on either chain, then on these assumptions the other member must be thymine; similarly for guanine and cytosine. The sequence of bases on a single chain does not appear to be restricted in any way. However, if only specific pairs of bases can be formed, it follows that if the sequence of bases on one chain is given, then the sequence on the other chain is automatically determined.

It has been found experimentally[3,4] that the ratio of the amounts of adenine to thymine, and the ratio of guanine to cytosine, are always very close to unity for deoxyribose nucleic acid.

It is probably impossible to build this structure with a ribose sugar in place of the deoxyribose, as the extra oxygen atom would make too close a van der Waals contact.

The previously published X-ray data[5,6] on deoxyribose nucleic acid are insufficient for a rigorous test of our structure. So far as we can tell, it is roughly compatible with the experimental data, but it must be regarded as unproved until it has been checked against more exact results. Some of these are given in the following communications. We were not aware of the details of the results presented there when we devised our structure, which rests mainly though not entirely on published experimental data and stereochemical arguments.

It has not escaped our notice that the specific pairing we have postulated immediately suggests a possible copying mechanism for the genetic material.

Full details of the structure, including the conditions assumed in building it, together with a set of co-ordinates for the atoms, will be published elsewhere.

We are much indebted to Dr. Jerry Donohue for constant advice and criticism, especially on interatomic distances. We have also been stimulated by a knowledge of the general nature of the unpublished experimental results and ideas of Dr. M. H. F. Wilkins, Dr. R. E. Franklin and their co-workers at King's College, London. One of us (J. D. W.) has been aided by a fellowship from the National Foundation for Infantile Paralysis.

J. D. WATSON
F. H. C. CRICK

Medical Research Council Unit for the
Study of the Molecular Structure of
Biological Systems,
Cavendish Laboratory, Cambridge.
April 2.

This figure is purely diagrammatic. The two ribbons symbolize the two phosphate—sugar chains, and the horizontal rods the pairs of bases holding the chains together. The vertical line marks the fibre axis

[1] Pauling, L., and Corey, R. B., Nature, 171, 346 (1953); Proc. U.S. Nat. Acad. Sci., 39, 84 (1953).
[2] Furberg, S., Acta Chem. Scand., 6, 634 (1952).
[3] Chargaff, E., for references see Zamenhof, S., Brawerman, G., and Chargaff, E., Biochim. et Biophys. Acta, 9, 402 (1952).
[4] Wyatt, G. R., J. Gen. Physiol., 36, 201 (1952).
[5] Astbury, W. T., Symp. Soc. Exp. Biol. 1, Nucleic Acid, 66 (Camb. Univ. Press, 1947).
[6] Wilkins, M. H. F., and Randall, J. T., Biochim. et Biophys. Acta, 10, 192 (1953).

Fig. 3.1 Watson and Crick's famous paper (*Nature*, **171**, 737 (1953)) on the structure of DNA. Reproduced courtesy of Macmillan Magazines Ltd.

DNA to its *function* of carrying the genetic information. There have subsequently been many advances in our understanding of the way in which DNA is packaged in both prokaryotic and eukaryotic cells.

3.2 *Transformation in* Diplococcus pneumoniae

In 1928 Griffith experimented with two different strains of the bacterium, *Diplococcus pneumoniae*, which can cause pneumonia, and subsequent death, in mice. It was known that this organism can exist in two forms, the so-called smooth (S) form and the rough form (R). The 'smooth' form of cell has a

☐ The current name for *Diplococcus pneumoniae* is *Streptococcus pneumoniae*.

(a)

(b)

Fig. 3.2 (a) Rough and (b) smooth colonies of *Diplococcus pneumoniae*. The smooth colonies have a glistening appearance because the cells are coated with a polysaccharide capsule which is essential for the pathogenicity of the bacteria. The mutants devoid of the polysaccharide coat are non-pathogenic and form colonies with a rough appearance. Photograph courtesy of Biophoto Associates.

surrounding capsule of polysaccharide and colonies growing on agar plates have a smooth, glistening appearance. In contrast, the rough cells lack this **polysaccharide capsule**, and colonies have a 'rough' appearance when grown on agar plates (Fig. 3.2). Several rough and smooth strains (designated, for example Types II S; II R; III S; III R and so on) were known, which differed in the chemical nature of their polysaccharide. All smooth strains caused pneumonia in mice, whereas rough strains did not. The polysaccharide capsule protects the bacteria against the host's immune system, and the cells can therefore multiply within the host, causing pneumonia. The rough cells are not so protected and consequently are not 'virulent'.

A change from S-type to R-type occurs as a result of spontaneous mutation in one S cell in about 10^7; the reverse change also occurs infrequently, but when it does the **reversion** is back to the same capsule type. Thus, the change Type II S → R → Type II S can occur, but not Type II S → R → Type III S.

Griffith showed that injection of a small dose of heat-killed S strain bacterial cells into mice was not fatal. The dead bacterial cells could not of course multiply within the host to produce pneumonia. Similarly, injection of a small dose of *live* R cells into mice is not fatal because the host animal is able to mount an immune response to them.

The curious thing was that if a mixture of (non-fatal) live R cells and (non-fatal) heat-killed S cells were injected, then the mice contracted pneumonia and many of them died (Fig. 3.3). It was possible to isolate *live* S cells from the diseased mice. More surprisingly, however, was a more detailed observation. This was that injection of Type II R cells (non-virulent) together with heat-killed Type III S not only produced pneumonia, but also that live Type III S cells could be isolated from the mouse. The change from Type II to Type III had never been observed to occur by mutation. Nevertheless these new virulent, Type III S cells reproduced true to type for many generations. Griffith concluded therefore that some of the dead Type III S cells had **transformed** the Type II R cells into living, virulent Type III S cells whilst in the mouse. These cells had acquired new genetic information, namely that required to manufacture Type III polysaccharide.

The mouse played only a passive part in this transformation: the transforming process could occur in a test tube with a mixture of live Type II R cells and heat-filled Type III S cells. Indeed, transformation could occur using a cell-free extract of dead S cells. The chemical nature of the 'transforming principle' was investigated in 1944 by Avery and his co-workers who showed it to be DNA and not protein.

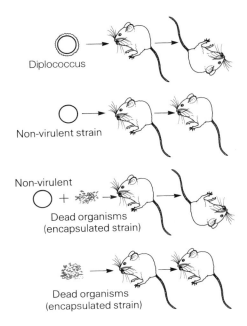

Diplococcus

Non-virulent strain

Non-virulent + Dead organisms (encapsulated strain)

Dead organisms (encapsulated strain)

Fig. 3.3 The experiments of Griffith in 1928 established the idea of transformation in bacteria, i.e. of 'something' from the heat-killed, virulent strain transforming the non-virulent strain so that the infected mice succumbed to pneumonia.

Exercise 1

How could it be demonstrated that a recently isolated (purified) transforming agent was composed of DNA?

---

reversion: a reversion is a back mutation.
transformation: the genetic modification of a bacterial cell as a result of the entry of extraneous DNA into that cell. It occurs naturally and is used experimentally in genetic engineering. The word

transformation is also used to describe the changes that occur in cultured cells after treatment with tumour viruses or treatment with carcinogens.

3.3 The Hershey and Chase experiment

Another series of experiments, reported in 1952 by Hershey and Chase, ha become famous as a direct proof that DNA is indeed the genetic material. At that time it was known that there were a number of viruses, called **bacteriophages**, that could infect bacterial cells. The structure of these viruses was rather simple: they consisted of a piece of DNA, protected by a protein coat forming a polyhedral *capsid* (Fig. 3.4). When such bacteriophages infect bacterial cells each virus acts like a 'molecular syringe': the bacteriophage attacks the bacterial cell and then 'injects' an infective component into the bacterium (Fig. 2.32). The infection with bacteriophage results in the cellular machinery being *'hijacked'* and turned over to the synthesis of new bacterio-phage particles. Later, the bacterial cells *lyse*, releasing several hundred new virus particles. The same principle applies as that in the transformation of *Diplococcus* described by Griffith. The transfer of DNA results in genetic information being transferred from one organism to the other. In this case it is the information to make new virus particles that is transferred to the bacterial cell.

See Chapter 2

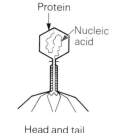

Protein
Nucleic acid

(a)   Head and tail

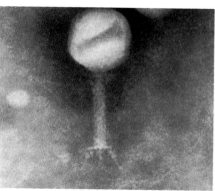

(b)

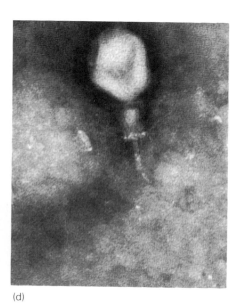

(d)

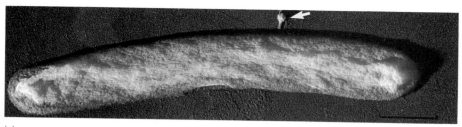

(c)

Fig. 3.4 (a) Structure of a typical T bacteriophage. The protein coat and associated structures (all protein) are quite complicated. (b) Electron micrograph of T bacteriophage. The DNA within the head of bacteriophages may be injected into a bacterial cell. The coat and all protein parts of the bacteriophage remain *outside* the cell. (c) A single bacteriophage (arrowed) caught in the act of injecting its DNA into a bacterial cell as seen by electron microscopy. (d) Electron micrograph showing appearance after injection of DNA. Electron micrographs (b) and (d) courtesy Dr G. Cleator, Department of Medical Virology, University of Manchester.

capsid: *the protein coat of a virus particle, usually made up of identical polypeptide subunits, forming a geometrical shape. Each of the subunits is called a capsomere.*
hijacked: *slang term meaning to steal shipment of goods. Said to have originated during prohibition*

in the USA when bootleggers fell prey to other criminals. When a hold-up took place, the command was 'Stick'em up high, Jack!'
lyse: *to break open a cell, releasing its contents. Originally derived from* **haemolysis***: when red blood cells are placed in water (lower osmotic pressure than cell contents) they burst.*

Hershey and Chase showed that it was bacteriophage DNA and not bacteriophage protein that was transferred from the bacteriophage to the bacterium during infection. In this case the information to make new virus particles was transferred to the bacterial cell. They knew that bacteriophage protein lacked phosphorus and that bacteriophage DNA did not contain sulphur. Using the radioisotopes, ^{32}P and ^{35}S, therefore, it was possible to label each of these separately. They prepared one batch of bacteriophage containing ^{32}P, i.e. its DNA contained radioactive label, and allowed this to infect some cells of the bacterium *Escherichia coli*. After a few minutes the resulting suspension was briefly agitated in a kitchen blender. This had the effect of shearing off the virus particles that were adhering to the outside of the bacteria, but by this time they had injected their DNA into the bacterial cells. Centrifugation first at low speed, and then at high speed, produced two pellets: one containing the bacteria; the other the bacteriophage. Hershey and Chase found that the radioactivity was mostly in the bacterial cells, confirming that it was the DNA that passed into the cells simultaneously with the genetic information (Fig. 3.5). To show that this was in fact the case they then did the experiment the other way round, namely, they labelled the bacteriophage's protein coat by using ^{35}S, and showed that following infection, this did *not* pass into the bacterial cells.

These experiments showed fairly conclusively that, at least in bacteria, as well as in the viruses that infected them, DNA was the carrier of the genetic information. Another line of evidence came from work with tobacco mosaic virus, although here the genetic information turned out to be carried in RNA. DNA and RNA are similar in structure, and it is not unreasonable that RNA should, in some instances, carry genetic information in a similar manner to DNA.

3.4 Tobacco mosaic virus

Tobacco mosaic virus (TMV) causes a debilitating disease in tobacco plants with serious economic consequences. Electron microscopy of the virus particles (Fig. 3.6) shows the virus to be a long cylinder of M_r about 40×10^6. Each virus particle contains a single-stranded RNA molecule of M_r 2×10^6, which is coiled, and around which are packed 2130 identical protein molecules ('capsomeres'), of M_r 18 000, to form a capsid.

In 1957 Fraenkel-Conrat and Singer experimented with two different strains of TMV which were distinguishable because of differences in the protein capsomeres. They were able to separate the proteins and RNA components and also recombine them to give infective virus particles. In this way they could make various combinations of protein and RNA, and infect tobacco plants and subsequently isolate new virus from the tobacco leaves. They found that whatever combination was used, the type of protein capsomere in the newly synthesized virus particles was always that corresponding to the *RNA* of the infecting particle not the protein (Fig. 3.7). Thus, in this virus too, it is the nucleic acid and not the protein that carries the genetic message.

(a)

INDEPENDENT FUNCTIONS OF VIRAL PROTEIN AND NUCLEIC ACID IN GROWTH OF BACTERIOPHAGE*

By A. D. HERSHEY AND MARTHA CHASE

(*From the Department of Genetics, Carnegie Institution of Washington, Cold Spring Harbor, Long Island*)

(Received for publication, April 9, 1952)

The work of Doermann (1948), Doermann and Dissosway (1949), and Anderson and Doermann (1952) has shown that bacteriophages T2, T3, and T4 multiply in the bacterial cell in a non-infective form. The same is true of the phage carried by certain lysogenic bacteria (Lwoff and Gutmann, 1950). Little else is known about the vegetative phase of these viruses. The experiments reported in this paper show that one of the first steps in the growth of T2 is the release from its protein coat of the nucleic acid of the virus particle, after which the bulk of the sulfur-containing protein has no further function.

(b)

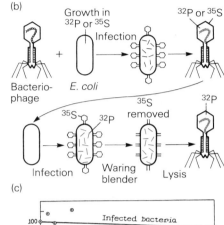

(c)

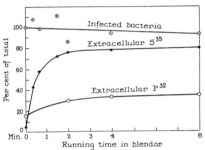

Fig. 1. Removal of S^{35} and P^{32} from bacteria infected with radioactive phage, and survival of the infected bacteria, during agitation in a Waring blendor.

Fig. 3.5 (a) Part of the original paper published by Hershey and Chase in which they showed that when a bacteriophage infects a bacterial cell only the bacteriophage DNA goes into the cell (*Journal of General Physiology*, **36**, 39 (1952) reproduced by kind permission of Rockefeller University Press). When this happens the bacterial cell acquires new genetic information, i.e. the information for making phage particles: the information must therefore be in the DNA of the bacteriophage not the protein. (b) Outline of the Hershey and Chase experiment. Either ^{32}P or ^{35}S would have been used in a single experiment, not both at the same time. (c) The original graph from the above paper. From the middle plot it can be seen that after bacteriophage infection, 2–3 min agitation in a blender separated 80% of the extracellular ^{35}S which was incorporated into the protein of the bacteriophage. Most of the ^{32}P incorporated into the DNA could not be released because it was in the bacterial cell.

Exercise 2

If protein was the carrier of hereditary information what results might be expected from the Hershey and Chase experiment?

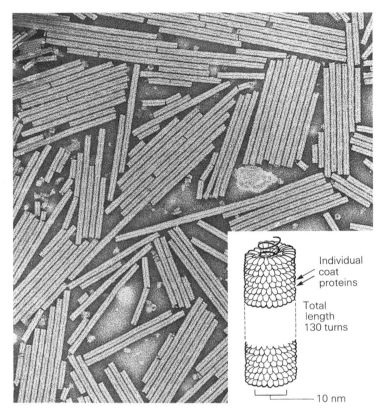

Individual coat proteins

Total length 130 turns

10 nm

Fig. 3.6 Electron micrograph of purified tobacco mosaic virus particles negatively stained with methylamine tungstate. Courtesy of C.M. Clay, Institute of Horticultural Research, Wellesbourne, UK. Inset: each particle is a long rod consisting of a coiled core of RNA surrounded by a tube made up of identical protein subunits. ×90 000.

Box 3.1
Mosaic diseases in plants

A celery leaf showing mosiac disease. Courtesy of Dr G.A. Walkey, Institute of Horticultural Research, Wellesbourne, UK.

The term mosaic disease refers to a number of plant diseases which can be of considerable economic importance (see table). All mosaic diseases produce characteristic symptoms in which some of the leaves are usually pale green or yellow because of loss or reduced production of chlorophylls in the infected areas. The shape and pattern varies considerably. If the discoloured areas are rounded, the symptoms are called a *mottle*. In some types of infection regular light and dark green banding may occur. In infected monocotyledonous plants mosaic symptoms usually appear as light and dark green striping.

Some mosaic infections result in plant hormone imbalances. The shape of the leaf lamina is affected leading to distortion. Tissue necrosis (death) may occur. Areas of necrosis can spread to stem and roots and eventually result in the death of the plant.

Yield reductions associated with some mosaic viral diseases

Crop	Viruses	Yield reduction (%)	Country
Beans	Bean yellow mosaic	33–64	USA
Cabbage	Turnip mosaic	36	England
Cassava	Cassava mosaic	24–75	Kenya
Lettuce	Lettuce mosaic	56	USA
	Cucumber mosaic	8–50	England
Raspberry	Raspberry mosaic	50	USA
Tobacco	Tobacco mosaic	5–16	USA

Adapted from Walkey, D.G.A. (1985) *Applied Plant Virology*, Heinemann, London, UK.

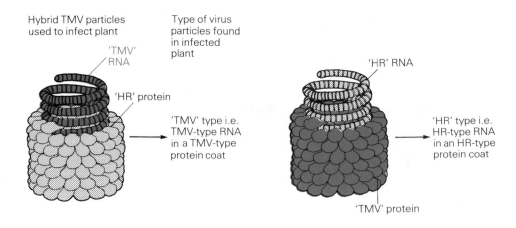

Hybrid TMV particles used to infect plant

'TMV' RNA

'HR' protein

Type of virus particles found in infected plant

→ 'TMV' type i.e. TMV-type RNA in a TMV-type protein coat

'HR' RNA

'TMV' protein

→ 'HR' type i.e. HR-type RNA in an HR-type protein coat

Fig. 3.7 When hybrid tobacco mosaic virus (TMV) is used to infect a tobacco plant, the new virus particles that can subsequently be isolated from the plant are not hybrids but are of the type dictated by the RNA of the infecting hybrid virus. Fraenkel-Conrat and Singer did these experiments in 1957, making hybrid virus particles by mixing the RNA of one strain (e.g. HR) with the protein coat of another strain (e.g. TMV) under the appropriate experimental conditions. Thus it is the nucleic acid not the protein that carries the genetic information.

3.5 Evidence that DNA is the genetic material in eukaryotes

From studies in biology a great deal of indirect evidence has accumulated that DNA is the genetic material in eukaryotic cells. For example it is known that each of the cells of a given organism contain the same amount of DNA, except for the haploid gametes, which contained exactly half this amount. This makes sense because when the gametes combined at fertilization a normal diploid cell with a full complement of DNA would result. Furthermore, although there are some exceptions, the more complicated the organism the more DNA its cells contain, which seems reasonable. (This is in fact not quite true, but there are probably reasons why this is not a stumbling block.)

Another line of indirect evidence came from a comparison of the likely response of DNA and protein to mutagenic agents. Both physical and chemical agencies that could modify DNA caused **mutations**; those that modified protein did not have this effect (see *Molecular Biology and Biotechnology*, Chapter 8). Thus, ultraviolet radiation of wavelength 260 nm (a wavelength absorbed by DNA) is more mutagenic than radiation of wavelength 280 nm (absorbed by proteins). Various highly mutagenic chemicals are also known. From a knowledge of chemistry these are much more likely to react with DNA than with proteins.

Whilst all this was known, experiments were in progress that demonstrated how information in prokaryote DNA could be transcribed to give mRNA and how this could be translated into protein. In addition, the genetic code was gradually elucidated. This identified the sequence of bases in a molecule of mRNA (and therefore DNA) which specified which amino acids should be joined to form a polypeptide chain. It was then possible to provide a chemical explanation for the phenomenon of mutations. A single base change in the DNA could be identified as being completely responsible for a single amino acid change in a polypeptide. Most significantly, the rules that applied to prokaryotes, by and large, also applied to eukaryotes.

Since Watson and Crick's paper in 1953, knowledge and capabilities have increased enormously. Techniques have developed which make it possible not only to sequence DNA, but also to make it by chemical synthesis and to transfer it between organisms. It is also now possible to isolate a gene such as

□ It might be expected that the more complicated the organism the more DNA each cell would contain (that is, more proteins therefore more genes) and this is broadly true. However, the precise correlation turns out to be less good. Thus a yeast cell, a fruitfly (*Drosophila*) cell and a human cell contain 0.05, 0.15 and 3.2 pg DNA respectively. Yet amphibians, which are usually thought of as being somewhat less complex than humans, are the vertebrate species with the most DNA (100 or more times as much as human beings). Even tulips have 10 times more than humans. The content of DNA per haploid cell is called the C value and the failure of C value to correlate with phylogenetic complexity is called the C value paradox. The reason for it is that much of a eukaryotic organism's DNA consists of non-coding sequences (sequences that do not contain information for making proteins) of unknown function.

See *Molecular Biology and Biotechnology*, Chapters 3 and 4

□ A single base change in the DNA can have very dramatic consequences. In sickle-cell anaemia in humans, a change from T to an A results in a change in the β-chain glutamic acid of haemoglobin at position 6 to a valine. This creates a 'sticky' patch on the protein molecule causing the sickle cell deoxyhaemoglobin molecules to tend to aggregate with serious clinical consequences for the individual.

mutation: *a change in the DNA. This can be as simple as a single base change, loss or addition, or as complicated as the loss of a large section of DNA or even a whole chromosome. A mutation is usually recognized by the resulting change in the* characteristics of a cell or organism. In fact a change may have no effect at all (silent mutation) at one extreme or may be fatal at the other.

the stretch of DNA carrying the information for making human insulin, and to sequence it and identify the order of bases. This piece of DNA may be put into another cell, such as a bacterium or a yeast, which will then express human insulin. This is a **transformation** (just as Griffith observed in 1928). DNA is transferred from cell to cell, and when this happens the recipient cell is found to have new genetic information, here the ability to produce a protein, human insulin. Such methodologies provide incontrovertible evidence that DNA is the genetic material in eukaryotes.

3.6 Exploiting DNA as the genetic material

Specific sequences of DNA may be looked for in the chromosome in order to find which part of a particular chromosome carries a particular gene (Table 3.1). The procedure of *in situ* hybridization depends entirely on the knowledge that DNA is the carrier of genetic information and provides evidence for this. The technique involves using a small piece of DNA whose sequence is complementary to all, or more usually part, of the sequence of the gene it is wished to identify. This piece of DNA, called a *probe*, is made radioactive.

DNA probes are used in a variety of ways. They may be used to answer the question: does a sample of DNA contain a sequence complementary to that of the probe? If it does then the probe and its radioactive label, **hybridize**, or stick to, this DNA. Probes may be used to find out on which chromosome, and indeed on which part of it, a particular gene lies. This is done by allowing the radioactive probe to hybridize with a preparation of cells or a histological section in which the chromosomes are visible. After a time, the excess probe is washed off, and, in the dark, a layer of photographic emulsion is applied. The preparation is stood in the dark (usually for weeks or even months because the amounts of radioactivity are small) and then developed in the same way that photographic film is developed. Black 'dots' appear where the radioactive probe has hybridized to DNA in the preparation. In the microscope it may be seen where black dots coincide with regions on the chromosome (Fig. 3.8) and hence where that particular gene lies. **Karyotyping** identifies which chromosome is which.

Table 3.1 *Location of genes on human chromosomes. The location of a particular gene on a particular chromosome may be established in a number of ways. Many characteristics are sex-linked which can point to a gene being situated on the X chromosome. A more direct way is to use in situ hybridization. Note that for haemoglobin the genes for the two types of chain that go to make the functional molecule are on different chromosomes: the same is true of the two types of chain that form the antibody immunoglobulin molecule*

Gene	Chromosome
Glutamate oxaloacetate transaminase	10
Thymidine kinase	17 (long arm)
Glucose phosphate isomerase	19
Haemoglobin α-chain	16
Haemoglobin β-chain	11
Antihaemophilic globulin (haemophilia)	X
Addison's disease*	X
Colour blindness	X
Ichthyosis	X
Muscular dystrophy	X

* In most of the cases of the X chromosome being involved the description refers to a medical condition for which the gene defect is unknown.

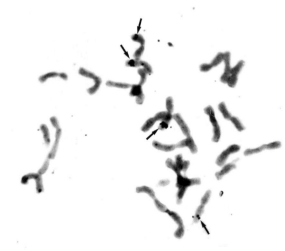

Fig. 3.8 *In situ* hybridization. Mitotic root tip chromosomes (from wheat plants) were treated with ribonuclease (since RNA would interfere) and then a radiolabelled DNA probe applied which recognizes (hybridizes to) the ribosomal RNA genes. The photograph shows the sites (arrowed) where the probe bound and which are therefore the sites of rRNA genes: 6–8 weeks of exposure was required. Courtesy of Dr J. Hutchinson-Brace, Department of Biological Sciences, The Manchester Metropolitan University, UK.

Box 3.2
Karyotyping

Karyotyping means representations of the chromosomal complement of a cell with individual mitotic chromosomes arranged in pairs in order of size.

A great deal is known about the fine details of chromosome structure. Some is at the molecular biological level; that is, the nucleotide sequences of individual genes, but a considerable body of information came from earlier cytological studies. Simple examples are the human X and Y chromosomes: the normal female has XX and the normal male XY. Karyotyping is valuable in medical diagnosis. Individuals with an extra copy of chromosome 21 (trisomy 21) have Down's syndrome, for example.

Karyotyping is carried out as follows. The chromosomes from one mitotic cell (usually a lymphocyte) are stained with a dye that reacts with DNA (orcein). The cell, 'squashed' on a microscope slide, is photographed. On the print, each pair of homologous chromosomes is identified (by length, shape, position of centromere), and these are cut out and mounted in order of decreasing size, and numbered. The entire set is the karyotype.

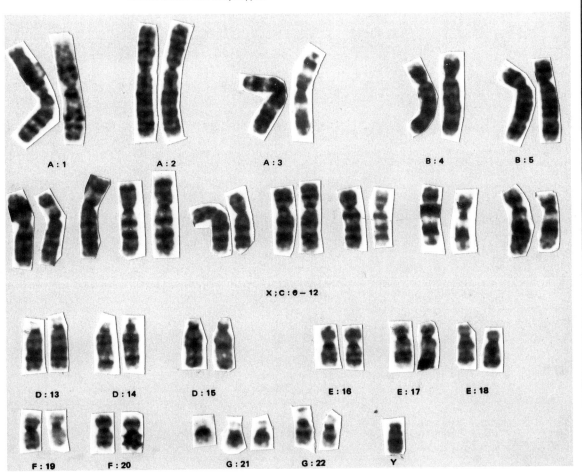

The karyotype of an individual with Down's syndrome (trisomy 21). Note that chromosome 21 appears three times. Because the chromosomes conventionally are placed in order of size, chromosome X appears in the second row between the pairs for chromosome 6 and 7. Courtesy of J.S. Haslam and K.P. O'Craft, Tameside General Hospital, Ashton-under-Lyne, UK.

Interestingly, it was not until about 1955 that it was firmly agreed that human beings have 46 chromosomes (see Wall, W.J. (1987) The chromosomes of man: an historical perspective, *Biologist*, **34**, 100–2).

Another way in which DNA probes may be used is to find out which bacterial cell out of one million cells carries a certain piece of DNA. This technique is indispensable in recombinant DNA technology where pieces of DNA have been inserted, at random, into plasmids which have been put into bacteria. The cells containing the plasmid are grown up on large agar plates at a dilution at which it can be certain that each individual colony originated from a single cell. The colonies are 'blotted' on to nitrocellulose paper and the cells broken open (lysed). Baking the filter at 80°C firmly fixes the DNA from the colonies on to the nitrocellulose paper as a replica of the plate. The paper may then be treated with a solution of the radioactive probe, washed, and placed against X-ray film. Any areas that take up the radioactive probe contain the DNA complementary to that of the probe. This allows the colony to be identified, picked off the original agar plate and grown up. This method (cloning) is a way of picking out a cell with a particular piece of DNA in it and of growing it up in a large volume of culture fluid, effectively **amplifying** that piece of DNA for further study.

These two methods have been described in some detail here because they show that molecular biology is based firmly on the knowledge that DNA is the genetic material in both eukaryotes and prokaryotes.

3.7 The nucleoid

In prokaryotes the DNA appears to exist free in the cytoplasm (Fig. 3.9) although there is some evidence that it is associated with the plasma membrane. Since the chromosomes of bacteria are over about 1 mm long, and a bacterial cell may be 1–2 μm in diameter, there must be some way of packaging or condensing the DNA. The chromosome of *E. coli* is a single, circular, double-helical molecule of DNA associated with positively charged protein molecules. This structure is often referred to as a **nucleoid**. The DNA is organized into a series of more than 50 negatively supercoiled loops held together by an RNA–protein core (Fig. 3.10) stabilized by a complex of positively charged polyamines and basic proteins. Bacterial chromosomes have been shown to contain two types of protein of M_r about 17 000 and 9000. Both are apparently necessary for chromosome formation. X-ray diffraction studies of the smaller protein have shown that it forms dimers, each subunit

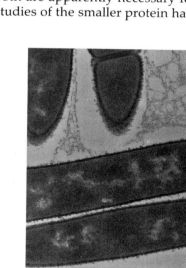

Fig. 3.10 Electron micrograph of an *E. coli* chromosome showing the multiple loops emerging from a central region. (Bluegenes © 1983 Designer Genes Posters Ltd. Available on T-shirts, posters and postcards from Carolina Biological Supply Co., Burlington, NC 27215, USA.)

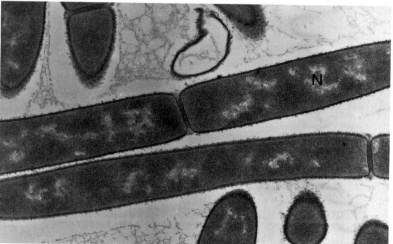

Fig. 3.9 Electron micrograph of a section through cells of *Bacillus subtilis* showing profiles of nucleoids (N). Courtesy of Dr I.D.J. Burdett, National Institute for Medical Research, Mill Hill, London, UK.

Reference Pettijohn, D.E. (1982) Structure and properties of the bacterial nucleoid, *Cell*, **30**, 667–9. A short, readable minireview of the organization of the bacterial chromosome.

having an extended 'arm'. The arms are rich in arginine residues which interact electrostatically with the sugar–phosphate backbones of the DNA, stabilizing it in its compact, superhelical conformation.

3.8 The nucleus

In eukaryotic cells, the DNA is protected in the nucleus which has a surrounding **envelope**. Electron microscopic studies show that the nuclear envelope consists of a double membrane separated by a **perinuclear space** (Fig. 3.11).

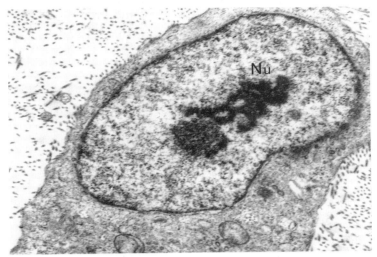

Fig. 3.11 Electron micrograph of a nucleus of a eukaryotic cell with enclosed nucleoli (Nu) (×75 000).

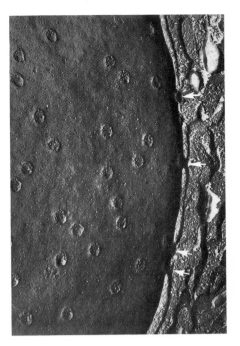

Fig. 3.12 Freeze-fracture preparation of nucleus and adjacent cytoplasm of an acinar cell of the pancreas. The plane of fracture was through inner membrane of nuclear envelope, showing fractured pores (arrowed). Micrograph by B. Gilula, from Fawcett, D.W. (1981) *The Cell*, 2nd edn, W.B. Saunders, with permission, courtesy of Professor D.W. Fawcett.

THE NUCLEAR ENVELOPE is interrupted by pores, found at regions where the outer and inner membranes fuse (Fig. 3.12 and 3.13a). The pores are surrounded by a highly organized **annulus** consisting of eight protein granules symmetrically arranged around the pore (Fig. 3.13b). The term **porosome** has been applied to the nuclear pore and its associated annulus.

The diameter of the pores, measured from electron micrographs, is about 70–80 nm. However, while proteins of M_r 15 000 or less freely enter the nucleus, serum albumin (M_r 67 000) enters slowly and larger proteins (M_r 120 000) are excluded. Nuclear proteins, like all proteins, are produced in the cytosol but must accumulate in the nucleus for functional reasons. Such proteins contain sequences of amino acid residues which act as signals, enabling them to be concentrated selectively in the nucleus. Glycoproteins present in the porosome appear to regulate the entry of material into the nucleus. The protein which regulates the transcription and replication of the DNA of the parasite SV40 virus contains a sequence rich in basic amino acid residues:

–Pro–Lys–Lys–Lys–Arg–Lys–Val–

which appears to be a typical nuclear-directing signal for proteins that must accumulate in the nucleus.

Nuclear pores can occupy up to 20% of the surface area of the nucleus, allowing material to be transported between the nucleus and cytoplasm. Indeed, material is often observable as an electron-dense plug within the pore (Fig. 3.13b).

□ Globular proteins of M_r 15 000 and 120 000 would have theoretical diameters of 1.63 and 3.26 nm, respectively, if they were perfect spheres.

□ A variety of signals consisting of sequences of amino acid residues target particular proteins to specific sites. The signals which allow proteins to cross the endoplasmic reticulum membrane are described in Chapter 1. Transit signals which 'direct' proteins to chloroplast and mitochondria are briefly reviewed in Section 5.5, while lysosomal-directing signals are covered in Section 1.6.

nuclear envelope: consists of the complex of two nuclear membranes, the perinuclear space and the nuclear pore complexes.

Reference Drlica, K. and Rouviere-Yaniv, J. (1987) Histone-like proteins of bacteria, *Microbiological Reviews*, **51**, 301–19. A detailed account of bacterial chromatin. Emphasis is on the biochemistry and genetics of bacterial histone-like proteins.

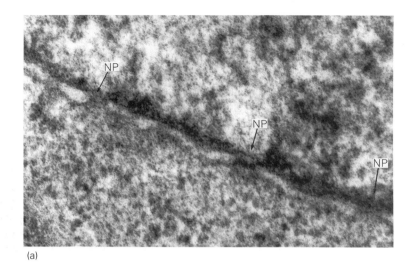

(a)

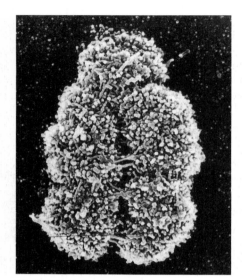

Fig. 3.14 Scanning electron micrograph of a human metaphase chromosome (×12 000). Courtesy of Dr C.J. Harrison, Christie Hospital, Manchester, UK.

☐ Protamines are yet another class of small basic proteins which bind to DNA. They are found in the heads of spermatozoa. Presumably space is so restricted that even further condensation of the DNA is required. Nearly two-thirds of the amino acid residues of protamines are basic (mostly Arg). Following their synthesis in the cytoplasm, protamines become phosphorylated and migrate to the nucleus where they simultaneously replace histones and are dephosphorylated. The DNA–protamine complex is stabilized by ionic interactions between the roughly equal numbers of positive charges carried by the protamines and the negative charges of the phosphates of the DNA.

☐ The DNA-binding protein of bacterial chromosomes which was described earlier (pp. 78) closely resembles histone proteins. For example, the protein from the *E. coli* nucleoid has 90 amino acid residues of which 9 are Lys and 5 Arg. Thus it particularly resembles the Arg-rich H3 and H4 histones.

See *Molecular Biology and Biotechnology,* Chapter 5.

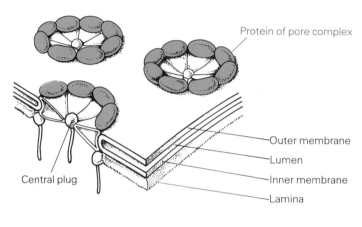

Protein of pore complex

Outer membrane
Lumen
Inner membrane
Lamina

Central plug

(b)

Fig. 3.13 (a) Nuclear pores (NP) of a rat liver cell as seen in the electron microscope (×120 000), courtesy of P.L. Carter, Department of Biological Sciences, The Manchester Metropolitan University, Manchester, UK. (b) Diagram showing the structure of the annulus of a nuclear pore. Redrawn from de Duve, C. (1984) *A Guided Tour of the Living Cell*, Scientific American Books, Inc., W.H. Freeman, USA, p. 282.

Chromatin

Eukaryotic chromosomes (Fig. 3.14) consist of about one-third DNA and two-thirds protein. The complex of chromosomal DNA and protein (a nucleo-protein) is called **chromatin**. The protein of chromatin consists of small, basic proteins known as *histones*, and additional proteins generally called some-what unimaginatively 'non-histone proteins'. The different histones may be separated by ion-exchange chromatography or electrophoresis (Fig. 3.15). All eukaryotic cells contain five types of histones, designated H1, H2A and H2B, H3 and H4, the sequences of which are highly conserved between species; that is, the amino acid sequences of comparable histones are similar in different species. All histones have a high content of the basic amino acid residues arginine and lysine (Table 3.2).

Histones are subjected to many post-translational modifications, including acetylation, methylation, phosphorylation and ADP-ribosylation. These modifications alter their size, charge and affinity for DNA, and are likely, therefore, to have a role in regulating transcription. The extent of these

histones: *basic proteins, rich in arginine and lysine, found in the nuclei of eukaryotic cells.*

Reference Jordan, G. (1988) The nucleus, *Biologist*, **35(5)**, 247–52. An interesting, easy-to-read account of the structure and functions of the nucleus and associated structures.

Table 3.2 *Types of histones from thymus of rabbit (H1) and calf (H2A–H4)*

Type	Location	Approx. M_r	Total	Residues Lys	Arg
H1	Linker	23 000	213	62	3
H2A	core	14 000	127	14	12
H2B	core	13 800	125	20	8
H3	core	15 300	135	13	18
H4	core	11 300	102	11	14

modifications varies during the cell cycle to reach a maximum during S phase when doubling of the DNA content of the nucleus occurs.

In interphase nuclei, chromatin is present as either heavily staining **heterochromatin** or more diffuse **euchromatin** (Fig. 3.11). There is evidence that heterochromatin is genetically inert, and replicates late in the S phase of the cell cycle. In contrast, euchromatin comprises those genes that the cell needs to transcribe actively, the so-called 'house-keeping' genes, producing RNA molecules. Since euchromatin is less tightly coiled than heterochromatin, perhaps not surprisingly, it is replicated early in S phase.

STRUCTURE OF CHROMATIN. Nuclei are generally 3–25 μm in diameter, although the size and shape varies considerably between organisms, the largest nuclei being associated with greater amounts of chromatin. The small volume of the nucleus can accommodate enormously long genomic DNA molecules. For example, the total length of DNA in human cells is about one metre! This is equivalent to packing a piece of string as long as the M1 motorway (about 187 miles) into a box of about 9 cubic feet volume. This packaging can only be achieved by an intricate folding of the DNA, in which the histones play a crucial role. The packing must be carried out so as to allow replication and expression of the genome.

Chromatin is made up of repeating structural units called **nucleosomes**. Each nucleosome consists of a length of DNA (Fig. 3.16b) about 200 base pairs long and an octomer of two each of histones H2A, H2B, H3 and H4. The histones form a core structure about $11 \times 11 \times 5.5$ nm with about 140 base pairs of the DNA wound around it in a left-handed helix. The remainder of the DNA connects the nucleosomes together in a flexible, jointed chain, rather like a string of beads (Fig. 3.16b). One molecule of histone H1 is located at the site where DNA enters and exits the core particle, effectively closing two full turns of DNA around the histone octomer (Fig. 3.16b). *In vivo*, the chromatin fibre occurs as a linear chain of closely packed nucleosomes forming a filament 10–11 nm in diameter. This chain of nucleosomes is arranged into successively higher ordered structures reducing the total length of the complex and allowing it to be packed into the nucleus. The next level of folding is a chromatin fibre of 25–30 nm diameter which appears to form through a continuous coiling of the 10–11 nm fibre, giving rise to a **'solenoid'**; with 6–7 nucleosomes per turn (Fig. 3.16c). The solenoid in turn is thought to be arranged in loops up to 250–400 nm long. These loops appear anchored to a system of fibrous proteins called the nuclear lamina (see later) (Fig. 3.16d). It is estimated that loops contain 10 000–180 000 bp, with a mean of about 63 000 bp per loop. Loops of this size would contain about 315 nucleosomes, with six nucleosomes per turn of the solenoid. Human cells, containing about a metre length of DNA, would have about 50 000 loops per nucleus.

The manner in which non-histone proteins associate with 10–11 and 25–30 nm diameter fibres is not known, nor have definite functional roles been ascribed to them. It is generally assumed that the non-histone proteins have biological functions in transcription and DNA replication and structural roles in the chromosome scaffold (see pp. 85).

Reference Isenberg, I. (1979) Histones. *Annual Review of Biochemistry*, **48**, 159–91. A rather old review, but still excellent to start reading about histone proteins.

Reference Kornberg, R.G. and Klug, A. (1981) The nucleosome. *Scientific American*, **244(2)**, 48–60. Old, but still the place to start learning about nucleosomes.

Exercise 3

What is the significance of the high content of basic amino acid residues in histone proteins?

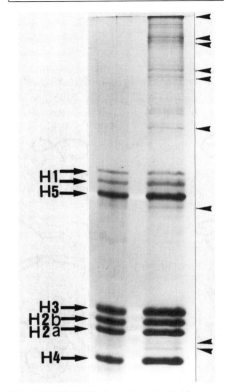

Fig. 3.15 SDS-PAGE of proteins extracted from chicken erythrocyte nuclei. The arrows on the track on the left show the designations and positions of histones H1–5 and those on the right (unpurified extract) show the non-histone proteins of 'total' nuclear protein. Courtesy of Dr J.E. Perez-Ortin, University of Valencia, Spain.

☐ Given their enormous sizes, the lengths of DNA molecules are often expressed in kbp (kilobase pairs): the length occupied by 1000 base pairs is 0.34 μm.

Exercise 4

A lysine residue (Lys 20) of histone H4 is frequently modified by methylation. What effect will this methylation have on the interactions of molecules of histone H4 and DNA?

Fig. 3.16 Sequence showing the stages in the packing of chromatin and the associated packing ratios. (a) Image of unstained DNA from calf thymus viewed by scanning tunnelling microscopy ($\times 1.76 \times 10^6$). Courtesy of Technical Information Department, Lawrence Berkeley Laboratory, California, USA. (b) Electron micrograph showing 'beads on a string' appearance of nucleosomes. The beads consist of octomers of histone proteins with associated DNA. Strands of DNA linking the nucleosomes together are apparent. (c) The 30 nm diameter solenoid fibre ($\times 78\,000$) formed by coiling of the beaded nucleosome. Both electron micrographs of fibroblasts. (d) Higher order loops of chromatin extending from a partially lysed interphase nucleus of *Bombyx mori*. Electron micrographs b–d courtesy of, Dr B. Hamkalo, Department of Molecular Biology and Biochemistry, University of California, Irvine, CA, USA. (e) Scanning electron micrograph of a human metaphase chromosome ($\times 12\,000$). Courtesy of Dr C.J. Harrison, Christie Hospital, Manchester, UK. Packing ratios are indicated.

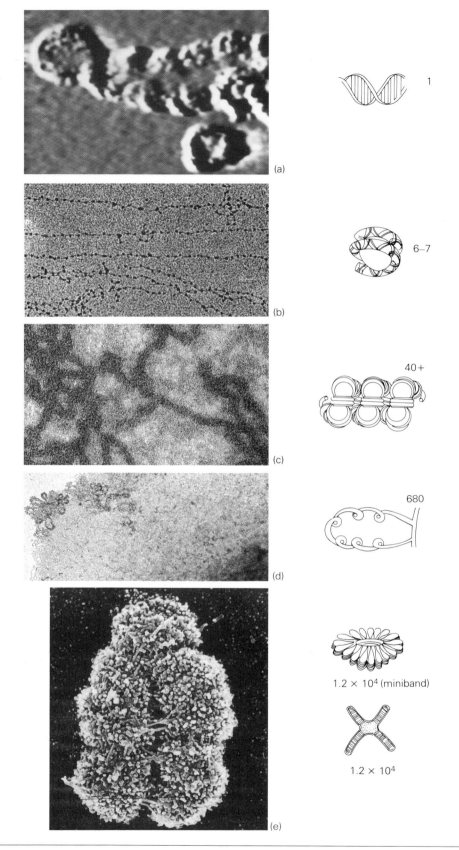

1

6–7

40+

680

1.2×10^4 (miniband)

1.2×10^4

☐ The ratio of the extended chromosomal DNA molecule to the length of a condensed form is described as the packing ratio of that condensation stage.

Reference Sperling, R. and Wachtel, E.J. (1981) The histones. *Advances in Protein Chemistry*, **34**, 1–60. A review article covering the dynamic structural role of histones in chromatin.

Box 3.3
The human genome project

In 1985 it was suggested that an ambitious project be undertaken to sequence the entire human genome. The idea was to produce a 'Book of Man'. This book would consist of 23 volumes, one for each chromosome, each of about 2000 pages, a page for each gene. Initially the idea was not widely supported for a variety of reasons. Sequencing of nucleic acids at that time was expensive and time-consuming. Also, the technologies for isolating DNA and preparing fragments to allow the complete genome to be sequenced did not exist. Furthermore, even if the sequence was completed analysis of the huge amount of information was not possible! However, the project was exciting for several reasons. Knowledge of the sequence of the human genome will help elucidate some of the problems of development; it will specify exactly what humans are, and why they differ. Further, the implications for medicine are profound. Many diseases have a genetic basis, either a gene is defective or part of its controlling mechanism is in error. Major human disorders often have a genetic element, for example, schizophrenia, heart disease. Knowing the sequence of human DNA should (eventually) enable the genes involved to be identified allowing individuals at risk to be treated at an early stage or advised on appropriate preventative measures. Such exciting possibilities meant that several senior researchers were enthusiastic and gradually support for the project was forthcoming.

Even now, however, the project still has detractors who argue that since about 95% of the sequence seemingly has no role but is evolutionary 'detritus', then determination of the whole sequence is a pursuit of questionable value. The point has also been made that the project will tie up a great deal of scientific resources, including scientists, which could perhaps be more usefully employed on other projects.

Before 1976 only 10–20 bp of DNA could be sequenced per person-year. The development of methods of sequencing nucleic acids by Sanger and by Maxam and Gilbert in 1976, increased this figure to about 5000 bp. Currently, something like 10^4 to 10^5 bp can be sequenced per person-year. However the human genome is about 3×10^9 bp in length, and probably codes for about 10^5 genes although the exact number is unknown. Clearly methods of sequencing will have to be improved to make the genome project feasible. The laboratory steps involved in sequencing are, however, amenable to automation, and the use of robotics and computer-controlled instrumentation should allow sequencing to proceed at a rate of about 5000 bp/day. Working at this speed, a team of 300 people would take 10 years to complete the sequence, although it is thought that a complete physical map of the genome should be completed in 5–7 years. The cost of the whole project is estimated at US$3 billion (approx. $1 a base). A large portion of the effort involved in the human genome project will be in preparing DNA, breaking it into pieces of a size suitable for analysis and separating them from each other. It is estimated that, realistically, the project is likely to take 15–20 years to complete.

The project will present a major challenge to information technology. The use of computers will be mandatory for two prime uses: (a) data management; that is, the storage, retrieval and distribution of information, and (b) analysis; that is, the identification of coding, non-coding and gene-controlling regions of the genome.

Large quantities of fundamental biological information (base sequences) will become available very quickly; trying to make sense of this information could well be the rate-limiting step in the project. Indeed, the interpretation of the sequence and the understanding of how genes interact and are controlled will probably not be completed until the twenty-first century.

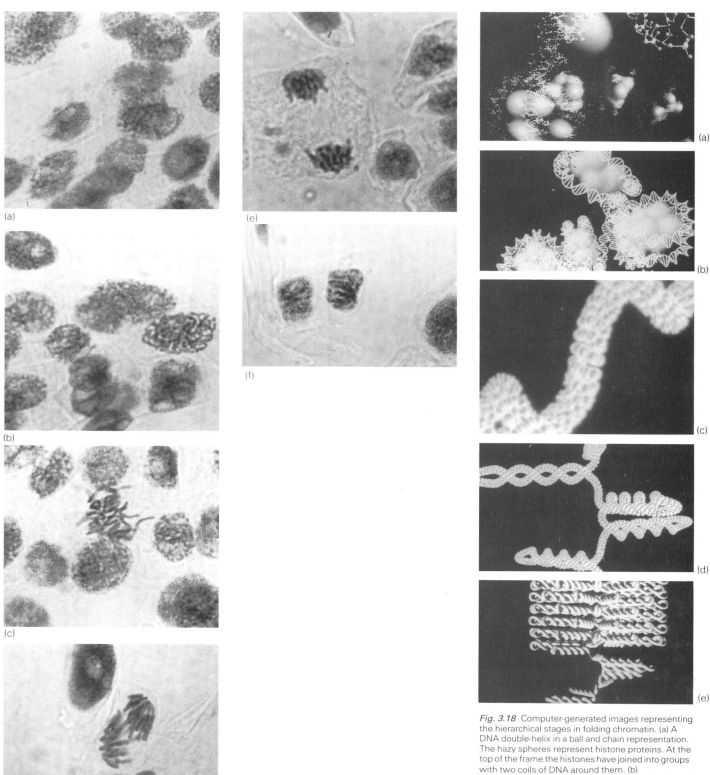

Fig. 3.17 Photomicrographs showing stages of mitosis in the meristem of the root cap of *Allium cepa*. (a) Interphase/early prophase; (b) prophase; (c) metaphase; (d) anaphase; (e) telophase; (f) late telophase.

Fig. 3.18 Computer-generated images representing the hierarchical stages in folding chromatin. (a) A DNA double-helix in a ball and chain representation. The hazy spheres represent histone proteins. At the top of the frame the histones have joined into groups with two coils of DNA around them. (b) Histone–DNA units (nucleosomes) arranging into (c) a chromatin fibre (solenoid). (d) Twisting of the solenoid to form portions of the minibands which orientate to give (e) an arm of a chromosome. Courtesy of Fujitsu Ltd from 'We Are Born Of Stars' ©.

MITOTIC CHROMOSOMES, that is, those in metaphase, are the most highly condensed form of chromatin (Fig. 3.16e and 3.17). During the transition from interphase to mitosis there does not appear to be any change in the structure of nucleosomes, nor in their organization into 10–11 nm and 25–30 nm fibres. However, there is a major rearrangement of the 25–30 nm fibres during prophase to form the highly condensed structure of the mitotic chromosome (see Box 3.2). This higher-ordered organization of chromatin has been remarkably difficult to determine. Since the organization of the chromatin fibre seems to be largely based upon successive coiling, many have suggested that metaphase chromosomes are the result of some forms of **supercoiling** of the chromatin (Fig. 3.18). The loops of chromatin appear to form twisted structures, 18 such loops forming a radial array, like the spokes of a wheel, around the core of the chromosome to give a so-called **miniband**. The continuous formation of minibands produces a stack of minibands which is the metaphase chromatid.

The existence of parallel arrays of loops of chromatin in chromosomes can be demonstrated by depleting chromosomes of their histone and non-histone proteins (Fig. 3.19). The remaining 2–8% of the chromosomal proteins form what appears to be a **scaffold** (Fig. 3.20) to which the loops of chromatin are presumably bound. Scaffolds consist of two main proteins of M_r 135 000 and 170 000. The larger is known to be the enzyme topoisomerase II, suggesting that the formation of superhelical coils is important in chromosome condensation.

Nuclear lamina

The nucleus and, indeed, nuclear pore complexes retain their shape even after lipids and proteins of the nuclear envelope have been extracted by treatment of isolated nuclei with non-ionic detergents. This retention of shape and nuclear pore structure is the result of a highly insoluble proteinaceous meshwork located on the **nucleoplasmic** side of the inner nuclear membrane

Exercise 5

A metaphase chromosome of length 4 μm contains a single DNA molecule 120 × 10^6 bases long. Calculate the packing ratio of the DNA.

See *Biological Molecules*, Chapter 8

☐ The *nuclear* lamina should not be confused with *basal* lamina (alternative term basement membrane), which is an extra-cellular structure supporting epithelial cells and which attaches them to underlying connective tissue.

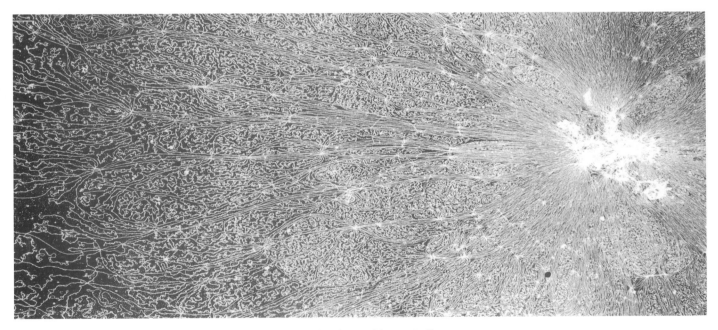

Fig. 3.19 Portion of chromatin spread from a condensed core after extraction of most of the protein. The structure shows numerous fibres forming loops. Reprinted from Mullinger, A.M. and Johnston, R.T. (1980) *Journal of Cell Science*, **46**, 61–86, courtesy of Dr A.M. Mullinger.

Reference Krachmarov, C. and Zlatanova, J. (1988) Nuclear skeletal structures. *Biochemical Education*, **16**, 122–7. A compact, yet wide ranging review of the packing of DNA and the protein skeletons associated with the nucleus, nucleolus and chromosomes.

nucleoplasm: the ground substance of the nucleus. The term is derived from Latin and Greek words meaning *a nut* and *form*.

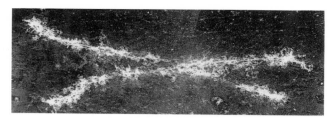

Fig. 3.20 A protein-depleted chromosome with a densely staining core or scaffold. Reprinted from Mullinger, A.M. and Johnson, R.T. (1980) *Journal of Cell Science*, **46**, 61–86, courtesy of Dr A.M. Mullinger.

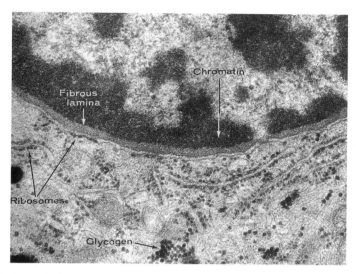

Fig. 3.21 Nucleus and adjacent cytoplasm of an epidermal cell from human skin showing the fibrous lamina. Micrograph by G. Szabo from Fawcett, D.W. (1981) *The Cell*, 2nd edn, W.B. Saunders, with permission, courtesy of Professor D.W. Fawcett.

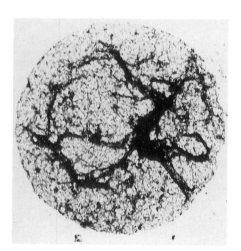

Fig. 3.22 Electron micrograph of nuclear lamina isolated from Ehrlich ascites tumour cells, obtained by extracting isolated nuclear envelopes with non-ionic detergent and high salt concentration. Courtesy of Dr C. Krachmarov, Institute of Molecular Biology, Bulgarian Academy of Sciences, Sofia, Bulgaria.

See Chapter 6

called the **nuclear lamina**. The nuclear lamina is about 30–100 nm thick (Fig. 3.21). It is often difficult to visualize in intact nuclei using electron microscopy because it is normally masked by peripheral chromatin. It is more easily observed following extraction of most proteins with detergent (Fig. 3.22).

The proteins that form the nuclear lamina are not well characterized, but it is generally composed of one to three prominent polypeptides called **lamins** A, B and C with M_r of 60–70 000 (Fig. 3.23). Lamins A and C are homologous with intermediate filaments of the cytoskeleton and presumably they all share a common evolutionary ancestor.

Both lamins A and C contain an internal region of M_r about 40 000, containing about 350 amino acid residues which form an α-helical rod-shaped domain (Fig. 3.24). This domain is flanked by non-helical regions at both the amino- and carboxyl-termini of M_r 4000 and 20–30 000 respectively. Intermediate filament proteins show a similar rod-shaped domain, although the lamins contain an insert of 42 amino acid residues in this region. Lamins A and C form dimers with a 50 nm long central portion with globular heads and tails (Fig. 3.24). These dimers aggregate to form long fibres with diameters of about 10 nm. These 10 nm filaments probably form the nuclear lamina network.

The third nuclear lamina protein (lamin B), although immunologically related to lamins A and C and therefore sharing some common structural features, is distinct from them. Its function in forming the nuclear lamina is less clearly understood. The nuclear lamina is closely associated with the inner nuclear membrane (Fig. 3.21). Indeed, some evidence indicates that lamin B has a specialized membrane-binding role and remains associated with membrane components during nuclear-envelope dissolution at mitosis. Furthermore, proteins of the inner nuclear membrane are relatively immobile in the membrane, possibly because of binding to the nuclear lamina (Fig. 3.25). Chromatin is also bound to the nuclear lamina. As described earlier, it is believed that loops of the 30 nm chromatin fibre are bound to the lamin network (Fig. 3.25) although the precise manner of such binding is unknown.

MITOTIC DIVISION is associated with several well-known distinct phases (Fig. 3.17). During late prophase the nuclear envelope disintegrates and its membranes become indistinguishable from those of the endoplasmic reticulum. The nuclear pores also disappear as morphologically defined units. In prophase, phosphorylation of nuclear proteins, including lamins, is known to occur. During disassembly, the lamins become 'hyperphosphorylated' and contain 2–3 mol P_i per mol of protein, compared to their normal content of only 0.25–0.4 mol P_i per mol lamin. The lamins are not degraded, but are reutilized in the formation of nuclear envelopes in the daughter cells; phosphorylation declines during reassembly. It is not, however, clear whether the disassembly/reassembly of the nuclei and the phosphorylation cycle of the lamins are causally related, although it has been suggested that the depolymerization and reassociation of the lamins, and therefore the dissolution and restructuring of the nucleus, may be controlled by the degree of phosphorylation.

The nuclear matrix

The nuclear matrix (Fig. 3.26) consists of the nuclear lamina, an extensive fibrogranular matrix and a residual nucleolar structure (see later). The nuclear matrix is believed to participate in several basic nuclear functions. Thus, in interphase nuclei chromatin is believed to be organized into loops (see earlier) attached to the matrix (Fig. 3.25). Each loop is thought to represent a discrete replicational and transcriptional unit, containing only one or, possibly, a few genes.

The nuclear matrix is enriched in genes that are actively being transcribed (that is, euchromatin, see earlier). Post-transcriptional activities also occur here. The matrix also contains binding sites for some hormones, drugs, etc., which can affect replication and transcription.

3.9 The nucleolus

Nucleoli are prominently staining regions of the nucleoplasm (Figs 3.11 and 3.27). They are composed of groups of ribosomal genes surrounded by their rRNA transcripts, together with many proteins.

Nucleoli are the sites of synthesis of rRNA molecules and for the assembly of ribosomal subunits using rRNA molecules, and the ribosomal proteins produced in the cytoplasm. The rRNA produced by transcription in the nucleus is cut into short lengths and the proteins bound to the RNA molecules. The shortest rRNA molecule is, however, made outside the nucleolus and migrates into the nucleolus during the formation of the ribosome subunits. The small ribosomal subunit takes about 30 minutes to be formed; assembly of the larger takes approximately twice as long.

In eukaryotes transcription of rRNA takes place on a 100–1000 repeated DNA sequence, the larger number being associated with cells that are highly active in protein synthesis and therefore have a large demand for ribosomes. Thus, the number and size of nucleoli depends upon the cell type and its

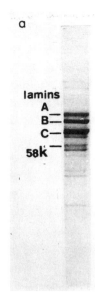

Fig. 3.23 SDS-PAGE of lamina proteins from Ehrlich ascites tumour cell. Lamins A, B and C are the major components with M_r 71, 68 and 64 000 respectively. Courtesy of Dr J. Zlatanova, Institute of Genetics, Bulgarian Academy of Sciences, Sofia, Bulgaria.

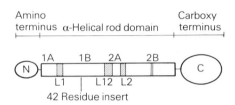

Fig. 3.24 Organization of molecular domains in lamins. In the 50 nm central rod domain the α-helical regions 1A, 1B, 2A and 2B are separated by the non-helical regions L1, L12 and L2. The rod domain is flanked by hypervariable non-helical domains.

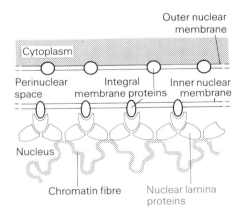

Fig. 3.25 Diagram showing possible arrangement of lamina protein subunits to form a fibrous network constituting the nuclear lamina. The structure organizes the nuclear envelope and binds to specific sites on the chromatin. Redrawn from Alberts *et al.* (1983) *Molecular Biology of the Cell*, Garland Publishing.

Reference Nelson, W.G., Peinta, K.J., Barrack, E.R. and Coffey, D.S. (1986) The role of the nuclear matrix in the organisation and function of DNA. *Annual Review of Biophysics and Biophysical Chemistry*, **15**, 457–75. A very clearly written article which lives up to its title!

Reference Scheer, U. and Benavente, R. (1990) Functional and dynamic aspects of the mammalian nucleolus. *BioEssays*, **12**, 14–21. Up-to-the-minute account of the roles of the nucleus, particularly in the formation of rRNA.

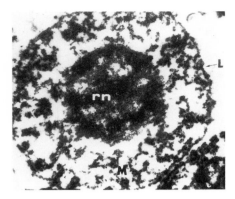

metabolic activity. Nucleoli are largest in active cells, but less prominent in metabolically sluggish types.

Each nucleolus is produced by a **nucleolar-organizing region** (NOR) restricted to a site on a **nucleolar-organizing chromosome** (NOC). All eukaryotic cells contain at least one such chromosome. However, since nucleoli often fuse, the number of nucleoli present is not a simple count of the number of NOCs. For example, in humans chromosomes 13, 14, 15, 21 and 23 are nucleolar-organizers. Fusion of the five nucleoli gives a single large nucleolus. NORs are typically located near the tips of NOCs (see, Fig. 3.8).

Fig. 3.26 Electron micrograph of thin section of nuclear matrix from an interphase nucleus. The fibrogranular matrix (M), nuclear lamina (L) and residual nucleolar skeleton (rn) are all clearly visible. Reproduced from Krachmarov, C. and Zlatanov, J. (1988) *Biochem. Educ.*, **16**, 122–7, with permission of Pergamon Press.

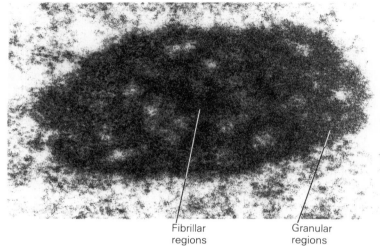

Fibrillar regions   Granular regions

Fig. 3.27 Nucleolus of a chorioallantoic membrane cell (×50 000). Courtesy of P.L. Carter, Department of Biological Sciences, The Manchester Metropolitan University, Manchester, UK.

□ In prokaryotes, transcription of rRNA is not compartmentalized. Indeed, in bacteria only about 10 repeated DNA sequences code for transcripts of ribosomal RNA. Nevertheless, the system is effective since active protein synthesis and therefore efficient ribosome production is maintained during periods of cellular proliferation.

Nucleoli consist of three major regions (Fig. 3.27). There are several **fibrillar centres** containing rRNA genes in the form of partially condensed chromatin. The fibrillar centres are surrounded by a dense **fibrillar component**, which in turn is surrounded by a **granular region** consisting of masses of ribosomal particles in the later stages of maturation.

The nucleolus contains many non-ribosomal proteins and RNA molecules. Some of these molecules form a fibrous **nucleolar skeleton** which is probably attached to the nuclear lamins. At present, the nature and functions of the nucleolar skeleton are ill-defined.

3.10 Origin of the nucleus

The cell biologist Cavalier-Smith has speculated on the biological advantages offered to evolving eukaryotic organisms by the presence of a nucleus. Since the macromolecules involved in DNA replication and repair, as well as transcription and processing of RNA, are concentrated in a relatively small volume, divorced from cytoplasmic activities, metabolic efficiency would be improved. Further, the DNA would be protected from shearing damage caused by cytoplasmic contraction and streaming. As endomembranes developed in the cell, the attachment of DNA to these membranes would be advantageous since the partitioning of DNA to the daughter cells during reproduction would also ensure that the new cells received an equal distribution of internal biomembranes.

It appears likely that mitochondria and chloroplasts originated as ***endosymbiotic*** prokaryotes, but there are several reasons why this would seem an unlikely origin for the nucleus. It is difficult to imagine how the

endosymbiont: *an organism which lives within another, the two organisms having a mutually beneficial relationship.*

Reference Cavalier-Smith, T. (1988) Origin of the cell nucleus. *BioEssays*, **9**, 72–8. A fascinating essay outlining the author's views on the evolutionary origins of the nucleus and its associated structures.

genome of the endosymbiont could take over from the host genome. Although nuclei, mitochondria and chloroplasts all have envelopes the envelopes are different in kind. Mitochondria and chloroplasts have topologically distinct outer and inner membranes held together by lateral adhesive regions. In contrast, the nuclear envelope is a single system: the outer and inner membranes are continuous with one another. Further, the nuclear envelope is continuous with the endoplasmic reticulum, the major endomembrane of the cell. It has therefore been suggested that the nucleus arose from an invagination of the plasma membrane around the host DNA. This invagination was not complete, but left pores allowing the DNA to communicate with the cytoplasm. If this was the case phagocytosis must have evolved before the cell nucleus.

It has been argued that phagocytosis originated in a protoeukaryotic cell which on 'losing' its cell wall replaced it with a glycoprotein surface coating and a rudimentary cytoskeleton to help prevent osmosis-induced lysis. Lamins are homologous to the intermediate fibres of the cytoskeleton; therefore the nuclear lamina may be a divergent specialization of an early cytoskeleton.

3.11 Overview

A considerable body of evidence shows that DNA is the hereditary material in cellular organisms. Viruses, however, use RNA or DNA as their genomic material. In prokaryotic organisms, the DNA is largely present as a single, circular chromosome, complexed with small basic proteins and associated with the plasma membrane. Eukaryotic cells are mainly defined by the presence of a nucleus, a discrete organelle which contains the genome. Nuclei are limited by an envelope consisting of a double membrane penetrated by pores allowing materials to pass between the nuclear contents and the cytoplasm.

Eukaryotic DNA occurs as linear molecules associated with basic proteins called histones which together form a nucleoprotein complex commonly called chromatin. Chromatin has a highly folded structure, its organization being essential to allow the elongated molecules of DNA to be folded and packaged into the confines of the nucleus. A variety of non-histone proteins are also found in the nucleus, and are presumed to be important in regulating transcription, translation and replication.

Chromatin is attached to the inner nuclear membrane by the nuclear lamina, a highly fibrous system of proteins which forms part of the nuclear skeleton. The major function of the nuclear skeleton is maintaining the shape and integrity of the nucleus and its contents.

Answers to Exercises

1. Treat the extract with deoxyribonuclease. If activity is lost then extract DNA.
2. That ^{35}S labelled protein would have entered the bacterial cell and the ^{32}P-labelled DNA would have been excluded.
3. The basic, positively charged proteins can form favourable ionic interactions with the negatively charged phosphates of DNA.
4. Decreased ionic interactions.
5. 10 200.

FILL IN THE BLANKS

1. _____ is the genetic material in organisms. In bacterial cells it is found in the _____ associated with basic proteins. _____ of the bacterial DNA shortens the overall length of the molecule, allowing the DNA to be packed into the cell. In eukaryotic organisms DNA is associated with basic proteins called _____ . These proteins are enriched in the amino acid residues _____ and _____ . The fundamental structural unit of chromatin is the _____ . These consist of about two and a half _____ of DNA wrapped around an _____ of _____ proteins. The _____ protein is found outside the _____ attached to _____ DNA. Further _____ of the chromatin produces greater packing of the DNA ultimately forming the condensed _____ present at _____ . Chromatin is attached to the _____ of the nuclear _____ , which in turn is attached to the _____ membrane of the nuclear _____ .

Choose from: arginine, chromosome, core, DNA, envelope, folding (coiling), H1, histone, histones, inner membrane, lamina, linker, lysine, metaphase (mitosis), nucleoid, nucleosome, octomer, skeleton, supercoiling, turns.

MULTIPLE-CHOICE QUESTIONS

2. Arrange the following two lists in the most appropriate pairings

nucleoid	nucleolus
nucleus	linker DNA
H1 histone	porosome
annulus	chromatin
nucleolar-organizing region	bacterial cell

3. State which of the following are true or false.
A. DNA is the carrier of genetic information in all prokaryotic organisms.
B. Mitotic chromosomes represent the most highly condensed form of chromatin.
C. Chromatin is found in the nucleoid.
D. Ribosomes are formed in the nucleolus.
E. The number of observable nucleoli is always equal to the number of nucleolar-organizing chromosomes.

SHORT-ANSWER QUESTIONS

4. (a) The enzyme responsible for the replicating *E. coli* chromosomal DNA has a turnover number of $750\,s^{-1}$ i.e. it adds 750 new nucleotides to the growing DNA strand every second. If the chromosomal DNA is 1.1 mm long what is the minimum time for replication to occur? (Remember replication of the *E. coli* chromosome involves two replication forks).
(b) In a rapidly growing culture of *E. coli*, the cell cycle, the time between successive cell divisions, takes about 20 minutes. How do you account for this given the time needed for DNA replication as determined in (a)?

5. The longest chromosome in *Drosophilia melanogaster* contains a single DNA molecule 21 mm long.
(a) How many bp does this molecule contain?
(b) Given a packing ratio of 1.2×10^4, how long would the molecule of DNA be in the mitotic chromosome?

Objectives

After reading this chapter you should be able to:

☐ describe the principles governing the formation of biological membranes from their components;

☐ explain how the fluid nature of biological membranes is attributable to the properties of its lipid and protein components;

☐ recognize the different constraints on membrane fluidity;

☐ relate membrane fluidity to the dynamic functions of membranes;

☐ outline transport across biological membranes.

4.1 Introduction

The plasma membrane separates the cell from its extracellular environment. Intracellular membranes isolate different parts of the cell from each other and give rise to multiple compartments, the endoplasmic reticulum, the Golgi complex, the nucleus, lysosomes, chloroplasts and mitochondria (Fig. 4.1),

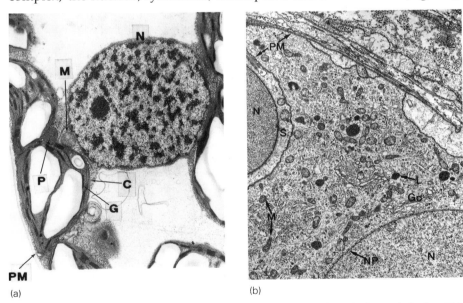

(a)   (b)

Fig. 4.1 Membranes of plant and animal cells. Electron micrographs of (a), tobacco leaf mesophyll cell, and (b), intramural ganglion in human urinary bladder. PM, plasma membrane; N, nucleus; NP, nuclear pore; P, peroxisome; M, mitochondrion; C, chloroplast; G, granum; L, lysosome; Go, Golgi body and S, adjoining satellite cell. Courtesy of Drs E. Sheffield, J.S. Dixon and C.J. Gilpin, University of Manchester, UK.

Reference Gennis, R.B. (1989) *Biomembranes: Molecular Structure and Function*, Springer-Verlag, Heidelberg, Germany. Good, up-to-date text covering a wide range of topics; particularly good on membrane structure.

each with their different functions. These biological membranes are sheet-like structures of considerable plasticity and are 6–10 nm in thickness.

Membranes serve a number of functions in the economy of the cell, but their major role is to act as a selective and responsive barrier between discrete masses of cytoplasm and their environments. The plasma membrane is a selective permeability barrier between the inside and outside of the cell. This allows the cytoplasm to be maintained in a homeostatic condition conducive to the biochemical reactions necessary to sustain life. The selective permeability of the plasma membrane depends on selective channels and pumps. Appropriate concentrations of specific ions and substrates are maintained in the cell, while substances that are not required or are toxic are excluded. The plasma membrane also acts as a responsive barrier allowing the cell to react to its environment. Specific receptors and *transducing* proteins in the membrane enable the cell to respond to signals generated by other cells.

Intracellular membranes provide sites for localization of enzymes and specialized proteins. Enzymes found in the rough endoplasmic reticulum and Golgi complex have a key role in cell secretion. Enzymes and carrier proteins found in the mitochondria and chloroplasts are necessary for energy transduction, resulting in ATP formation.

Membranes are therefore are vital in cell function. To understand cell function it is necessary to know how cell membranes are constructed and to appreciate their activities at the molecular level.

4.2 Chemical components of membranes

All cell membranes contain two principal components: lipids and proteins. These are associated non-covalently and vary in their relative compositions depending on the origin and function of the membrane (Table 4.1, see also Table 1.2). The other component in some membranes is carbohydrate, linked covalently either to lipids (glycolipids) or to proteins (glycoproteins). These carbohydrates are especially abundant in the plasma membrane of eukaryotic cells but absent from many intracellular membranes, including the inner mitochondrial membrane and the chloroplast lamellae. Glycoproteins are not found in prokaryotic cell membranes.

□ Major sugars found in the oligosaccharides of mammalian cell surfaces include L-fucose, galactose, mannose, N-acetyl-galactosamine, N-acetylglucosamine and N-acetylneuraminate, assembled in a variety of combinations.

Exercise 1

Density-gradient centrifugation can be used to separate subcellular structures which differ in density by as little as a few thousandths of a g cm^{-3}. Can this method be used to separate the outer and inner mitochondrial membranes from osmotically disrupted mitochondria (Table 4.1)? The mean densities of proteins and lipids are 1.2 and 0.92 g cm^{-3} respectively.

Table 4.1 Composition of some biological membranes (as percentages of dry weight)

Biological membrane	Lipid	Protein	Carbohydrate
Myelin	18	79	3
Mitochondrial:			
outer membrane	50	50	0
inner membrane	76	24	0
Chloroplast lamellae membrane	70	30	0
Mouse liver plasma membrane	44	52	4
Human erythrocyte plasma membrane	49	43	8
Halobacterium purple membrane	75	25	0

Table 4.2 Lipid composition of the plasma membrane of rat hepatocytes

Lipid	Amount (%)
Phosphatidylcholine	18
Phosphatidylethanolamine	11
Phosphatidylserine	9
Phosphatidylinositol	4
Phosphatidate	1
Sphingomyelin	14
Cholesterol	30

MEMBRANE LIPIDS. These are the fundamental components of biological membranes and form a double layer. All the major lipids found in membranes are **amphipathic**, i.e. their structure has a water-soluble hydrophilic region and a water-insoluble hydrophobic region (Fig. 4.2). Examples are: the phospholipids, phosphatidylinositol, phosphatidylcholine, phosphatidyl-serine and sphingomyelin; glycolipids such as cerebrosides and gangliosides; and sterols such as cholesterol (Table 4.2).

transduction: transfer of energy or a signal (information) from one system to another.

Reference Bretscher, M.S. (1985) The molecules of the cell membrane. *Scientific American*, **253**, 100–9. Good review of membrane structure and function.

Practically all the fatty acyl chains in membrane lipids of eukaryotic cells contain an even number of carbon atoms, mainly 14–24. Glycerophospholipids, for example, contain 16, 18 or 20 carbon atoms in their fatty acyl chains. Many fatty acyl chains also have one or more double bonds, nearly always in the *cis* configuration.

Dispersion in water of a mixture of dried phospholipids, such as those found in biological membranes, produces structures called **liposomes**. These are multilayered phospholipid vesicles which are effectively membranes within membranes (Fig. 4.3). They form because of hydrophobic interactions occurring between the fatty acyl chains. Liposome structure is further stabilized by electrostatic and hydrogen-bonding interactions between the polar-head groups of the lipid molecules. Further dispersion of liposomes by ultrasonication can produce vesicles having only one bilayer of lipid molecules. **Bilayers** are the structural unit of biological membranes, and form spontaneously from lipids in an aqueous environment.

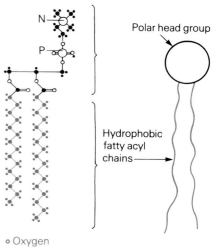

- o Oxygen
- • Hydrogen
- · Carbon

Fig. 4.2 Structure of phosphatidylcholine; the hydrophilic polar head group is represented as a circle while the two hydrophobic fatty acyl chains are shown as tails.

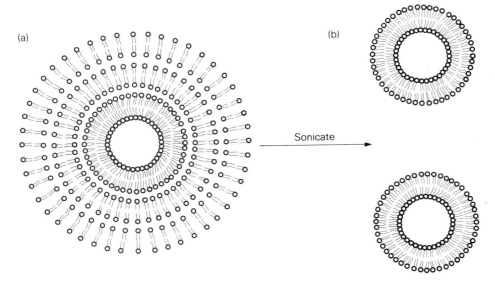

Sonicate

Fig. 4.3 (a) Dispersion of polar lipids by shaking in water at a water concentration greater than 40% by weight leads to formation of multilayer liposomes. (b) Sonication of this mixture leads to the formation of vesicles containing a single lipid bilayer. (c) Liposomes formed by sonicating phosphatidyl ethanolamine in aqueous buffer, and observed by freeze-fracture electron microscopy (×14 220). Courtesy of P.F. Knowles, Department of Biophysics, University of Leeds, UK.

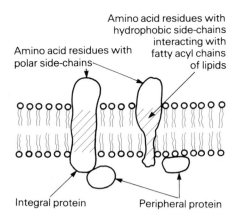

Amino acid residues with polar side-chains

Amino acid residues with hydrophobic side-chains interacting with fatty acyl chains of lipids

Integral protein

Peripheral protein

Fig. 4.4 The interaction of integral and peripheral membrane proteins with the lipid bilayer. Hydrophobic portions are shown in red.

MEMBRANE PROTEINS. These function as: transport proteins, carrying substances in and out of the cell; receptor proteins, transmitting signals from the exterior to the interior of the cell; or anchor proteins, for the attachment of structural proteins both inside and outside of the cell.

Membrane proteins are characteristically *integral* when embedded in the lipid bilayer such that part of their structure interacts directly with the fatty acyl chains of the lipid bilayer, or *peripheral* in which case they are bound to the bilayer surface (Fig. 4.4).

4.3 *Organization of membrane components*

Early ideas on cell membrane structure did not take into account the presence of proteins and viewed membranes as fairly rigid structures. Gorter and Grendel in 1925 suggested that the cell membrane was lipoidal and consisted of a lipid bilayer. These early pioneers extracted membrane lipids from **erythrocyte ghosts** and measured the area of a compacted monolayer after floating the extracted lipids as a layer on the surface of water (Fig. 4.5). The monolayer was found to be twice the area of the equivalent cell membrane area. Gorter and Grendel concluded that the red cell was surrounded by a lipid membrane two molecules thick, i.e. a lipid bilayer. (The experiments of Gorter and Grendel contained two major errors: the lipid extraction procedure was incomplete, leading to underestimation of total lipid; and the red cell was assumed to be spherical rather than biconcave, resulting in underestimation of total red cell surface area. Fortunately these errors cancelled each other out!)

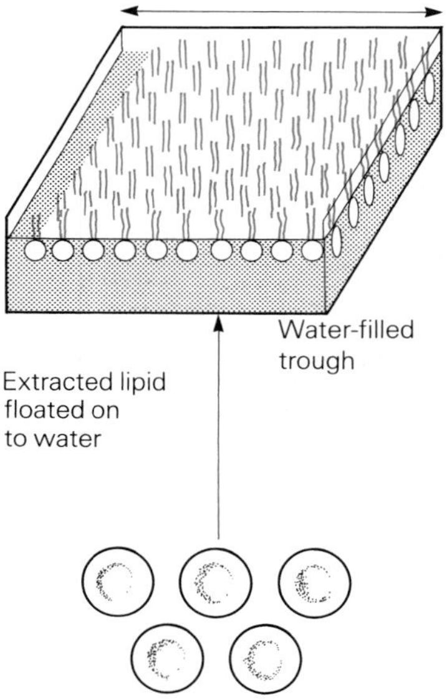

Fluid mosaic model for the dynamic properties of membranes

The 'fluid mosaic' model was proposed by Singer and Nicholson in 1972 (Fig. 4.6). In it proteins are envisaged as being either 'adsorbed' to the surface of the membrane or partially or wholly embedded in the lipid bilayer. The proteins maintain a globular structure and possess a high proportion of α-helical and random coil configurations rather than extended β-structures.

Both lipid and protein molecules are arranged in a tightly packed, water-excluding mosaic. In the core of the membrane there occur interactions between the hydrophobic regions of both lipids and proteins, while at the surface, the hydrophilic regions of the amphipathic molecules are solvated in the aqueous environment. Both the proteins and the lipids of the membrane are free to move laterally in the plane of the membrane.

This model is now generally accepted, being consistent with many experimental findings (for example, with observed patterns of electron micrographs following freeze-fracture of membranes (Fig. 4.7)). The various

Surface area of lipid monolayer measured

Water-filled trough

Extracted lipid floated on to water

Erythrocyte ghosts extracted with acetone

Fig. 4.5 The early experiments of Gorter and Grendel which first suggested that the cell membrane was composed of a lipid bilayer.

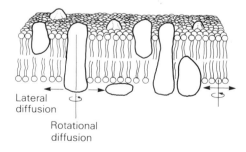

Lateral diffusion

Rotational diffusion

Fig. 4.6 The 'fluid mosaic' model of membrane structure (after Singer and Nicolson). The allowed movements for membrane proteins and membrane lipids are indicated by the arrows.

erythrocyte ghost: *purified erythrocyte plasma membrane obtained by breaking open the intact red cell and washing away the cytoplasm.*

Reference Jennings, M.L. (1989) Topology of membrane proteins. *Annual Review of Biochemistry*, **58**, 999–1027. An advanced review; worth reading.

arrangements of membrane proteins are visible within the membrane (Fig. 4.8). Dynamic studies confirm that both proteins and lipids can move about.

The existence of both peripheral and integral membrane proteins is also consistent with the conditions required to extract proteins from biological membranes. Peripheral proteins are bound to the surface of membranes (Fig. 4.9) by weak ionic interactions. They are relatively easily removed with aqueous solutions, for example, by changing the ionic strength. In contrast, integral proteins (the major proportion of membrane proteins) may only be removed with detergents. Non-ionic detergents, such as Triton X100 and octylglucosides (Fig. 4.10), have amphipathic properties, like membrane lipids, but are more hydrophilic. By substituting partially for lipids in association with integral proteins, they cause the solubilization of the proteins. Detergents have been useful in studying the biophysical properties of integral membrane proteins, a number of which have been isolated and examined in detail.

GLYCOPHORIN. This is the major sialoglycoprotein of the human erythrocyte membrane and is thought to play a role in antigenicity. Glycophorin is a glycoprotein (Fig. 4.11), containing 131 amino acid residues,

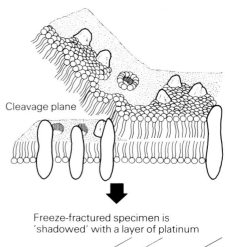

Cleavage plane

Freeze-fractured specimen is 'shadowed' with a layer of platinum

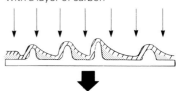

The 'shadowed' specimen is coated with a layer of carbon

The organic material is dissolved away and the remaining replica is then ready for examination by electron microscopy

Fig. 4.7 The technique of freeze-fracture electron microscopy involves rapidly freezing cells or membrane fragments at liquid nitrogen temperatures. The frozen membrane is then fractured by impact of a microtome knife. The cleavage plane usually occurs along the hydrophobic middle of the lipid bilayer exposing extensive regions within the bilayer. The exposed fracture faces are then shadowed with a layer of platinum which is evaporated from a filament and then coated with carbon. The organic material is dissolved away and the platinum/carbon replica is viewed by electron microscopy.

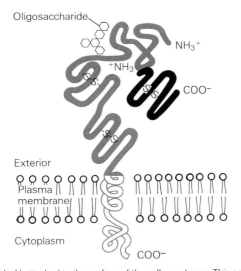

Fig. 4.8 Electron micrograph showing placental membranes following freeze-fracture.

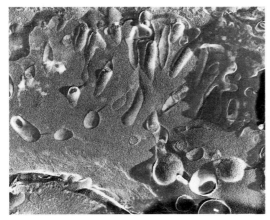

Oligosaccharide

NH₃⁺

⁺NH₃

COO⁻

Exterior

Plasma membrane

Cytoplasm

COO⁻

Fig. 4.9 β-microglobin (black) attached to the surface of the cell membrane. This peripheral protein is non-covalently bound to an integral protein (red), the HLA class 1 surface antigen. Redrawn from Ploegh, H.L. *et al.* (1981) *Cell*, **24**, 287–399.

Reference Gorter, E. and Grendel, F. (1925) Biomolecular layers of lipoids on the chromocytes of the blood. *Journal of Experimental Medicine*, **41**, 439–43. Original article.

Reference Singer, S.J. and Nicolson, G.L. (1972) The fluid mosaic model of the structure of cell membranes. *Science*, **175**, 720–31. Good starting place for reading about membrane structure.

Box 4.1
Characterization of membrane proteins using SDS gel electrophoresis

The ionic detergent sodium dodecyl sulphate (SDS) can be used to characterize membrane proteins. When present at high concentrations (1–2% w/v) this detergent binds to hydrophobic regions of proteins, denaturing and solubilizing them (see Section 6.3). Membrane proteins which have been solubilized with SDS may then be separated according to size by SDS polyacrylamide gel electrophoresis. Using this separation procedure also allows the M_r of the membrane proteins to be determined by comparison with proteins of known M_r treated in the same way.

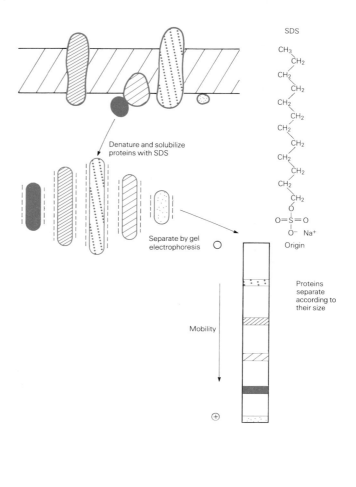

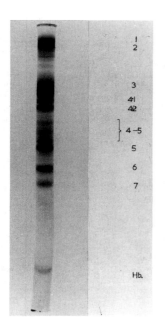

SDS-PAGE separation of proteins of erythrocyte membranes. Courtesy of Dr J.B.C. Findlay, Department of Biochemistry and Molecular Biology, University of Leeds, UK.

of which 34 are embedded in the lipid bilayer (residues 62–95). One segment of this portion (residues 73–95) contains only uncharged, mainly hydrophobic amino acid residues, such as phenylalanine, leucine, isoleucine, valine, tryptophan and tyrosine. The membrane-associated portion of the protein

Reference Dalby, R.E. (1990) Positively charged residues are important determinants of membrane protein topology. *Trends in Biochemical Sciences*, **15**, 253–7. Short readable article on the orientation of membrane proteins.

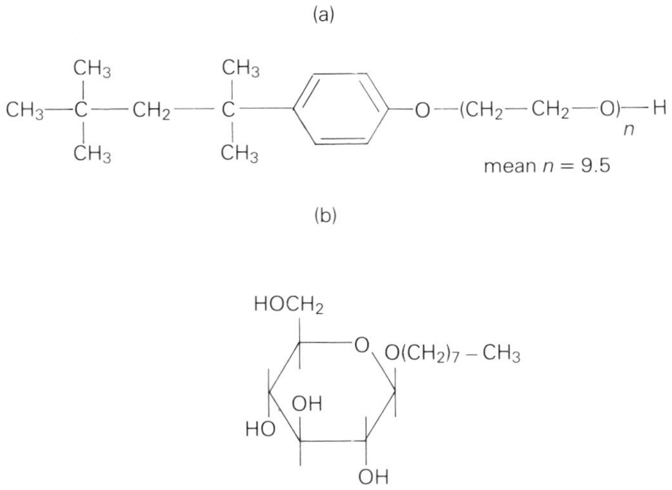

(a)

$$CH_3-\underset{\underset{CH_3}{|}}{\overset{\overset{CH_3}{|}}{C}}-CH_2-\underset{\underset{CH_3}{|}}{\overset{\overset{CH_3}{|}}{C}}-\underset{}{\bigcirc}-O-(CH_2-CH_2-O)_n-H$$

mean $n = 9.5$

(b)

Fig. 4.10 The non-ionic detergents (a) Triton X100 and (b) octylglucoside.

□ Sialate or *N*-acetylneuraminate is a principal terminal sugar on oligosaccharide chains and makes a major contribution to the negative charge found on many cell surfaces.

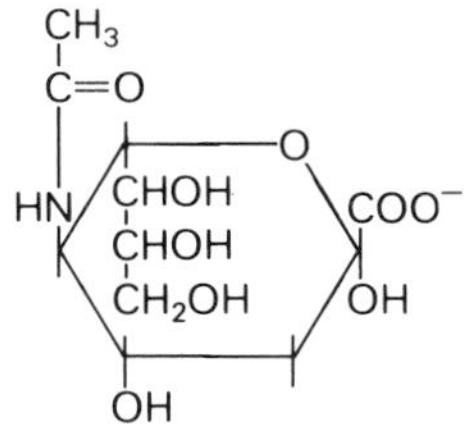

apparently adopts an α-helical conformation with the hydrophobic side-chains pointing outwards and forming hydrophobic interactions with the surrounding fatty acyl chains. The amino-terminal portion of the protein (residues 1–61) is on the extracytoplasmic side of the membrane. This portion is heavily glycosylated and rich in sialate (*N*-acetylneuraminate) residues. The carboxy-terminal part of the molecule is inside the cell. It is rich in charged amino acid residues (96–131). Amino acid residues 96 and 97 (positively charged arginine residues) and residues 100 and 101 (positively charged lysine residues) appear to interact with the negatively charged phospholipid head groups at the cytoplasmic side of the membrane.

□ Circular dichroism is an optical absorption method which measures the differences in absorption by a test sample of left- and right-circularly polarized light. When related to the wavelength of the light this produces a curve of circular dichroism (CD). Differences in absorption may be related to protein conformation by comparison with polypeptides and proteins of known structure.

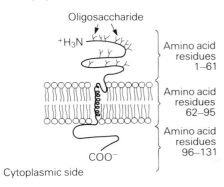

Extracytoplasmic side

Oligosaccharide

Amino acid residues 1–61

Amino acid residues 62–95

Amino acid residues 96–131

Cytoplasmic side

Fig. 4.11 Glycophorin showing its association with the lipid bilayer.

Fig. 4.12 Computer-generated model of the nicotinic acetylcholine receptor embedded in a lipid bilayer. This integral membrane protein consists of five subunits each of which have several membrane-spanning helices. Courtesy of Dr D. Osguthorpe, Molecular Graphics Unit, University of Bath, UK.

Reference Viitalia, J. and Jamefeltt, J. (1985) The red cell surface revisited. *Trends in Biochemical Sciences*, **10**, 392–5. Good coverage of the carbohydrate content of the erythrocyte membrane.

Reference Paulson, J.C. (1989) Glyco-proteins: what are the sugar chains for? *Trends in Biochemical Sciences*, **14**, 272–6; Wickner, W. (1989) Secretion and membrane assembly. *Trends in Biochemical Sciences*, **14**, 280–3. Both from *TIBS* special issue on structure and function of proteins.

□ Optical rotatory dispersion is an optical method which measures the extent of rotation of monochromatic linearly polarized light by the sample under test. When rotation effects are expressed as a function of wavelength a curve of optical rotatory dispersion is obtained (ORD). The degree of rotation can be related to protein conformation, by using 'standard' proteins of known conformation.

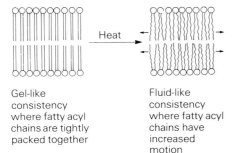

Gel-like consistency where fatty acyl chains are tightly packed together

Fluid-like consistency where fatty acyl chains have increased motion

Fig. 4.13 The fluid and gel-like states of the lipid bilayer.

□ Electron spin resonance spectroscopy is a technique for detecting the change in energy as an unpaired electron oscillates between two energy states (paramagnetism). The technique can be used to observe the mobility of compounds containing these unpaired electrons in lipid bilayers.

Other membrane proteins are composed of polypeptide subunits and have several membrane-spanning helical regions (Fig. 4.12).

MEMBRANE PROTEIN STRUCTURE. The globular nature of these proteins is consistent with results from physical measurements, such as **circular dichroism** and **optical rotatory dispersion**, which indicate how much α-helix and β-pleated sheet is present.

Experimental investigations on liposomes composed of different glycerophospholipids suggest two possible consistencies for the lipid bilayer: *gel-like*, where the acyl chains of the polar lipids packed tightly in a crystalline array; and *fluid-like*, where the fatty acyl chains have increased motion about their –C–C– bonds and assume a more random configuration (Fig. 4.13).

This change in consistency or **phase transition** may be explained by a 'melting' of the fatty acyl chains. The temperature at which this occurs depends on a number of factors. In general, fluidity is favoured by the presence of fatty acyl chains with short chain lengths or one or more *cis* double bonds. Lipids with short acyl chains have less surface area for van der Waals' interactions. Similarly, because unsaturated acyl chains of *cis* configuration have 'kinks' which hinder stable van der Waals' contacts with other acyl chains, the bilayer tends to adopt a more random fluid state. Biological membranes contain a mixture of fatty acids. Cells adjust their fatty acid content and maintain membrane fluidity appropriate to their environment.

Cholesterol, a major component of mammalian plasma membranes, can affect membrane fluidity. By intercalating within the lipid bilayer, with its polar-hydroxyl group in contact with the aqueous environment, the rigid steroid nucleus interacts with the fatty acyl chains near their polar-head groups. This dampens the random movement of the fatty acyl chains but also prevents them from forming stable bonds with one another. Cholesterol therefore prevents the fluid to gel phase transition of the membrane normal at low temperatures but, in addition, has an overall dampening effect on membrane fluidity.

Experimental determination of membrane fluidity

Various techniques have been used to measure the motion of lipids and proteins in artificial and biological membranes. For example, lipid molecules can be synthesized which carry a 'spin label' such as a nitroxide group. The nitroxide group has an unpaired electron whose spin creates a *paramagnetic* signal that can be detected by **electron spin resonance** (ESR) spectroscopy. A lipid labelled in this way can be introduced into a biological membrane and its motion within the bilayer easily measured (Fig. 4.14). This technique has revealed that a lipid molecule may change places with its neighbour in the membrane 10^7 times every second.

The lateral motion of the proteins in biological membranes can be measured by first labelling membrane proteins with a *fluorescent chromophore*. An intense laser beam is then used to bleach the fluorescent label irreversibly in a small patch of the cell membrane. Recovery of fluorescence in this patch, as unbleached proteins migrate into the bleached area, provides a measure of protein mobility. Diffusion coefficients for proteins in cell membranes vary from 10^{-9} to 10^{-11} cm^2 s^{-1}. This is 10 to 1000 times slower than the diffusion coefficients for membrane lipids.

Experimental evidence is therefore consistent with membrane proteins and lipids being in a dynamic state and undergoing both rapid lateral movement in the plane of the membrane and rotational motion about their perpendicular axis.

paramagnetism: *a property of unpaired electrons, whereby they oscillate between low- and high-energy states when placed in a magnetic field.*
fluorescent chromophore: *molecule which emits visible light when irradiated with ultraviolet. Examples include fluorescein, which emits green light, and rhodamine, which emits red light.*

Reference Robertson, R.N. (1983) *The Lively Membranes*, Cambridge University Press, UK. Emphasizes membrane dynamics. Inexpensive, useful text.

Local anaesthetics such as Procaine and Lignocaine are thought to exert their effects by increasing the lipid fluidity of the plasma membrane of the nerve axon. This leads to blockage of the protein channels responsible for the flux of K^+ and Na^+ across the membrane during depolarization (see Section 9.3). The transmission of the action potential along the sensory nerve fibres is thus blocked leading to alleviation of pain.

(a)

(b)

Structures of (a) lignocaine and (b) procaine.

Fig. 4.14 A derivative of phosphatidylcholine containing a nitroxide spin label. This derivative can be used to determine the movements of lipids in artificial and biological membranes.

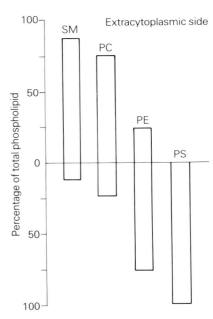

Fig. 4.15 Lipid asymmetry in the human erythrocyte membrane. SM, sphingomyelin; PC, phosphatidylcholine; PE, phosphatidylethanolamine; PS, phosphatidylserine.

Percentage of total phospholipid

Constraints to membrane fluidity — membrane asymmetry and 'boundary layers'

Movement of lipid components from one side of the bilayer to the other, sometimes called 'flip-flop', is thermodynamically unfavourable since it requires solvation of the polar-head group of the lipid in the inner hydrophobic core of the bilayer. Such restricted movement means that biological membranes maintain an asymmetric distribution of lipid components between the sides of the bilayer. For example, the plasma membrane of eukaryotes has a predominance of phosphatidylethanolamine, phosphatidylserine and phosphatidylinositol at the cytoplasmic surface, while choline-based glycerophospholipids and sphingomyelin are concentrated at the extracytoplasmic surface (Fig. 4.15). Glycolipids are

Reference Houslay, M.D. and Stanley, K.K. (1982) *Dynamics of Biological Membranes*, Wiley, Chichester, UK. Provides a general background with considerable detail relating to the functional aspects of cell membranes.

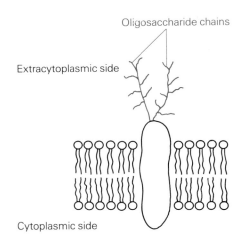

Oligosaccharide chains

Extracytoplasmic side

Cytoplasmic side

Fig. 4.16 Glycosylation of membrane proteins at the extracytoplasmic side of the membrane. (See also Fig. 4.11.)

Exercise 2

A number of methods may be used to label selectively proteins exposed at the cell surface. One method is to label tyrosine residues in membrane proteins with ^{125}I using the enzyme lactoperoxidase. Because of its high M_r, lactoperoxidase cannot penetrate the cell so only proteins exposed at the cell surface are labelled. Devise a method to determine the orientation of proteins in the plasma membrane of isolated human erythrocytes using this technique.

---

restricted to the outer face of the plasma membrane. Lipid asymmetry may have a functional role, such as supporting and regulating the properties of membrane proteins, maintaining a charge distribution between the two leaflets of the bilayer, and providing an intracellular sink for essential cations such as Ca^{2+} and Mg^{2+}.

Membrane proteins also show asymmetric distribution. For example, cell surface proteins such as hormone receptors, and proteins involved in cell recognition are often heavily glycosylated on the extracytoplasmic side of the membrane (Fig. 4.16) but not on the surface facing the cytoplasm.

As with membrane lipids, the polar nature of this glycosylation serves to maintain a particular asymmetric conformation of the protein within the membrane since movement of the glycosylated portion through the hydrophobic core of the bilayer is thermodynamically unfavourable.

It appears that integral membrane proteins may be surrounded with a distinctive 'boundary layer' or 'microdomain' of membrane lipid that is in a different physical state to that in the bulk of the membrane. This microdomain may influence the biological activity of the membrane proteins in some way. Indeed, this is in keeping with the finding that some isolated membrane enzymes, and membrane transport proteins, require specific lipids for biological activity. Physical measurements on the mobility of lipids in microdomains indicate that they are still mobile, and it is not clear whether they have any substantial effect on the function of membrane proteins in the physiological state.

Lateral and rotational movement of some membrane proteins is also limited by other constraints. Some plasma membrane proteins may be partially immobilized by their interactions with the cytoskeleton or with structural proteins or **proteoglycans** in the extracellular matrix. Other plasma membrane proteins may be involved in the different junctions between cells.

4.4 Junctions between cells

The structural and functional integrity of animal tissues and organs depends on the organization of their cells. This organization is mediated by specialized regions of the plasma membrane called **cellular junctions**. Plant cells are also connected by junctions called plasmadesmata.

There are three functional categories of junction between animal cells:

(1) **desmosomes** (zonulae adherens) whose prime purpose is to hold cells together;
(2) **tight junctions** (zonulae occludens) which not only hold cells together but also seal the space between them so that molecules cannot pass;
(3) **gap junctions** which form channels of communication between cells by allowing passage of small molecules or ions.

All these types of junction are found in the intestinal epithelium (Fig. 4.17).

DESMOSOMES occur in three different structural forms: spot desmosomes, hemidesmosomes and belt desmosomes. **Spot desmosomes** act like rivets by holding cells together (Fig. 4.18). The cytoplasmic surface of the plasma membrane in the region where spot desmosomes occur contains dense plaques. These plaques serve to link cells together mechanically through interconnecting filamentous structures. In this way the plasma membranes of the two cells are held parallel to each other at a fixed distance of approximately 30 nm. On the cytoplasmic side of the membrane the dense plaques are also connected to **tonofilaments**. Tonofilaments help dissipate stress on mechanical stretching of the tissue. **Hemidesmosomes** (half desmosomes)

---

proteoglycans: *glycoproteins found in the extracellular matrix which are rich in carbohydrate, the amount of carbohydrate often exceeding 90–95% (see Chapter 7).*

tonofilaments: *a class of cytoskeletal filaments also known as intermediate filaments or stress fibres. These filaments, with a diameter of 8–10 nm, are built from various non-motile structural proteins, among them keratins, vimentin and desmins (see Chapter 6).*

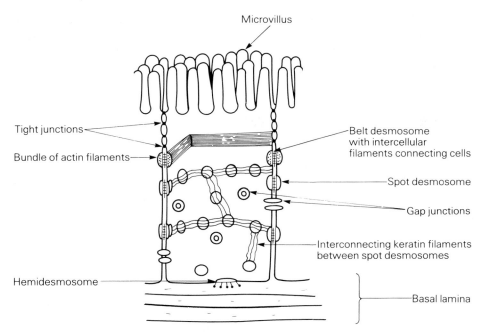

Microvillus

Tight junctions

Bundle of actin filaments

Belt desmosome
with intercellular
filaments connecting cells

Spot desmosome

Gap junctions

Interconnecting keratin filaments
between spot desmosomes

Hemidesmosome

Basal lamina

Fig. 4.17 Arrangement of the different types of cell junction found in the epithelial cells of the small intestine.

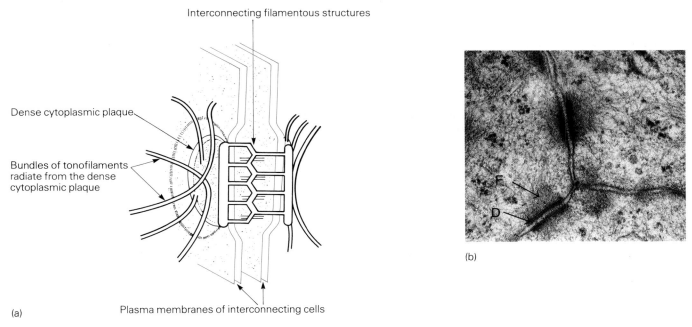

Interconnecting filamentous structures

Dense cytoplasmic plaque

Bundles of tonofilaments
radiate from the dense
cytoplasmic plaque

Plasma membranes of interconnecting cells

(a)

(b)

Fig. 4.18 (a) A spot desmosome. The dense cytoplasmic plaque is thought to be associated with the plasma membrane of each interconnecting cell. Poorly characterized intercellular filaments connect adjacent plaques while tonofilaments either pass along the surface or terminate within the plaque. (b) Electron micrograph showing spot desmosomes in the epithelial cells of human small intestine (×49 200). D, dense cytoplasmic plaque associated with the plasma membrane of each interconnecting cell; F, tonofilaments which radiate into the cytoplasm from these plaques. Courtesy of Dr R. Griffin, Department of Science, University of the West of England, UK.

link the cell membrane to extracellular connective material known as the *basal lamina* which is secreted by some epithelial cells. This serves to anchor the cells to the underlying matrix. **Belt desmosomes** form a continuous band

---

basal lamina: *a network or matrix of extracellular connective material secreted by some cell types. It is composed of proteoglycans and other glycoproteins, reinforced by collagen fibres, and serves as an attachment site for sheets of epithelial cells (see Chapters 6 and 7).*

Reference Macdonald, C. (1985) Gap junctions and cell–cell communication. *Essays in Biochemistry*, **21**, 86–118. High-level but worth reading.

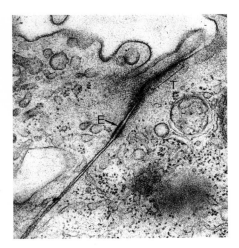

Fig. 4.19 Electron micrograph showing a tight junction (T) and part of a belt desmosome in the epithelial cells of human small intestine (E) (×40 000). Courtesy of Dr R. Griffin, Department of Science, University of the West of England, UK.

between interconnecting cells. They lack the dense plaques found in spot desmosomes. Instead they are anchored to loosely connected mats of actin filaments on the cytoplasmic side of the membrane.

Desmosomes are found in all animal tissues, particularly those that undergo lateral or shear stress, such as skin and linings of the stomach, intestine (Figs 4.18 and 4.19), bladder and uterus. When these tissues are stressed the desmosomes are thought to distribute shear forces amongst the cells via the tonofilaments beneath the membrane and the filamentous network which connects the cells together.

TIGHT JUNCTIONS are commonly found in epithelial cell sheets such as those lining the intestine (Fig. 4.19) and the bladder. At tight junctions the plasma membranes of interconnecting cells are so closely opposed that there is no intercellular space. Treatment of cells with **proteases** destroys tight junctions, implicating linkage proteins as essential structural units in these junctions. Another view is that the tight junction is formed by fusion of the two outer lipid layers of the plasma membrane (one from each cell) to form a continuous leaflet.

Tight junctions serve at least two functions. They seal layers of epithelia so that molecules cannot pass across the epithelium in either direction. In the intestine this prevents nutrients that have been transported into the bloodstream by the epithelial cells from leaking back into the intestine. Tight junctions also polarize cell function by preventing lateral diffusion of membrane proteins from one side of the cell to the other. In the epithelial layer of the intestine this ensures that membrane transport systems required for uptake of nutrients from the intestine and those required for secretion of nutrients into the bloodstream are kept on separate faces of the cell.

GAP JUNCTIONS (Fig. 4.20) are probably the commonest type of animal cell junction. They are also called **communicating junctions** because they allow the passage of small molecules and ions (up to M_r 1200) between interconnecting cells. This means that metabolic substrates (for fuels and building blocks, and secondary messengers such as cAMP and Ca^{2+}) may

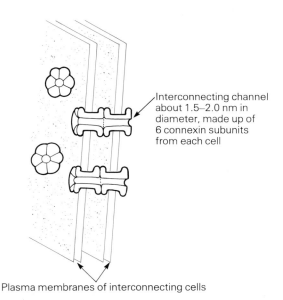

Interconnecting channel about 1.5–2.0 nm in diameter, made up of 6 connexin subunits from each cell

Plasma membranes of interconnecting cells

Fig. 4.20 Gap junctions. The plasma membranes of the interconnected cells are penetrated by channels composed of hexamers of connexin which allow the free movement of small molecules and ions between the two cells.

proteases: enzymes that break proteins down into peptides and amino acids by catalysing the cleavage of peptide bonds.

pass freely from one cell to another. The metabolism of interconnecting cells can be coupled or coordinated such that cells of a particular tissue act in synchrony. For instance, during tissue development, metabolic coupling may distribute nutrients where there is no functional circulatory system and deliver regulatory signals such as cAMP during cell differentiation. In more specialized tissues, such as heart and smooth muscle, gap junctions permit electrical coupling. This allows the free flow of Ca^{2+} between cells that is necessary to maintain smooth muscular contraction.

These communicating channels are composed of a single protein (M_r 25 000) called **connexin**. Six connexin molecules from the plasma membrane of one cell are aligned with six connexin molecules of the adjoining cell arranged in the same symmetrical array (Fig. 4.20) to form an hexagonal cylinder. Small molecules can pass from one cell to another through this cylinder. Ca^{2+} plays an important role in controlling the permeability of these gap junctions. When Ca^{2+} levels are abnormally high, the gap junctions seal as a result of a conformational change in the connexin molecules. This protects normal cells during tissue injury by closing them off from damaged or dying neighbouring cells.

PLASMADESMATA. These are the major intercellular junctions of higher plants. They allow transport of materials of relatively low M_r between neighbouring cells. They are thus like the gap junctions of animal cells, but have a totally different structure. Plasmadesmata are fine strands of cytoplasm, limited by plasma membrane. They connect a living plant cell to adjacent cells through holes 30–50 nm in diameter which puncture the

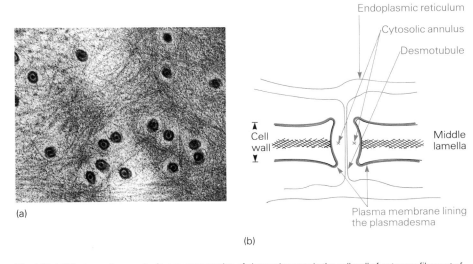

(a)

(b)

Fig. 4.21 (a) Electron micrograph of transverse section of plasmadesmata in the cell wall of a stamen filament of *Triticum aestivum* (×55 300). Courtesy of Biophoto Associates. (b) Schematic view of plasmadesma.

Box 4.3
Impaired metabolic coupling in cancer cells

The metabolic and ionic coupling which occurs in normal cell types is lacking in some cancer cells. For example, cells of liver tumours and from thyroid and stomach cancers have been shown to be non-coupled. This is due to a large reduction in the numbers of gap junctions in these cancer cells. This loss of metabolic coupling may be important since molecules involved in growth regulation normally pass through gap junctions. In cancer cells the loss of this coupling may contribute to their unrestricted growth.

intervening cell walls (Fig. 4.21). A desmotubule runs through the centre of each plasmadesma and is continuous with the endoplasmic reticulum of the connected cells. The ends of plasmadesmata are constricted by overlaps of adjacent cell walls (Fig. 4.21).

Microinjection of peptides of different sizes linked to fluorescent dyes has demonstrated that molecules of M_r up to about 800 pass freely through plasmadesmata. Movement is subject to regulation, and can be shut off. Some evidence suggests Ca^{2+} and protein phosphorylation may be involved in this regulation. Plasmadesmata ensure that multicellular plants are not merely aggregates of cells separated by cell walls, but are organisms in which the metabolic activities of different parts are regulated and integrated.

4.5 *The membrane as a dynamic entity*

Biological membranes are neither static nor inert. Rather, they are dynamic and asymmetric structures in which defined movements of proteins and lipids take place. These dynamic properties are vital for membrane functions.

Endocytosis and exocytosis

Cells can take in materials from their environment by **endocytosis**, a process entirely different from membrane transport. There are three types of endocytosis: **pinocytosis**, **phagocytosis** and **receptor-mediated endocytosis**.

PINOCYTOSIS. In this process cells continually infold parts of their plasma membrane to form small endocytic (pinocytotic) vesicles (Fig. 4.22). Extracellular fluid, which may contain dissolved nutrients, is trapped inside the vesicles and transported into the cytoplasm. Since pinocytosis is continuous, large volumes of extracellular fluid are ingested by cells. For example, in macrophages up to 25% of the cell's volume can be ingested in an hour.

PHAGOCYTOSIS. This endocytic process is used to ingest insoluble material (Fig. 4.22). In protozoa it is a form of feeding. Large particles are ingested in phagocytic vacuoles. These vacuoles fuse with lysosomes where the ingested material is digested. The products then pass into the cytosol to be utilized as food. In mammals, polymorphonuclear leucocytes and macrophages (phagocytes) are the two major blood cells that carry out phagocytosis. Both cells use phagocytosis as a means of ingesting insoluble material such as microorganisms and this process has a role in defending the body against infection. When the foreign material touches the cell membrane the cell may respond with a forward flow of cytoplasm around the point of contact (Fig. 4.23). In phagocytes, stimulation of the phagocytic process is controlled by specialized cell receptors, the best characterized of which recognize antibody molecules. When foreign particles coated with antibody bind to the receptors on the phagocytic cells, the cell responds by engulfing the particle and forming a phagocytic vacuole.

Unlike pinocytosis, very little extracellular fluid is taken in with the vacuole during phagocytosis. This is because, once the insoluble material has been surrounded, the vacuole is constricted to extrude any excess fluid before it is closed off to release the vacuole into the cytoplasm.

RECEPTOR-MEDIATED ENDOCYTOSIS. This process entails binding of macromolecules to specific cell surface receptors before endocytosis. Before the macromolecule and receptor are internalized, the occupied receptors are

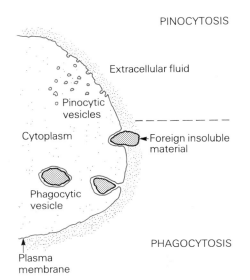

Fig. 4.22 Overview of pinocytosis and phagocytosis in an animal cell.

PINOCYTOSIS

Extracellular fluid

Pinocytic vesicles

Cytoplasm

Foreign insoluble material

Phagocytic vesicle

Plasma membrane

PHAGOCYTOSIS

Reference Tarkatoff, A.M. (ed.) (1985) Membrane dynamics and specificity. *Trends in Biochemical Sciences*, **10**, 413–64. Excellent collection of review articles covering both membrane structure and function.

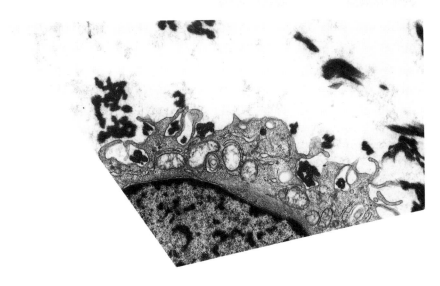

Fig. 4.23 Electron micrograph showing phagocytosis of fibrin clots at the surface of a human macrophage (×10 880). Courtesy of Dr R. Griffin, Department of Science, University of the West of England, UK.

thought to concentrate in areas of the plasma membrane rich in proteins, forming **endocytic pits**. This selective concentrating mechanism allows cells to internalize large amounts of specific ligands without the need to ingest large volumes of extracellular fluid. Examples of macromolecules internalized in this way are *transferrin* and *low-density lipoprotein* (LDL).

In the electron microscope, endocytic pits appear to be coated with bristle-like structures on the cytoplasmic side of the membrane. These pits are called **coated pits**. The bristle-like structures contain some major proteins the best characterized of which is **clathrin**. This is a fibrous protein composed of a large polypeptide (M_r 180 000) associated with several smaller polypeptides (M_r 35 000) (Fig. 4.24). Polymerization of clathrin produces a characteristic polyhedral coat, resulting in a cage-like structure on the intracellular surface of the coated pit and subsequently formed coated vesicle. Seconds later, the endocytosed coated vesicle sheds its coat forming an uncoated vesicle known as an endosome.

Pinocytosis, phagocytosis and receptor-mediated endocytosis are all rapid processes, taking from seconds to minutes to internalize vesicles. Since a cell's surface area and volume remain constant during these processes, new membrane must be continually added to the cell surface at the same rate that it is removed by endocytosis. Considering that these processes are extremely rapid and there is little evidence for comparable rates of membrane degradation, the endocytosed membrane must be returned to the cell surface by membrane recycling. How this membrane recycling occurs is unclear; it is likely to differ in the different endocytic processes. Many endosomes resulting from receptor-mediated endocytosis are thought to deliver their contents to lysosomes where the receptor-bound ligands are degraded by lysosomal enzymes. In the uptake of low-density lipoproteins (LDL) prior dissociation of the receptor from LDL is thought to take place in the endosome. Only LDL is then delivered to the lysosomes in small vesicles which bud off from the endosome (Fig. 4.25). The unoccupied receptors and most of the vesicle membrane are then recycled to the cell surface by exocytosis (see later). The LDL is delivered to the lysosomes and acted upon by the lysosomal enzymes to release free cholesterol from the cholesterol esters. The free cholesterol is then available to the cell for new membrane synthesis. If too much free cholesterol accumulates in the cell, both the

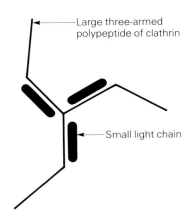

Large three-armed polypeptide of clathrin

Small light chain

Fig. 4.24 Clathrin has the form of a triskelion composed of one large three-armed polypeptide associated with three smaller light chains. Association or polymerization of these triskelions leads to the formation of a polyhedral lattice-like structure composed of hexagons and pentagons which is found on coated pits and coated vesicles.

---

transferrin: a plasma glycoprotein of M_r *80 000 which binds iron (Fe^{3+}). There are two binding sites for Fe^{3+} per molecule of transferrin. Transferrin is used to transport iron in the blood because free Fe^{3+} will form $Fe(OH)_3$ and precipitate.*

low-density lipoprotein: a complex of about 1500 esterified cholesterol molecules encapsulated by a lipid bilayer containing several copies of a single large protein.

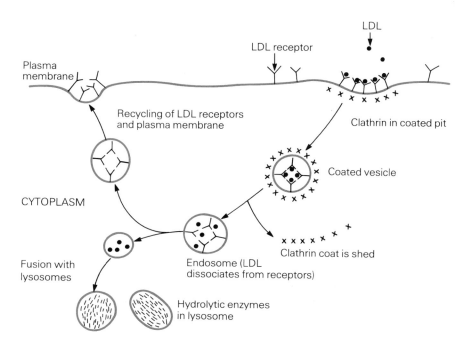

Fig. 4.25 Overview of receptor-mediated endocytosis of low-density lipoprotein (LDL) via clathrin-coated pits and the recycling of plasma membrane and LDL receptors.

synthesis of the cell's own cholesterol and the synthesis of LDL receptor proteins are shut off. This regulatory mechanism means that less cholesterol is made by the cell and less is taken up from the LDL circulating in the blood.

EXOCYTOSIS. This involves fusion of vesicles from the interior of the cell with the plasma membrane. The vesicles then expel their contents into the surrounding medium (Fig. 4.26). Exocytosis is important in the secretion of hormones by endocrine glands, release of neurotransmitters by nerve cells, and a variety of other secretory processes. It is also the major process whereby membrane is returned to the cell surface after endocytosis.

Cell signalling and cell recognition

Since the plasma membrane is at the surface of the cell it is on the front line for receiving messages, such as those carried by hormones, from other cells. Many of these hormones cannot cross the plasma membrane. In these cases, the message must be transmitted to the interior of the cell by some other mechanism. This dynamic function of the plasma membrane, referred to as **transmembrane signalling** or **signal transduction** allows the message to be transmitted across the membrane by the association and dissociation of

Reference Steinman, R.M. *et al.* (1983) Endocytosis and the recycling of plasma membrane. *Journal of Cell Biology*, **96**, 1–27. Wide-ranging review but at a high level.

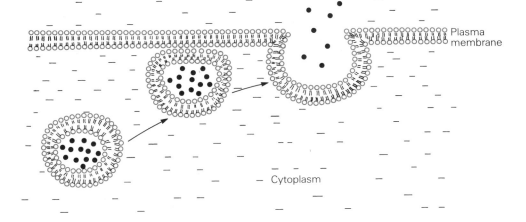

Fig. 4.26 Bilayer adhesion and fusion in exocytosis.

membrane proteins. This process is important in understanding how cells communicate with one another and respond to the requirements of the organism.

The plasma membrane is also important in intercellular cell recognition. Most animal cells are surrounded by a coat, the **glycocalyx** (Fig. 4.27). This coat is made up of oligosaccharide chains of membrane glycoproteins and glycolipids. It is important in cell recognition since the highly specific structures of oligosaccharide chains allows one cell to recognize another.

The ABO blood groups are based on specific combinations of terminal carbohydrate residues on the erythrocyte cell coat. If an individual receives a blood transfusion of the wrong blood group the immune system will recognize the erythrocytes as 'foreign' and destroy them. The histocompatibility antigens are membrane glycoproteins found on the surfaces of all human cells. Their recognition by the immune system as either 'self' or 'foreign' is extremely important to the success or otherwise of organ transplantation and tissue graft operations. Such operations often employ immunosuppressive drugs to prevent the new tissue from being rejected.

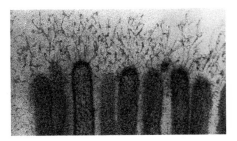

Fig. 4.27 High-power electron micrograph showing a well-developed glycocalyx extending from microvilli of intestinal cells of the bat *Myotis lucifugus*. Reprinted from Fawcett, D.W. (1965) Surface specializations of absorbing cells. *Journal of Histochemistry and Cytochemistry*, **13**, 75–91. Photograph courtesy of Prof. D.W. Fawcett.

4.6 *Membrane transport*

Membrane transport is also a dynamic function. The cell must maintain an internal environment conducive to the biochemical reactions that sustain life. The pH and ionic environment within the cell must be maintained within narrow limits. The cell must also acquire molecules for energy (fuels) and biosyntheses (building blocks) from the exterior, while at the same time excreting toxic material and releasing secretory products. Biological membranes must therefore be **selectively permeable**.

Three processes allow molecules and ions to be transported across biological membranes: passive diffusion, facilitated diffusion and active transport.

Passive diffusion

Passive diffusion is the free movement of molecules across a membrane down a concentration gradient. If the molecule is uncharged, then its movement is in accordance with Fick's First Law of Diffusion. The net flux of molecules

Reference Neubig, R.R. and Thomsen, W.J. (1989) How does a key fit a flexible lock? Structure and dynamics in receptor function. *BioEssays*, **11**, 136–41. Excellent essay on structure and functions of membrane receptor proteins.

Reference Bonting, S.L. and de Pont, J.J.H.H. (eds) (1981) Membrane transport, *New Comprehensive Biochemistry*, Vol. 2, Elsevier, Amsterdam, Holland. For reference only.

(a) Lattice theory

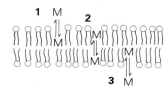

(b) Pore theory

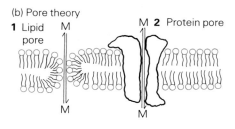

Fig. 4.28 Diffusion of molecules through biological membranes. (a) Lattice theory; (1) The molecule (M) enters the fatty acyl core on one side of the bilayer; (2) The molecule moves through to the other side of the bilayer; and (3) The molecule enters the aqueous compartment on the other side of the membrane. (b) Pore theory; molecules (M) can pass through the membrane either via lipid pores or protein pores without entering the hydrophobic core of the membrane.

through the membrane per unit area of the membrane (J) is dependent on the diffusion coefficient of the molecule (D) and its concentration gradient $(dc)/(dx)$ across the membrane. Thus

$$J = -D\frac{dc}{dx}.$$

Since discrete binding sites on the membrane are not involved during this process, saturation does not occur at high solute concentrations. However, Fick's law does not apply to charged molecules since the rate of diffusion of a charged molecule across the membrane is not *only* determined by the concentration gradient but also by any electrical potential gradient (the difference in charge) existing across the membrane.

Fick's law assumes that no constraints are imposed on a molecule during its passage through the membrane. However, in order to pass through the membrane, molecules must first dissolve in the fatty acyl core. In general this means that the more hydrophobic the molecule, the more easily it will pass through the membrane.

The **lattice theory** adequately accounts for the ability of many small uncharged molecules to diffuse through biological membranes. This theory states that diffusion can occur by the movement of the molecule into a vacancy in the lattice structure of the lipid bilayer (Fig. 4.28a). Such vacancies are thought to arise from random lateral diffusion of lipid molecules in the bilayer. Conditions which increase lateral diffusion of lipids, such as elevated temperatures and decreased cholesterol content, do indeed increase the rate of diffusion of many substances.

However, this simple theory does not account for the diffusion of all molecules through biological membranes. Molecules like water and the hydrophobic gases O_2, CO_2 and N_2 cross biological membranes at a higher

Box 4.5
Permeability of cell membranes and fresh-water green algae

In many early studies on the permeability of cell membranes, fresh-water green algae such as *Chara* and *Nitella* spp were used. These algae are composed of giant, multinucleate cells with a surface area of several square centimetres and a central vacuole of $25 \, cm^3$ or more in volume! Using cells of this size it is possible to assess the permeability of the membranes directly. Individual cells are placed in solutions of the solute under test for various lengths of time. The amount of solute accumulated by the cell can be estimated by simply squeezing out the cell sap and measuring the amount of solute absorbed.

Experiments of this type indicated that penetration of the membrane by non-electrolytes is largely due to simple diffusion through the cell membrane.

Chara sp. (magnification ×8). Courtesy of M.J. Hoult, Department of Biological Sciences, The Manchester Metropolitan University, UK.

rate than expected. This suggests that aqueous pores exist in the membrane. Such pores may be formed by rearrangement of the lipid bilayer or by transmembrane proteins (Fig. 4.28b). In both cases solutes pass into the cell without entering the hydrophobic core of the membrane. For larger molecules such as sugars, amino acids and small charged species such as Ca^{2+}, Na^+, K^+ and H^+, passage through biological membranes by simple diffusion is either very slow or totally restricted.

Facilitated and active transport

Lipid bilayers are largely impermeable to polar and ionic compounds. Thus, polar molecules and ions that pass through biological membranes are transported with the aid of specific carrier proteins embedded in the lipid bilayer. Biological membranes from different sources and different cellular localities may possess different permeabilities according to which carrier proteins they contain.

The transport of molecules and ions across biological membranes by protein carriers differs from simple diffusion in a number of ways. Firstly, it is highly specific for different molecules or ions. Secondly, it is faster than expected from passive diffusion. Thirdly, a maximum rate of transport is reached when all sites on the proteins become saturated. The kinetics of the process can be described by the equation for simple enzyme-catalysed reactions:

$$ J = \frac{J_{max}[S]}{K_m + [S]} $$

where J is the rate of transport of the species into the cell, J_{max} is the maximum rate achievable, $[S]$ is the concentration of the substance to be transported, and K_m is the concentration of the substance which gives half-maximal transport across the membrane.

Transport across biological membranes by protein carriers may be passive or active. Passive transport is referred to as facilitated diffusion and occurs when an ion or molecule crossing a membrane moves *down* its electrochemical or concentration gradient until equilibrium is attained. Metabolic energy is not expended since the free energy change, ΔG, for this process is negative given that the particle is moving down its concentration gradient. For example, the free energy change (ΔG) for the movement of one mole of substance down its concentration gradient from concentration C_2 to concentration C_1 would be:

$$ \Delta G = -2.3\ RT \log \frac{[C_2]}{[C_1]} $$

At equilibrium $[C_2] = [C_1]$ and ΔG is zero.

For charged substances, movement across a membrane by facilitated diffusion must also take into consideration the electrical potential gradient that exists across the membrane such that:

$$ \Delta G = -2.3\ RT \log \frac{[C_2]}{[C_1]} + zF\Delta\psi $$

where z is the valence of the ion, F is the Faraday, and $\Delta\psi$ is the potential difference across the membrane.

Active transport, unlike facilitated diffusion, uses metabolic energy to transport ions or molecules *against* an electrochemical or concentration gradient.

Box 4.6
Ionophores

Some microorganisms synthesize compounds called **ionophores** that are lipid-soluble and increase the permeability of bilayers to certain ions. Some ionophores have been used as antibiotics. They have been extensively used in studies on the permeabilities and functions of biomembranes.

There are two classes of ionophores: mobile ion carriers (e.g. **valinomycin**) and channel formers (e.g. **gramicidin A**).

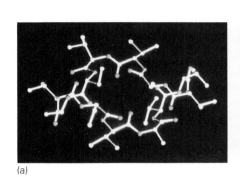
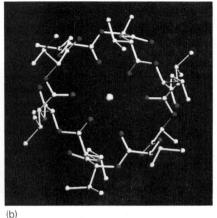

(a)   (b)

Computer-drawn models of (a) valinomycin and (b) valinomycin with a centrally bound K^+. Courtesy of M.J. Hartshorn, Polygen, University of York, UK.

Valinomycin is obtained from some species of *Streptomyces*. It is a cyclic compound shaped rather like a ring doughnut. The interior of the ring is hydrophilic and has six oxygen atoms that coordinate preferentially with a K^+. In contrast, the exterior of the ring is hydrophobic, conferring lipid solubility on the complex. Consequently, if the concentration of K^+ at one surface of a membrane is high, a K^+

(a)

HN–L-Val–Gly–L-Ala–D-Leu–L-Ala–D-Val–L-Val–D-Val–L-Trp–D-Leu–L-Trp–D-Leu–L-Trp–D-Leu–L-Trp—C—N—(CH₂)₂—OH

(b)

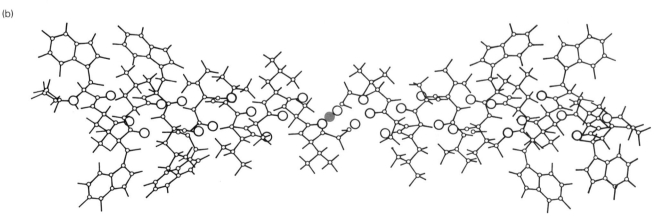

(a) Primary structure of gramicidin A (note the modified termini) and (b) a computer-generated model of a dimer of gramicidin A with an internal K^+ (red). Courtesy of M.J. Derham, Department of Chemistry, The Manchester Metropolitan University, UK.

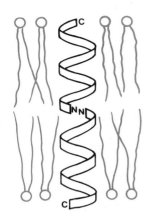

is picked up by a valinomycin molecule, shuttled across the membrane and released at the other surface where the concentration is low. These features allow K$^+$ to be transported into mitochondria, interfering with oxidative phosphorylation, presumably to the advantage of the *Streptomyces* species.

Gramicidin A is made by the bacterium *Bacillus brevis*. Unlike valinomycin, it is an open-chain peptide of 15 amino acid residues, forming a β-helix of about 3½ turns. Two molecules of gramicidin A associate at their amino-termini giving a dimer of sufficient length to span a biomembrane. The structure has a hydrophobic exterior and forms a central hydrophilic channel across the membrane. Transport is not specific: H$^+$, Na$^+$ and K$^+$ can all move through the channel. This can dissipate the H$^+$, Na$^+$ and K$^+$ gradients across the cell membranes of other bacteria leading to their death. Dimers of gramicidin A are unstable, and the channel is only formed for an average of one second. However, during this time, as many as 2×10^7 cations per channel can be transported, a rate about 10^3 faster than with a mobile carrier.

FACILITATED DIFFUSION. In some cell types molecules such as sugars and amino acids and small ions such as HCO$_3^-$ and Cl$^-$ are transported through biological membranes by facilitated diffusion.

An example is the transport of glucose across the erythrocyte plasma membrane into the cell, where it is rapidly metabolized. The concentration of glucose in plasma is higher than that in the erythrocyte. The glucose transporter protein found in erythrocytes has a M_r of 45 000, is an integral protein and probably spans the membrane. Glucose transport is thought to occur by a 'gated pore' mechanism rather than involving carrier molecules moving through the membrane. In the 'gated pore' mechanism (Fig. 4.29), the binding of glucose to a specific site on the exterior surface of the transporter induces a conformational change in the polypeptide which in turn generates a passage (a pore) through which the protein-bound glucose molecule passes to the cytosol. Conformational changes in the membrane depend on the fluid nature of the environment which surrounds the transporter.

ACTIVE TRANSPORT. The selective permeability of biological membranes is important in cellular homeostasis. Cells maintain an intracellular environment of high K$^+$ and low Na$^+$ concentrations, establishing ionic gradients which are used to drive other transport processes. Similarly, since Ca^{2+} is an important secondary messenger, the free cytosolic concentration is kept as low as 10^{-6}–10^{-7} mol dm^{-3} under non-stimulatory conditions.

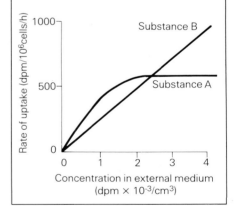
Extracytoplasmic side

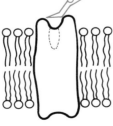 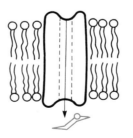 Glucose

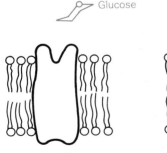

Cytoplasmic side

Fig. 4.29 The 'gated pore' mechanism of the glucose transporter.

Exercise 5

In an aerobically respiring cell, 50% of the ATP produced is used to support Na^+–K^+ ATPase activity. How many moles of Na^+ can be exported from the cell per mole of glucose catabolized?

To maintain this closely defined ionic environment, active transport systems are used. These systems consume up to 50% of the cell's available energy and are therefore an important factor in cellular metabolism. For instance, the active transport of Na^+ and K^+ ions across the membrane in many eukaryotic cells is mediated by Na^+–K^+ ATPase, which involves direct hydrolysis of ATP. This active transport system is present in the plasma membranes of most animal cells. In many cell types it functions by pumping three Na^+ out of the cell while simultaneously transporting two K^+ into the cell, although different stoichiometries are known for other cell types. This keeps the intracellular concentration of K^+ and Na^+ appropriate for cellular metabolism. The Na^+ pumped out of the cell is also used to drive active **cotransport systems**, since the concentration of Na^+ outside of the cell is much greater than that found inside owing to the impermeability of the cell membrane to Na^+.

The Na^+–K^+ ATPase is also important in generating the membrane potential of many animal cells. This potential is brought about by the ability of K^+ to leak out of the cell down its steep concentration gradient. As a result of net efflux of K^+ the inside of the cell becomes electrically negative relative to the outside.

Experimental investigations of Na^+–K^+ ATPase from various tissues, including the mammalian kidney and the human erythrocyte membrane, indicate that this transport system is made up of a tetramer of two polypeptide chains: a glycosylated α-subunit of M_r of 50 000 closely associated with a non-glycosylated β-subunit of M_r 100–120 000. The larger polypeptide is the catalytic subunit responsible for ATP hydrolysis. It has binding sites for ATP and Na^+ on its cytoplasmic surface and binding sites for K^+ on its external

Box 4.7
Specific inhibitors of the Na^+–K^+ ATPase are useful in cardiac heart disease

Drugs which inhibit Na^+–K^+ ATPase, in particular the cardiac glycosides, have been useful in elucidating the mechanism of Na^+–K^+ ATPase. For example, ouabain specifically inhibits Na^+–K^+ ATPase by binding to the outer surface of the larger subunit. Cardiac glycosides, such as ouabain and digoxin, also have considerable clinical application in the treatment of congestive heart failure. The major pharmacological effect of these drugs is to increase the Na^+ concentration in heart cells by inhibiting the Na^+–K^+ ATPase. The resulting lower Na^+ concentration gradient then leads to a less effective Na^+-linked export of Ca^{2+} so that the intracellular level of Ca^{2+} is raised in the heart cell. This in turn leads to greater and more frequent contraction of the cardiac muscle.

Lactone rings

General structure of cardiac glycosides.

Reference Finean, J.B., Coleman, R. and Michell, R.H. (1984) *Membranes and their Cellular Function*, 3rd edn, Blackwell Scientific, Oxford, UK. Particularly good on membrane transport mechanisms.

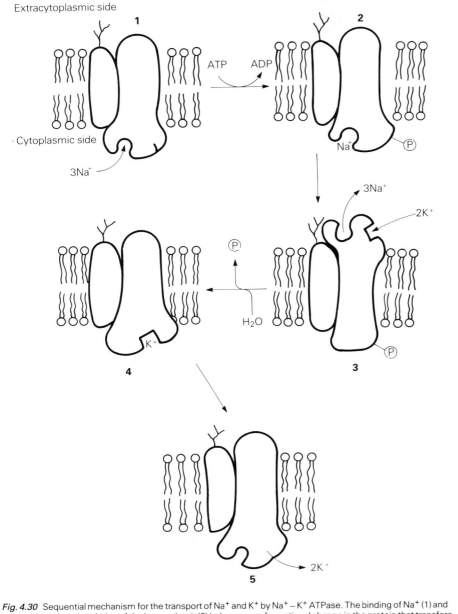

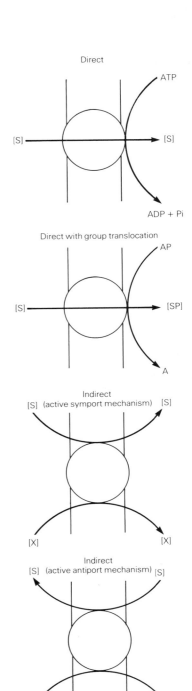

Fig. 4.30 Sequential mechanism for the transport of Na⁺ and K⁺ by Na⁺ – K⁺ ATPase. The binding of Na⁺ (1) and subsequent phosphorylation of the large subunit (2) induces a conformational change in the protein that transfers the Na⁺ across the membrane (3) and releases it outside the cell. The binding of K⁺ on the external surface and subsequent dephosphorylation return the large subunit to its original conformation, resulting in the transfer of K⁺ across the membrane (4) and its release into the cytoplasm (5). The diagram shows the Na⁺ – K⁺ ATPase as a dimer, although in the membrane it is thought to exist as a tetramer of two large and two small subunits.

surface. It is thought that the binding of Na⁺ results in the auto-phosphorylation of an aspartate residue of the large catalytic subunit at the cytoplasmic surface (Fig. 4.30). This produces a conformational change in the protein that transfers Na⁺ across the membrane and releases it on the outside. The binding of K⁺ on the external surface of the catalytic subunit then results in its dephosphorylation and returns the protein to its original conformation while at the same time transporting K⁺ to the inside of the cell where it is released. The function of the smaller glycosylated subunit in this sequential mechanism is unknown.

Fig. 4.31 Modes of active transport.

Mechanisms of transport

Carrier systems involving active transport may be divided into two broad categories (Fig. 4.31). The first involves enzyme systems where the free energy of hydrolysis of ATP or an alternative 'high energy' phosphate donor is directly coupled to the transport of molecules or ions. The Na^+–K^+ ATPase is one example of this type of active transport. Also included is the mechanism of **group translocation** commonly found in bacteria where active uptake of sugars involves conversion of the sugar to its phosphate derivative *during* transfer across the membrane. The phosphorylation of the sugar gives it a negative charge and prevents its passive diffusion back across the membrane. In this way bacteria are able to accumulate sugars and other nutrients from their surroundings against very high concentration gradients. Phospho-enolpyruvate, rather than ATP, is normally the high-energy phosphate donor in this mechanism.

The other category of active transport, the indirect or active cotransport mechanism, is where movement of one molecule or ion *down* its electrical or concentration gradient is coupled to movement of another molecule *up* its concentration gradient. In some animal cells, for example, intestinal epithelia, Na^+ gradients drive the transport of glucose into the cell. In these indirect active transport processes the energy stored in the molecule or ion moving down its gradient is used to power the transport of the other molecule up its gradient (Fig. 4.31).

If the transported molecule and cotransported ion move in the same direction, the process is a **symport** mechanism. If the two substances move in

Table 4.3 *Examples of carrier transport systems*

Substance transported	Type of transport	Membrane type
ATP out, ADP in	Facilitated diffusion	Inner mitochondria
Glucose	Facilitated diffusion	Most animal cell plasmalemma
Glucose	Indirect active (Na^+ symport)	Some animal cell plasmalemma
H^+	Direct active (uses ATP)	Stomach epithelial cell membrane
Na^+ out, K^+ in	Direct active (uses ATP)	Animal cell plasmalemma
Glucose	Group translocation (uses phosphoenol pyruvate)	Many bacterial cell membranes
P_i in, PGE or DHAP out	Facilitated diffusion	Inner chloroplast membrane
HCO_3^-	Facilitated diffusion	Inner chloroplast membrane
Exchange of dicarboxylates e.g. oxaloacetate and malate	Facilitated diffusion	Inner chloroplast membrane

---

Box 4.8
Hartnups disease, an inherited disorder of amino acid transport

Hartnups disease is an inherited genetic disorder named after the family in which it was first recognized. The disease is characterized by the inability of renal and intestinal epithelial cells to absorb neutral amino acids from the lumen. In the kidney this inability manifests itself as the excretion of amino acids in the urine (aminoaciduria). The intestinal defect results in maladsorption of free amino acids from the diet. Therefore the clinical symptoms are those due to essential amino acid deficiencies. In particular, a deficiency in nicotinamide occurs because of the failure to adsorb its precursor, tryptophan. Investigations with patients suffering from Hartnups disease have revealed different transport systems for free amino acids and dipeptides or tripeptides. Since the genetic lesion does not affect the transport of peptides, the disease can be treated with a high-protein diet supplemented with nicotinamide.

---

lumen: cavity or tubular part of an organ or vessel.

opposite directions, it is an **antiport** mechanism (Fig. 4.31). An example of a symport mechanism is the absorption of glucose and amino acids linked to the entry of Na^+ in the *lumen* of the small intestine. An example of an antiport mechanism is the entry of Na^+ linked to the export of Ca^{2+} in cardiac muscle cells.

Some examples of facilitated diffusion and active transport used in different cell types are listed in Table 4.3.

—————— *Exercise 6* ——————

If the facilitated uptake of leucine and isoleucine by a cell was inhibited by the non-standard amino acid nor-leucine,

$$^-OOC-\overset{\overset{\displaystyle H}{|}}{\underset{\underset{\displaystyle NH_3^+}{|}}{C}}-CH_2-CH_2-CH_3$$

which other standard amino acid is likely to be carried across the membrane by the same permease as leucine and isoleucine?

4.7 Overview

Biological membranes are sheet-like structures, 6–10 nm thick, composed principally of lipid and protein molecules held together by non-covalent interactions. Many membranes also contain carbohydrate units linked covalently to proteins and lipids. Membranes act as selective and responsive barriers either between discrete masses of cytoplasm or between the cytoplasm and the extracellular space. They form boundaries between compartments of different composition.

Membrane lipids are amphipathic: they contain both hydrophilic and hydrophobic moieties. Examples are glycerophospholipids (e.g. phosphatidylcholine), sphingolipids (e.g. sphingomyelin), sterols (e.g. cholesterol) and glycolipids (e.g. gangliosides and cerebrosides). These spontaneously form lipid bilayers when dispersed in aqueous solution. The lipid bilayer is a fluid-like structure where both lateral and rotational movements of molecules can take place. The fluidity is maintained by a number of factors, including degree of unsaturation, length of fatty acyl chains, and presence of sterols such as cholesterol.

Membrane proteins normally have distinctive functional properties, depending on the source and cellular locality of the membrane. They function as transport proteins, hormone receptors and transducers of external signals and energy. Many membrane proteins, like their lipid counterparts, are amphipathic molecules which allows them to be embedded in the lipid bilayer. These proteins, unless restricted by specific interactions, are capable of both rotational and lateral movement in the lipid bilayer.

The ability of some membrane proteins to move within the lipid bilayer is restricted by their interaction with the cytoskeleton, the extracellular matrix or cellular junctions. There are three categories of cellular junctions: adhering junctions (desmosomes), sealing junctions (tight junctions) and communicating junctions (gap junctions).

The fluid nature of biological membranes is important in many functions, including endocytosis and exocytosis, transmembrane signalling, and membrane transport. The passage of some small uncharged molecules through membranes occurs by passive diffusion. However, most ionic or polar molecules that cross membranes do so with the aid of specific carrier proteins embedded in the lipid bilayer. This type of carrier-mediated transport is called facilitated diffusion or active transport.

Facilitated diffusion is a passive process, the ion or molecule passing through the membrane moves down its electrical or chemical concentration gradient until equilibrium is attained. An example is the transport of glucose into red blood cells. The specific binding of glucose to the carrier protein is thought to induce a conformational change in the protein which generates a specific passage way or pore for glucose to pass through.

Active transport of molecules across membrane uses metabolic energy enabling ions or molecules to move against an electrical or chemical concentration gradient. An example is the pumping of Na^+ out and K^+ into the cell, by Na^+–K^+ ATPase using the free energy of hydrolysis of ATP.

1. The densities of the outer and inner mitochondrial membranes are 1.060 and 0.9872 g cm^{-3} respectively. Therefore they can be separated.

2. The orientation or 'sidedness' of membrane proteins can be determined by labelling the exposed proteins on the surface of the intact cell. The membrane proteins are then solubilized with SDS and fractionated by SDS polyacrylamide gel electrophoresis. Labelled proteins, that is those exposed at outer surface, may then be detected by their radioactivity. It is also possible to label the surface of sealed 'inside-out' red cell ghosts by the same technique. Thus one can determine which proteins are transmembrane proteins (that is, those proteins exposed both at the extracytoplasmic and cytoplasmic side of the membrane).

3. Vesicles loaded with radiolabelled glucose and ethanol are first separated from their surrounding medium by centrifugation and then resuspended in medium free of both ethanol and glucose. Any release of radiolabelled ethanol or glucose from the vesicles could be measured by sampling the bathing medium for radioactivity at appropriate time intervals. This is possible since the radioactivity emitted from the ^{3}H-ethanol may be distinguished from that from ^{14}C-glucose.

Ethanol is smaller and more hydrophobic than glucose which is relatively large and polar. Therefore ethanol should pass through the lipid bilayer at a much faster rate.

4. The high initial rate of uptake of substance A and the subsequent plateau indicate that the red blood cell membrane contains a specific carrier for substance A, which initially increases the rate of transport but becomes saturated at high glucose concentrations. In contrast, the rate of transport of substance B is linear up to high concentrations, suggesting that this transport is not mediated by a specific carrier but that it is able to diffuse freely across the membrane. In the red blood cell, substances such as glucose are transported into the cell by specific protein carriers whereas lipid-soluble substances like hydrophobic gases (O_2, N_2 and CO_2) or ethanol enter the cell by passive diffusion.

5. One mole of glucose produces 38 moles of ATP. Three Na^+ are transported per ATP. Therefore the number of moles of Na^+ is $38 \times 0.5 \times 3 = 570$.

6. Methionine, which has a similar overall structure, shape, and hydrophobicity to nor-leucine.

$$^-OOC-\underset{\underset{NH_3^+}{|}}{\overset{\overset{H}{|}}{C}}-CH_2-CH_2-S-CH_3$$

QUESTIONS

FILL IN THE BLANKS

1. Biological membranes contain protein and lipid molecules which are associated _____ . The weight ratio of these components varies according to the cellular locality and _____ of the membrane.

A characteristic feature of the lipids found in biological membranes is that they are _____ . Examples of these lipids are glycerophospholipids such as _____ _____ and sphingolipids such as _____ _____ .

Proteins found in biological membranes generally have specific functions. These functions include membrane _____ , hormone _____ and _____ signalling. Some proteins are deeply embedded in the lipid bilayer and are referred to as _____ proteins. Extraction of these proteins requires the use of _____ .

Both membrane proteins and membrane lipids may diffuse rapidly in the _____ of the membrane. Important in maintaining the fluid nature of biological membranes is the _____ and degree of _____ of the constituent fatty acyl chains. In mammals the presence of _____ is also important in modifying membrane fluidity.

Choose from: amphipathic, cholesterol, detergents, function, integral, length, non-covalently, phosphatidylethanolamine, plane, reception, sphingomyelin, transport, transmembrane, unsaturation.

MULTIPLE-CHOICE QUESTIONS

State which of the following statements are true or false:

2. Lipids which are found in biological membranes:
A. are amphipathic.
B. are commonly referred to as triacylglycerols.
C. contain only unsaturated fatty acyl chains.
D. are normally covalently associated with proteins.

3. The Na^+–K^+ ATPase:
A. is an example of a facilitated transporter.
B. contains four subunits of equal M_r.
C. is responsible for the transport of Na^+ out of and K^+ into the cell.
D. is capable of catalysing the synthesis of ATP.

4. Transferrin is taken up by cells using:
A. pinocytosis
B. phagocytosis
C. receptor-mediated endocytosis
D. exocytosis

5. Gap junctions:
A. contain the protein connexin.
B. are connected to tonofilaments.
C. attach cells to basal lamina.
D. are communicating junctions between interconnected cells.

SHORT-ANSWER QUESTIONS

6. Describe an electrophoretic technique which can be used to investigate the protein composition of biological membranes.

7. It is necessary that biological membranes maintain a fluid-like structure in order to carry out a number of functions. Make a list of these functions.

8. The inner mitochondrial membrane is a highly specialized membrane which differs markedly from the plasma membrane. Using a mammalian cell as an example make a list of the common and the dissimilar features of the two membranes.

9. In an imaginary cell the uptake of glucose and four amino acids is driven by a Na^+ gradient. The inward transport of each of the amino acids and glucose is mediated by five separate transport proteins. If the internal Na^+ concentration is maintained at 5% of the external concentration, what ratio of internal to external concentration will be maintained for each of the five solutes?

5

Mitochondria and chloroplasts

Objectives

After reading this chapter you should be able to:

☐ describe the detailed ultrastructure of mitochondria and chloroplasts;

☐ relate the structures of these energy-transducing organelles to their biological functions;

☐ outline the biosynthesis and possible evolutionary origins of these two energy-transducing organelles

5.1 Introduction

In 1771, Joseph Priestley showed that a sprig of mint was needed to 'revitalize' the air 'spoilt' by the respiration of a mouse trapped within an air-tight bell jar. This was the first demonstration of the chemical interrelationship between plants and animals. Today, Priestley's observation is understood as the exchange of carbon dioxide, a product of tissue respiration in the mitochondria of the mouse (and mint), and dioxygen (O_2), a product of photosynthesis in the chloroplasts of the mint. This exchange of gases is necessary for the capture and utilization of the energy of sunlight.

The conversion of the energy of sunlight into a biologically useful form requires the co-operation of chloroplasts and ***autotrophic*** prokaryotic cells with mitochondria and ***heterotrophic*** prokaryotic cells (Fig. 5.1). The chloroplast imports the readily available simple inorganic molecules, carbon dioxide and water, and produces carbohydrate. This ***endergonic*** reaction, driven by the input of the energy of sunlight, is called **photosynthesis**. Dioxygen is the by-product of photosynthesis and is released into the atmosphere. The mitochondrion imports fuel molecules and oxidizes them to H_2O and CO_2. This is an ***exergonic*** process and the released energy is used, in part, to drive the phosphorylation of adenosine diphosphate (ADP) to ATP. This process produces most of the ATP formed during tissue respiration.

Thus, chloroplasts and mitochondria are mutually dependent. The mitochondria rely on chloroplasts for the provision of chloroplast-derived metabolites and dioxygen, and the chloroplasts depend on mitochondria for replenishing the supply of atmospheric CO_2. However, the fundamental importance of chloroplasts in biology lies in their ability to capture the energy of sunlight and transform it into the chemical potential energy of ATP.

Chloroplasts and mitochondria are not merely structural features of a eukaryotic cell with a role in energy transduction. They possess DNA, and can replicate independently of the cell's nucleus. Moreover, chloroplasts have characteristics in common with autotrophic prokaryotic organisms, and mitochondria resemble some heterotrophic prokaryotic organisms. This

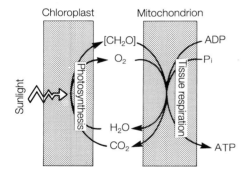

Fig. 5.1 The interaction between chloroplasts and mitochondria in the conversion of the energy of sunlight into the chemical potential energy of ATP.

☐ Because mitochondria and chloroplasts obey the Second Law of Thermodynamics it is erroneous to describe them as energy-producing organelles. ATP production in these organelles is the result of energy-*transduction*.

autotrophic: *a term applied to those organisms which can manufacture their own organic constituents from inorganic material. Autotrophic organisms can be phototrophic or chemotrophic.*
heterotrophic: *a term applied to those organisms which require a supply of organic material from which they can make most of their organic constituents.*

endergonic reactions: *chemical reactions driven by the input of energy.*
exergonic reactions: *chemical reactions which yield energy.*

The interrelationship in the anabolic activity of chloroplasts and the catabolic activity of mitochondria mediated by the exchange of carbon dioxide and dioxygen between the atmosphere and organisms is shown in Figure 5.1. The concentration of atmospheric CO_2 has been relatively constant, at about 270 parts per million, over the past few millennia. Since the Industrial Revolution, however, it has increased by about 25%, and is predicted to reach 600 parts per million by the year 2050. The increased atmospheric CO_2 concentration is the result of the burning of fossil fuels such as coal and oil, although deforestation has an exacerbating effect. CO_2 is called a 'greenhouse' gas because it absorbs infrared radiation reflected from the Earth's surface. An increase in atmospheric CO_2 concentrations is predicted to result in warming of the planet by a couple of degrees by the year 2050. Although the higher atmospheric CO_2 concentration may slightly improve photosynthetic efficiency, the planet's ecosystem will be threatened by major alterations in climate which could accompany the global warming.

suggests that chloroplasts and mitochondria may have evolved from free-living prokaryotic organisms which were engulfed by a host cell. Indeed, as Thomas wrote in *The Lives of a Cell*, 'it is usual to view chloroplasts and mitochondria as enslaved creatures captured to supply ATP for cells unable to respire, or to provide carbohydrate and dioxygen for cells unequipped for photosynthesis'. Irrespective of their evolutionary origins, mitochondria and chloroplasts are in a fundamental sense the most important things on Earth. To an alien visiting Earth, they would be identified as responsible for maintaining the carbon-based life process on this planet.

5.2 Energy-transduction pathways in mitochondria and chloroplasts

To understand mitochondrial and chloroplast ultrastructure, it is necessary to know about the energy transduction reactions of tissue respiration and photosynthesis.

The process of photosynthesis in chloroplasts involves two distinct sets of

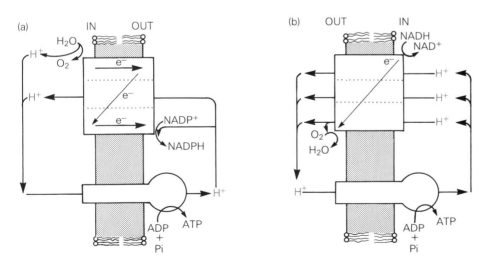

Fig. 5.2 Chemiosmotic coupling of photosynthetic and respiratory electron transport to ATP synthesis in (a) chloroplasts and (b) mitochondria, respectively.

Reference Thomas, L. (1974) *The Lives of a Cell*, Bantam Books, London, UK. A collection of interesting short articles on various aspects of biology by an excellent essayist.

□ The chemiosmotic coupling of electron transport to ATP synthesis was proposed by Mitchell in 1961. Although his ideas were initially met with scepticism, they could be tested experimentally and several studies in the 1960s and 1970s supported Mitchell's proposal. Today, researchers are concentrating on the mechanistic details of H^+ pumping by the electron transfer chains and ATP synthetase. Mitchell was awarded the Nobel Prize in Chemistry in 1978.

Dr Peter Mitchell, Nobel Prize Laureate in Chemistry (Courtesy of Dr P. Mitchell, Glynn Research Institute, Bodmin, UK).

See *Energy in Biological Systems*, Chapter 3–7.

□ A common textbook error is to show flavin adenine dinucleotide (FAD) as the 'mobile' mediator of hydrogen between succinate and ubiquinone of the respiratory chain. Unlike NAD^+, which is a cofactor, FAD is a prosthetic group tightly attached to the membrane-bound succinate dehydrogenase. It is therefore correct to consider the whole protein, and not a component part, as the mediator of hydrogen between succinate and ubiquinone (just as one would describe haemoglobin, and not the haem, as an oxygen carrier of erythrocytes).

Exercise 1

How many mitochondria (treat as cylinders $10\,\mu m$ in length and $0.5\,\mu m$ in diameter) would occupy 20% of the volume of a liver cell (treat as a sphere of $30\,\mu m$ diameter)?

---

enzyme-catalysed reactions. The initial 'light' (or **photophosphorylation**) reactions are summarized in Figure 5.2. They involve the transfer of hydrogen atoms from water to nicotinamide adenine dinucleotide phosphate ($NADP^+$) and are coupled to the production of ATP. The photosynthetic electron transport chain consists of two types of photosystems (PSI and PSII) and a cytochrome b/f complex embedded within the **energy-transducing thylakoid membrane**. Electron transfer, driven by photochemical charge transfer reactions within the photosystems, results in translocation of H^+ across the thylakoid membrane and generation of an electrochemical potential gradient. Dissipation of this gradient through membrane-bound ATP synthetase results in phosphorylation of ADP to ATP. The products of the 'light' reactions of photosynthesis, ATP and NADPH, are consumed in the 'dark' reactions involving the Calvin cycle enzymes within the chloroplast.

Mitochondria are the power stations of the cell. They import fuels from the cytosol and export energy as ATP to the cell. These are: pyruvate from the glycolytic breakdown of carbohydrate; oxoacids from the breakdown of protein; and acyl units from the breakdown of fats.

The organic carbon residues are fed into the tricarboxylic acid pathway (fatty acid derivatives first have to undergo β-oxidation), and are completely oxidized to CO_2. The released hydrogen ions and electrons enter the respiratory electron transport chain either indirectly, via NADH, or directly upon the oxidation of succinate by succinate dehydrogenase. Electron transfer, driven by the redox potential difference of the initial reductant (for example, NADH) and oxidant (that is, oxygen) through the respiratory electron transport chain, results in the translocation of H^+ across the inner mitochondrial membrane and the generation of an electrochemical potential gradient of H^+ (Fig. 5.2). ATP is generated by ATP synthetase upon dissipation of the electrochemical potential gradient of H^+ through the ATP synthetase.

The reactions of both respiration and photosynthesis involves two spatially separate sets of reactions. The set of enzymes involved in either the formation (as in chloroplasts) or breakage (as in mitochondria) of carbon–carbon bonds are functional in the aqueous compartment of the organelle, whereas the set of enzymes involved in the production of ATP are membrane-bound in 'energy-transducing' membranes. Although the chloroplasts and the mitochondria possess their own distinctly unique set of enzymes for processing carbon–carbon bonds (that is, the catabolic reaction pathway of mitochondria is dissimilar to the anabolic reaction pathway of chloroplasts), both organelles produce ATP using a similar reaction mechanism (see Figs 1.27 and 1.31).

5.3 *Mitochondria*

A few eukaryotic cell types lack mitochondria and obtain ATP from glycolytic fermentation. Some, such as erythrocytes, lose their mitochondria during development. Others, such as the *petite* yeast mutant have lost the ability to synthesize ATP through complete, or partial, loss of the mitochondrial genome, and some, such as obligately anaerobic flagellates, have poorly developed mitochondria. Most eukaryotic cells contain several hundred mitochondria, although many zooflagellates (e.g. trypanosomes and leishmanias) and algae possess only a single mitochondrion which must divide in synchrony with cell division. Liver cells, for example, can contain up to 2500 mitochondria occupying a substantial part (for example up to 20%) of the cell's volume.

---

catabolism: *from the Greek kata, down, and bolism, to throw.*
mitochondrion: *introduced by Benda in 1898, this term originates from the Greek for 'thread-like' granules and was used to describe their appearance during spermatogenesis.*

Reference Harold, F.M. (1986) *The Vital Force*, W.H. Freeman, New York, USA. A clear and cogent examination of the physical principles and experimental studies of bioenergetics.

Mitochondria were identified as subcellular structures by light microscopy studies of animal cells in the 1840s; plant mitochondria were first observed in 1904. They can be distinguished from other internal organelles by their ability to be stained with the dye Janus Green. This redox dye can be oxidized to a coloured form by cytochrome c oxidase of the respiratory electron transport chain. These early studies showed that mitochondria were similar in size and shape to bacteria. In 1890, Altmann speculated that mitochondria (then called **bioblasts**) were the basic units of cellular activity. He also proposed that mitochondria were capable of an independent existence and that, by their association with the 'host cell', they conferred the properties of life to the eukaryotic cell.

Mitochondrial ultrastructure

The first extensive electron microscopic study of mitochondria was made by Palade in 1953. His studies of thin sections of different mammalian tissues showed that mitochondria were rod or thread-shaped organelles with hemispherical ends, of diameter about 0.5 μm and length about 2 μm. Figure 5.3a shows an electron micrograph of a typical mitochondrion from a muscle cell. All mitochondria have an outer and an inner membrane. The two membranes enclose and separate two compartments within the mitochondrion: the narrow intermembrane space between the outer and inner membranes, and the internal **matrix** enclosed by the inner mitochondrial membrane. Figure 5.3b is a schematic to show the structure of a typical mitochondrion.

Figure 5.4 shows freeze-etch micrographs of heart mitochondria. The fracture face of the outer membrane is smooth, but as can be seen, the fracture face of the inner membrane is covered with particles. This observation is consistent with the different functions of the two membranes of mitochondria.

Mitochondria are not immobilized within the cell: microscopy in combination with time-lapse microcinematography reveals that not only are they mitochondria motile but also that their shape is plastic (see Fig. 1.25). Mitochondria are probably associated with the microtubules of the cytoskeleton which may determine the orientation and spatial distribution of

□ Supravital staining uses stains to identify living cells, or specific organelles within living cells.

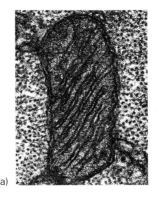

(a)

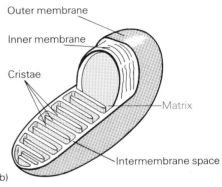

(b)

Fig. 5.3 (a) Electron micrograph of a mitochondrion from rat soleus muscle (×72 000). Courtesy of M.J. Cullen, Muscular Dystrophy Research Labs, Newcastle General Hospital, UK. (b) Schematic mitochondrion, showing the lamellar arrangement of cristae.

Box 5.2
Looking at membranes in the electron microscope

Specimens for examination by transmission electron microscopy (see Section 1.2, also Boxes 6.1 and 7.2) can be treated by freeze-fracture, freeze-etching and negative staining techniques.

Freeze-fracture enables the interior of a cell membrane to be visualized. Cells are frozen in liquid nitrogen (at −196°C) in the presence of a cryoprotectant to prevent the formation of large ice crystals. The frozen block is cracked with a knife and the fracture plane passes through the hydrophobic core of a membrane. Two fracture faces of the membrane are exposed (the 'f' and 'e' faces) which can be shadowed with platinum.

Freeze-etching enables the exterior surface of cell membrane to be visualized. The technique is similar to freeze-fracture, except that cryoprotectants are not included and the cracked block is freeze-dried allowing some ice-sublimation. This causes the ice level to be lowered around and within the cells, exposing unfractured membranes. The exposed parts of the cell are then shadowed with platinum.

In **negative-staining** the specimen is coated with the salt of a heavy metal, for example, phosphotungstate, which cannot penetrate into the material, but which creates a contrast between the free areas of the copper grid on which the specimen is mounted and the specimen itself.

Reference Tzagoloff, A. (1982) *Mitochondria*, Plenum, New York, USA. Now somewhat dated, this excellent text covers structure, function and biogenesis of mitochondria in great detail.

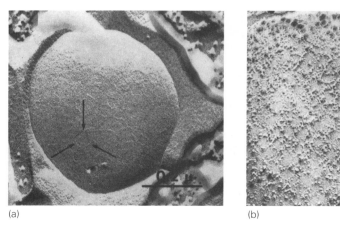

(a)　　　　　　　　　　　　　　　　　(b)

Fig. 5.4 Freeze-etch heart mitochondria: (a) smooth fracture face of the outer membrane; (b) particle-covered inner membrane. Courtesy of J. Wrigglesworth, Department of Biochemistry, Kings College London, UK.

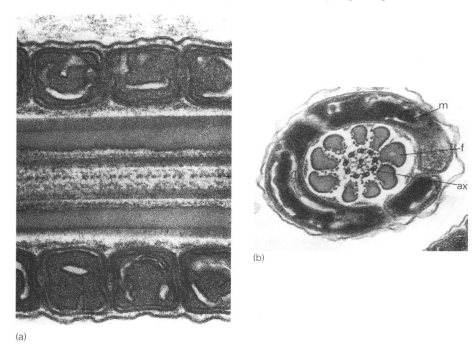

(a)

(b)

Fig. 5.5 Midpiece section of an intact spermatozoon from hamster. Within the plasma membrane one can detect the mitochondria (m) which surround the outer dense fibres (f) enveloping the central axoneme (ax). (a) Longitudinal section; (b) transverse section. Courtesy of G. Olson, Department of Cell Biology, Vanderbilt University, USA.

Fig. 5.6 Freeze-fracture through a single crista of a rat heart mitochondrion showing dense aggregation of membrane particles (×51 000). Courtesy of F. Sjostrand, Department of Biology, University of California, USA.

these organelles within a cell; mitochondria form long sinuous chains within some cells (see Fig. 1.19). In other cells, the mitochondria appear to be permanently positioned near a site of high ATP consumption; examples are the mitochondria wrapped around the flagellum in the midpiece of spermatozoa (Fig. 5.5), or the mitochondria aligned in the plane of the myofibrils of muscle cells (see Chapter 10). In many cells, such as those of muscle, the mitochondria are often close to fat droplets in the sarcosol. These droplets are the fuel reserve for the mitochondria, and their close proximity ensures rapid utilization of the released triglycerides.

Electron micrographs of mitochondria show that the inner membrane is deeply folded inwards at a number of places, so as to project into the matrix compartment. These invaginations, termed **cristae**, increase the surface area of the 'energy-transducing' membrane permitting many electron transfer chain units and ATP synthetase units to be packed into a mitochondrion. The cristae are usually lamellar (in Fig. 5.6 note the dense aggregation of membrane particles in this single crista). There are, however, exceptions. In some mitochondria, especially those of protozoa, the cristae may be branched, tubular structures, or may be arranged parallel to the long axis of the organelle. Figure 5.7 shows a mitochondrion from a fungal cell which has a tubular network of cristae.

The inner membrane of mitochondria has a larger surface area than that of the outer membrane and some mitochondria have more closely packed cristae than others. The observed difference in cristae content between cell types is probably indicative of differences in their energy demand. Thus, tissues which demand a greater output of ATP, such as cardiac muscle, have a more developed inner mitochondrial membrane than, for example, liver mitochondria. In some cases, such as yeast cells growing in anaerobic conditions, the number of cristae is very small.

Mitochondrial morphology can be grossly affected by the age of the organism or by certain conditions of physiological stress. Riboflavin or copper deficiency results in enlarged mitochondria (about 10-fold increase in volume) with a paucity of cristae. The cristae of blowfly muscle mitochondria appear to degenerate as the insect ages. Abnormalities in mitochondrial morphology in human tissue have been observed and may be indicative of inborn errors of metabolism.

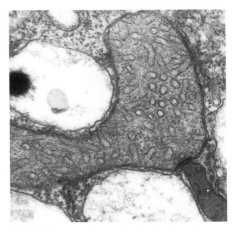

Fig. 5.7 Electron micrograph of a mitochondrion from *Saprolegnia* sp. (water mould), showing a tubular arrangement of cristae ($\times 57\,600$). Courtesy of G. Beakes, Department of Biology, The University, Newcastle upon Tyne, UK.

Isolation and purification of mitochondria

To study how functional activities are related to mitochondrial ultrastructure, mitochondria must be separated from the other cellular material for biochemical investigation. Intact mitochondria were first isolated from rat liver by Hogeboom, Schneider and Palade in 1948 and the method used today is much the same. Fresh liver is chopped, and gently homogenized in an isotonic, buffered medium using a glass mortar and a Teflon pestle. Sucrose ($0.25\,\mathrm{mol\,dm^{-3}}$) is usually used as the osmotic stabilizer. With tougher tissue, such as heart, pretreatment with a proteolytic enzyme ensures adequate cell breakage. With plant tissues and yeast, it is usually necessary to use a carbohydrase to digest the cell wall (Box 7.1) before the plasma membrane is

☐ Consider an isolated organelle surrounded by its semipermeable membrane suspended in a support medium. The solution is: **hyper**tonic if water passes by osmosis from the solution to organelle; **hypo**tonic if water passes from organelle to solution; and **iso**tonic if a net flow of water does not cross the semipermeable membrane.

Box 5.3
Mitochondrial myopathies

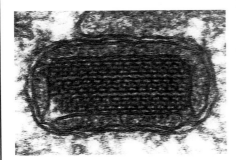

Mitochondrial **myopathies** describes a heterogeneous class of rare clinical disorders affecting different tissues and with a variable onset of symptoms. Initial diagnosis is often by observation of elevated concentration of lactate in the blood. Electron microscopy of biopsy samples may reveal abnormal mitochondrial morphology. For example, in some disorders the cristae of the inner mitochondrial membrane may form honeycomb patterns of concentric whirls. In other disorders, the mitochondria may have a vacuolated appearance with a paucity of cristae.

The identification of mitochondrial myopathies can be followed up using sensitive measurements of enzymatic activity and immunological techniques to determine the enzymatic location of the disorder. With a few patients this has led to successful treatment. In general, the structural changes observed in the mitochondria appear to result from disorders in either the respiratory electron transport chain or ATP synthetase, and not in the decarboxylation reactions.

Electron micrograph of a mitochondrion from a patient exhibiting a mild form of mitochondrial myopathy ($\times 105\,000$). Courtesy of M.J. Cullen, Muscular Dystrophy Research Labs, Newcastle General Hospital, UK. The patient has partial deficiency of cytochrome *c* oxidase (complex IV) activity. Abnormal inclusions within the inner mitochondrial membrane have a regular lattice structure.

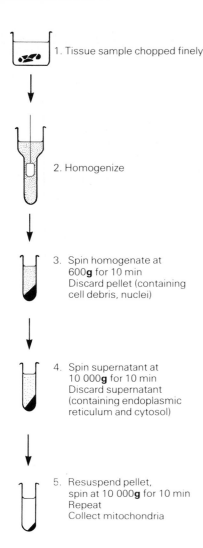

1. Tissue sample chopped finely

2. Homogenize

3. Spin homogenate at 600**g** for 10 min Discard pellet (containing cell debris, nuclei)

4. Spin supernatant at 10 000**g** for 10 min Discard supernatant (containing endoplasmic reticulum and cytosol)

5. Resuspend pellet, spin at 10 000**g** for 10 min Repeat Collect mitochondria

Fig. 5.8 The procedures used in the isolation and purification of mitochondria.

Exercise 2

Non-respiring mitochondria were suspended in either 150 mmol dm^{-3} KCl or 150 mmol dm^{-3} KSCN. No decrease in light-scattering at 520 nm could be observed in either case. When valinomycin was added (an ionophore which makes the inner membrane permeable to K$^+$), light scattering was observed with the mitochondria suspended in KSCN but not KCl. Is the inner membrane permeable to Cl$^-$ or to SCN$^-$ or to both anions?

---

broken. The homogenized material is centrifuged at a low speed (600g) for 10 min, and the pellet containing the nuclei and unbroken cells is discarded. The supernatant is then spun at 8000–10 000g for 10 min. The resulting supernatant which contains ribosomes and fragments of endoplasmic reticulum is discarded. The pellet contains intact mitochondria and is normally resuspended and recentrifuged to increase their purity. The procedure is outlined in Figure 5.8.

Suspensions of isolated mitochondria are turbid because of differences in refractive index between the mitochondrial matrix and the buffering medium. An increase in the volume of the matrix as a result of an influx of a permeant solute accompanied by water can therefore readily be monitored by a decrease in the amount of light scattering. However, for swelling of the mitochondria to be observed either the inner membrane must be permeable to both the cation (C$^+$) and anion (A$^-$) species of the major osmotic component, or the overall charge balance across the membrane must be unaffected. This technique has been used to detect various transporters across the inner membrane of mitochondria.

Compartmentation of mitochondrial enzymes

Early electron microscopic studies of fixed and stained tissues failed to distinguish outer and inner mitochondrial membranes. However, in 1962, Fernandez-Moran, working with negatively stained samples, showed that the inner membrane was morphologically distinct from the outer membrane.

There are four distinct compartments within the mitochondrion in which an enzyme could be located (Fig. 5.3b). The protein could be embedded in either the inner or the outer membrane, or it could be present in either the intermembrane space or the internal matrix. The locations of mitochondrial enzymes have been established by fractionating broken mitochondria into their component parts.

The outer membrane can be removed either by swelling mitochondria in hypotonic buffer, or by treatment with low concentrations of the detergent digitonin. These treatments rupture the outer membrane without breaking the inner one. The constituents of the intermembrane space are simultaneously released. Osmotically intact mitochondria devoid of their outer membranes are called **mitoplasts**. The broken pieces of the outer membrane reseal to form small vesicles and may be separated from the mitoplasts by sucrose density gradient centrifugation. The soluble enzymes of the matrix are released upon rupture of the inner membrane by a strong detergent or by sonication. Following sonication, the pieces of inner membrane may reseal to form inverted vesicles called submitochondrial particles.

Whereas the outer membrane is smooth in appearance, the inner one is covered with small spherical particles of 9 nm diameter connected to the membrane by 'stalks'. These spheres can also be detected on the outer surfaces of submitochondrial particles (Fig. 5.9a). Racker and colleagues were able to develop mild techniques to remove the spheres from submitochondrial particles (Fig. 5.9b). The isolated spherical particles were shown to possess ATPase activity, that is, they catalysed the reaction:

$$H_2O + ATP \rightarrow ADP + P_i$$

The depleted inner mitochondrial membranes were unable to produce ATP during electron transfer. Reconstitution of ATP synthetase activity to the submitochondrial particles required re-binding of the spherical particles (Fig. 5.9c). The spherical particles were called 'coupling factors'. Today, the 4 nm diameter spherical particles are recognized as the F_1 head-group component of the F_0F_1 ATP synthetase.

Reference Racker, E. (1976) *Mechanisms in Bioenergetics*, Academic Press, New York, USA. The 'whys' and 'hows' of scientific research in the field of bioenergetics written by one of its leading figures.

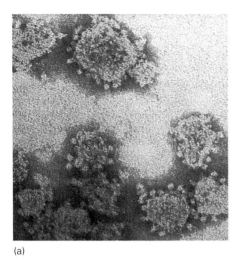
(a)

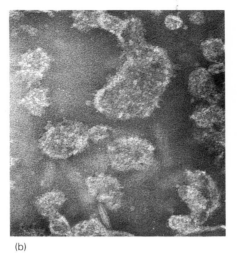
(b)

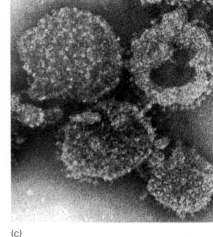
(c)

Fig. 5.9 Electron micrographs of negatively stained submitochondrial particles: (a) untreated 'control' particles (×153 600); (b) after treatment with Sephadex and urea (×153 600); (c) Sephadex–urea particles reconstituted with coupling factor (×153 600). Courtesy of Professor E. Racker, Cornell University, USA.

THE OUTER MEMBRANE. This resembles most other eukaryotic membranes in its lipid composition. It contains only about 5% of the total mitochondrial protein. The major protein is **porin** which forms large, non-specific, aqueous channels through the outer membrane. The channels ensure that the outer mitochondrial membrane is freely permeable to molecules of M_r of 10 000 or less. Several other enzymes are located in the outer mitochondrial membrane, such as NADH cytochrome b_5 oxido-reductase and monoamine oxidase.

THE INNER MEMBRANE. This contains about 20% of the total mitochondrial protein, accounting for the increased particulate nature observed in electron micrographs of inner compared with outer mitochondrial membranes (Fig. 5.4). It is the energy-transducing membrane, and is impermeable to most small ions. This is thought to be due to the presence of bisphosphatidylglycerol (also called cardiolipin; Fig. 5.10). This phospholipid, which has only a low abundance in the outer mitochondrial membrane, accounts for about 10% of the total lipid content of the inner membrane (Table 5.1). Three major types of enzyme complex are found in the inner mitochondrial membrane: (1) components of the respiratory electron transport chain, namely complexes I, II, III and IV, cytochrome c and ubiquinone (Fig. 5.11); (2) F_0F_1 ATP synthetase; (3) specific transport proteins (Table 5.2). These transporters regulate the movement of metabolites into and out of the mitochondrial matrix.

The number and type of transporters in a mitochondrion vary according to

Fig. 5.10 Bisphosphatidylglycerol (cardiolipin).

See *Energy in Biological Systems*, Chapter 3

□ Ubiquinone and plastoquinone are the mobile carriers of hydrogen atoms (H^+ and e^-) in the respiratory and photosynthetic electron transport chains respectively).

Table 5.1 *Lipid composition of mitochondrial membranes*

	Outer membrane	Inner membrane
	(mg/mg protein)	
Cholesterol	30.1	5.6
Phospholipids	0.9	0.3
	(% total phospholipids)	
Phosphatidylcholine	55	45
Phosphatidylethanolamine	25	28
Phosphatidylinositol	14	4
Bisphosphatidylglycerol (cardiolipin)	3	22

Which transporters are involved in the synthesis of glucose from pyruvate in rat liver mitochondria?

Table 5.2 Some specific transporter proteins

Carrier	Out (C)*		In (M)†	Role
Adenine nucleotide	ADP^{3-}	$\rightleftarrows$	ATP^{4-}	ATP synthesis
Phosphate	$H_2PO_4^-$	$\rightleftarrows$	OH^-	ATP synthesis
Pyruvate	Pyr^-	$\rightleftarrows$	OH^-	TCA cycle substrates
Dicarboxylate	$malate^{2-}$	$\rightleftarrows$	HPO_4^{2-}	Allow net export of TCA intermediates from matrix to be used for gluconeogenesis and fatty acid synthesis
Tricarboxylate	$H^+ + citrate^{3-}$	$\rightleftarrows$	$malate^{2-}$	
Glutamate	$glutamate^{2-}$	$\rightleftarrows$	OH^-	
Carnitine	$acylcarnitine^+$	$\rightleftarrows$	$carnitine^+$	β-oxidation TCA cycle substrates
Glutamate–aspartate	$glutamate^{2-}$	$\rightleftarrows$	$aspartate^{2-}$	Malate/aspartate shuttle allowing
2-oxoglutarate–malate	$2\text{-}oxoglutarate^{2-}$	$\rightleftarrows$	$malate^{2-}$	oxidation of cytosolic NADH
Ca^{2+}	Ca^{2+}	$\rightarrow$		

* Cytosol
† Mitochondrial matrix

Table 5.3 Cell-type distribution of inner mitochondrial membrane transporters

Carrier	Liver	Heart	Blowfly muscle
Adenine nucleotide	+	+	+
Phosphate	+	+	+
Dicarboxylate	+	+	−
Tricarboxylate	+	−	−
α-oxoglutarate	+	−	−
Glutamate	+	+	−
Ca^{2+}	+	+	−

From Lehninger, A.L. (1971) in *Biomembranes* (ed. L.A. Manson), Vol. 2, pp. 147–64, Academic Press, New York, USA.

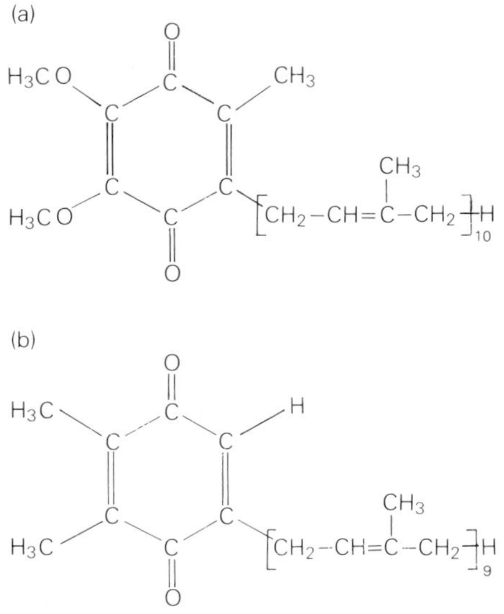

(a)

(b)

Fig. 5.11 (a) Ubiquinone and (b) plastoquinone.

tissue- and cell-type. For example, liver mitochondria need a greater variety of transporters than muscle mitochondria (Table 5.3), since they are involved in the initial reactions of several anabolic pathways, while muscle mitochondria are dedicated to ATP-generation to power contraction.

THE INTERMEMBRANE SPACE. This contains only about 5% of the mitochondrial protein. It is the location of cytochrome *c*, an extrinsic protein of the respiratory electron transport chain involved in shuttling electrons from complex III to IV. It can be displaced from the inner mitochondrial membrane by treatment with high concentrations of salt. Most dehydrogenases of the inner mitochondrial membrane have access to the matrix side only. However, some, such as α-glycerophosphate dehydrogenase which feeds reducing equivalents from cytosolic NADH into the respiratory electron transport chain, have access to the intermembrane space.

Box 5.4
The role of mitochondrial autoantigens in primary biliary cirrhosis

Primary biliary cirrhosis (PBC) is a chronic cholestatic liver disease characterized by inflammation of the intrahepatic bile ducts. This progressive disease is predominantly detected in middle-aged women. It is the most common disease for which liver transplantation is carried out. It is thought that PBC is an autoimmune disease since anti-mitochondrial antibodies can be detected. The autoantibodies are directed against at least four antigens of the inner mitochondrial membrane. Recent immunological data indicate that an autoantigen (M2a) of M_r 70 000 is a component of the pyruvate dehydrogenase multienzyme complex. This enzyme has a structurally similar analogous form in bacteria. Although the aetiology of primary biliary cirrhosis is not known, it is possible that it is due to bacterial infection.

THE MATRIX. The majority of the mitochondrial proteins are found in the matrix, which contains the enzymes that catalyse the reactions of the tricarboxylic acid cycle and fatty acid oxidation. The mitochondrial matrix of some cell types contains some of the enzymes involved in urea formation and gluconeogenesis. The matrix contains ribosomes, and the other enzyme systems responsible for synthesis of mitochondrial DNA, RNA and protein.

5.4 Chloroplasts

Priestley's pioneering studies on the exchange of gases between animals and plants were followed up first by Ingenhousz, who showed that the illumination of the green tissue of plants was needed for photosynthesis.

Chloroplasts were first observed as subcellular structures within plant cells by von Mohl in 1837 and they were finally confirmed as the site of photosynthesis by Engelmann in 1894. Microscopic examination of the alga *Spirogyra* (Fig. 1.29b) showed Engelmann that motile bacteria moved towards the part of the chloroplast that was illuminated thereby generating oxygen. In a second experiment, he illuminated the alga with different coloured lights. More bacteria surrounded those parts of the algal filament in the blue and red compared with the green regions of the spectrum (see Fig. 5.12). Engelmann correctly concluded that the green chlorophyll pigment was the photoreceptor of photosynthesis. It is now known that all the chlorophyll pigments are associated with photosystems. Most of the chlorophyll has a light-harvesting role (absorbing incident sunlight), but within each photosystem a few chlorophyll molecules (probably only four) are essential for the photochemical charge transfer reaction which 'traps' the energy of sunlight.

Chloroplast ultrastructure

Chloroplasts, which belong to a group of organelles called **plastids**, are generally ellipsioids, 4–10 μm long and about 1 μm in width (Fig. 5.13a, Figs 5.21 and 5.23d,e). They are therefore significantly larger than mitochondria. The number of chloroplasts in a plant cell can vary from one to over 100, depending on the type of plant and the growth conditions.

Chloroplasts are surrounded by an outer and an inner membrane which together are called the **envelope**. Each membrane is about 6–8 nm thick and they are separated by a gap of 10–20 nm. The envelope encloses an aqueous compartment called the **stroma**, in which (depending on plant type) starch granules can be found. The most distinctive feature of a chloroplast is, however, the single internal membrane which extends throughout the organelle from pole to pole. This is the **thylakoid** membrane, named by the German botanist Menke, which encloses an internal aqueous compartment called the **lumen** (Fig. 5.13b).

Isolation of chloroplasts

Photosynthetically active chloroplasts were first isolated by Hill in 1939. These early preparations consisted of 'broken' chloroplasts in which the outer envelope membrane had been ruptured during the isolation procedure and the contents of the stroma lost. The chloroplasts were thus unable to fix CO_2, and were only active in the 'light reactions' when supplied with artificial hydrogen acceptors (the so-called Hill Reaction). These are often termed 'Hill acceptors'. Today's modern procedures allow the isolation of 'intact' chloroplasts with an unbroken envelope and therefore the full retention of the contents of the stroma.

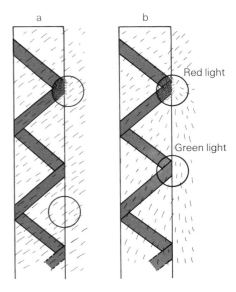

Fig. 5.12 Engelmann's experiment for studying photosynthesis in the alga *Spirogyra* using motile bacteria (which migrate towards regions of high dioxygen concentration). (a) Illumination of the alga's chloroplast and not its cytosol results in the production of dioxygen. (b) Red light (or blue light) is more effective than green light in promoting the photosynthetic production of dioxygen.

☐ A photosystem comprises a set of light-harvesting pigment proteins involved in the absorption of sunlight, along with the photochemical reaction centre which traps the energy of sunlight by driving electron transport between donor and acceptor. The various photosystems found in photosynthetic organisms differ in the number and composition of light-harvesting pigment proteins and the chemical nature of the donor and acceptor molecules of the reaction centre.

☐ All plastids comprise a double outer membrane and an internal membrane system. Some act as storage organelles. (For example, amyloplasts contain starch; elaioplasts contain lipid.) Chloroplasts are the most complex in structure.

☐ 'Broken' chloroplasts cannot assimilate CO_2 to carbohydrate because the Calvin cycle enzymes are lost during the isolation procedures. These chloroplast preparations also lose their complement of $NADP^+$, which acts as the ultimate hydrogen acceptor of the photosynthetic electron transport chain. To detect the 'light reactions' of photosynthesis in 'broken' chloroplasts, measured as light-dependent oxygen evolution, it is necessary to supply $NADP^+$, or some other suitable (but non-physiological) hydrogen acceptor, to the chloroplasts. These artificial acceptors, for example ferricyanide or oxalate, are termed Hill acceptors.

---

thylakoid: *from the Greek* thulakos, *empty pouch.*

Reference Govindjee and Coleman, W.J. (1990) How plants make oxygen. *Scientific American*, **262**(2), 50–8. Clear, concise synopsis of the water-splitting reactions of photosynthesis.

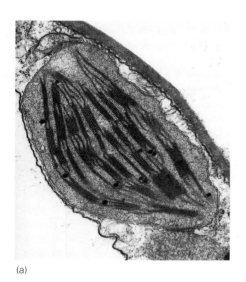

(a)

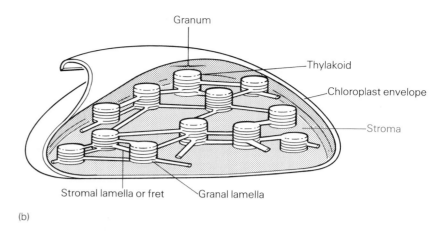

(b)

Fig. 5.13 (a) Electron micrograph of a chloroplast from pea. Note the presence of starch granules. (×21 000). Courtesy of G. Beakes, Department of Biology, The University, Newcastle upon Tyne, UK. (b) Representation of a chloroplast showing the complex architecture of the single thylakoid membrane.

Exercise 4

The chlorophyll content of a single chloroplast is about 2.5×10^{-12} g. Assuming that the chlorophyll content of a leaf is about 0.1% wet weight, calculate the number of chloroplasts in a leaf weighing 2.5 g.

Exercise 5

Estimate the percentage of intact chloroplasts from the following experimental observations. A low light-dependent oxygen evolution rate of $20\,\mu\text{mol O}_2\,\text{mg chlorophyll}^{-1}\,\text{h}^{-1}$ was obtained when an aliquot of freshly prepared chloroplasts was added to a buffered, isotonic medium. A higher rate of $200\,\mu\text{mol O}_2\,\text{mg chlorophyll}^{-1}\,\text{h}^{-1}$ was obtained, however, if the chloroplasts were first osmotically shocked in distilled water before being assayed.

Isolation of 'intact' chloroplasts involves the suspension of plant material in an isotonic, buffered medium followed by mechanical breakage of the cell wall. Cell debris is removed by filtering the homogenate through several layers of muslin. The chloroplasts are then sedimented by centrifugation at 3000g for 90 s. The degree of intactness of the chloroplast preparation can be tested by using potassium ferricyanide, which can replace $NADP^+$ as the terminal electron acceptor. However, because ferricyanide cannot penetrate the outer envelope of chloroplast, its use is restricted to 'broken' chloroplast preparations. This inability to penetrate the envelope can be used to estimate the population of intact chloroplasts within a preparation.

Inverted thylakoid membrane vesicles, with the normally inward membrane face exposed to the suspension medium, can be isolated by **phase separation**. In this procedure ruptured thylakoid membranes are allowed to reseal to form vesicles which can be either 'right side out' (that is, with the usual membrane orientation) or 'inside out'. Since the normally 'inside' thylakoid membrane surface has a different charge density to the 'outer' surface, the 'inverted' thylakoid membrane vesicles can be separated from 'right side out' vesicles. This procedure involves low-speed centrifugation (1000**g** for 3 min) following suspension in a buffered medium containing 6% (w/w) dextran and 6% (w/w) polyethylene glycol. The centrifugation separates the two phases; the inverted vesicles partition with the dextran phase. This procedure needs to be repeated several times to obtain 'clean' separation of the vesicles.

Chloroplast enzyme distribution

Like mitochondria, chloroplasts are subdivided into several distinct compartments, each with its characteristic components and functions. The **chloroplast envelope membranes** are enriched in the two galactosyl-glycerides: mono- and di-galactosyldiacylglycerol (Fig. 5.14). Although both envelope membranes contain carotenoids, neither contains chlorophyll. The outer envelope membrane is freely permeable to most ions and metabolites. The inner stromal-facing envelope is, like the inner mitochondrial membrane, highly selective. The transport of specific solutes across this membrane involves transporters. The chloroplast, unlike the mitochondrion, is not used by the cell as a major source of ATP. Indeed, the envelope membrane lacks an ATP/ADP transporter, and any ATP 'transferred' from chloroplast to cytosol crosses the envelope indirectly as triose phosphate and 3-phosphoglycerate.

The **stroma** contains the enzymes of the Calvin cycle, including the most abundant protein in the chloroplast (and purportedly the most abundant protein in nature) **ribulose bisphosphate carboxylase (rubisco)**. The stroma contains chloroplast DNA, ribosomes and the enzymes needed for the synthesis of DNA, RNA, pigments, proteins and (unlike mitochondria) lipids (although desaturation of fatty acids must occur in the cytosol).

The **lumen** of the thylakoid membrane is thought to contain few proteins, of which the best characterized are extrinsic proteins of the photosynthetic electron transport chain. These include the water-splitting enzyme of photosystem II and plastocyanin, which acts by transferring electrons from the cytochrome *b*/*f* complex to photosystem I.

The **thylakoid membrane** is rich in glycolipids, the major lipid being monogalactosyldiacylglycerol (Table 5.4). The enzymes involved in photophosphorylation are located in the thylakoid membrane. The three major components of the electron transport chain are the photosystems and the cytochrome *b*/*f* complex. The cytochrome *b*/*f* complex and the ATP synthetase are analogous in both structure and function to their mitochondria counterparts.

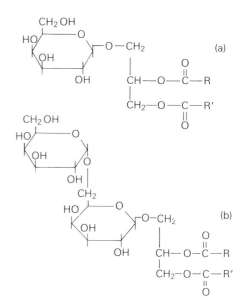

Fig. 5.14 Structure of the galactolipids. (a) Monogalactosyldiacylglycerol and (b) digalactosyldiacylglycerol. R and R' represent the fatty acid residues.

Table 5.4 *Polar lipid composition of thylakoid membranes (from spinach)*

	Outer envelope	Inner envelope	Thylakoid membrane
Lipid/protein (mg/mg)	2.5	0.8	0.4
Lipid class (% weight)			
monogalactosyldiacylglycerol	17	49	52
digalactosyldiacylglycerol	29	30	26
phosphatidylglycerol	10	8	9.5
phosphatidylcholine	32	6	4.5
phosphatidylinositol	5	1	1.5
sulphoquinovosyldiacylglycerol	6	5	6.5

Appression and photosystem II pigment proteins

Although the basic structure of chloroplasts is similar in different plants, subtle differences occur in the number and pigment composition of the photosystems, and the degree of ***appression*** of the thylakoid membrane.

Electron micrographs of chloroplasts reveal that the thylakoid membrane can be arranged as a series of flattened lamellae either tightly appressed to form stacks, called **grana**, or non-appressed as single lamellae (stromal lamellae or frets) interconnecting the grana. Figure 5.15 shows that the different regions of the thylakoid membrane have different surroundings. The non-appressed membranes at the sides of the grana (termed margins), or at the 'top' and 'bottom' of the grana stacks, or the interconnecting stromal lamellae, are in direct contact with the stroma. In contrast, the appressed lamellae within the grana stacks are not in direct contact with the stroma.

Freeze-fractured thylakoids of chloroplasts reveal four types of fracture face (Fig. 5.16). Two are associated with appressed thylakoid lamellae (EFs and PFs), and two with the non-appressed lamellae (EFu and PFu). The appressed and non-appressed thylakoid lamellae have a different distribution of intramembrane particles (Fig. 5.17). Fracture faces from appressed lamellae have a higher density of large intramembrane particles (Table 5.5). Moreover, whereas the PF particles appear to be evenly distributed throughout the thylakoid membrane, the EF particles are concentrated in the appressed lamellae. It has not yet been possible to identify the particles on the freeze-fracture micrographs to photosystems I and II, cytochrome *b*/*f* complex, or the

☐ The 1988 Nobel Prize in Chemistry was awarded to Huber, Deisenhofer and Michel for their work on the crystal structure of the reaction centre of the photosystem from the eubacterium *Rhodopseudomonas viridis*. Although this reaction centre cannot oxidize water to dioxygen (see Section 5.2) it is believed to resemble an ancestral form of photosystem II in its similarity in protein structure, amino acid sequences, ability to photoreduce quinone and sensitivity to triazine-based herbicides.

☐ The surfaces of membranes in electron micrographs of freeze-fracture thylakoids are denoted by their contact with the stroma (PS face) or lumen (ES face), and by their fractured surfaces (PF and EF). Membranes from the appressed regions of the grana are designated by the subscript 's', and those from the non-appressed region by 'u'.

Exercise 6

Calculate how many spinach leaves (average weight 5 g and having a 0.1% chlorophyll content), undergoing photosynthesis for 10 h at a rate of 100 μmol CO_2 mg chlorophyll^{-1} h^{-1}, would trap the equivalent of 10^6 J of energy. Assume that all the CO_2 is converted into sucrose (17 kJ/g).

appressed: *closely pressed together but not united.*

Reference Hall, D.O. and Rao, K.K. (1987) *Photosynthesis*, 4th edn, Edward Arnold, London, UK. An inexpensive but comprehensive account of the biochemical reactions of photosynthesis.

Fig. 5.15 Electron micrographs of chloroplast grana of *Zea mays*. (a) 'Top view' (×90 200) and (b) 'side view' (×123 000). Unappressed stromal thylakoids or frets (F) are observable in both views. Numerous frets (white arrowheads) can be seen at the top right of a granum. Since the section is only thick enough to accommodate about three granum discs, each disc must develop many connections to frets. (A few ribosomes (circled) can be seen in the stroma (S)). Reprinted from Gunning, B.E.S. and Steer, M.W. (1975) *Ultrastructure and the Biology of Plant Cells*, Edward Arnold, London, UK, with permission. Photographs courtesy of Professor M.W. Steer, University College, Dublin, Eire. (c) Representation of a granum, showing that the different regions of the thylakoid membrane have different surroundings.

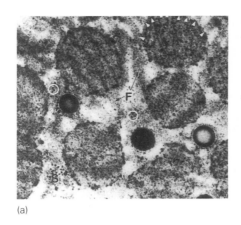

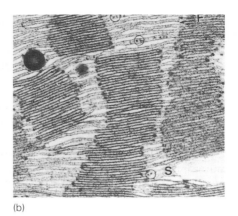

(a)

(b)

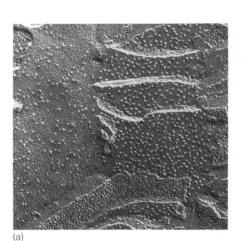

Margin membrane

End membrane — Lumen — Appressed lamellae

(c)

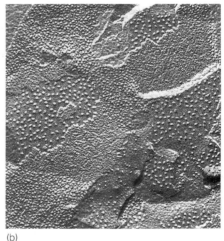

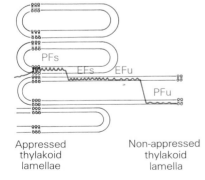

Appressed thylakoid lamellae

Non-appressed thylakoid lamella

Fig. 5.16 The four fracture faces of a thylakoid membrane.

Fig. 5.17 Freeze-fractured chloroplasts. (a) Spatial relationship between granal and stromal lamellae (×42 000). (b) The four fracture faces of the thylakoid membrane (×51 120). Courtesy of L.A. Staehelin, University of Colorado, USA.

Table 5.5 *Size distribution of intramembranous particles of thylakoid membranes*

Particle distribution	Unstacked membranes	Stacked membranes
PF particles/μm^2	3400	3600
EF particles/μm^2	570	1500
% total length	40	60
EF particles	20	80

From Staehelin, L.A. (1976) *Journal of Cell Biology*, **71**, 136–58.

membrane-spanning F_0 of the ATP synthetase. However, biochemical data indicate that photosystem II is located in the appressed regions of the thylakoid membrane and is spatially segregated from photosystem I and the ATP synthetase is located in the non-appressed thylakoid lamellae. It is therefore tempting to assign photosystem II to the EF particles. The PF particles are probably a mixture of photosystem I, cytochrome b/f complexes, CF_0 and free light-harvesting pigment proteins. The proposed heterogenous distribution of the major enzyme complexes within the thylakoid membrane is shown in Figure 5.18.

The thylakoid membrane is a **dynamic** structure, and the degree of membrane appression in isolated thylakoid membranes can be regulated by the ionic composition of the suspending medium. At low salt concentrations, the granal stacks dissociate and photosystems II and I intermingle. Under these conditions, energy absorbed by photosystem II can be transferred to photosystem I. Reformation of grana and the separation of photosystems II and I into their respective regions of the thylakoid membrane can be achieved by adding appropriate concentrations of monovalent, divalent or trivalent cations. The cations screen the charges on the surface of photosystem II causing aggregation within the plane of the membrane to form photosystem II patches. The membrane then folds up to form appressed lamellae, and photosystem I and ATP synthetase are squeezed into the non-appressed regions of the thylakoid membrane.

The **number of grana** found within a chloroplast and the number of appressed lamellae within the granum can vary with cell type. For example, algae which contain phycobilins as secondary light-harvesting pigments to photosystem II appear either to lack grana completely, or to have grana comprised of a pair of appressed membranes. The absence, or restricted size, of grana in algal chloroplasts containing phycobilins might be because phycobilisomes are extrinsic proteins sitting on the thylakoid membrane surface and probably sterically hinder membrane appression. Chloroplasts of diatoms (a class of algae) (Fig. 5.19) contain chlorophyll c as the secondary pigment and lack phycobilins. However, the thylakoid membrane can be seen as a pair of appressed membranes with a low density of granal stacks consisting of a number of appressed pairs of membrane.

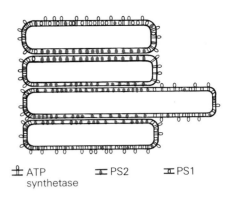

⊥⊥ ATP synthetase   ⊞ PS2   ⊞ PS1

Fig. 5.18 The putative heterogenous distribution of the membrane-embedded photosystems (PS) I and II and ATP synthetase complexes within the thylakoid membrane.

☐ Diatoms are algae of the class Bacillariophyceae within the division Chrysophyta. They are microscopic, unicellular plants whose outer cell wall is impregnated with silica.

Box 5.5
Physiological control of photosynthetic electron transport

Photosystems I and II possess different pigments, and therefore have different absorption spectra. Thus, illumination with light which is preferentially absorbed by one type of photosystem will have a serious effect on the intersystem electron transport involving plastoquinone and the cytochrome b/f complex. For example, an over-active photosystem II will ultimately lead to complete reduction of the plastoquinone pool and hence inactivation of photosystem II. However, chloroplasts can regulate the photochemical activities of photosystems II and I, via a protein kinase which transfers a phosphate group from ATP to a threonine residue at the amino-terminus of a light-harvesting chlorophyll a/b protein (LHC2). In the dephosphorylated state, LHC2 is associated with the photosystem II reaction centre in the appressed lamellae. However, when phosphorylated, the LHC2 migrates from the appressed lamellae to the non-appressed lamellae and becomes associated with photosystem I reaction centres. There may also be a decrease in the surface area of the granal stacks. The dephosphorylation of LHC2 is catalysed by a phosphoprotein phosphatase. Interestingly, the activity of the protein kinase (but not the phosphatase) is controlled by the redox state of the plastoquinone pool. Thus, the distribution of the energy of sunlight between photosystems II and I is constantly fine-tuned by an enzyme that can monitor the flow of electrons through the intersystem electron transport.

Fig. 5.19 Electron micrograph of a chloroplast from a diatom Bacillariophyceae (×13 760). Note the outer cell wall impregnated with silica and composed of two halves, one of which overlaps with the other. Courtesy of G. Beakes, Department of Biology, The University, Newcastle upon Tyne, UK.

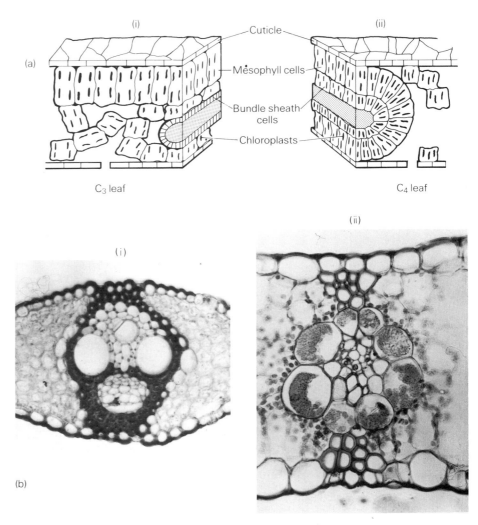

Fig. 5.20 Comparison of the anatomies of C_3 and C_4 plants. In (a, i) the C_3 leaf, palisade mesophyll cells typically form a layer in the upper portion, in (a, ii) the C_4 leaf, they are arranged in a ring around the bundle sheath (Kranz anatomy). (b) Photomicrographs showing (i) the C_3 arrangement in *Molinia caerulea* (×350) and (ii) Kranz anatomy in *Zea mays* (×350). Courtesy of M.J. Hoult, Department of Biological Sciences, The Manchester Metropolitan University, UK.

See *Biosynthesis*, Chapter 2

Not all higher plant cells have chloroplasts like that shown in Figure 5.13a. Those of the bundle sheath cells of C_4 plants (Fig. 5.20) are also devoid of appressed thylakoid lamellae (Fig. 5.21). The absence of grana is not due to the presence of phycobilisomes, but to depleted levels of photosystem II. The mesophyll cell chloroplasts of C_4 plants fix atmospheric CO_2 by its condensation with phosphoenolpyruvate to form oxaloacetate. The oxaloacetate is then reduced to malate by the oxidation of NADPH. Malate is transported to the chloroplast of the bundle sheath cell where it is decarboxylated to pyruvate which migrates back to the mesophyll cell. The CO_2 released is re-fixed by the action of rubisco. $NADP^+$ is reduced to NADPH. However, since the bundle sheath chloroplasts have no need to produce NADPH using electron transport, there is therefore little need for photosystem II. ATP generation in bundle sheath chloroplasts is the result of a cyclic electron transport process involving photosystem I. As expected, freeze-fracture electron micrographs of bundle sheath chloroplasts indicate depletion of the large EF particles found in the chloroplasts of C_3 plants.

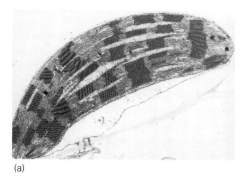 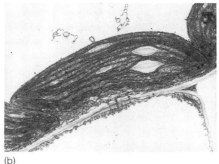

(a)   (b)

Fig. 5.21 Electron micrographs of chloroplasts from (a) a mesophyll cell (×9165) and (b) a bundle sheath cell (×9400), both of *Zea mays*. Note the differences in the structures of the thylakoid membranes between the two cells. Courtesy of Dr R. Basso, Department of Biology, Universita di Padova, Italy.

5.5 Biogenesis of mitochondria and chloroplasts

New organelles need to be continually synthesized in the cell to keep abreast of cell growth and division, or to replace those which have been degraded. Some are formed *de novo*, such as the lysosome which is made by 'pinching off' part of the Golgi apparatus. Mitochondria and chloroplasts are, however, formed by the growth and division of existing organelles. Although both organelles contain DNA, the size and gene composition of their genome is not sufficient to support an independent existence within the host cell. Their biogenesis requires the co-operation of the nuclear and organelle genomes. How the expression of the two genomes is coupled is poorly understood.

The mitochondrial genome is quite small compared with the chromosomes. It consists of a circular DNA molecule with an M_r about 12×10^6. Only about 1% of the cellular DNA is found in mammalian mitochondria. The chloroplast genome is larger, with an M_r of $85–95 \times 10^6$, and can account for as much as 15% of the total cellular DNA. Although mitochondrial and chloroplast genomes encode a relatively few proteins (only 13 in mitochondrial DNA), these organelles are able to carry out DNA replication, DNA transcription and polypeptide synthesis (Fig. 5.22). However, the protein synthesizing machinery of the organelles more closely resembles that of prokaryotes than that of eukaryotes. In fact, most of the proteins found in the mitochondria and chloroplasts are encoded in the nuclear genome and are synthesized on cytosolic ribosomes. Consequently, they have to be imported into the organelle and this is an energy-requiring process.

Most of the mitochondrial and chloroplast proteins encoded in the nuclear genome and synthesized on cytosolic ribosomes contain an **amino-terminal leader sequence**. This facilitates targeting to, and uptake by, the appropriate organelle and transfer to the relevant intra-organelle location. The amino-terminal sequence is removed during the import process by a specific peptidase in the mitochondrial matrix or the chloroplast stroma. Molecular biologists are trying to identify the residues involved in the targeting and cleavage of the precursor proteins. In one study, the first 22 residues of the cytochrome *c* oxidase subunit IV precursor were fused in-frame to cytosolic dihydrofolate reductase (DHFR). The resulting fusion protein was found to be imported into, and cleaved to 'mature' (that is, the cytosolic form of) DHFR by mitochondria. However, if only the first 12 residues of the cytochrome *c* oxidase subunit were fused to DHFR, the amino-terminal sequence was not removed because the cleavage site is no longer present in the fusion protein. It is presumed that the protein is transferred into the organelle in an unfolded state.

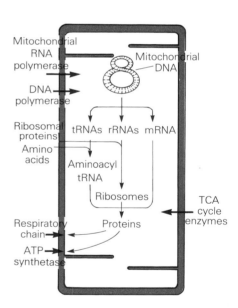

Fig. 5.22 Mitochondrial enzymes are encoded either by the organelle DNA or by nuclear DNA. Note that the operation of the mitochondrial genetic system requires proteins synthesized in the cytosol.

The division of meristematic cells results in passive segregation of the mitochondria of the mother cell, each daughter cell receiving approximately equal numbers. However, during the differentiation of the daughter cells the numbers of mitochondria increase considerably. For example, in the root cap of *Zea mays*, new cells contain only about 200 mitochondria. During subsequent growth this number increases to about 2000.

The manner of this increase is not fully understood. Newly formed meristematic cells contain numerous particles termed **initials** which have diameters of about 50 nm and are limited by a double membrane. During the growth and development of the cell, the initials rapidly increase in size and their inner membrane extends to form small folds. At this stage the initials are called **promitochondria**. The promitochondria develop into mature mitochondria.

Initials are thought to arise by budding from, or division of, existing mitochondria.

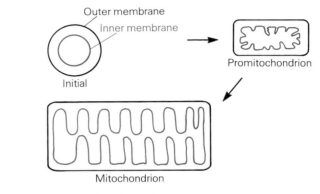

Mitochondrial development in meristematic cells.

Recent studies indicate that the subunits of some oligomeric proteins, such as rubisco of chloroplasts, cannot spontaneously assemble into their correct structural form. The primary structure of the subunits does not carry sufficient information to specify the final structure of the holoenzyme. These proteins require the post-translational assistance of a ubiquitous class of conserved proteins called *chaperonins*. These proteins, found in bacteria,

Fig. 5.23 Electron micrographs illustrating the development of chloroplasts from proplastids. (a/b), (c) and (d/e) show normal development in the light, while (b), (c'), (d'), (e'), (f') and (d/e) show the effects of, and recovery from, etiolation. (a) Meristematic cell of stem apex from oat *Avena sativa* (×5300). Proplastids (P) are clearly visible, together with other subcellular structures: nucleus (N), vacuole (V), and cuticle (C). (b) Enlarged view (×77 000) of proplastid of root tip. The envelope (circled) and invaginations of the inner membrane (small arrows) are apparent. Starch grains (S), small particles possibly plastid ribosomes (boxed) and nucleoid areas (large arrows) are present in the stroma. (c) Differentiation of the meristematic tissue of the stem apex in *A. sativa* (×8000). The plastids have developed complex internal membranes although grana have yet to form. Vacuoles (V), nuclei (N) and starch grains (S) are clearly distinguishable. Some intercellular spaces (IS) are visible. (d/e) Mature chloroplasts from *A. ventricosa* (×33 000) and *Zea mays* (×18 500) respectively. The 'side' (S) and 'top' (T) views of the grana connected together by frets (F) are strikingly visible. The envelope (E), ribosomes (circled), starch grains (G) and plastoglobuli (P) of typical chloroplasts are apparent. (c') Two etioplasts in leaves of *A. sativa* seedlings (×47 000). The envelope is arrowed (black) while the prolamellar body (PLB) can be seen to be a semicrystalline lattice. The left-hand prolamellar body (above white arrow) is obviously different in structure from the neighbouring lattice. (d') Shows the rapid effect of illumination on the morphology of the prolamellar body (×36 000). A remnant of the body is evident (PR). The plastid ribosomes occur mainly as clusters and chains suggestive of polyribosomes. Invaginations of the inner plastid envelope membrane (stars) and a nucleoid (N) are visible. (e') The effects of rapid biosynthesis within the organelle. Overlapping regions of the primary thykaloids (large arrowheads), the first stages of grana formation, are visible after 2–4 hours of illumination. The membranes of the primary thykaloids are continuous with those of the remnant of the prolamellar body (small arrows). Chains of polyribosomes are marked with asterisks (×64 000). (f') Extended illumination (10 hours) has led to the development of nearly mature chloroplasts. (×36 000). The small overlaps seen in (e') have been extended and developed giving small, but recognizable grana with interconnecting frets. A mass of plastoglobuli marks the remnants of the prolamellar body. Electron micrographs reprinted from Gunning, B.E.S. and Steer, M.W. (1975) *Ultrastructure and the Biology of Plant Cells*, Edward Arnold, London, UK, with permission. Photographs courtesy of Professor M.W. Steer, University College, Dublin, Eire.

chaperonins: protective proteins whose presence is necessary for some large proteins to fold appropriately. From the French chape, *a hood or covering protector.*

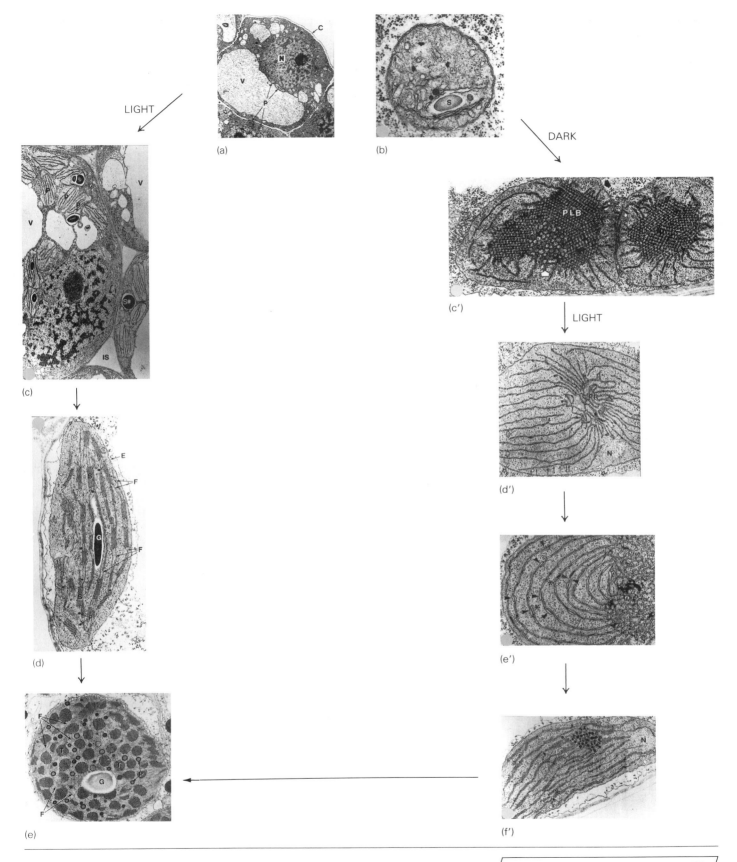

(a)

(b)

LIGHT

DARK

(c)

(c')

LIGHT

(d)

(d')

(e')

(e)

(f')

chloroplasts and mitochondria, are homologous and are thereby presumed to be functionally related to, the groE 'heat-shock' protein. In the case of the holoenzyme form of rubisco, consisting of eight large and eight small subunits, the chaperonins may assist in the folding of the monomeric form of both subunits, or in the construction of the homodimers, or in the assembly of the complete holoenzyme, or in all three stages.

The development of chloroplasts

Chloroplasts develop from proplastids (Fig. 5.23a,b) which are found in plant embryonic tissues, young tissues of roots, shoots and leaves and in the meristems of mature plants. They are smaller than chloroplasts having diameters of only 1–3 μm. Proplastids contain organelle DNA and can replicate by division, but they lack the elaborate internal membranes of chloroplasts.

Light stimulates the enlargement of proplastids and the biosynthesis of chloroplast proteins and pigments and these processes are accompanied by an expansion of the inner proplastid membrane (Fig. 5.23c). The inner membrane buds off vesicles which become arranged into stacks. Stacking of these membranous vesicles appears to be mediated by **binding proteins** associated with photosystem II. Eventually, these morphological changes bring about the formation of a mature thylakoid system containing the proteins and pigments essential for photosynthesis (Fig. 5.23d,e).

Plants subjected to extended periods of darkness become etoliated. Their recovery from etoliation has been studied extensively as a model of chloroplast development. Although these studies have helped in the understanding of normal development, recovery from etoliation differs from the usual maturation of chloroplasts in several respects.

In complete darkness, proplastids develop to form **etioplasts** (Fig. 5.23c'). Etioplasts are larger than proplastids ($4–8 \times 2–4$ μm). They lack grana, but contain a semicrystalline array of internal membranes called a **prolamellar body** (Fig. 5.23c'), which contains a yellow protochlorophyll pigment. When exposed to light, etioplasts develop rapidly (Fig. 5.23d'). The protochlorophyll is converted to chlorophyll. Other pigments, photosynthetic enzymes and components of the electron transport chain are also synthesized. These biosyntheses are paralleled by the development of the usual internal chloroplast structure. The prolamellar body loses its crystalline nature and is changed into extended flattened sacs called **primary thylakoids**, which have overlapping portions. These stages, intermediate between etioplasts and chloroplasts (Fig. 5.23d',e) are often called **etiochloroplasts**. The initial overlapping sections become more extensive (Fig. 5.23f'), eventually becoming fully developed granal stacks (Fig. 5.23d,e).

5.6 Evolutionary origin of mitochondria and chloroplasts

According to the *endosymbiont* hypothesis, mitochondria and chloroplasts are the evolutionary descendants of free-living prokaryotes, which during evolution have been stripped of practically all functions except those related to energy transduction. Some prokaryotic organisms have features similar to those of energy-transducing organelles: *Paracoccus denitrificans* has an electron transport chain like that in mitochondria, and *Prochloron* has a photosynthetic electron transport chain (with chlorophyll *b*) like that found in chloroplasts from the higher plants.

It is envisaged that the modern eukaryotic animal cell arose from the

endosymbiont: symbiosis (Greek symbios, a companion) is when two species live together to their mutual benefit. An endosymbiont lives within its companion.

Reference Gunning, B.E.S. and Steer, M.W. (1975) *Ultrastructure and Biology of Plant Cells*, Edward Arnold, London, UK, pp. 109–11, 130–1, 249–59. Marvellously illustrated textbook, full of splendid electron micrographs.

emergence first of a nucleated cell dependent on glycolytic fermentation for its ATP supply. This was followed by the emergence of an anucleated cell, possessing a respiratory chain and therefore a more efficient method of producing ATP. Early in the evolutionary timescale, the nucleated cell either absorbed, or was invaded by, an anucleated cell. Initially, a stable endosymbiotic relationship was established which conferred advantages over the free-living nucleated cells or anucleated cells. Since that time, the nuclear genome has slowly absorbed (or been infiltrated by) most of the genome of the anucleated organism (Fig. 5.24). The eukaryotic plant cell is probably a descendant of this cell, which some time early on in evolution, absorbed a second prokaryotic cell capable of carrying out photophosphorylation.

An alternative hypothesis is that the ancestral cell could carry out oxidative phosphorylation, but its genomic material was contained in a number of plasmids. Modern prokaryotic cells followed from the integration of the plasmid DNA into a single chromosome, whereas modern eukaryotic cells followed from the segregation of the genes into separate membrane-enveloped compartments.

Modern mitochondria and chloroplasts are not, as Altmann suggested, capable of independent existence outside of the host cell. However, it is interesting to speculate as to why these energy-transducing organelles retained so little of their genetic material. To quote from Thomas's, *The Lives of a Cell*, on mitochondria (but equally applicable to chloroplasts): 'instead of evolving as we have done, manufacturing longer and ever more elaborately longer strands of DNA and running ever increasing risks of mutating into evolutionary cul-de-sacs, they elected to stay small and stick to one line of work. To accomplish this, and to ensure themselves the longest possible run, they have got themselves inside all the rest of us'. It may be more appropriate to consider mitochondria and chloroplasts not as servants to the nucleus master, but as non-predatory parasites within the cell!

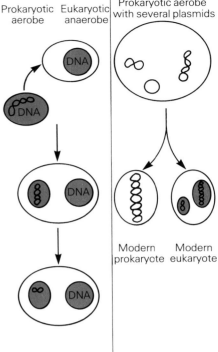

Fig. 5.24 Possible evolutionary origin of mitochondria. (a) Endosymbiont hypothesis, (b) cluster-clone hypothesis.

5.7 *Overview*

A basic tenet of molecular biology is that biological function depends on structure. This is obviously true for a relatively simple protein such as cytochrome *c* with its clearly defined activity in electron transport. However, it also holds for complex structures, such as the mitochondrion or chloroplast. For example, the degree of convolution of the internal 'energy-transducing' membranes in these organelles correlates with the metabolic activity of the cell. Thus, muscle mitochondria differ from liver mitochondria in having more cristae, and bundle sheath chloroplasts have fewer granal stacks than mesophyll chloroplasts in C_4 plants. The chemical structure of the internal membrane is important because it has to store energy (as an electrochemical potential gradient of protons across itself), and to use this energy in measured amounts to synthesize ATP. Thus, the inner mitochondrial membrane is impermeable to protons but does permit the import of cytosolically synthesized proteins.

Answers to Exercises

1. 7214 mitochondria per liver cell.
2. Permeable to SCN$^-$ but not to Cl$^-$.
3. Dicarboxylate and tricarboxylate carriers.
4. 10^9 chloroplasts per leaf.
5. 90% intact.
6. 406 leaves.

FILL IN THE BLANKS

1. The mitochondrion has two membranes: the _____ membrane, which is permeable to small ions, and the _____ membrane, which is impermeable to small ions. The internal aqueous phase is termed the _____ and contains the enzymes of the _____ cycle. The _____ membrane contains the electron transport chain, the ATP synthetase and several _____ . Submitochondrial particles have a _____ membrane orientation.

The chloroplast is a _____ found in plants. The function of the chloroplast is to assimilate carbon dioxide into carbohydrate in a reaction process called _____ . This consists of two sets of reactions. The initial _____ reactions use the energy of _____ to transfer hydrogen atoms from _____ to $NADP^+$, and to generate ATP. The enzymes of this set of reactions are embedded within the _____ membrane.

Choose from: inner (2 occurrences), light (2 occurrences), matrix, outer, photosynthesis, plastid, TCA, thylakoid, transporters, transverse, water.

MULTIPLE-CHOICE QUESTIONS

Which of the following statements are true or false:
2. The inner membrane of mitochondria:
A. appears smooth on freeze-etch electron micrographs.
B. contains the protein porin.
C. is permeable to protons.
D. binds the 9 nm diameter coupling factors.
E. is rich in the lipid, bisphosphatidylglycerol.

3. The chloroplast of a C_3 plant:
A. has a thylakoid membrane which can be appressed to form granal stacks.
B. is similar to chloroplasts from the mesophyll cells of C_4 plants.
C. possesses a single envelope membrane.
D. can synthesize all its own proteins.
E. is larger than a typical mitochondrion.

SHORT-ANSWER QUESTIONS

4. Draw a mitochondrion. Label the locations of the major enzymes involved in the degradation of pyruvate.

5. List the differences in the structure and function of chloroplasts of mesophyll and bundle sheath cells.

6

The cytoskeleton

Objectives

After reading this chapter you should be able to:

☐ describe the methods which have been used to study the cytoskeleton;

☐ explain the roles of microfilaments in non-muscle and muscle cells;

☐ appreciate that microfilaments and microtubules are highly conserved and ubiquitous labile aggregates in eukaryotes;

☐ appreciate that intermediate filaments are relatively stable but vary in composition according to the tissue in which they are found;

☐ outline the range of structures formed by aggregated microtubules and their different functions in the cell;

☐ describe the mechanisms of cellular movement and how these contribute to embryonic development.

6.1 Introduction

The cytosol of eukaryotic cells is supported by a cytoskeleton formed from three highly organized filamentous protein networks. These are composed of **microfilaments**, **intermediate filaments** and **microtubules**. They are distinguished by their appearance and dimensions when examined using electron microscopy, and by differences in their chemical nature including their reactions to specific antibodies. Prokaryotes (excepting mycoplasma) do not possess such a cytoskeleton.

See Chapter 2

Microfilaments are 6–7 nm in diameter and are composed of the protein **actin**. They are found in all eukaryotic cells, usually in association with a second protein, **myosin**. Intermediate filaments are 8–11 nm in diameter; they are intermediate in diameter between microfilaments and microtubules. In different cell types intermediate filaments are composed of different proteins. Thus ***vimentin***, ***desmin***, **glial fibrillary acidic protein** (GFAP), three **neurofilament** proteins (NF) and over 20 **cytokeratin** proteins have been detected in different intermediate filaments. Usually one cell type contains only one type of intermediate filament, although there are exceptions. Microtubules appear as double-edged hollow tubules, 25 nm in diameter. They are made up of subunits of the protein **tubulin**, and are found in all eukaryote cell types.

These three filamentous networks perform the functions of a skeleton for the cell, which are to allow movement and to provide support. Cell movement includes changes in positions of organelles within the cytosol, changes in shape of the cell as a whole and migration of the cell. Filamentous elements of the cytoskeleton are long and often extend right across the cell so that they can spread forces of stress throughout the cytoplasm.

vimentin: *an intermediate filament protein. From the Latin* vimineus, *made of strong pliable strands.*

desmin: *an intermediate filament protein that binds microfilaments together. From the Greek* desmo, *bond ligament.*

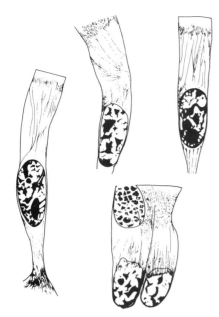

Fig. 6.1 A drawing made in 1895 of frog intestinal epithelial cells clearly contains cytoplasmic filaments. Reproduced by permission of Rockefeller University Press, from Porter, K.R. (1984) The cytomatrix: a short history of its study. *Journal of Cell Biology*, **9**, 3s–12s.

6.2 A brief history

In the last 150 years different kinds of microscopy have provided extensive evidence for the existence of a cytoskeleton. Early in the nineteenth century many scientists using primitive microscopes examined both animal and plant cells and probed cytoplasm with dissecting needles. The cytoplasmic matrix was described as a 'clear, glutinous, diaphanous substance', which stuck to dissecting tools like mucus. Cytoplasmic filaments were identified in 1895 (Fig. 6.1). In 1908 Dahlgren tried to visualize the very viscous organized nature of protoplasm by means of a diagram (Fig. 6.2).

Experiments showed that cytoplasm could be manipulated and organelles pushed aside with needles, but that after a while the organelles returned to their original positions. During centrifugation of cells organelles appeared to move freely in a cytoplasm which seemed less viscous. After a brief period the cytoplasm became viscous again and organelle movement was restricted. These experiments revealed two properties of cytoplasm, namely the ability to change its viscosity and the ability to recoil after deformation, that is, elasticity.

The use of the transmission electron microscope in the 1950s enabled the filamentous arrangement of muscle fibres to be appreciated, and later provided the first evidence that there were three main components of the cytoskeleton in non-muscle cells. These were distinguished by their different diameters in thin sections of cells. Other electron microscope techniques have provided three-dimensional views of the cytoskeleton.

Box 6.1
The contribution made by electron microscopy

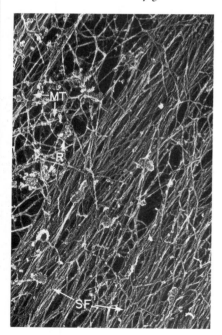

In conventional transmission electron microscopy (see Section 1.2, Boxes 5.2 and 7.2) the electron beam is accelerated using a potential of 40–100 kV. This enables the beam to pass only through those specimens that are less than 0.2 μm thick. High-voltage electron microscopy (HVEM) accelerates electrons to 1000 kV enabling electrons to pass through specimens as much as 1–2 μm thick. If cells are spread thinly whilst growing in culture, they do not need to be cut into sections for observation by HVEM.

For HVEM, cells are frozen in liquid propane at −185°C, left unfixed or fixed in glutaraldehyde, then dried under a vacuum while still frozen (freeze-dried). The cytoskeleton appears as a *three-dimensional* lattice, composed not only of the three main cytoskeletal elements, but also of fibres 3–15 nm in diameter called microtrabeculae. It is not certain whether these are real structures or artefacts produced during cell preparation.

Similar studies can be made in the conventional transmission electron microscope using ion-etching. Cells are frozen, then etched by an ion beam to remove overlying cytoplasm. The underlying material, including the cytoskeleton is coated with a layer of carbon. The cellular material is then dissolved away in acid leaving the carbon layer as a replica of the ion-etched surface. Further coating of the replica with a heavy metal directed from an angle, casts shadows which give contrast and a three-dimensional appearance to the replica.

The cytoskeleton viewed as a platinum-shadowed replica of a freeze-dried cell. MT, microtubules; SF, stress fibres; R, ribosomes. Reprinted from Heuser, J.E. and Kirschner, M. (1980). Filament organization revealed in platinum replicas of freeze-dried cytoskeletons (1980). *Journal of Cell Biology*, **86**, 212–34. Photograph courtesy of Dr M. Kirschner.

6.3 The isolation and characterization of cytoskeletal proteins

Actin and myosin, the two main protein components of microfilaments in striated muscle, were isolated as long ago as 1941. By 1979, several proteins of striated muscle had been found also in non-muscle cells. These proteins included actin, myosin, tropomyosin (but not troponin), α-actinin, fascin and **gelsolin**. Since 1980 there has been a tremendous increase in the number of new proteins isolated from non-muscle cells and characterized by their association with the cytoskeleton.

No doubt many more such proteins remain to be discovered. Proteins occurring in different tissues often share similar functions. Some of them may be found in striated and smooth muscle and non-muscle cells while others may be restricted to non-muscle cells. Unfortunately the nomenclature is non-systematic and idiosyncratic.

Biochemical studies on cytoskeletal proteins

Although it is relatively easy to identify enzymes and other proteins with biological activity, it is difficult to identify structural proteins if they have no measurable biological activity. Cytoskeletal proteins have been extracted from cells using detergents of various kinds, sometimes followed by other procedures such as solubilization with $6\,mol\,dm^{-3}$ guanidinium chloride. Once solubilized they are typically identified by their apparent M_r or isoelectric points as measured by gel electrophoresis (Fig. 6.3) and may be further investigated using antibodies (Fig. 6.4). Antibodies may be used to identify, and then to localize, the protein antigen to which they were raised.

Some molecules which are not antibodies (for example, cytochalasin B, phalloidin, colchicine, taxol), nevertheless bind specifically to cytoskeletal proteins. These may be used to study movement *in vivo* by following changes in their intracellular distribution.

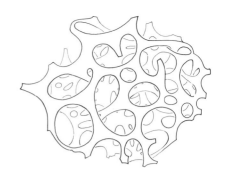

Fig. 6.2 Drawing made in 1908 to illustrate the reticular theory of protoplasm; it closely resembles the photographs of living cytoplasm taken on far more sophisticated microscopes in the 1970s. Reproduced by permission of Rockefeller University Press, from Porter, K.R. (1984) The cytomatrix: a short history of its study. *Journal of Cell Biology*, **9**, 3s–12s.

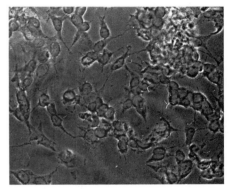

(a)

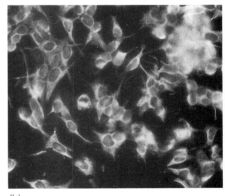

(b)

Fig. 6.4 Tissue-cultured cells from a muscle tumour. (a) Viewed in the phase-contrast light microscope. (b) The same cells stained for vimentin using an antivimentin antibody bound to fluorescein and viewed in the fluorescence microscope.

Fig. 6.3 Two-dimensional SDS-polyacrylamide gel electrophoresis (SDS-PAGE) of a whole cell extract prepared from endothelial cells (courtesy of Mr M. Clarke, Christie Hospital, Manchester, UK). SDS renders proteins more soluble by conferring negative charges on them. The electrophoresis separates proteins in the first dimension by charge (isoelectric focusing, left-to-right in the illustration) and in the second dimension according to relative molecular masses because of the sieving effect of the gel (top-to-bottom in the illustration).

gelsolin: *a protein which polymerizes soluble G-actin into insoluble F-actin gel. From the Latin* gelare, *to congeal and* solutus, *dissolved.*

Reference Taylor, D.L. and Wang, Y.-L. (1980) Fluorescently labelled molecules as probes of the structure and function of living cells. *Nature*, **284**, 405–10. Interesting aspects of methods used to study the cytoskeleton. Useful for reference.

Box 6.2
Specific antibodies may be used to determine the distribution of a cytoskeletal protein

Cytoskeletal proteins have no intrinsic biological activity by which they may be identified. Instead, specific antibodies with attached marker molecules may be used to locate them.

Antibodies are first raised to a purified cytoskeletal protein. Then in order to be made visible when bound to the protein against which it was raised, an antibody must be bound to a marker molecule. Marker molecules used for light microscopy include fluorescent dyes (such as fluorescein or rhodamine) or enzymes (for example peroxidase, alkaline phosphatase or β-galactosidase), which can catalyse the formation of a coloured reaction product.

For electron microscopy the markers must be electron dense. Examples of electron-dense markers are ferritin, an iron-containing protein, colloidal gold, or enzymes (for example, peroxidase) which react with their substrates to produce an electron-dense product.

The distribution of the antigen in other cells and tissues can be determined by using the antibody to stain the cells and tissues. These procedures are called **immunocytochemistry** and **immunohistochemistry**, respectively.

In the brain, for example, the intermediate filament protein, glial fibrillary acidic protein (GFAP) is found in **astrocytes** but not neurons, whereas neurofilament protein (NFP) is characteristic of neurons but not astrocytes. A particular antibody, e.g. anti-NFP, can then be used to demonstrate the neuronal nature of uncharacterized cells in other tissues.

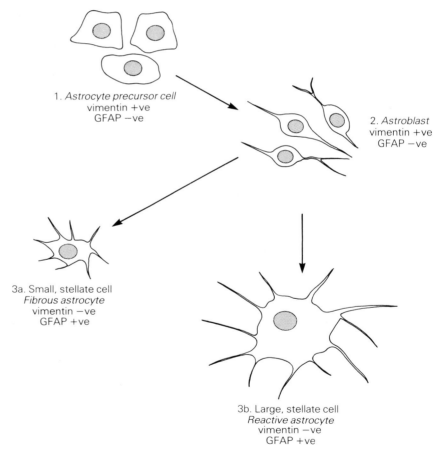

1. *Astrocyte precursor cell*
vimentin +ve
GFAP −ve

2. *Astroblast*
vimentin +ve
GFAP −ve

3a. Small, stellate cell
Fibrous astrocyte
vimentin −ve
GFAP +ve

3b. Large, stellate cell
Reactive astrocyte
vimentin −ve
GFAP +ve

The changes in cell shape during astrocyte differentiation. The expression of intermediate filament proteins (vimentin and glial fibrillary acidic protein (GFAP)) changes during differentiation as shown by stages 1 to 3a/b.

immunohistochemistry: *the study of antigens, antibodies and their interactions in tissues. From the Latin* immunes, *free from, and the Greek* histos, *tissue and* chemeia, *alchemy.*

astrocyte: *a specialized supporting cell with branching processes, found in the central nervous system. From the Greek* astros, *star and* kytos, *cell.*

Cytoskeletal proteins vary during development and it is important to know at which stage of fetal or post-natal development a particular protein appears in a cell or disappears from it. For example, a switch occurs from a fetal type of myosin to an adult type during muscle development. This was realized when an antibody with specificity for adult muscle did not stain fetal muscle tissue and *vice versa*.

The disaggregation of tissues into individual cells for growth in tissue culture *in vitro* enables a study of alterations in the cytoskeleton to be made during differentiation and indicates how this may be related to changes in cell shape or motility.

Astrocyte precursor cells in fetal brain are capable of dividing. In culture they are flat and polygonal like epithelial cells and contain vimentin intermediate filaments. After some time in culture they become astroblasts, less flat with long projections but also containing vimentin. Eventually astroblasts develop into stellate cells containing not vimentin but glial fibrillary acidic protein (GFAP) (the equivalent of mature astrocytes *in vivo*). Factors affecting the 'switch off' of vimentin production and 'switch on' of GFAP may be studied by adding to the culture substances which induce differentiation.

Box 6.3
Microinjection techniques

Cytoskeletal proteins tagged with relatively non-toxic fluorescent dyes such as fluorescein may be injected into living cells to study dynamic movements of the cytoskeleton. A video camera may be attached to the light microscope to film the movements of such tagged molecules inside a cell. Colloidal gold is a useful marker in the video camera system, because it appears dense and black on the TV screen in contrast to the grey of the cytoplasm. Such minute amounts of injected material necessitate TV image intensification technology.

Often, the function of a particular protein suggested by experiments *in vitro*, can be confirmed to be the same as that *in vivo* by injecting the protein into a living cell in this way.

The influence of molecular biology

Sometimes it is possible to isolate and purify cytoskeletal proteins to homogeneity despite their lack of easily measurable biological activity. In such cases the proteins may be sequenced and subjected to other investigations in order to reveal their structure. In many cases, however, this has not been possible and for some of the cytoskeletal proteins it is only the development of the techniques of molecular biology in the last few years that has enabled their sequences to be determined. For example, it was not possible to purify keratin polypeptides. Eventually mRNA coding for keratins was isolated and used to manufacture cDNA. This was cloned in bacterial plasmids and clones containing keratin genes isolated. This DNA was then sequenced and the cDNA 'translated' into the amino acid sequence by reference to the genetic code.

See *Molecular Biology and Biotechnology*, Chapter 9

The amino acid sequences of many cytokeratins are now known, even though the proteins themselves have never been isolated in pure form. Even possessing the amino acid sequence has so far not allowed many useful predictions of the secondary and tertiary structures of the proteins *in vivo*.

6.4 Microfilaments

Actin is found in every eukaryotic cell type. In Protozoa, slime moulds, nervous tissue or blood platelets, it can represent as much as 10–20% of the total cellular protein. Six kinds of actin are known. All are globular proteins,

Reference Birchmeier, W. (1984) Cytoskeleton structure and function. *Trends in Biochemical Sciences*, **9**, 192–5. Covers research into the cytoskeleton between 1975 and 1984.

5.5 nm in diameter and have an M_r of 42 000. Four different α-actins are known from striated, cardiac, smooth vascular and smooth enteric muscle tissues. The other two actins β and γ, are found in non-muscle cells, with every cell type having its own characteristic ratio of β to γ actin. The amino acid sequences of the different actins from different species are nearly identical, showing that actin genes have been highly conserved during evolution. Muscle α-actins differ from one other by only four amino acid residues while four β- and γ-actins differ from α by only 24 or 25 amino acid residues of a total of 374 in the whole molecule. The presence of the modified amino acid residue, **3-methylhistidine** is characteristic of all kinds of actin and may be used as a marker in its structural identification (Fig. 6.5).

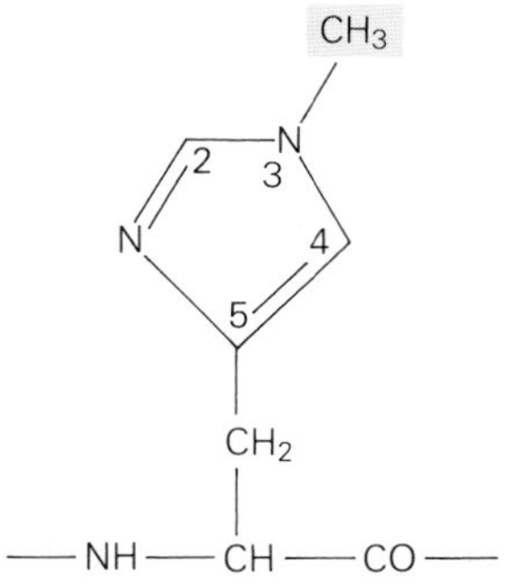

Soluble **globular (G) actin** monomers are capable of polymerizing to form insoluble **filamentous (F) actin** both *in vivo* and *in vitro*. F-actin as viewed in the transmission electron microscope is a beaded double right-handed helical microfilament with a repeat of 37 nm and a diameter of 7 nm (Fig. 6.6). One model suggests that the filament is a double helix while another suggests it is single. All beads are identical G-actin monomers, which are polymerized head-to-tail. The head-to-tail arrangement of monomers imparts a polarization to the actin filament which is reflected in its treadmilling.

Fig. 6.5 N-γ-methylhistidine.

Box 6.4
Effects of drugs on microfilaments

It is difficult to demonstrate an equilibrium between G- and F-actin in living cells because the equilibrium is dynamic. However, two drugs with different mechanisms of action on microfilaments have been widely used to study actin polymerization and depolymerization. The fungal products called **cytochalasins**, particularly cytochalasin B, have a specific inhibitory effect on G-actin polymerization, inhibiting cell movement, cell division, phagocytosis and formation of microspikes. Microspikes are made of cytoplasm extending outwards as part of a larger pseudopodium. However, these compounds do not affect mitosis or muscle contraction. In contrast a fungal alkaloid, **phalloidin**, specifically inhibits depolymerization of F-actin and stabilizes microfilaments. Living cells cannot take up phalloidin but once injected it inhibits locomotion of amoebae and vertebrate cultured cells. It has a second use, in that injected fluorescently labelled phalloidin beautifully highlights the arrangement of microfilaments in cells.

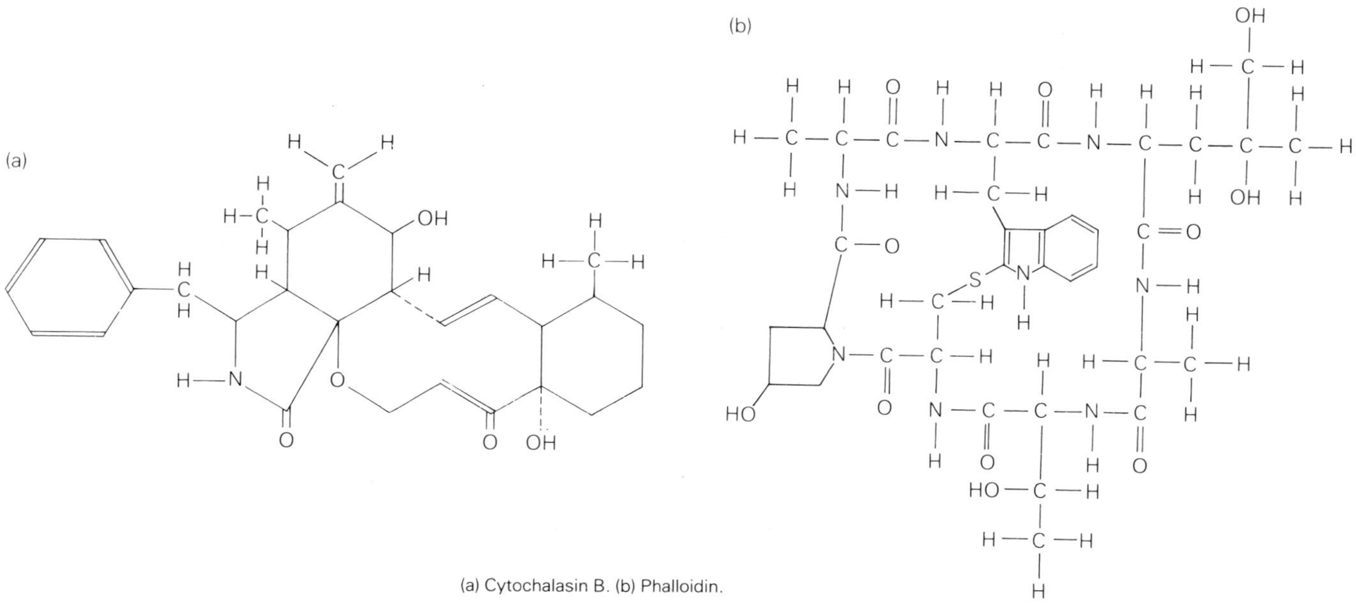

(a) Cytochalasin B. (b) Phalloidin.

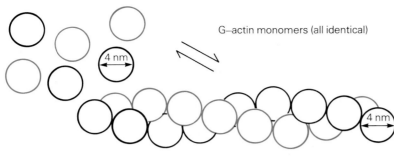

G–actin monomers (all identical)

4 nm

4 nm

F–actin polymer = 2 intertwined chains of G–actin monomers
(13 monomers in 1 complete turn)

Fig. 6.6 The structures of G- and F-actin.

□ Treadmilling is the process by which a
monomer incorporated into one end of an
F-actin filament by polymerization moves
(treadmills) along the filament until it is
eventually lost at the other end by
depolymerization. An equilibrium exists
between polymerization at one end
(dependent on the concentration of free
monomers) and depolymerization at the
other end (also probably dependent on
free monomer concentration) (Fig. 6.7).
This change in position of the growing and
disintegrating ends of actin filaments
allows some flexibility in the shape and
position of actin filament networks within
a cell.

Mg^{2+}-dependent G- to F-actin polymerization is possible only at one end
while Ca^{2+}-dependent depolymerization can only occur at the other end.
These are called the (+) and (−) ends respectively (Fig. 6.7). Treadmilling
is controlled by actin-binding proteins and is linked to the hydrolysis of ATP
to ADP.

(a)   (−) end   (+) end

(b)   (−) end

(+) end

(c)   (−) end

(+) end

Fig. 6.7 Treadmilling, showing the relationship between (+) and (−) ends of F-actin filaments. The monomers
highlighted gradually 'treadmill' along the polymer away from the (+) end towards the (−) end where they will
eventually be depolymerized.

Myosin in non-muscle cells

Myosin is an asymmetric hexamer M_r 500 000. There are two heavy chains
(each M_r 230 000) and four light chains (M_r 16–20 000). Two of the light chains
were originally called 'essential' because their presence was thought to be
necessary for ATPase action of the myosin head while the other two were
called 'regulatory' as they seemed to regulate the Ca^{2+}-mediated binding

of calmodulin to myosin. It is now known that essential light chains are not needed for ATPase activity and that all four light chains interact to regulate the binding of Ca^{2+}–calmodulin.

Myosin molecules have the same basic structure whether they occur in muscle or non-muscle tissue. Like actins, they vary only slightly in amino acid sequence.

Starting from the carboxy-terminal ends, about 50% of the pair of heavy chains are twined about each other in a coiled coil of α-helices called the **rod** (see Fig. 6.8). These rods contain the binding sites for myosin molecules so that bipolar filaments with a diameter of 8–12 nm can form. Assembly into bipolar filaments, necessary for the action of myosin *in vivo*, is thought to involve phosphorylation of the light chains.

The other half of each heavy chain coils with one essential light chain and one regulatory light chain to form a globular head. There are two globular heads in every myosin molecule. Each head region contains a site for binding ATP, with associated ATPase activity, an actin-binding site and other sites which bind divalent cations such as Ca^{2+}.

Actin–myosin interactions in non-muscle cells

A great deal is known about the interaction of actin with myosin in skeletal muscle tissue. The mechanism of contraction in non-muscle tissues may be similar to that in skeletal muscle in that it is activated by Ca^{2+} and involves the

---

Box 6.5
Investigation of actin filament distribution using myosin fragments

Myosin subfragments may be formed by digestion with proteolytic enzymes. Trypsin cleaves myosin between the rods (called light meromyosin, LMM) and the heads. A head with a rod connecting piece and two light myosin chains is called heavy meromyosin (HMM). Digestion with papain results in the heads plus light chains splitting clearly from the rod or tail. The former are called subfragment-1 (S-1). Subfragment-2 is the rod connecting piece which forms part of HMM (Fig. 6.8).

Both HMM and S-1 myosin subfragments have been used in binding studies *in vitro*. Microfilaments that bind to myosin subfragments may be identified as actin. The structure of the actin–myosin binding complex and its relevance to cytoplasmic contraction *in vivo* may be investigated in this way as well.

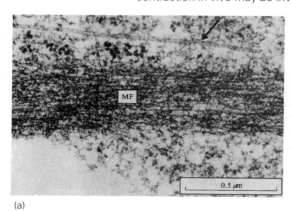

(a)

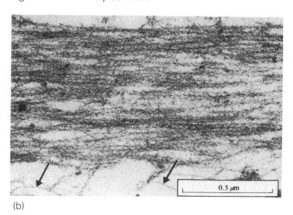

(b)

(a) Electron micrograph of actin microfilaments (MF) in a mouse 3T3 fibroblast. The arrow denotes a microtubule. Reproduced by permission of Academic Press, from Goldman, R.D. *et al.* (1975). The distribution of actin in non-muscle cells. *Experimental Cell Research*, **90**, 333–44. (b) A mouse 3T3 cell treated with heavy meromyosin (HMM) fragments giving microfilaments a barbed appearance. Intermediate filaments do not bind HMM (arrowed). Reproduced by copyright permission of the Histochemical Society, from Goldman, R.D. (1975) The use of heavy meromyosin binding as an ultrastructural cytochemical method for localizing and determining the possible functions of actin-like microfilaments in nonmuscle cells. *Journal of Histochemistry and Cytochemistry*, **23**, 529.

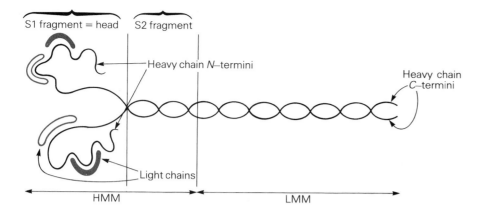

Fig. 6.8 Myosin. Cleavage sites for heavy meromyosin (HMM), light meromyosin (LMM) and subfragment 1 (S1) are indicated.

Table 6.1 *A comparison of microfilament proteins in muscle and non-muscle cells*

Striated muscle tissue	Non-muscle cells
Contains α-actin	Contain β- and γ-actins
Actin filaments stable and permanent	Actin filaments unstable—treadmilling
Actin filaments in a rigid network	Actin filament network changes shape
Contains myosin	Contain myosin
Contains tropomyosin and troponin	Contain tropomyosin, calmodulin but not troponin
Ca^{2+} regulate contraction	Ca^{2+} regulate contraction
Myosin bound to actin in definite sarcomeres	Myosin bound to actin in structures reminiscent of sarcomeres
Binding of actin to myosin is strong	Binding of actin to myosin is weak and may need a protein of M_r 110 000 as a co-factor
Actin binding stimulates myosin ATPase activity 50–200 fold	Actin binding stimulates myosin ATPase activity only 2–5 fold

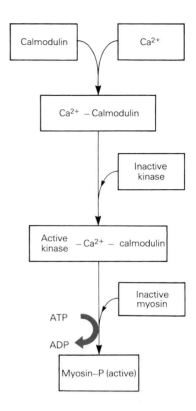

Fig. 6.9 Calmodulin regulation of actin–myosin interaction.

binding of tropomyosin to actin. However, in non-muscle cells tropomyosin–actin binding is regulated, not by troponin as in skeletal muscle, but by calmodulin (Table 6.1, Figs 6.9 and 6.10). The binding of actin to myosin appears to be much weaker than that in skeletal muscle; and the non-muscle cell cytoskeleton, unlike that in skeletal muscle, is not arranged for maximum efficiency of contraction. Nevertheless, structures reminiscent of muscle sarcomeres may be observed by electron microscopy of non-muscle cells using antibodies specific to actin and myosin subfragments.

Microfilaments in both muscle and non-muscle cells are closely associated with other molecules called **microfilament accessory proteins**. These are mainly actin-binding proteins necessary to effect changes in the molecular forms of actin in a cell and can be classified into four groups according to their functions (Table 6.2).

GROUP 1 PROTEINS INFLUENCE ACTIN–MYOSIN INTERACTIONS. Non-muscle cells seem to contain most of the proteins found in striated muscle, with the exception of troponin. Troponin is replaced in non-muscle cells by the Ca^{2+}-binding proteins, calmodulin, and possibly caldesmon.

GROUP 2 PROTEINS REGULATE POLYMERIZATION OF G-ACTIN TO F-ACTIN. Polymerization or depolymerization is controlled by the protein gelsolin and Ca^{2+}. This is considered to be the mechanism for reversible conversion of endoplasm to more viscous ectoplasm in amoebae to produce

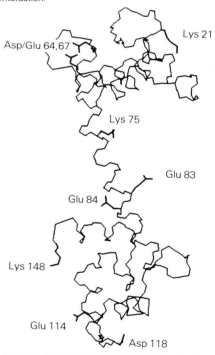

Fig. 6.10 A molecular model of calmodulin. Courtesy of Dr K. Brew, Department of Biochemistry, University of Miami, USA.

movement (Fig. 6.11). Directional movement of the animal may depend on uneven Ca^{2+} concentrations in the cytosol. Low concentrations would increase the viscosity (endoplasm to ectoplasm) and cause a pseudopod to extend, while high concentrations would decrease the viscosity (ectoplasm to endoplasm) and cause pseudopod retraction.

Table 6.2 *Actin-binding proteins in vertebrate non-muscle cells*

	M_r	Function
Group 1		
Myosin	500 000	May cause actin filaments to move (as in muscle)
Tropomyosin	70 000	Regulates binding of actin to myosin heads
Caldesmon	140 000	
Group 2		
Gelsolin	91 000	Polymerizes actin monomers into filaments. Low [Ca²⁺] Depolymerizes actin filaments into monomers. High [Ca²⁺]
Capping proteins	various	Bind to one end of filament, preventing loss or addition of actin monomers
Profilin	16 000	Binds to actin monomers to prevent polymerization (inhibits nucleation)
Group 3		
$\alpha; \beta$-Spectrin	260 000; 225 000	
Filamin	250 000; 250 000	Proteins like spectrin which cross-link adjacent
Fodrin	265 000; 260 000	actin filaments
TW 260/240	260 000; 240 000	
Fimbrin	68 000	Cross-links actin filaments to form parallel actin fibres
Fascin	58 000	Cross-links actin filaments to form parallel actin fibres
Villin	95 000	Cross-links actin filaments to form parallel actin fibres
Group 4		
α-Actinin	100 000	Binds actin filaments to cell membrane. Cross-links actin filaments?
β-Actinin		
Talin	225 000	Binds ends of actin filaments to cell membrane
Vinculin	130 000	Binds ends of actin filaments to cell membrane
+ Meta-vinculin		
Microvillus protein	110 000	Links sides of actin filaments to microvilli membranes
+ Calmodulin	17 000	
Synapsin I	82 000	Binds actin filament bundles to synaptic vesicles

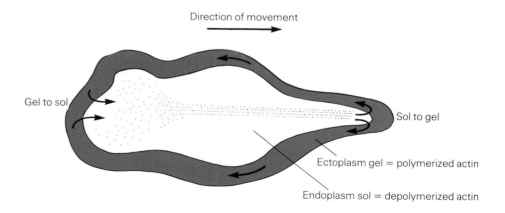

Fig. 6.11 Amoeboid movement.

GROUP 3 PROTEINS ARE CROSS-LINKING PROTEINS OF F-ACTIN FILAMENTS. Actin filaments cross-linked into **stress fibres** make important skeletal struts (Fig. 6.12). These appear to be lost and reformed during movement of *in vitro* cultured cells. Myosin, tropomyosin and α-actinin are arranged along the stress fibres in a manner reminiscent of striated muscle sarcomeres (Table 6.1 and Fig. 6.13). Movements of pseudopodia at the

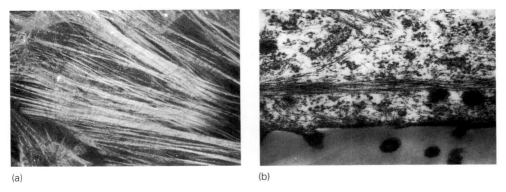

(a)

(b)

Fig. 6.12 Actin stress fibres in tissue-cultured endothelial cells. (a) Stained using an antibody bound to fluorescein and viewed in a fluorescence microscope. Courtesy of Dr S. Kumar, Christie Hospital, Manchester, UK. (b) An electron micrograph revealing a broad band of microfilaments near the cell surface.

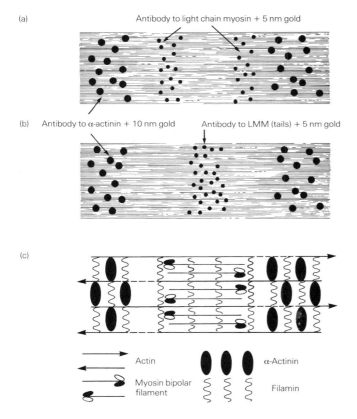

Fig. 6.13 (a) and (b) Drawing of stress fibres in non-muscle cells showing the localization of antibodies specific to light chain myosin (heads), light meromyosin (tails) and α-actinin. (c) The distribution of proteins in non-muscle deduced from patterns similar to those in (a) and (b). Staining with antibodies at the electron microscope level revealed the distribution of different muscle proteins and particularly the head—tail arrangement of myosin molecules. This allowed (c) to be drawn up.

leading edge of moving cells *in vitro* and of the retraction fibres at the trailing edge all need microfilament contraction. The free energy necessary for contraction is supplied by the hydrolysis of ATP.

The long microvilli of the intestinal brush border have cores of actin filament bundles stiffened by the proteins villin and fimbrin. Actin is arranged with its barbed ends towards the tips (Fig. 6.14). Myosin is found only at the base of microvilli in the apical cytoplasm, where it provides a mechanism for movement of the microvilli.

A pool of unpolymerized actin molecules (kept together by spectrin, in the absence of a limiting membrane) lies in the perinuclear cap in *Thyone* sperm,

☐ When heavy meromyosin binds to actin filaments the complex resembles the flights on an arrow $(+) \rightarrow \rightarrow \rightarrow \rightarrow (-)$. One end of the filament is pointed and corresponds to the $(-)$ end while the other is barbed and corresponds to the $(+)$ end.

(a)

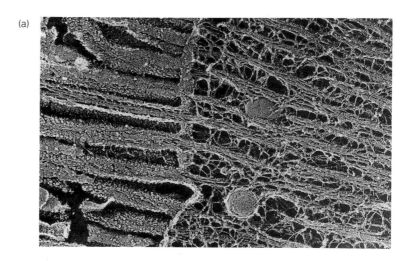

(b)

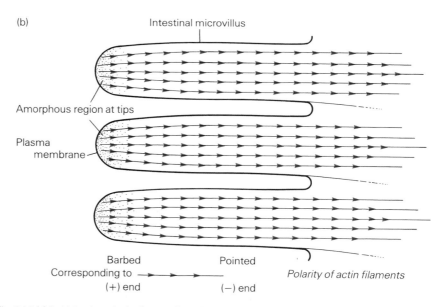

Fig. 6.14 (a) A chicken intestinal cell treated by quick-freeze, deep-etch technique to demonstrate the terminal web of the microvilli in the brush border. From Mooseker, M.S. (1984) *Journal of Cell Biology*, **99**, 1046–1125. Specimen prepared by Dr N. Hirokawa, Tokyo University, Japan. Courtesy of Dr M.S. Mooseker, Department of Biology, Yale University, USA. (b) Drawing of similar cell treated with heavy meromyosin fragments to show the polarity of F-actin in microvilli. Redrawn from Darnell, J.E. *et al.* (1986) *Molecular Cell Biology*, Scientific American Books Inc, p. 834.

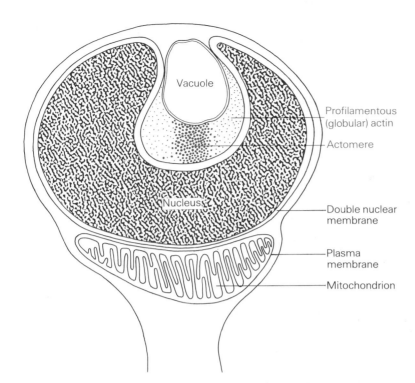

Vacuole

Profilamentous
(globular) actin

Actomere

Nucleus

Double nuclear
membrane

Plasma
membrane

Mitochondrion

Fig. 6.15 Representation of the head of a sea cucumber spermatozoon (*Thyone* sp). The actin fibres (red) are packed together in the actomere, the site of their polymerization. Following polymerization the stiff actin fibres are extruded (upwards in diagram) and penetrate the egg allowing fertilization to occur.

Exercise 1

How may actin filaments in non-muscle cells be identified?

ready for extrusion as stiff polymerized filaments from the sperm during fertilization (Fig. 6.15).

GROUP 4 PROTEINS MEDIATE ATTACHMENT OF ACTIN FILAMENTS TO THE PLASMALEMMA. Stress fibres would not function properly if they were not anchored at one or both ends to the plasmalemma. Vinculin and α-actinin are known to play a part in this anchorage.

6.5 Intermediate filaments

Intermediate filaments (IFs) are so called because they are thinner than microtubules but thicker than microfilaments. They are typically 10–11 nm in diameter, but vary between 7 and 15 nm.

They are the least soluble skeletal elements of a cell. The different components of the cytoskeleton can be extracted from a cell sequentially using detergents or solutions of high or low ionic strength or low pH. A non-ionic detergent solubilizes and removes most cellular proteins, leaving behind actin, myosin and intermediate filaments. Actin and myosin can then be extracted using a buffer with a high salt concentration, leaving behind only intermediate filaments which may be solubilized by sodium dodecyl sulphate (SDS) or pH 2.6 (Fig. 6.16).

Intermediate filaments appear to be the most stable skeletal elements, since at physiological pH and ionic strength, they always seem to exist in the form of insoluble polymers (filaments). In contrast, microfilaments and micro-

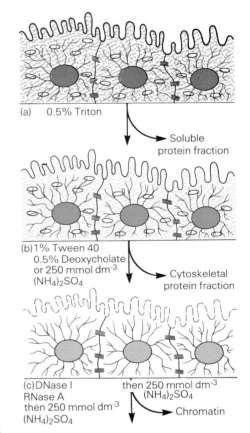

(a) 0.5% Triton

Soluble protein fraction

(b) 1% Tween 40
0.5% Deoxycholate
or 250 mmol dm^{-3}
$(NH_4)_2SO_4$

Cytoskeletal protein fraction

(c) DNase I
RNase A
then 250 mmol dm^{-3}
$(NH_4)_2SO_4$

then 250 mmol dm^{-3}
$(NH_4)_2SO_4$

Chromatin

(d)
Nuclear matrix + intermediate filaments only

Fig. 6.16 Detergent extraction of different cytoskeletal components. (a) Whole cells, (b) cells after removal of soluble proteins. (c) Shows the presence of intermediate filaments and the nuclear skeleton with associated chromatin. (d) Intermediate filaments and nuclear skeleton only.

Reference Gilula, N. and Wolpert, L. (eds) (1989) *Current Opinion in Cell Biology*, Vol. 1, No. 1, *Cytoplasm and Cell Motility*, Current Science, London, UK. A collection of up-to-date short reviews of all aspects by experts in the field.

tubules exist as insoluble polymers (filaments) and soluble monomers (separate globular molecules) and undergo cycles of polymer breakdown to monomers and monomer reformation to polymers (e.g. treadmilling; Fig. 6.7).

Structure of intermediate filaments

The proteins making up microfilaments and microtubules are the highly conserved molecules, actin and tubulin. In contrast, there seem to be families of polypeptide chains of intermediate filaments which are variable in size (Table 6.3).

The most recent evidence indicates that intermediate filaments are composed of two double polypeptide α-helices coiled about one another into a cable with a left-handed twist (Fig. 6.17). There are three regions where the polypeptide chains are randomly coiled. These regions may allow projection of side arms from intermediate filaments for interaction with other cytoskeletal elements or cellular organelles. Since there is no strict coiling requirement in these regions, they may also be sites of amino acid sequence variation in intermediate filaments. The carboxyl- and amino-terminal regions of each polypeptide are also randomly coiled with variable amino acid sequences. These are probably responsible for end-to-end (head or tail) or side-to-side aggregation of the molecules into filaments.

Table 6.3 *Tissue-types and their associated intermediate filaments*

Name of filament	M_r	Tissue	Name and number of proteins
Tonofilaments	40–70 000	Epithelial	Cytokeratins (~20)
Vimentin	57 000	Mesenchymal	Vimentin (1)
Desmin	53 000	Muscle	Desmin (1)
Neurofilaments			
NFL	70 000	Neurons	Neurofilament triplet
NFM	150 000		(NFT) (3)
NFH	200 000		
Glial filaments	52 000	Astrocytes	Glial fibrillary acid protein (GFAP) (1)

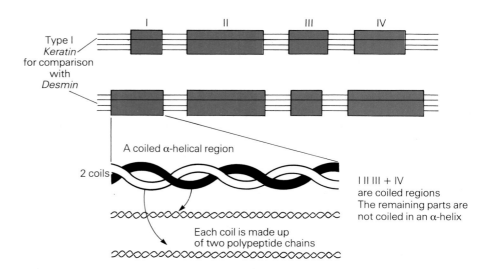

Fig. 6.17 A model for intermediate filament structure.

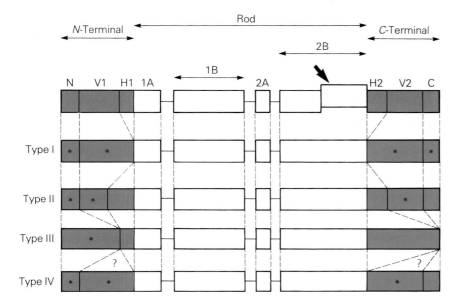

Fig. 6.18 The molecular structures of intermediate filaments. The rod is made up of four segments 1A, 1B, 2A, 2B joined together by three linker segments. 1A, 1B, 2A, and 2B are composed of repeating heptad peptides. H1 and H2 are homologous (same sequence) regions of different lengths in N and C termini respectively. V1 and V2 are non-homologous (variable in sequence) regions of different lengths in N and C termini respectively. The dotted lines indicate how length of each region compares with the same region in other types of intermediate filament. The asterisks indicate variability in sequence among members of the same type. (See also keratins in Table 6.4.) The black arrow indicates a reversal in polarity in the heptad peptides making up 2B. Redrawn from Steinert, P.M. and Parry, D.A.D. (1985) *Annual Reviews of Cell Biology*, **1**, 41–65.

Intermediate filaments as indicators of tissue type

Intermediate filaments are widely distributed in eukaryotic cells, occurring in both invertebrates and vertebrates. Their presence in plant cells has also been reported. The five main families shown in Table 6.3 may be reclassified as four types according to the amino acid sequences at their amino and carboxy termini (Fig. 6.18):

Type I	acidic cytokeratins
Type II	basic cytokeratins
Type III	vimentin, desmin, GFAP
Type IV	neurofilaments

Molecules of types III or IV can co-polymerize with other molecules of the same type; thus vimentin may assemble with desmin to form filaments. In the assembly of keratin subunits into filaments, co-polymerization of one type I molecule with one type II molecule seems to be essential to form a filament. The ability of molecules to co-polymerize indicates that they are more homologous (closely related) than molecules which lack this ability.

Function of intermediate filaments

The functions of intermediate filaments are as yet unknown. These may include maintaining the nucleus in an appropriate position, definition of cell shape and movement of organelles. Every one of the five intermediate filament families is characteristic of different tissue types. This tissue specificity might eventually shed some light on the function of intermediate filaments during differentiation.

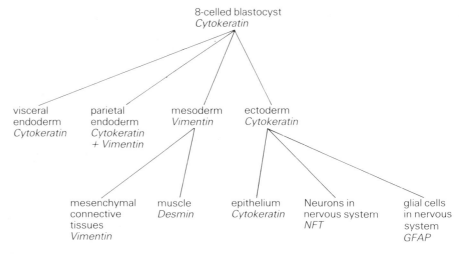

Fig. 6.19 The relationship of intermediate filament type to differentiation in mammalian cells.

Taken individually, the intermediate filament proteins also show interesting associations with certain stages of embryonic development (Fig. 6.19).

See Chapter 12

CYTOKERATINS. This first intermediate filament type to appear during cleavage of the **zygote** is detectable as early as the eight-cell stage. Later, during **embryogenesis**, the cytokeratin family exists in epithelial cells. Simple, single-layered epithelia or rapidly dividing epithelial cells contain the cytokeratins of lower M_r, while complex epithelia with highly differentiated cells contain larger cytokeratins with hydrophobic amino and carboxy termini. These interact with a closely associated cytoplasmic protein, **filaggrin**, to form an insoluble but pliable complex providing a protective barrier in the skin (Table 6.4). Cytokeratin is also associated with desmosomes (Fig. 6.20), the junctions that hold epithelial cells together.

See Chapter 4

Table 6.4 *Expression of keratin intermediate filaments in different epithelial tissues*

Type of epithelium	Type I (acidic) M_r	Type II (basic) M_r	Length and composition of V subdomain in *N*- or *C*-terminal regions
Simple	40 000 45 000 46 000	52–54 000 52 000 54 000	Short or absent?
Stratified	48 000	56 000	50–90 residues long,
Squamous	50 000	58 000	40% glycine
Hyperproliferating stratified squamous	48 000	56 000	48 residues long, low glycine content
Terminally differentiated:			
Cornea	55 000	64 000	104–130 residues long, 60% glycine
Skin epidermis	56 500	65–67 000	104–130 residues long, 60% glycine
Wool, hair	48 000	54 000	25–55 residues long, cysteine rich

THE NEUROFILAMENT PROTEINS. These three proteins, NF-L, NF-M and NF-H, are synthesized in neurons. They are not equally distributed since axons preferentially contain NF-M and NF-H, while cell bodies contain more NF-L and NF-H. They have highly charged carboxy-terminals, which might transport molecules along axons.

zygote: *a cell resulting from the fusion of oocyte and sperm which can develop to give an organism. From the Greek zygon, a yoke.*
embryogenesis: *the development of the embryo from the zygote.*

Reference Osborn, M. and Weber, K. (1982) Intermediate filaments: cell type specific markers in differentiation and pathology, *Cell*, **31**, 303–6. Applications of intermediate filament typing in pathology.

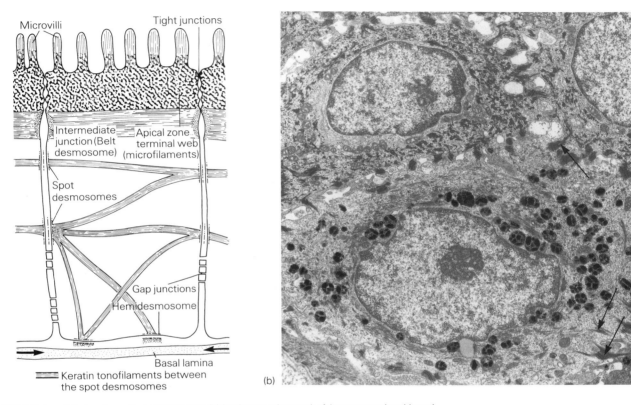

Fig. 6.20 (a) Intercellular junctions of gut epithelial cells. (b) An electron micrograph of desmosomes in epidermal cells of skin (×9625). Several desmosomes are indicated by arrows. Courtesy of Drs T.D. Allen and C. Potten, Christie Hospital, Manchester, UK. See also Figs 4.17 and 4.19.

VIMENTIN. This is expressed in **mesenchymal** tissues and is typical of primitive cells. Cells in tissue culture revert to expressing vimentin even if this is not their mature intermediate filament component, perhaps as a result of being stimulated to migrate or divide rapidly.

DESMIN. This protein is expressed in skeletal muscle cells, increasing in the later stages of differentiation, as the earlier expression of vimentin decreases. Desmin and vimentin seem to help microfilaments to align in register in skeletal muscle fibres and both remain as minor components of mature skeletal muscle. Desmin occurs in smooth muscle and also in non-muscle cells, such as those lining blood vessels, the endothelial cells. Thus it is not entirely muscle-specific.

mesenchymal: *relating to connective tissue, composed of stellate cells embedded, with fibres, in a gelatinous ground substance. From the Greek* mesos, *middle and* egchyma, *infusion.*

Reference Steinert, P.M. and Parry, D.A.D. (1985) Intermediate filaments: conformity and diversity of expression and structure, *Annual Review of Cell Biology*, **1**, 41–65. Fairly technical, for reference only.

Normally when patients with cancer are treated by radiation or drugs the decision as to the type of treatment is based on the diagnosis of the tumour. For instance, some tumour types are susceptible to radiation treatment while others are resistant. As tumours grow, the normal characteristic tissue structure breaks down and the individual cells often lose their characteristic surface proteins, making their identification very difficult. However, tumours of unknown origin may be diagnosed by identifying their intermediate filaments using specific antibodies. This is because intermediate filament expression is cell specific, each intermediate filament protein being associated with a particular cell type (Table 6.3). For example, tumours expressing cytokeratins can be classified as carcinomas (of epithelial origin) and distinguished from sarcomas (of mesenchymal origin) which express vimentin.

The success of this method depends on the absolute specificity of an antibody for one intermediate filament. Confusion has existed in the past because antibody specificity was not stringently checked before use. If intermediate filament proteins are subjected to SDS-electrophoresis, and subsequently added antibody binds to only the relevant protein on the gel, the antibody has appropriate specificity. Intermediate filaments share common rod domains, which means that an antibody with a specificity for part of a rod domain might well react, for example, with both vimentin and desmin and be useless in detecting either separately in tissues.

Another complicating factor is that tumour cells do not always obey the 'rules' of tissue expression (Fig. 6.19). It has been hotly debated, but now finally agreed, that in contrast to normal non-cancerous cells, tumour cells can co-express two intermediate filament types, for example, cytokeratin and vimentin. These do not co-polymerize into filaments but exist as two separate networks. Providing the limitations mentioned above are taken into account, intermediate filament expression in tumours remains a method for their diagnosis.

Several intermediate filament-associated proteins have been identified, including filaggrin, paranemin, plectin and synemin. Possibly they perform the same functions as actin-binding proteins: for example, filaggrin is reported to cross-link keratin into bundles.

Intermediate filaments and nuclear lamins

Recently it has been discovered that intermediate filaments are homologous to the proteins making up the scaffold of the nucleus called nuclear lamins. A receptor for vimentin filaments has been demonstrated on a nuclear lamin and it seems that vimentin filaments stretch from the nucleus, where their carboxy-termini are located, to the plasmalemma, where there are receptors for their amino-termini. Since they form a connection between the plasmalemma and the nucleus, intermediate filaments could possibly transduce extracellular signals received at the plasmalemma directly to the nucleus, or signals from the nucleus directly to the plasmalemma.

6.6 Microtubules

Microtubules were first identified as hollow tubes 25 nm in diameter using electron microscopy (Fig. 6.21). In the 1960s it was observed that juniper meristem cells contained microtubule cross-sections apparently made up of circular subunits. Techniques of rotating photographs one relative to another, demonstrated a periodicity of 13 subunits in every complete turn (Fig. 6.22). Isolated microtubule fragments negatively stained and examined sideways-on have subunits arranged in rows called **protofilaments** (Fig. 6.23).

Biochemical studies have demonstrated that microtubules are formed by

Reference Gull, K. (1990) Microtubules the cells most dynamic organelles? *Biological Sciences Review*, **2**(5), 18–22. A nicely illustrated starting place for learning about microtubules.

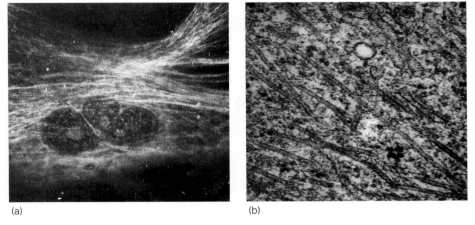

(a)   (b)

Fig. 6.21 (a) Tissue-cultured cell stained with antimicrotubule antibody bound to fluorescein, showing as lighter filaments against a dark background and viewed in a fluorescence microscope. Courtesy of Dr S. Kumar, Christie Hospital, Manchester, UK. (b) Electron micrograph of microtubules in an endothelial cell.

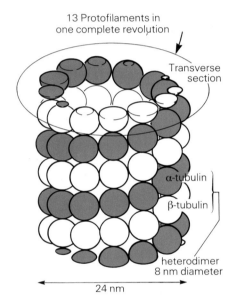

13 Protofilaments in one complete revolution

Transverse section

α-tubulin
β-tubulin

heterodimer 8 nm diameter

24 nm

Fig. 6.23 A three-dimensional view of microtubule structure.

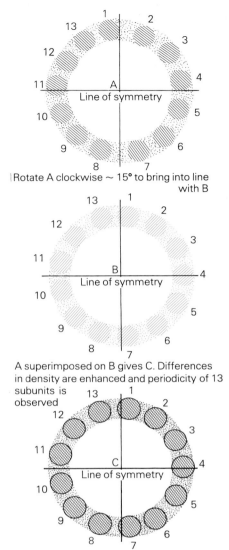

Rotate A clockwise ~ 15° to bring into line with B

A superimposed on B gives C. Differences in density are enhanced and periodicity of 13 subunits is observed

Fig. 6.22 How the use of image rotation indicated the presence of 13 subunits in each microtubule.

polymerization of tubulin dimers. Each dimer consists of one α-tubulin polypeptide of M_r 50 000 and one β-tubulin polypeptide of M_r 50 000. The dimers are polymerized in a head-to-tail fashion which imparts polarity to the microtubule (compare microfilaments).

Microtubules and actin filaments: a comparison

Microtubules are capable of treadmilling in a similar way to F-actin. They are likewise polarized, having polymerizing (+) and depolymerizing (−) ends. Polymerization is dependent on GTP hydrolysis, unlike actin which relies on ATP as a free energy source. Two GTP molecules bind to a tubulin dimer and remain bound after polymerization. One becomes dephosphorylated to GDP while the other remains as GTP with, as yet, unknown function. Like actin, Mg^{2+} promotes polymerization while Ca^{2+} and calmodulin cause

---

Exercise 3

Given that the diameter of the microtubules visible in Figure 6.21b is 25 nm, calculate the magnification of the electron micrograph.

Table 6.5 *Microtubule proteins and microtubule-associated proteins*

Protein	M_r	Tissue origin	Year isolated	Function
Tubulin α	50 000	Pig brain	1968	Main microtubule component
β	50 000			
Dynein	300–600 000	*Tetrahymena* cilia	1963	Microtubule sliding movement
MAP 1	310 000	Brain	1974	Cross-links microtubule bundles to actin *in vitro*
2	270 000	HeLa cells	1978	
Tau proteins	55–70 000	Brain	1975	Stimulates microtubule formation
Tektins	47 000	Sea urchin sperm	1982	} Outer doublet of flagella
Secretin	51 000	Neuronal synaptic vesicles	1984	
	55 000		1987	
Plectin	300 000	Tumour cells containing vimentin	1980	} Bound to vimentin in outer doublet of flagella
			1987	
Kinesin	134 000	Sea urchin embryo	1985	Translocates microtubules in spindle
	110 000	Squid axoplasm	1985	Translocates microtubules in spindle

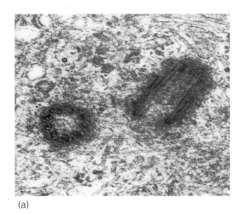

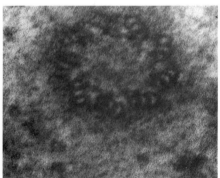

(a)

(b)

Fig. 6.24 (a) An electron micrograph of a pair of centrioles ($\times$57 000). (b) In one centriole nine triplet microtubules are visible. ($\times$175 000). Courtesy of Dr A. Sattar, St Mary's Hospital, Manchester, UK.

depolymerization. Polymerization is also controlled by binding to **micro-tubule-associated proteins** (MAPs) some of which are the so-called **tau** proteins (see Table 6.5).

Whereas the polarization of F-actin was demonstrated using heavy meromyosin fragments, the polarization of microtubules was demonstrated by polymerization of short microtubule fragments *in vitro*. Newly polymerized molecules formed 'curved hooks' at the ends, always with a clockwise arrangement at one end and an anticlockwise arrangement at the other. If microtubules *in vivo* are depolymerized using a drug such as colchicine and allowed to regrow, regrowth is faster at the (+) end than is depolymerization at the (−) end. A lack of depolymerization at the (−) end may be the result of anchorage of this end. The direction of microtubule growth is always away from the centrosome or centriolar region of a cell, where it is initiated *in vivo*.

Specialized structures formed by microtubules

Several cellular structures are derived from microtubules. These include centrioles, cilia and flagella and the mitotic spindle.

Centrioles are short parallel condensations of microtubules 0.3 μm long and 0.1 μm diameter composed of nine triplets arranged in a circular fashion (Fig. 6.24). They are not present in unfertilized eggs but are visible after fertilization probably after polymerizing from electron-dense material in the cytosol. In animal cells, **microtubule organizing centres (MTOC)** are observable by electron microscopy. These consist of two *centrioles*

centriole: a short hollow cylinder of microtubules. From the Greek kentron, *centre.*

Reference Karrenti, E. and Mars, B. (1986) Centrosomes and the spatial distribution of microtubules. *Trends in Biochemical Sciences,* **11**, 460–3. A readable account.

surrounded by electron-dense material (Fig. 6.25). Each centriole of a pair in the MTOC divides and migrates to opposite ends of the cytoplasm where they initiate microtubule polymerization to form the **mitotic spindle**. In higher plant cells where centrioles are absent, microtubules are polymerized in electron-dense material called **asters**.

Single centrioles lying just inside the plasmalemma, called basal bodies (Fig. 6.26), are sites for the generation of cilia or flagella. **Basal bodies** and centrioles are interconvertible. For example, when flagella of *Chlamydomonas* are reabsorbed at a certain stage in its life cycle, the basal bodies are converted into centrioles. Vertebrate cells with many cilia have centrioles capable of migrating towards the plasmalemma, where they bud off multiple modified cilia (**satellites**) instead of dividing. Each satellite gives rise to a cilium.

Cilia and flagella are specialized microtubular structures associated with cellular movement in Protozoa, larvae of invertebrates and gametes of animals and plants. Microtubules extend in the core or **axoneme** along the whole length of a cilium or flagellum (up to 200 μm!) in a 9 + 2 arrangement (Fig. 6.27). Nine outer doublets surround two central singlets. Every doublet consists of an inner A subfibre with 13 subunits and an outer B subfibre with only 10. Arms made of dynein, a protein having ATPase activity, project from the doublets at regular 24 nm intervals along the axoneme (Fig. 6.28). These make contacts with neighbouring doublets when ATP is depleted. When ATP is restored, the contacts are broken, and the dynein arms change shape to point 'downwards' towards the basal body. Cycles of movement in dynein arms generate a force which is dependent on hydrolysis of ATP and causes microtubules to slide past one another (as in actin–myosin interactions in striated muscle) lengthwise along the cilium. Cross-links and radial spokes (Fig. 6.29) made up of the protein nexin prevent lengthwise movement and convert it into ciliary bending, visible as a whiplash movement in one direction (Figs 6.30 and 6.31). They also regulate activation of dynein side arms. Mutations leading to loss of axoneme components often result in non-motile cilia. For example, some non-motile human sperm have been found to lack central microtubules; that is, they have a 9 + 0 axonemal structure.

In multiciliated cells, rows of cilia are co-ordinated to beat together a fraction of a second after the row in front or before the row behind. This co-ordinated beating is called **_metachronal rhythm_** (Fig. 6.32). The ciliated epithelial cells of the human upper respiratory tract are an example. Debris trapped in mucus is swept upwards out of the lung by co-ordinated movements of the cilia.

The mitotic spindle

Microtubules appear to be essential in the formation of a spindle and chromosome movement during mitosis of higher animals. There are two kinds of microtubule in the spindle. Polar fibres join the equator to one or other pole. Polar fibres from one pole overlap at the equator with polar fibres from the other pole. **Kinetochore** fibres join the centromeric region of a chromosome to one or other pole.

At early metaphase, chromosomes jostle together so that they are arranged on a metaphase plate at the equator, with each centromere facing one or the other pole. During anaphase the two chromatids are pulled apart by the action of kinetochore microtubules and by polar microtubules sliding past one another to elongate simultaneously the spindle axis and push the poles apart. Thus the chromosomes separate, with the chromatids of a pair migrating to opposite poles (Fig. 6.33). What molecule generates the force enabling

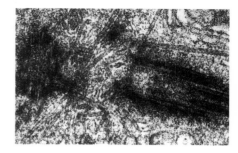

Fig. 6.25 Electron micrograph of a microtubule organizing centre. (×80 600). Note the surrounding electron-dense material. Courtesy of Dr A. Sattar, St Mary's Hospital, Manchester, UK.

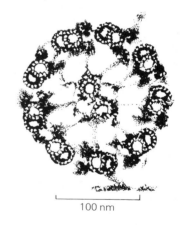

Fig. 6.26 An electron micrograph of a basal body (×150 500). Note the radial spokes (examples arrowed). Courtesy of Dr A. Sattar, St Mary's Hospital, Manchester, UK.

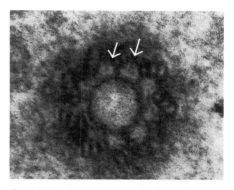

|⟵   100 nm   ⟶|

Fig. 6.27 A flagellum in cross-section. Redrawn from Alberts, B. *et al.* (1989) *Molecular Biology of the Cell*, 2nd edn, Garland, New York, p. 645.

Exercise 4

Why are abnormal sperm with 9 + 0 flagellar structure non-motile? How is this linked to human infertility?

metachronal rhythm: *the pattern set up by rows of beating cilia: From the Greek* meta, *change of position and* chronos, *time.*
cytokinesis: *the division of the whole cell after nuclear division. From the Greek* kinesis, *movement and* kytos, *cell.*

Reference Hepler, P. (1985) The plant cytoskeleton, in *Botanical Microscopy* (ed. A.W. Robards), Oxford University Press, Oxford, UK. One of few reviews on the plant cytoskeleton.

Exercise 5

A 20 μm² surface of *Paramecium* plasma membrane contains 300 profiles of cilia in cross-section. What proportion of the membrane area is made up of cilia, if the diameter of a cilium is 250 nm?

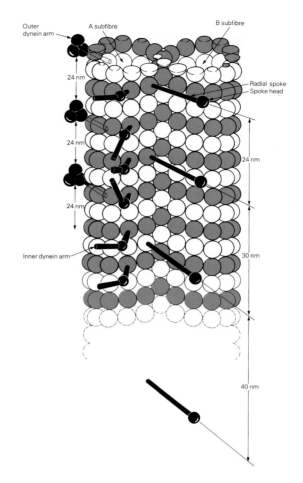

Fig. 6.28 Axoneme with side arms of dynein (red) projecting from microtubules in a cilium.

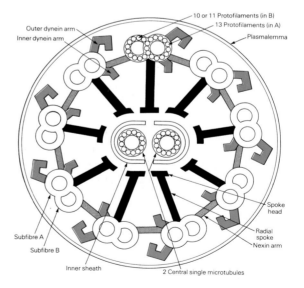

Fig. 6.29 Cross-section of a cilium/flagellum.

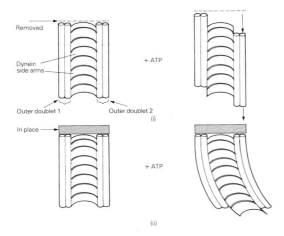

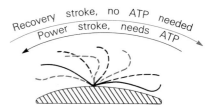

(a) *Cilium*

Sperm is *pushed* away from direction of waves of bending

Movement of sperm

Waves of bending of flagellum, 30-40/s

(b) *Flagellum of sperm*

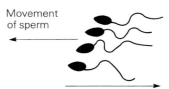

Trypanosome is *pulled* towards direction of waves of bending

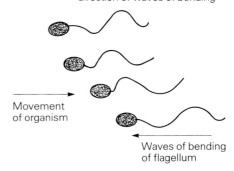

Movement of organism

Waves of bending of flagellum

(c) *Flagellum of trypanosomes*

Fig. 6.31 The whiplash motion in (a) cilia (b) flagella of spermatozoa and (c) flagella of trypanosomes.

Fig. 6.30 A demonstration of how microtubule sliding and interactions with central microtubules generate bending movements in cilia/flagella. (i) If radial spokes are hydrolysed away, then doublet 2 can slide relative to doublet 1 when ATP is added. (ii) If radial spokes are present, then the only movement doublet 1 can make relative to doublet 2 results in bending.

microtubules to slide past one another is not known, but a recently discovered molecule, kinesin, has been tentatively called the spindle 'motor'. At telophase, the nucleus which had earlier disintegrated to allow mitosis to occur, reforms. Cytoplasmic division or **cytokinesis** follows. This occurs by furrowing caused by contraction of a ring of microfilaments running around the equatorial region (Fig. 6.34).

Microtubules also play a significant role in plant cell division. They lie beneath the cell wall parallel to cellulose fibrils in the wall (Fig. 6.35) and appear to influence the position and direction of new wall synthesis after

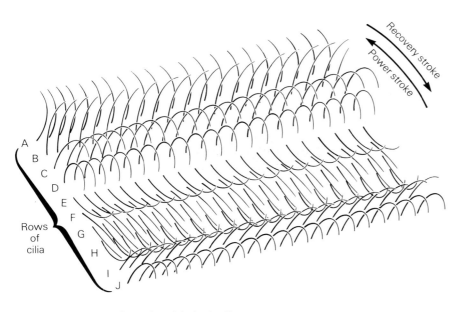

Fig. 6.32 The appearance of metachronal rhythm in cilia.

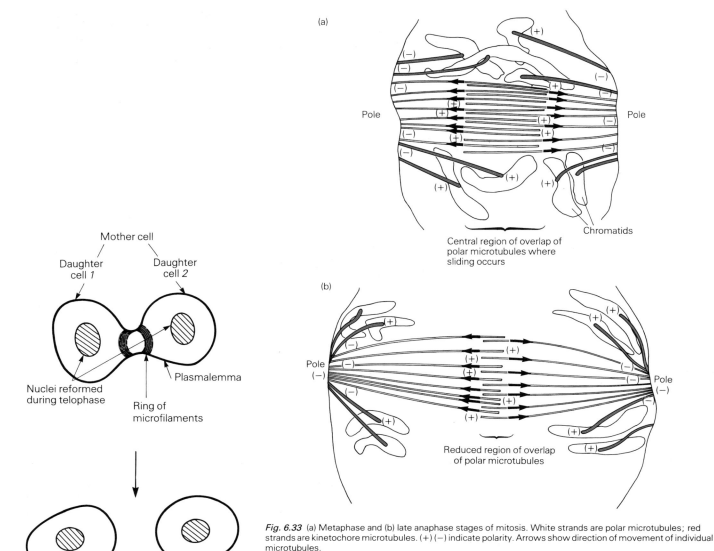

(a)

Pole

Pole

Chromatids

Central region of overlap of polar microtubules where sliding occurs

(b)

Pole (−)

Pole (−)

Reduced region of overlap of polar microtubules

Fig. 6.33 (a) Metaphase and (b) late anaphase stages of mitosis. White strands are polar microtubules; red strands are kinetochore microtubules. (+) (−) indicate polarity. Arrows show direction of movement of individual microtubules.

Mother cell

Daughter cell *1*

Daughter cell *2*

Nuclei reformed during telophase

Ring of microfilaments

Plasmalemma

Two new separate cells

Fig. 6.34 A diagram of cytokinesis in an animal cell.

Reference McIntosh, R. (1984) Mechanisms of mitosis. *Trends in Biochemical Sciences*, **9**, 192–5. Self-explanatory title; a readable account.

division. This is significant because plant cell division is not always symmetrical (Fig. 6.36).

The action of the drug colchicine (Fig. 6.37) in depolymerizing microtubules has been exploited in **karyotyping**, the process of analysing the chromosomes of a cell.

☐ Cells treated with colchicine arrest at the metaphase stage of division. They are gently ruptured and spread on a slide to release their condensed chromosomes which can be photographed, counted and arranged in order of size and shape.

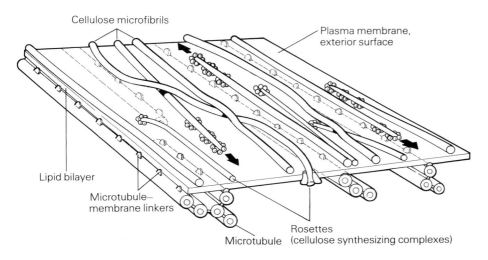

Fig. 6.35 Plant cell microtubules as organizers of cell wall orientation. Reproduced by permission of A.R. Liss, from Staehelin, L.A. and Giddings, T.H. (1982) in *Developmental Order: Its Origin and Regulation*, (Eds S. Subtelny and P.B. Green), p. 133–47.

Microtubules are in overall charge of the cytoskeleton

Microtubules bind to actin microfilaments and intermediate filaments and may control the distribution of the latter. They are involved with actin filaments in coated vesicle and synaptic vesicle invagination from the plasmalemma into the cell. Microtubules also bind to cellular organelles and probably act as guidelines for, and translocators of, organelles moving within the cell. It has been suggested that microtubules control the cytoskeleton.

The cytoskeleton of most cells is complicated and the interactions between all of its constituent proteins are not well understood. The most thoroughly studied, because it is simpler and easier to isolate, is the cytoskeleton of erythrocytes (red blood cells).

Exercise 6

How was it demonstrated that the cytoskeleton consists of three main components?

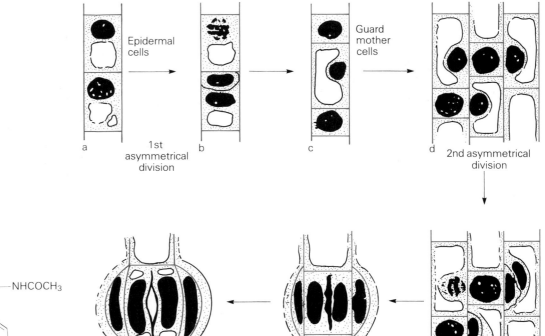

Fig. 6.37 Colchicine.

Fig. 6.36 Plant cell divisions are not always equal. Redrawn from Pickett-Heaps, J.D. and Northcote, D.M. (1966) Cell division in the formation of the stomatal complex of the young leaves of wheat. *Journal of Cell Science*, **1**, 121–8.

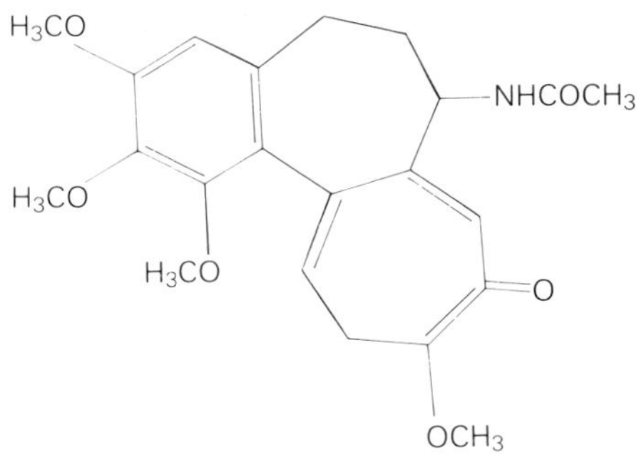

6.7 The erythrocyte cytoskeleton

The erythrocyte cytoskeleton is not typical since it does not criss-cross the whole cell but is confined to a peripheral thin cage lying just inside the plasmalemma. The shape of red blood cells is more rigidly controlled than in most eukaryotic cells which often continually change shape rather than having a fixed morphology. Erythrocytes are noted for their ability to be deformed and then to regain their original shape. Indeed, this deformability is necessary for them to be able to circulate through small capillaries.

Their cytoskeleton consists of five main proteins: actin, spectrin, ankyrin and **Band** 4.1 and 4.9 proteins (Table 6.6 and Fig. 6.38). These proteins are not integrated into the plasmalemma but still remain after extraction of erythrocyte ghosts with the detergent Triton X-100 which extracts most non-integral proteins.

There is evidence that actin filaments (albeit very short ones) co-exist with myosin and tropomyosin and that Ca^{2+} and calmodulin regulate actin-myosin interactions. Microfilament contractions thus may bring about some change in erythrocyte shape. Several other actin-binding proteins have been identified. Band 4.1 protein promotes actin polymerization while Band 4.9

☐ Suspending erythrocytes in dilute buffered solution causes them to rupture as water passes into them by osmosis. The internal contents leak out leaving a 'ghost', consisting only of an external membrane and its associated cytoskeletal proteins.

band: *some of the cytoskeleton proteins of erythrocytes are named according to their position on the gel following SDS-PAGE.*

(a)

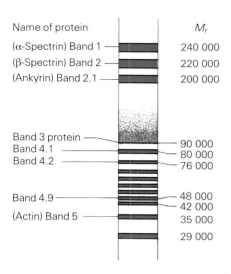

Name of protein		M_r
(α-Spectrin) Band 1		240 000
(β-Spectrin) Band 2		220 000
(Ankyrin) Band 2.1		200 000
Band 3 protein		90 000
Band 4.1		80 000
Band 4.2		76 000
Band 4.9		48 000
		42 000
(Actin) Band 5		35 000
		29 000

(b)

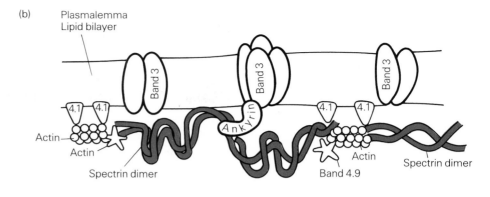

Fig. 6.38 (a) A representation of erythrocyte cytoskeletal proteins analysed by polyacrylamide gel electrophoresis. (b) How the cytoskeletal proteins are linked together. Reproduced by permission.

Table 6.6 *The major proteins of the erythrocyte cytoskeleton*

	M_r	Function
Actin	42 000	Contraction?
α-Spectrin	240 000	Binds all other proteins
β-Spectrin	220 000	Gives strength and elasticity to the cytoskeleton
Band 4.1 protein	82 000	Promotes actin polymerization and stabilizes filaments
Band 4.9 protein	48 000	Controls lengths of actin filaments polymerized from monomers

protein controls actin filament length. Band 4.1 protein also mediates actin binding to the ends of spectrin tetramers (Fig. 6.38). Spectrin is a long fibrous protein. Heterodimers composed of one α- and one β-spectrin aggregate end-to-end into tetramers. The extent of aggregation of spectrin is possibly controlled by phosphorylation. There is a binding site near the centre of tetramers for ankyrin, a peripheral protein that anchors spectrin to the integral membrane Band 3 protein. When ankyrin changes shape, it behaves as though hinged. Rotation about these hinges and folding/unfolding of the spectrin polypeptide chain probably account for most of the erythrocyte strength and deformability.

In spite of their unique deformability, erythrocytes resemble other cells in containing intermediate filaments composed of vimentin. Most erythrocytes, except those in mammals, contain a peripheral circular band of microfilaments lying just inside the plasmalemma. Erythrocytes have been extensively studied in molecular terms but it should be remembered that their unique properties of rigidity and deformability make them far less dynamic than most eukaryotic cells and therefore atypical.

Patients with the hereditary diseases elliptocytosis and spherocytosis have lower than normal amounts of spectrin in their erythrocytes or produce defective spectrins. As a result, their erythrocytes are less deformable than normal and become trapped during circulation through the spleen and prematurely digested, releasing bilirubin. The overload of bilirubin causes jaundice and gall stones, while immature erythrocytes are sent into the blood by the bone marrow to compensate for the anaemia caused by erythrocyte destruction. Luckily the patient's symptoms may be 'cured' by removing the spleen, even though the primary defect of the disease remains unknown.

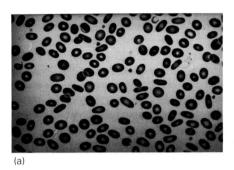 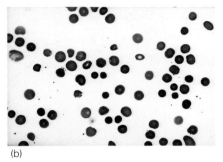

(a)   (b)

Photomicrographs of blood smears from patients with (a) elliptocytosis, note the presence of abnormal elliptical erythrocytes, and (b) spherocytosis, the abnormal erythrocytes (spherocytes) lack the usual biconcave shape and appear a uniform dark colour. Courtesy of Dr D.I.K. Evans, Royal Manchester Children's Hospital, UK.

6.8 Movements of cells during the embryonic development of animals

Animal cells are one of two main types, epithelial or mesenchymal. Epithelial cells are tightly joined together in layers. Mesenchymal cells tend to occur singly or in loose arrangements.

The formation of tissues and organs during embryogenesis often depends upon changes in curvature of layers of epithelial cells including folding, invagination/evagination and branching (Fig. 6.39). In these cells microfilaments are arranged such that their contraction produces the desired change in shape. For example, some cells contain microfilament bundles circumferentially arranged around their base and apex which contract enabling the cells to become taller and thinner, a process called palisading (Fig. 6.39).

Microfilament contraction is also a passive tensile force to prevent cells being stretched or broken by opposing forces. Microfilament bundles 'locked' in this way around the base or apex of a cell could prevent that part of the cell from changing shape. Grooves in epithelial layers might arise by constriction of the bases of cells while their apices remain anchored. Conversely, cells spread by an increase in width × decrease in height, which seems to involve depolymerization of microfilament bundles, for example, in the formation of the embryonic neural plate (Fig. 6.39e).

All kinds of cells migrate during embryonic development. A Ca^{2+}- and ATP-dependent interaction of actin and myosin microfilaments may be responsible for cell migration as well as shape changes. Migration should not be regarded as a characteristic of cells in isolation from their surroundings.

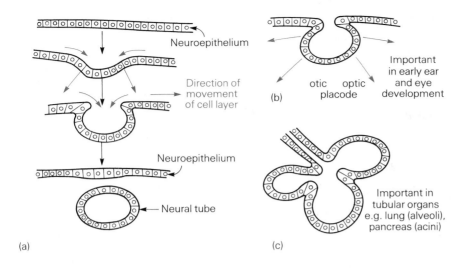

Neuroepithelium

Direction of
movement
of cell layer

Neuroepithelium

Neural tube

(a)

(b) otic optic
placode

Important
in early ear
and eye
development

(c) Important in
tubular organs
e.g. lung (alveoli),
pancreas (acini)

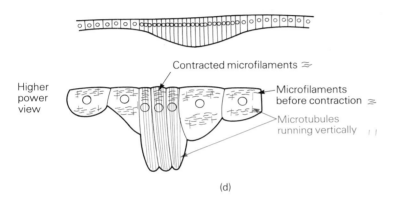

Contracted microfilaments ≈

Higher
power
view

Microfilaments
before contraction ≈

Microtubules
running vertically ⊥⊥

(d)

Cell spreading/flattening
occurs during neural plate
formation

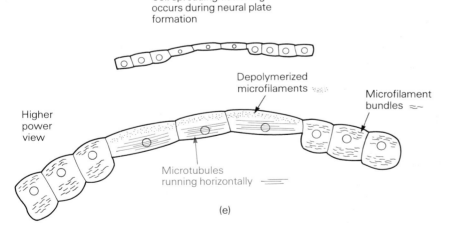

Depolymerized
microfilaments

Microfilament
bundles ≈~

Higher
power
view

Microtubules
running horizontally ≡

(e)

Fig. 6.39 Movements of cell layers: (a) folding, (b) invagination, (c) branching, (d) constriction. (e) Diagram of cell spreading as in neural plate formation. Constriction palisading is probably involved in movements of cell layers in a–c.

Indeed, an interaction of the cell with its extracellular matrix, the complex supportive matrix made up of glycoproteins found between cells, is important in determining how fast and in which direction a cell will travel.

Microtubule bundles run from apex to base of elongated epithelial cells as if forming a supportive backbone. The support afforded by microtubules in animal cells occurs also in plant cells. However, higher plant cells have additional features to help maintain their rigidity, namely a cellulose cell wall and the turgor pressure of the cell sap.

Cell movements in the adult animal, embryo and in tissue culture should not be regarded as a prerogative of microfilaments alone. It is unlikely that change in shape affects only this element of the cytoskeleton and it is possible that the different intermediate filament molecules might also be necessary to produce the cell shapes and movements characteristic of different tissue types.

6.9 Concluding remarks

The ability of the cytoskeleton to alter its form and function is governed by the biochemistry of its many constituent proteins, particularly their state of phosphorylation by protein kinases or dephosphorylation by phosphatases. The proteins making up microfilament, intermediate filament and microtubule networks interact with one another using ATP- and GTP-dependent mechanisms, where Ca^{2+} and Mg^{2+} play a major role. Their interactions are further modified by accessory proteins which co-exist in close proximity. Self-assembly of proteins into filaments, aggregation into larger fibrous structures, anchorage to the plasmalemma and co-ordinated cytoskeletal movements are all controlled by these accessory proteins. The arrangements of microfilaments in muscle and microtubules in cilia or flagella are extreme specializations of basic functions that are common to all eukaryotic cells.

6.10 Overview

The cytoskeleton comprises three types of filament, microfilaments, microtubules and intermediate filaments, possibly arranged in a micro-trabecular lattice. Actin (microfilaments) and tubulin (microtubules) are found in all cells and are highly conserved. Actin and tubulin genes belong to two multiple gene families (their DNAs are sufficiently similar to have arisen from a single ancestral gene). The minor differences in amino acid sequences probably account for slight variations in these proteins (isoforms) in different cells.

Actin microfilaments interact with myosin to produce cytoplasmic contraction and cellular movement in a manner resembling that in striated muscle, although less efficiently and with some different regulatory proteins. Microfilaments contribute to structures such as microvilli, stress fibres and sperm acrosomes where they aid penetration of the ovum. Microtubules guide movements of organelles and substances within the cell. They form the mitotic spindle and specialized structures such as cilia, flagella and centrioles. Microtubules share properties with microfilaments in that both can exist as soluble globular subunits or insoluble filamentous forms, interchangeable by treadmilling. Sliding of microfilaments and microtubules to produce movement depends on the hydrolysis of ATP catalysed by specific ATPases. In contrast, intermediate filaments are more stable, are harder to extract with detergents and do not undergo treadmilling. Intermediate filaments are represented by five gene families showing varying degrees of homology and are more tissue specific. Different intermediate filaments are characteristic of different cell types. Differentiation often involves a change from one intermediate filament type to another. The function of intermediate filaments is not yet known.

The accessory proteins that exist for microfilaments, intermediate filaments and microtubules are less well characterized; some show evidence of belonging to gene families (e.g. spectrin) while others appear to be more tissue specific (e.g. synapsin).

The cytoskeleton is responsible for movement within cells, changes in cell shape, and cell movement, all of which are important in embryogenesis as well as in adult life.

Answers to Exercises

1. By their diameter in the electron microscope, their binding to heavy meromyosin and their biochemistry. Once solubilized, they can be purified by two-dimensional polyacrylamide gel electrophoresis and identified by their M_r. Antibodies raised to the protein can be used to identify the protein wherever it occurs in cells and tissues. Phalloidin will bind actin specifically.

2. Intermediate filaments have a region in their rod domain which has the same or very similar sequence to that in nuclear lamins. This implies they have evolved from a common ancestral protein.

3. Measurement of a microtubule gives a diameter of 2 mm.

   At a magnification of × 1000

   1 μm becomes 1 mm

At a magnification of × 1000

   2 μm becomes 2 mm

   or 2000 nm becomes 2 mm

At a magnification of

   $$1000 \times \frac{2000\,\text{nm}}{25} \text{ becomes 2 mm}$$

   $$= \times 80\,000$$

4. Flagella with a 9 + 0 structure lack central microtubules and therefore cannot form cross-links to regulate sliding of microtubules and convert it into bending (Figs 6.29, 6.30). Since they are immobile, they cannot swim up the female reproductive tract to fertilize an ovum. Thus fertilization cannot occur.

5. Area of cross-section of one cilium (πr^2)

   $$= (125)^2 \times 3.142\,\text{nm}^2$$

   $$= 49\,093\,\text{nm}^2$$

   Area of cross-section of 300 cilia

   $$= 14\,727\,900\,\text{nm}^2$$

   $$= 14.728\,\mu\text{m}^2$$

   Therefore, proportion of membrane occupied by cilia

   $$= \frac{14.728}{20} \times 100\% = 73.6\%$$

6. They could be seen to have different diameters in sections of cells examined by transmission electron microscopy.

FILL IN THE BLANKS

1. The six different kinds of actin in cells do not differ very much in amino acid composition. Thus they are _____ molecules with important roles in the cell. All are globular proteins, _____ _____ in diameter with an M_r of _____ . _____ . Actins are found in muscle tissues, while _____ and _____ actins occur in non-muscle cells. _____ occurs in all actins and may be used as a marker in its structural identification.

Group 2 proteins control actin polymerization e.g. _____ is involved in amoeboid movement. Capping proteins stabilize polymerized filaments and allow fairly permanent extensions of cytoplasm e.g. _____ . Group 3 proteins _____ actin into stress fibres for cell migration, into microvilli and into stiff actin rod associated with fertilization of an ovum by a spermatozoon.

_____ cross-links actin to form the rigid erythrocyte cytoskeleton. Group 4 proteins anchor actin filaments to _____ . Without anchorage, stress fibres would not function to allow cells to change shape and particularly to spread. Signals received at the plasmalemma often stimulate changes in cell shape via these anchoring molecules.

The proteins making up microfilaments and microtubules are the highly _____ molecules _____ and _____ . Intermediate filaments are coded for by multiple _____ families and are composed of two double _____ coiled about one another into a cable with a _____ twist. The amino and carboxy termini are _____ coiled with _____ amino acid sequences as are three regions linking four blocks of the _____ domain. There are more than 20 kinds of _____ intermediate filaments. Type I are _____ while type II are _____ proteins. _____ exist as a triplet of proteins in _____ . Cells that are stimulated to divide rapidly in culture, re-express _____ .

Choose from: 5.5 nm, 42 000, α-, α-helices, acidic, actin, β-, basic, conserved, cross-link, cytokeratin, γ-, gelsolin, gene, highly-conserved, left-handed, membranes, N-γ-methylhistidine, microvilli, neurofilaments, neurons, randomly, rod, spectrin, tubulin, variable, vimentin.

MULTIPLE-CHOICE QUESTIONS

State which of the following are true or false.

2. SDS-PAGE electrophoresis:
A. separates proteins on the basis of M_r.
B. separates proteins on the basis of charge.
C. all proteins in the gel have a net negative charge.
D. the proteins are not denatured.
E. can only be used in one dimension (to run proteins in one direction only).

3. The myosin molecule may be digested into fragments.
A. One molecule has two 'heads'.
B. S-1 fragments are 'heads'.
C. There are no light chains associated with 'heads'.
D. The amino acid termini of the myosin heavy chains are in the 'heads'.
E. Light meromyosin (LMM) is part of the 'tail' only.

4. Several drugs stabilize and destabilize microtubules and microfilaments in cells
in vitro.
A. Colchicine causes depolymerization of microtubules.
B. Colchicine prevents microtubule polymerization.
C. Taxol causes microtubule depolymerization.
D. Phalloidin binds only to actin.
E. Cytochalasin B depolymerizes actin filaments.

SHORT-ANSWER QUESTIONS

5. What is the advantage of (a) high-voltage electron microscopy, (b) freeze-etching, compared with the 'ordinary' transmission electron microscopy in studying the cytoskeleton?

6. Explain briefly how an antibody may be used to visualize the distribution of a cytoskeletal protein in a cell.

7. How do microfilaments of non-muscle cells resemble those in striated muscle and how do they differ?

8. What are stress fibres and what is their function?

9. How does the demonstration of the intermediate filament content of tumours help tumour diagnosis and the treatment of cancer?

10. Explain how microtubule polymerization and depolymerization bring about movements of chromosomes along a mitotic spindle.

11. There are five intermediate filament families. Match each of the families in (A) with one of the tissues in (B).

A	B
desmin	mesenchymal cells
vimentin	glial cells
glial fibrillary acidic protein	muscle
cytokeratins	neurons
neurofilament proteins	epithelial cells

ESSAY QUESTIONS

12. Compare the structure, formation and mechanism of movement of microfilaments, microtubules and intermediate filaments.

7

The extracellular matrix

Objectives

After reading this chapter you should be able to:

☐ describe the mechanical properties of connective tissues and their extracellular matrix;

☐ discuss the mechanisms involved in cell–extracellular matrix interactions;

☐ Explain how the composition/structure of the extracellular matrix modulates cellular processes.

7.1 Introduction

At some period during their existence, all cells in multicellular animals come into contact with an intricate meshwork of interacting **fibrous proteins**, **proteoglycans** and **structural glycoproteins** that make up the **extracellular matrix** of most tissues. The cell wall of plants (Box 7.1), and the bacterial capsule may also be regarded as an extracellular matrix. This matrix not only serves to maintain tissue structure but also influences the differentiation, migration, proliferation and shape of the cell with which it is in contact. The structural, or mechanical, properties of the extracellular matrix components are essential for the normal functioning of the *connective tissues*, such as cartilage, skin (dermis), tendon, bone, the intervetebral disc, teeth and the basal laminae.

Most, if not all, of the components of the extracellular matrix appear to be able to interact with cells either indirectly through proteins such as fibronectin, or via surface receptors.

7.2 Composition and structural diversity

The macromolecules that constitute the extracellular matrix are secreted locally by all cells, with the exception of the mature blood cells. The exact composition of the matrix secreted depends on the cell-type, its state of differentiation and its metabolic status. The basic components of all matrices are: the fibrous proteins, **collagens** and **elastin**; the **glycosaminoglycans** and proteoglycans, and the structural, or adhesion, proteins such as **fibronectin** and **laminin**. The glycosaminoglycans, proteoglycans and structural proteins trap water molecules to form a highly hydrated gel-like 'ground substance' in which the insoluble fibrous proteins are embedded, giving the matrix strength, rigidity and resilience (Fig. 7.1).

Exercise 1

Can you name some modern materials used in the construction of buildings automobiles or household objects, that use the principle of 'fibres in matrix' to give rigidity, resilience and strength.

☐ The term **matrisome** has been proposed for the basic functional complex of extracellular matrixes. This is seen as a complex of collagen, glycoproteins (for example, laminin in basement membranes) and proteoglycans and a tissue-specific organizing protein (that is, **nidogen** in some basement membranes).

---

connective tissues: tissues which provide structural support for other tissues and organs throughout body, i.e. cartilage, tendon, bone, walls of blood vessels. Adipose tissue is also included in some definitions.

Reference Hay, E.D. (1991) *Cell Biology of Extracellular Matrix*, Plenum, New York, USA. Good reviews on most aspects of extracellular matrix and development.

Box 7.1
Cell walls

Bacterial, algal, fungal, protozoan and plant cells do not have the protective homeostatic environment afforded animal calls. They are in direct contact with an unregulated, potentially hostile, environment and possess a reinforced cell outer coat or cell wall. The latter may be considered to be analogous to an extracellular matrix.

The bacterial plasmalemma resembles that found in animal cells in that it is a double membrane system. However, in bacteria a peptidoglycan layer (which imparts rigidity) is interposed between the inner and outer membranes. The peptidoglycan is composed of long polysaccharide chains of alternating *N*-acetylglucosamine and *N*-acetylmuramate residues (see Section 7.7). Linked to the muramate residues are short peptide chains (4–5 amino acids) containing some D-amino acids. Cross-linking between the peptide side-chains is extensive and the entire inner wall can be considered as one large bag-shaped peptidoglycan complex.

Many bacteria are able, under appropriate conditions, to produce a capsule which consists of high M_r polysaccharides and polypeptides. The capsule of *Bacillus anthracis* is composed of a polypeptide of glutamate residues. That of *Acetobacter xylinum* is pure cellulose, while that of the filamentous *Streptomyces* bacterial species is hyaluronate. The function of such capsules may be as a protection against the immune system.

The plant cell wall is a specialized form of extracellular matrix. In plants, the newly formed cells are small in relation to their ultimate size. To accommodate future enlargement the cell walls of young growing plant cells are thinner and only semi-rigid (**primary cell walls**). The mature cell may retain its primary cell wall, thickening it considerably, or deposit new tough wall layers of a different composition forming **secondary cell walls**. In higher plants the cell wall is composed of **cellulose** fibres embedded in a protein/polysaccharide matrix, predominantly **hemicellulose** and **pectin**.

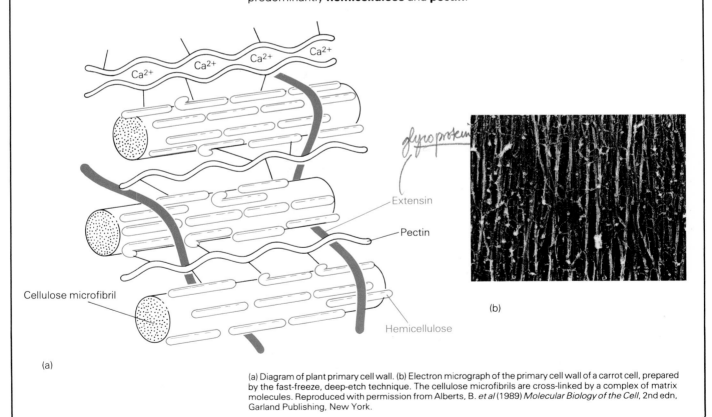

glyroprotein

Extensin

Pectin

Cellulose microfibril

Hemicellulose

(a)

(b)

(a) Diagram of plant primary cell wall. (b) Electron micrograph of the primary cell wall of a carrot cell, prepared by the fast-freeze, deep-etch technique. The cellulose microfibrils are cross-linked by a complex of matrix molecules. Reproduced with permission from Alberts, B. *et al* (1989) *Molecular Biology of the Cell*, 2nd edn, Garland Publishing, New York.

Reference (1987) *Methods in Enzymology*, **144, 145**. Advanced texts but give overviews on elastin, collagens, proteoglycans and electron microscopy of matrix molecules. Worth reading.

Box 7.1 cont'd

Cellulose is a linear polymer of glucose units linked by a characteristic β1–4 glycosidic bond. This linkage imparts a flat ribbon-like structure to the molecules, which adhere strongly to one another in parallel arrays of 70 chains to form **microfibrils**. Hemicelluloses are a heterogeneous group of branched polysaccharides that bind tightly, but non-covalently, to the surface of cellulose microfibrils and covalently to pectin molecules. The exact composition of plant hemicelluloses varies with species and developmental stage. They all have a β1–4-linked polysaccharide backbone of a single sugar-type from which short side-chains of other sugars protrude.

The third major constituent polysaccharide, the **pectins**, are heterogeneous, branched and highly hydrated polysaccharides made up of an α1–4-linked galactouronate backbone interrupted by occasional 1–2-linked rhamnose residues. The negatively charged pectins bind Ca^{2+} to form a semi-rigid gel. Pectin is abundant in the **middle lamella**, a region that serves to cement together the cell walls of adjacent cells.

One of the major glycoprotein components of primary cell walls in higher plants and algae are **hydroxyproline-rich glycoproteins** or **extensins**. They bind tightly to cellulose microfibrils. The large number of repeating amino acid sequences, Ser-$Hyp)_4$ is typical of structural proteins (for example, collagen; see *Biological Molecules*, Chapter 3, page 70), and they have been shown to be essential for normal wall development.

Cell wall maturation occurs by deposition of more primary cell wall constituents or when new components are laid down between the plasma membrane and the primary cell wall to form a secondary cell wall. In some cases spiral macromolecules are deposited within the existing wall or on its outer surface: lignin in xylem cells and cutin and waxes in epidermal cells, for example.

(a)

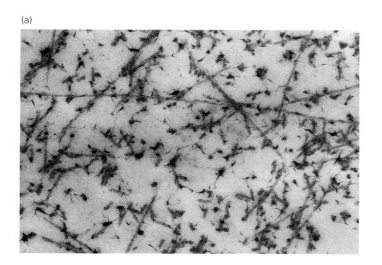

(b)

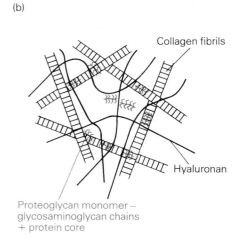

Collagen fibrils

Hyaluronan

Proteoglycan monomer –
glycosaminoglycan chains
+ protein core

Fig. 7.1 (a) Electron micrograph of the extracellular matrix produced by cultured chick-embryo chondrocytes stained with alcian blue. It shows the rod-like fibrils of Type II collagen with dark-stained bodies which are collapsed proteoglycan aggregates ($\times$35 100). Courtesy of Prof. A.I. Caplan, Dept of Biology, Case Western Reserve University. Cleveland, Ohio, USA. (b) Diagram of such a matrix.

7.3 *The fibrous proteins*

Mammalian connective tissues contain two essentially hydrophobic proteins which aggregate to form a fibrous scaffolding for the tissue.

See *Biological Molecules*, Chapter 3

COLLAGENS are a heterogenous family of proteins which occur in all eukaryotic animal phyla, except for the protozoans, and form the major

Reference Woodhead-Galloway (1980) *Collagen: the Anatomy of a Protein.* Institute of Biology Study Series in Biology, No. 117. Edward Arnold, London, UK. An inexpensive account of collagen as a protein.

Table 7.1 *General properties of collagens*

Type	No. of unique chains	Fibrous (F)/ non-fibrous (NF)	Gp*	Cartilaginous	Location
I	2	F	1		Interstitial tissues of organs bone, skin, etc.
II	1	F		Yes	Cartilage
III	1	F	1		Co-distributes with Type I
IV	2	NF	2		Basement membranes – some tumours
V	3	F	1		Minor component interstitial tissues
VI	3	Beaded fibrils	2		Most interstitial tissues
VII	1	Laterally-associated dimeric units	2		Placenta – anchoring fibrils
VIII	1	?	2		Cultured endothelial cells Descemets membrane
IX	3	F‡	3	Yes	Cartilage
X	1	F?	3	Yes	Cartilage
XI	3†	F	1	Yes	Mineralizing cartilage

* The collagens have been ordered into three groups with respect to size and proportion of triple helix in tropocollagen: M_r Gp1 larger than 95 000/high percentage helix; Gp2 large/low percentage triple helix, and Gp3 small/high percentage triple helix.
† One chain may be a highly glycosylated Type II α-chain.
‡ One α-chain has covalently-linked chondroitin sulphate chain.

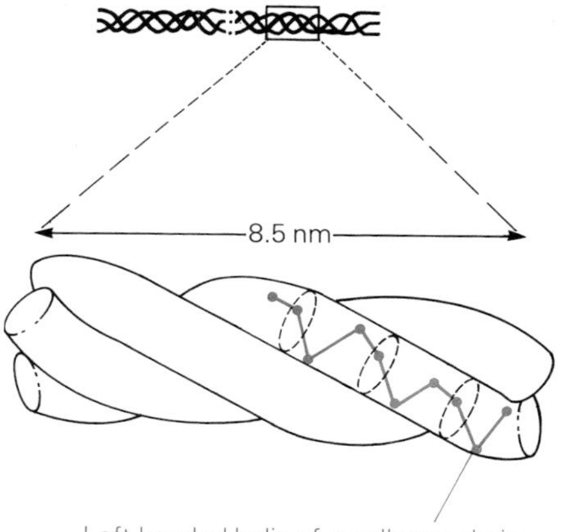

Left-handed helix of α-collagen chain

Fig. 7.2 Schematic representation of the triple helical region of a collagen molecule showing the internal GXY helix of one chain. Every third residue is glycine, the only amino acid small enough to occupy the internal site in the triple helix. The tropocollagen molecule shown is representative of the fibrous collagens, Types I, II, III, V and Xi, which have a 300 nm long helical region.

component of the extracellular matrix. More than eleven types of collagen are now recognized and characterized to varying degrees, and it is probable that others will be found (Table 7.1).

All collagen molecules are composed of three polypeptide chains, each containing considerable lengths of the repeating Gly–X–Y tripeptide sequence which allows triple-helix formation (Fig. 7.2). After being secreted into the extracellular space as pro-collagen molecules, carboxyl and amino terminal extension peptides are cleaved off by procollagen peptidases with the formation of insoluble, cross-linked, extracellular aggregates which function primarily as supporting elements within the tissue. Types I, II, III, V and XI collagens contain a long unbroken region of triple helix and assemble into ordered polymers called fibrils (10–300 nm in diameter) or fibres several μm in diameter (Fig. 7.3).

When viewed by electron microscopy, collagen fibrils exhibit a characteristic banding pattern due to the staggered arrangement of tropocollagen molecules. The fibrils are strengthened and stabilized by both intramolecular and intermolecular covalent cross-linking between lysine residues (Fig. 7.3). The minor cartilage collagens, types IX and X, also form heterogenous fibrils with type II collagen. Types IV, VI, VII and VIII collagens have large globular domains, either at one end (the carboxyl terminus in type IV collagen) or at both the carboxyl and amino ends (type VI) and do not associate into these simple fibrillar structures (Fig. 7.4). Type IV collagen appears to be stabilized by both lysine-derived cross-links and intermolecular disulphide bonds making it difficult to extract from its normal location in **basement membranes**.

ELASTIN AND ELASTIC FIBRES fulfil a requirement in the vertebrate body for tissues that can bend, twist and stretch reversibly during normal movement. The protein elastin, found in most connective tissues in conjunction with collagen, provides this property. High levels of elastin are present in ligaments, vessel walls (especially arteries) and lung tissue with lesser amounts in skin, tendons and other connective tissues.

☐ Two non-collagenous proteins possess collagen-like, triple-helical regions, namely **acetycholinesterase** and **complement component C1q**. Acetylcholinesterase hydrolyses acetylcholine to acetate and choline thus depolarizing the synaptic membrane (Section 9.8). It is located in the synaptic cleft where it is bound to a collagen/ proteoglycan network.

The action of the complement system results ultimately in the lysis of microorganisms and infected cells by creating holes in their plasma membranes. The process may be initiated by the formation of antibody complexes on the cell surface. Complement component C1q binds to the Fc portion (Section 11.3) of the antibody molecules and initiates the multistep cascade. C1q comprises six globular binding regions, each on a triple-helical, collagen-like stalk.

basement membrane: *a sheet of extracellular matrix seen in light microscopy underlying epithelial and endothelial cells and surrounding adipocytes (fat storage cells), Schwann cells and muscle cells. Usually separates these cells from the mesenchymal or connective tissue cells.*

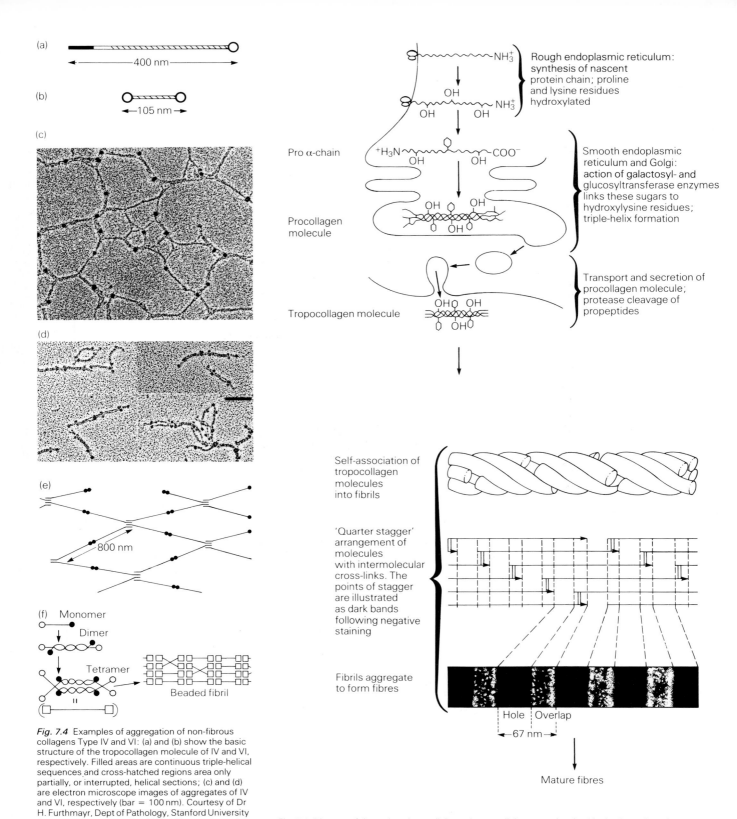

(a)

←———400 nm———→

(b)

←—105 nm—→

(c)

(d)

(e)

←800 nm→

(f) Monomer

Dimer

Tetramer

Beaded fibril

Rough endoplasmic reticulum:
synthesis of nascent
protein chain; proline
and lysine residues
hydroxylated

Pro α-chain

Smooth endoplasmic
reticulum and Golgi:
action of galactosyl- and
glucosyltransferase enzymes
links these sugars to
hydroxylysine residues;
triple-helix formation

Procollagen
molecule

Transport and secretion of
procollagen molecule;
protease cleavage of
propeptides

Tropocollagen molecule

Self-association of
tropocollagen
molecules
into fibrils

'Quarter stagger'
arrangement of
molecules
with intermolecular
cross-links. The
points of stagger
are illustrated
as dark bands
following negative
staining

Fibrils aggregate
to form fibres

Hole ┊ Overlap
←—67 nm—→

Mature fibres

Fig. 7.4 Examples of aggregation of non-fibrous collagens Type IV and VI: (a) and (b) show the basic structure of the tropocollagen molecule of IV and VI, respectively. Filled areas are continuous triple-helical sequences and cross-hatched regions area only partially, or interrupted, helical sections; (c) and (d) are electron microscope images of aggregates of IV and VI, respectively (bar = 100 nm). Courtesy of Dr H. Furthmayr, Dept of Pathology, Stanford University School of Medicine, USA; (e) and (f) are schematic representations of these micrographs, (e) showing the 'chicken-wire' structure of Type IV collagen and (f) the multimeric beaded fibrils of Type VI collagen.

Fig. 7.3 Diagram of the various intracellular and extracellular events involved in the formation of a mature collagen fibre. Some conversion of procollagen to tropocollagen is known to occur within the secretory vesicle. Furthermore, while intermolecular cross-links are shown (II), intramolecular and interfibrillar cross-links also form with maturation of collagen fibres.

The amino acid composition of elastin in certain respects resembles that of collagen. One-third of the residues are glycines and it is also rich in proline (10–13%), but not hydroxyproline or hydroxylysine residues. A high proportion of the remaining residues are non-polar (60%) making elastin a very hydrophobic protein. Elastin is found in all vertebrate species, with the exception of the *Agnatha* (for example, lampreys or hagfish).

Elastin is synthesized as a monomer of M_r 72 000, called **tropoelastin**. This is thought to have little secondary structure and to exist largely as a random coil. Intermolecular cross-linking occurs extracellularly and is catalysed by **lysyl oxidase** giving several different types of cross-link. **Desmosine** and **isodesmosine** are the most common. In mature elastin 28 of 34 possible lysine residues may be involved in intermolecular cross-linking.

The microscopic appearance of elastic fibres in tissue is 'amorphous' (Fig. 7.5), although small fibrils are associated with these fibres in most tissues. Analysis of these fibrils suggests that they could be type VI collagen. Fibre formation occurs close to the cell surface, which may allow the cell to control the orientation of the fibres. Also, the associated fibrillar proteins are thought to play a role in the organization of elastin molecules into fibres.

The amorphous appearance of elastin in tissues and the probable random coil structure of tropoelastin, suggests a cross-linked, random-coil structure for the fibres that allows the network to stretch and recoil like a rubber band (Fig. 7.6). Hydrophobic attraction between the elastin molecules appears to play a part in the recoil process. The associated collagen fibrils may serve to limit stretching.

□ In highly cellular tissues, such as the liver, muscle and embryonic cartilage, the extracellular matrix is a relatively minor component, with a microscopically less organized structure. It helps maintain tissue integrity and has been likened to an intercellular 'glue'.

See *Biological Molecules*, Chapter 3

□ Elastin from reptiles, amphibians and fish has a lower content of hydrophobic amino acid residues than that of mammals and birds. This correlates with a parallel change in systolic blood pressure, which is 4000 Pa in amphibians but 16 000 Pa in humans. The degree of hydrophobicity may be related to the elasticity of the elastin molecules.

Fig. 7.5 Electron micrograph of adult mouse trachea showing amorphous elastin fibres (e), and collagen fibres (c). Bar = 1 μm. Courtesy of Dr C. Frangblau, Dept of Biochemistry, Boston University School of Medicine, USA.

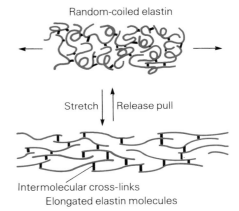

Random-coiled elastin

Stretch ⇅ Release pull

Intermolecular cross-links
Elongated elastin molecules

Fig. 7.6 Extensively cross-linked, randomly-coiled elastin molecules can elongate allowing the fibre to stretch and recoil.

7.4 *The ground substance*

Glycosaminoglycans (Table 7.2) form an important component of the extracellular matrix. The high negative charge on these polysaccharides imparted by the presence of sulphate and carboxyl groups, ensures they bind large amounts of water and cations. However, with the exception of *hyaluronan*, the glycosaminoglycans are not found as the free polysaccharide chains and are synthesized directly onto a protein core.

agnatha: *a class of fish-like primitive jawless vertebrates. It comprises the cyclostomes (lamprey and hagfish) and their extinct relatives, the ostracoderms.*

hyaluronan: *this term is increasingly applied to emphasize the macromolecular nature of the molecule, which was formerly known as hyaluronic acid.*

Table 7.2 *Properties of the glycosaminoglycans*

Glycosaminoglycan	Molecular weight	Repeating disaccharide	Sulphate/ disaccharide	Linked to protein	Tissue distribution
Hyaluronan (hyaluronic acid or hyaluronate)	4000–8 million	Glucuronate/*N*-acetylglucosamine	0	−	Various connective tissues, umbilical cord, rooster comb Synovial fluid, cartilage, vitreous body, skin, kidney, heart valves
Chondroitin-4-sulphate	5000–50 000	Glucuronate/*N*-acetylgalactosamine	0.2–1.0	+	Cartilage, cornea, bone, skin, arteries, young nucleus pulposus
Chondroitin-6-sulphate	5000	Glucuronate/*N*-acetylgalactosamine	0.2–2.3	+	Cartilage, cornea, bone, skin, arterial wall, umbilical cord, tendon, intervertebral disc
Dermatan sulphate	15 000–40 000	Iduronate (or glucuronate)/*N*-acetylgalactosamine	1.0–2.0	+	Skin, tendon ligaments, umbilical cord, heart valves
Heparin	6000–25 000	Glucuronate (or iduronate)/*N*-acetylgalactosamine	1.0–2.0	+	Arterial wall, lung, liver, skin, mast cells
Heparin sulphate	5000–12 000	As heparin	0.2–3.0	+	Lung, arteries, cell surfaces, basement membranes
Keratan sulphate	4000–12 000	Galactose/*N*-acetylglucosamine	0.9–1.8	+	Cartilage, cornea intervertebral disc, bone

HYALURONAN, ALSO CALLED HYALURONIC ACID OR HYALURONATE

is an 'honorary' proteoglycan, because the polysaccharide chain is not covalently bound to a protein core, although it does form non-covalent complexes with these protein cores. In common with the other members of the glycosaminoglycan family, it is an unbranched polysaccharide chain composed of a regular repeating disaccharide unit (glucuronate-*N*-acetyl-glucosamine). Hyaluronan is unique amongst the glycosaminoglycans for several reasons: its chain length is much greater than that of other glycosaminoglycans; it is not sulphated and does not undergo *epimerization* during synthesis; it is not covalently bound to a protein core, and it is not synthesized in the Golgi apparatus as are the other glycosaminoglycans. Hyaluronan is synthesized directly through the plasma membrane and elongation of the chain takes place by addition of sugars to the reducing end (Fig. 7.7). It is normally found in adult tissues in small amounts but is present in higher amounts during embryonal development, wound healing and in certain specialized tissues such as cartilage, the vitreous body of the eye, the umbilical cord and synovial fluid.

Hyaluronan is present in tissues as a coat surrounding the cell surface (Fig. 7.8), as part of larger aggregates with proteoglycans (see below) or as a seemingly free polysaccharide in synovial fluid and the vitreous body. Due to both repulsion of negatively charged groups and hydrogen bonding between the adjacent sugar residues there is an inherent stiffness in the molecule resulting in a polydisperse, high M_r, expanded coil structure (Fig. 7.9). However, at higher concentrations the chains can entangle and form loose double- or triple-helical structures forming a continuous network entrapping water molecules (compare with the agars of algae and the pectins of higher plants). The ability of even low concentrations of hyaluronan to form a continuous network is dependent on the high molecular weight of the poly-saccharide. Limited cleavage of the molecules results in a considerable reduction in the viscosity of hyaluronan solutions. In synovial fluid and the vitreous body of the eye the high concentration of hyaluronan serves two purposes: (i) it produces a highly viscous gel which retards water movement and also dampens the effects of rapid loading in joints; (ii) it excludes cells or large particulate matter to keep the light path clear in the eye.

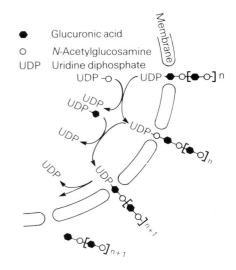

Fig. 7.7 Hyaluronan synthesis, showing the sequential addition of glucuronate from uridine diphosphate glucuronate (UDP-O) and *N*-acetylglucosamine from uridine diphosphate *N*-acetylglucosamine (UDP–●) to the reducing end of the chain. The chain is synthesized directly through the plasma membrane.

See *Biological Molecules*, Chapter 6

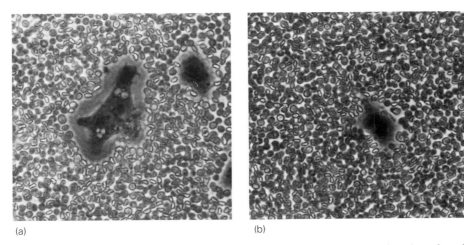

Fig. 7.8 (a) The surrounding coat of hyaluronan is apparent by its exclusion of erythrocytes from the surface of the cultured cell. (b) Fibrosarcoma cells from which the hyaluronan has been removed by digestion with hyaluronidase. The erythrocytes are no longer excluded and are closely associated with the cell surface.

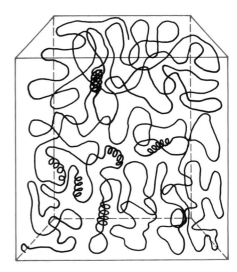

Fig. 7.9 The expanded coil model of hyaluronan. A molecule of M_r 8×10^6 occupies a cube of 300 nm^3.

epimerization: *inversion of configuration about one chiral centre, i.e. conformational change from one isomer to another: here the inversion is required to produce D-iduronate.*

The vastly greater resolution (0.5–1.5 nm) of the electron microscope over that of the light microscope (see Section 1.2 and Boxes 5.2 and 6.1) has been used to advantage in studying large aggregates, such as collagen, microtubules and also small isolated molecules such as fibronectin and laminin. Generally, large particles are examined by **negative staining** procedures.

In negative staining, heavy metal deposits (uranyl acetate or phophotungstate) surround the specimen and collect in cavities within the specimen. Thus, the particles appear as a bright image on a dark background, with open irregularities shown as dark patches. For smaller particles or large molecules, a technique called **rotary shadowing** is employed. The sample is placed on a grid, dried and placed in an evacuated chamber. Also in the chamber is a filament of a heavy metal (usually platinum and carbon) which is heated, causing it to evaporate. The filament and sample are positioned in such a manner that the metal vapour deposits on the grid on only one side of the specimen; that is, it casts a 'shadow' on the uncoated grid. This produces a 'three-dimensional' image of the sample.

See *Molecular Biology and Biotechnology,*
Chapter 5

Proteoglycan complexes

The glycosaminoglycans other than hyaluronan are synthesized directly onto a core protein *via* a linking oligosaccharide sequence (Table 7.3). Knowledge of the synthesis of proteoglycans is derived largely from experimental work on normal or tumour **chondrocytes**. The process closely resembles the biosynthesis of *N*- and *O*-glycosylated proteins. The core protein is synthesized on the rough endoplasmic reticulum. Addition of *N*-linked high-mannose-containing oligosaccharide commences in the endoplasmic reticulum, as does the addition of xylose residues. The further assembly of the glycosaminoglycan chains, addition of *O*-linked oligosaccharides and subsequent modifications of the glycosaminoglycan chains, such as sulphation and the epimerization of D-glucuronate to L-iduronate,

Table 7.3 *Examples of linkage regions of proteoglycans*

Chondroitin sulphate, dermatan sulphate, heparan sulphate and heparin
(*O*-linked)

–Gal–Gal–Xylose–Serine

Keratan sulphate
1. Cartilage (*O*-linked)

2. Cornea (*N*-linked)

Gal, Galactose; GlcNAc, *N*-acetylglucosamine; GalNAc, *N*-Acetylgalactosamine; Fuc, Fucose; Sia, Sialic acid.

chondrocyte: *cartilage-forming cell.*

take place in the Golgi apparatus. The completed cartilage proteoglycan is a large protein core containing covalently-linked chondroitin sulphate and keratan sulphate chains, together with other *N*- and *O*-linked oligosaccharides (Fig. 7.10). This is secreted by the cell and forms an aggregate with hyaluronan. This is stabilized by a second protein, the **link protein**, which binds to both hyaluronan and the proteoglycan molecules (Fig. 7.11).

The high negative charge density of the aggregated glycosaminoglycan chains causes them to repel each other, giving rise to a highly extended structure in solution, occupying a much larger volume than would be expected from their molecular weights, that is, 30–50 times their dry weight. This ability to bind and control large amounts of water is essential for the normal functioning of tissues such as cartilage, aorta and tendon, which must act as cushions for repetitive, variable, compressive loads. The molecules are reversibly compressible. When subjected to a compressive load, water is displaced from the molecules, decreasing the domain occupied by the proteoglycan aggregate, but increasing the charge density and thus the repulsion between glycosaminoglycan chains. When the load is removed, the proteoglycan molecules expand, imbibing water molecules. The concentration of proteoglycans in most cartilage is three to five times higher than would be possible if the molecules were fully extended and they occupy restricted spaces as small as 10% of their maximum. They are analogous to partially compressed coiled springs which can resist loading with less deformation than an uncompressed spring. In cartilage the swelling of the proteoglycan aggregates is limited by the network of collagen fibrils. The

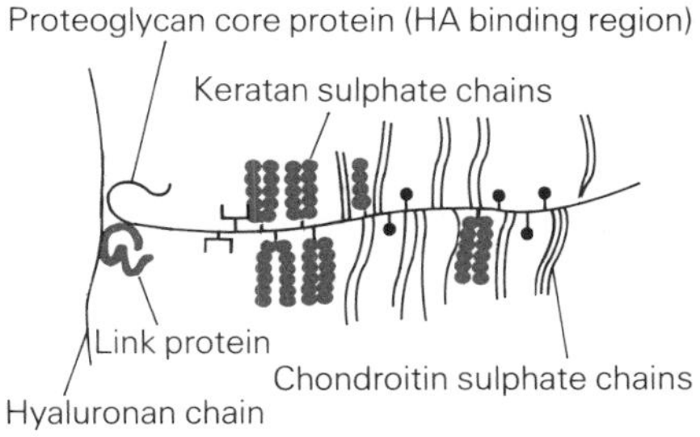

Fig. 7.10 Detailed drawing of a proteoglycan monomer aggregated with link protein and hyaluronan-*N*-linked oligosaccharides (Ψ) and *O*-linked oligosaccharides (↑).

Exercise 2

Age changes seen in human cartilage proteoglycans are: a decrease in the size of hyaluronan; reduced efficiency and stability of aggregate formation, and a decrease in the size of the glycosaminoglycan chains on the core proteins. What direct effects would you expect such changes to have on the cartilage matrix?

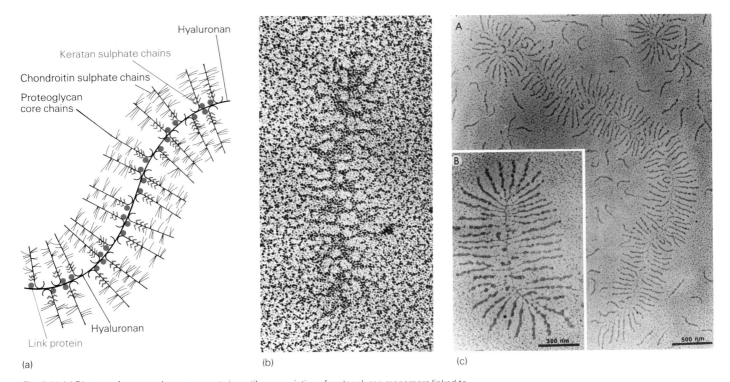

Fig. 7.11 (a) Diagram of a proteoglycan aggregate in cartilage consisting of proteoglycan monomers linked to hyaluronan and stablized by link proteins. (b) Electron micrograph of a chondrosarcoma proteoglycan subunit (×80 000). (c) Electron micrograph of (A) a large and (B) smaller proteoglycan aggregate from bovine foetal cartilage. Courtesy of Dr J.A. Buckwater, Dept of Orthopaedics, Veterans Medical Center and University of Iowa, USA; from *Journal of Bone and Joint Surgery* (1983) **65A**, 958–74 and *Collagen Related Research* (1983) **3**, 489–504.

resulting swelling pressure exerted by the proteoglycans gives cartilage its compressive stiffness and the ability to sustain loading during normal articulation of joints.

Three other, possibly more abundant, forms of proteoglycan have been recognized in recent years. Most tissues, including cartilage, contain smaller non-aggregating proteoglycans which have a core protein of M_r 40 000, with one to three covalently-linked chondroitin, dermatan or keratan sulphate chains. Their function is unclear but appears to be associated with the ordering and regulation of the fibril diameter of collagen.

The third form of proteoglycan is that present on the cell surface intercalated in the lipid bilayer of the plasma membrane. Such proteoglycans usually contain heparan sulphate or heparin chains. The latter are normally found on the surface of vascular endothelial cells. The exact structure of these proteoglycans has yet to be elucidated. However, the number of heparan sulphate chains per molecule is only three to eight and these are located at the extracellular, amino-terminal portion of the core protein.

Several functions have been proposed for these cell-surface proteoglycans. They are known to be able to bind types I, III, IV and V but not type II collagen. Some cellular specificity for individual collagens has been reported. Also, their ability to bind to fibronectin and cell–cell adhesion molecules, and to self-associate is thought to play a role in regulating cell–extracellular matrix and cell–cell communication. Heparin/heparan sulphate proteoglycans have also been shown to control surface coagulation (blood vessels), lipoprotein metabolism (blood vessels), smooth muscle cell growth, neurite formation, and basement membrane permeability (blood vessels and kidney). Mixed proteoglycans containing chondroitin sulphate and heparin sulphate on the same core have recently been reported; also an increasing number of cell surface glycoproteins may contain chondroitin sulphate chains. The significance of this is unknown.

Extracellular non-aggregating heparan sulphate proteoglycans are present in small amounts in most tissues and are major components of basement membranes and some tumours. They form a fourth group of proteoglycans, since protein analysis indicates that they are not derived from the cell-surface heparan sulphate proteoglycans. In the same way, core protein studies suggest that the large aggregating and small non-aggregating proteoglycans are not related and perform different functions. The main function of these heparan sulphate proteoglycans appears to be that of filtration.

☐ Small dermatan sulphate-containing proteoglycans (DS–PG) are bound to Type I collagen at the junction of the fibrils. Small DS–PG can self-associate and this property may be related to its ability to inhibit fibril formation.

Exercise 3

Cultured brain cells synthesize two forms of heparan sulphate proteoglycan (HS–PG): one an integral membrane form present on the cell surface and a small non-aggregating form found free in the culture medium. Suggest a metabolic experiment to test the theory that the small HS–PG is derived from the membrane-type HS–PG.

7.5 Extracellular matrix diversity

The term extracellular matrix encompasses the microscopically amorphous pericellular layer close to the surface of most cells; the more fibrous interstitial matrix of most tissues; the calcified or hard matrix of bone and teeth, and the basement membranes. The structural, or mechanical, functions of the extracellular matrix are well known from the study of the 'classical' connective tissues such as mature cartilage, tendon, bone and intervertebral disc. These comparatively acellular tissues have an extracellular matrix which has a structure and molecular composition adapted to perform a largely mechanical role (Fig. 7.12a, b).

The calcified matrix of bone is rigid and mechanically strong. It is formed by the deposition of calcium phosphate, or carbonate, in a cartilage-like collagen–proteoglycan network (Fig. 7.12c). The matrix of tendon is composed of longitudinally-orientated bundles of collagen which are strong and relatively inelastic to withstand longitudinal stress (Fig. 7.13). Cartilage

Reference Gallagher, J.T., Lyon, M. and Steward, W.P. (1986) Structure and function of heparan sulphate proteoglycans. *Biochemical Journal*, **236**, 313–25. Advanced review article but very readable.

Reference Poole, A.R. (1986) Proteoglycans in health and disease: structures and functions. *Biochemical Journal*, **236**, 1–14. Advanced but good introduction to different forms of proteoglycans.

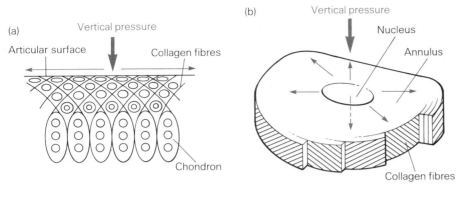

(a)

Vertical pressure

Articular surface

Collagen fibres

Chondron

(b)

Vertical pressure

Nucleus

Annulus

Collagen fibres

☐ Humans are tallest in the morning. The pressure exerted on the intervertebral disc results in loss of tissue fluid during the day. This results in a decrease of 3 cm in height during the day. Thinning of the intervertebral disc with age also results in a decrease in height of several centimetres.

(c)

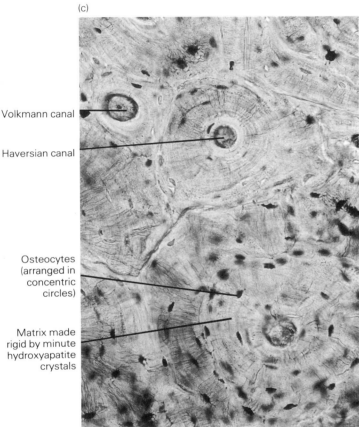

Volkmann canal

Haversian canal

Osteocytes (arranged in concentric circles)

Matrix made rigid by minute hydroxyapatite crystals

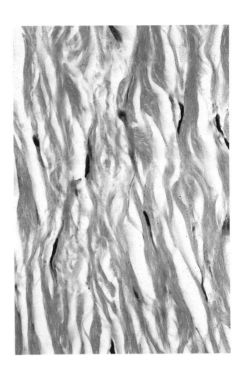

Fig. 7.13 LS tendon: closeup of the wavy bundles of collagen fibres. × 320. Courtesy of Biophoto Associates.

Fig. 7.12 (a) A hypothetical view of articular cartilage. The basic structural/functional unit, the chondron, is a single chondrocyte (or cell group) surrounded by an hydrated proteoglycan extracellular matrix. The osmotic pressure exerted by the proteoglycan keeps the collagen framework taut. Point loads are rapidly converted into tensile strain throughout the tangential surface collagen fibrils and eventually result in an increase of the pressure within the chondron, thus increasing the tissue's mechanical resistance. (b) The intervertebral disc is made up of 15–20 concentric fibrous lamellae (the annulus fibrosis) which enclose a gel-like central region. Each lamella is composed of coarse bundles of collagen fibres which run an oblique course between adjacent vertebrae. The direction of the fibres alternates between lamellae. The lamellae contain a sparse cellular population embedded in a hydrated proteoglycan 'ground substance' which acts as a lubricant. The deformable, but incompressable centre redirects vertical comprehensive forces into tensile strain on the collagen corset, or annulus. The annulus also allows about three degrees of rotation per vertebra. Under prolonged loading, fluid loss from the nucleus serves to increase the mechanical resistance. (c) A section of human compact bone. The bone cells (osteocytes) appear as concentric layers of black-stained cells around a central cavity (Haversian canals) containing blood vessels and other cells. The matrix is composed of strong collagen fibres impregnated with calcium phosphate (hydroxyapatite) crystals. A second series of cavities (Volkmann canals) run perpendicular to the length of the bone and convey periosteal vessels into the Haversian canals (×375). Courtesy M.J. Hoult, Dept of Biological Sciences, The Manchester Metropolitan University, UK.

Reference Brown, C.H. (1975) *Structural Materials in Animals*. Edward Arnold, London, UK. Old, but gives good account of comparative skeletal structures in animals.

Reference Caplan, A.I. (1984) Cartilage. *Scientific American*, **251**(4), 82–90. Basic text on structure and function of cartilage.

Mature corneal stroma contains a small keratan sulphate proteoglycan (KS–PG) the molecules of which are thought to be small enough to fit between the uniformly arranged collagen fibrils and contribute to the refractive properties of the tissue. The embryonic cornea is opaque and contains large DS–PG (dermatan sulphate proteoglycan) which alters the interfibrillar distance between the collagen fibrils. Similar opacity is seen in corneal scars and in the genetic deficiency **corneal macular dystropy**, a degenerative genetic disorder characterized by progressive degeneration of the cornea marked by the appearance of grey punctate opacities within the stroma.

☐ Load-bearing connective tissues are normally avascular since the presence of tissue blood vessels would disrupt the architecture and mechanical properties of these tissues.

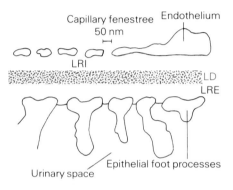

Fig. 7.14 Part of the kidney glomerulus showing the periphery of a capillary where filtration takes place. The filtration surface consists of a thin capillary endothelial cell containing fenestrae, or pores, about 50 nm in diameter. The endothelial cell sits on the *lamina rara interna* (LRI) of the underlying glomerula basement membrane. The *lamina densa* (LD) separates this from the *lamina rara externa* (LRE) on which sits the epithelium. The epithelial foot processes are separated by 25 nm filtration slits, which allows the glomerular filtrate to pass into the urinary space.

--- *Exercise 4* ---

The classical weight-bearing connective tissues are normally avascular (i.e. lacking blood vessels). Suggest how their normal functions allow them to obtain nutrients and dispose of waste products?

and intervertebral disc possess highly structured networks of collagen fibres which are both prestressed and lubricated by a water-holding proteoglycan–glycoprotein 'ground substance', which enables these tissues to dampen and redistribute considerable transient loads through the collagen matrix. In skin, collagen fibrils form a wickerwork of interlacing bundles providing a stretchable and tough barrier whereas its counterpart covering the eye, the cornea, is a transparent lattice of collagen fibrils of uniform diameter and distance apart. The alternating layers of collagen are orientated essentially perpendicular to each other, with a gradual clockwise shift in the orientation. This structural arrangement serves both barrier and optical functions.

Another specialized form of extracellular matrix is the basement membrane. Microscopically these are seen as thin amorphous structures at the basal or abluminal surface of epithelial and endothelial cells respectively, and have a different chemical composition and organization to the underlying mesenchymal matrix (see Table 7.1). Such membranes undoubtedly impart mechanical strength to the attached epithelial or endothelial cell layer but they also have important functions as molecular filters. An important example of this is filtration in the glomerulus of the mammalian kidney (Fig. 7.14).

Older studies emphasized the mechanical functions and properties of the extracellular matrix and resulted in a view of the extracellular matrix as an inert supporting material synthesized by the cells as a mere scaffolding, on (or in) which to reside. Two important observations contradict this view:

1. The loss of normal differentiated functions, encountered when cells are placed in culture, can be reversed if the cells are grown on the naturally occurring matrix, or on isolated components of the matrix (Table 7.4).
2. Removal of certain matrix components has profound effects on organ development (Table 7.5).

Cell–extracellular matrix interactions, it was realized, are important in processes such as embryogenesis, differentiation, growth, cell migration and tissue repair. It is now apparent that most cells possess specific cell surface receptors for many components of the extracellular matrix but the way in which these control cellular responses is not well understood.

Table 7.4 *Influence of matrix components on cultured cells*

Component	Cell	Effect
Collagen (Type I) individual α1, α2 chains, peptides	Myoblasts	Promotes muscle-fibre formation
Collagen (Type I and II) + procollagen	Chick embryonic sclerotomal cells	Chondrocyte maturation
Floating collagen gels (Type I)	Rat hepatocytes Mammary epithelium	Morphological and biochemical differentiation (i.e. cytochrome P450 induction, secretion milk proteins, lumen formation)
Type (IV) kidney epithelium	Mammary epithelium Endothelial cells	Stimulates growth Inhibits growth
Hyaluronan	Myoblasts Chondroblasts Endothelial cells Adrenal cortical cells	Inhibits muscle-fibre formation Inhibits chondrocyte maturation Inhibits growth Inhibition of corticosterone secretion, induced motility
	Fibroblasts	Concentration-dependent migration
Fibronectin	Cancer, or transformed, cells	Increased adhesion 'normal morphology'

Table 7.5 *In vivo effects of changes in extracellular matrix alteration*

Stimulus	Cells	Effect
Implanted demineralized bone matrix	Subepidermal fibroblasts	Conversion to chondrocytes and osteocytes
Collagenase treatment (also inhibition of collagen synthesis)		Perturbed lung and salivary gland development
Collagenase treatment (also inhibition of collagen synthesis)	Chick embryonic skin	Feather morphogenesis
Hyaluronidase treatment		Neural crest development inhibited
Implanted hyaluronan		Blood vessel formation inhibited

7.6 Focal adhesions: specialized cytoskeleton–extracellular matrix associations

Many components of the extracellular matrix promote the adhesion, aggregation and spreading of tissue-cultured cells. In electron microscopy, sites of cell matrix interaction appear as discrete areas of thickened cell surface adjoining the extracellular matrix (Fig. 7.15). These have become known as adhesion plaques, focal contacts or focal adhesions, and closely resemble cellular junctions of the adherens type. Ultrastructural examination shows that microfilaments and intermediate filaments are present in high frequency at these attachment sites. The microfilaments are often present as bundles, known as stress fibres, extending from the focal adhesion to the nucleus. Studies with specialized light microscopy techniques have revealed a close association between **actin** bundles (stress fibres) and contact points (Fig. 7.16). This close association between the cytoskeleton and the extracellular matrix may well mediate the metabolic control exerted on the cell by its extracellular matrix.

See Chapter 6

Much of our current knowledge about the composition of focal adhesions comes from immunological localization studies and analysis of the material remaining attached to the substrate after disrupting cultured cells. Such

actin: *major constituent of microfilaments.*

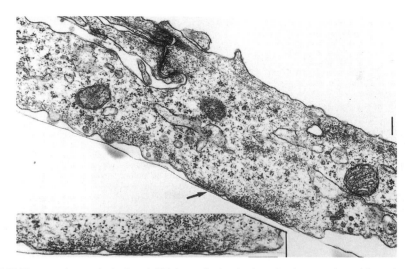

Fig. 7.15 Electron micrograph of cultured chick lens cells showing large focal contacts (arrow) formed with the substratum, as well as adherens-type junctions (Chapter 4) with neighbouring cells. The insert shows the microfilaments associated with focal contacts and bundles of intermediate filaments which are enriched in these areas (bar = 0.2 μm). Courtesy of Dr B. Geiger, Dept of Chemical Immunology, The Weizmann Institute of Science, Israel; from *Journal of Cell Science*, Suppl. 8, 251–72.

analyses have shown that these sites contain a variety of cytoskeletal and extracellular matrix components (summarized in Figure 7.17). The exact composition is known to vary with the length of time the cells are in contact with the substrate and also between normal and cancer cells. One protein, **talin**, seems to be uniquely associated with focal adhesions and is not present in cell–cell adherens junctions.

7.7 *Molecules that mediate cell adhesion*

□ Cell-cell interactions also play a major role in regulating differentiation and cell growth. They are mediated by **cell adhesion molecules** (CAMs) which are present in adherens junctions. A CAM on one cell binds to a CAM of the same type on an opposing cell. This cell specificity leads to cell ordering in the developing embryo.

One of the most obvious consequences of cell–extracellular matrix interaction is the adhesion, spreading and migration of cells cultured *in vitro*. Many types of cells will rapidly adhere to and spread on surfaces coated with extracellular matrix components. The degree of spreading is dependent on the concentration of matrix component coated on the surface; that is, the number of focal contacts the cell can make. Also, when a cell encounters limiting concentrations of an adhesive matrix component, it tends to migrate to regions where the concentration is sufficient to allow cell spreading. It appears that cell migration occurs when cell–matrix interactions are few in number but sufficient to impart the traction needed for motility; increasing the number of cell adhesion sites immobilizes the cell. Such a mechanism may also be valid *in vivo*, directing cell migration during embryogenesis and tissue repair.

Two approaches have been used to characterize the proteins involved in cell adhesion. One is based on the production of antibodies which are capable of inhibiting or disrupting cellular adhesion, spreading and migration. The molecules with which these antibodies interact are then identified by immunoprecipitation, immunoblotting, or antibody affinity chromatography. The advent of monoclonal antibodies, with their greater specificity, has greatly increased the usefulness of this approach.

See Chapter 11

The second approach employs a purified extracellular matrix component immobilized on a chromatographic matrix such as agarose or polyacrylamide. Affinity chromatography of a suitable cell or tissue extract is then performed to identify proteins which interact with the immobilized component. This approach identifies possible cell surface receptors for the immobilized extra-

talin: *a link protein between integrins and the actin cytoskeleton. Present in adhesion plaques.*

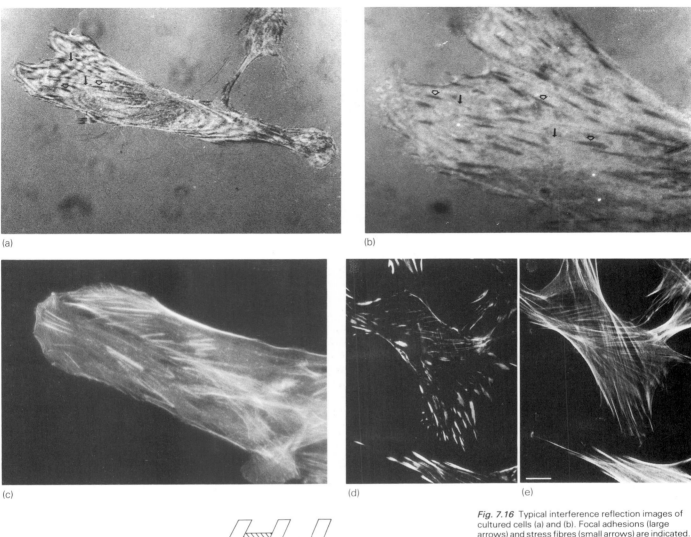

(a)

(b)

(c)

(d)

(e)

Fig. 7.16 Typical interference reflection images of cultured cells (a) and (b). Focal adhesions (large arrows) and stress fibres (small arrows) are indicated. (c) Typical pattern of actin stress fibres, stained with phalloidin, terminating at adhesion contacts, courtesy of Drs A. Sattar, and J. Aplin, Dept of Obstetrics and Gynaecology, University of Manchester, St Mary's Hospital, Manchester, UK. Double fluorescent staining of chick fibroblasts for (d) vinculin and (e) actin showing typical association of vinculin with end of actin stress fibres (bar = 10 μm). Courtesy of Dr B. Geiger, Department of Chemical Immunology, The Weizmann Institute of Science, Israel, from *Journal of Cell Science*, Suppl. 8, 251–72).

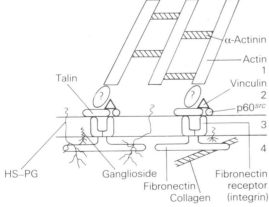

Fig. 7.17 Some of the components identified in focal adhesions. (1) The actin filaments are shown cross-linked by α-actinin. Filamin and myosin have also been detected. (2) On the cytoplasmic side of the plasma membrane are the proteins talin and vinculin, which link some receptors (integrins) to actin. There is an undetermined linkage component(s) between vinculin and actin. Other possible components at the same site include regulating proteins such as clatharin, Ca^{2+}-dependent proteases and the *src*-oncogene product (a protein kinase). (3) Integral membrane components include the integrin group of receptors (see text), intercalated heparan sulphate proteoglycan (HS–PG) and gangliosides which appear to play a regulatory role. (4) The extracellular compartment has been shown to contain fibronectin, heparan sulphate and chondroitin sulphate proteoglycans, and hyaluronan.

Reference Burridge, K., Molony, L. and Kelly, T. (1987) Adhesion plaques: sites of transmembrane interaction between the extracellular matrix and the actin cytoskeleton. *Journal of Cell Science*, suppl. 8, 211–29. Good review although in some places rather advanced. The whole supplement is of interest.

cellular matrix component; confirmation requires that the antibodies to the affinity purified 'matrix receptor' perturb cellular adhesion or localization to focal adhesions. Further validation can be given by showing that these antibodies, or peptides derived from the matrix component, inhibit receptor–ligand (extracellular matrix component) binding.

The application of these techniques has shown that two different classes of molecules are involved in cell–matrix adhesion: membrane receptors and cell–surface/extracellular matrix glycoproteins. The latter have also been termed **adhesive** or **structural glycoproteins**.

Structural glycoproteins

Structural glycoproteins are multifunctional molecules. They usually interact with several components of the extracellular matrix and with cell surfaces, through specific binding domains, and appear to play an important role in cell–matrix interactions as cell-anchoring proteins. Their ability to self-associate and bind to other components of the matrix suggests that they also mediate cellular organization of the surrounding matrix. **Fibronectin** and **laminin** are the best characterized examples of this group of proteins.

FIBRONECTINS are large glycoproteins (M_r 450 000) present in most body fluids and connective tissues, and are synthesized by most animal cell types in culture (Table 7.6). Fibronectin was originally discovered in the early 1970s as a cell surface protein present on normal fibroblasts but not on cancer cells. It was subsequently shown to be identical with 'cold insoluble globulin', a plasma protein known to bind to fibrin and fibrinogen. The discovery that fibronectin promotes cell attachment has stimulated intensive study. Indeed, it has become the classical example of a cell–substrate adhesion protein. Fibronectins have been implicated in a variety of biological activities, most of

Table 7.6 *Distribution of fibronectins*

Major component	Cells and tissues Absent or minor component	Body fluids (mg dm^{-3})	
Connective tissues	Lymphocytes, erythrocytes	Plasma	300
Glial tissue	Myelocytes	Amniotic fluid	170
Basement membranes	Epithelial cells (intestinal	Cerebrospinal fluid	1–3
Fibroblasts	glomerular, liver and	Seminal fluid	100
Myoblasts	mammary)	Urine	10
Undifferentiated chondrocytes			
Endothelial cells			
Amniotic epithelial cells			
Astroglial and Schwann cells			
Platelets and macrophages			

Reference Hynes, R.O. (1986) Fibronectins. *Scientific American*, **254**(6), 32–4. Good basic introduction.

Table 7.7 Properties of fibronectin binding sites

Site (physiological importance)	Properties
Fibrin(ogen) I (clot formation, wound healing)	Binding weak at 37°C, probably stabilized by transglutaminase cross-linking to fibrin(ogen)
Heparin I (cell binding, cell migration)	Much weaker than heparin II, at 37°C, physiological significance uncertain
Staphylococcal-binding (infection)	Thought to opsonize *Staphylococci*, but more probable that strains have evolved the ability to bind fibronectin to aid infection
Transglutaminase Cross-linking site (clot formation)	Involves glutamine residue three residues from amino terminus
Collagen-binding (gelatin) (cell-matrix binding and ordering)	Composed of two Type II repeats. Binds gelatin more tightly than native collagen. Also binds denatured collagen portions of complement C1q and acetylcholinesterase
DNA-binding (?)	Next to collagen binding site towards carboxyl terminus. Significance unknown
Cell attachment sites (cell migration, matrix interaction)	Contains Arg–Gly–Asp (RGD) sequence which bind to cell surface receptor(s)
Heparin II (proteoglycan aggregation, cell-surface attachment)	Major function is the binding of proteoglycans
Fibrin(ogen) II (clot formation)	As fibrin(ogen) I

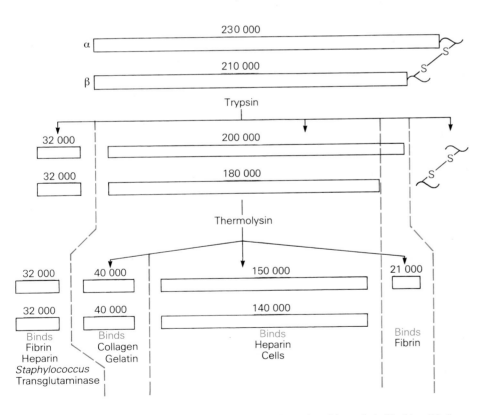

Fig. 7.18 An illustration of proteolytic dissection of fibronectin with trypsin and thermolysin. The M_r and binding affinity of the digestion product is indicated at each stage.

☐ Staphyloccocal cells bind to fibronectin. It is thought to aid the microorganism in establishing infection. The earlier notion that this was a host defence mechanism seems improbable given the greater ability of bacteria to adapt and shed their fibronectin binding site.

which involve adhesive or ligand-binding functions (Table 7.7). They are composed of subunits of M_r 220–250 000 linked by disulphide bonds into dimers (Fig. 7.18).

Fibronectin contains specific binding domains for fibrin, heparin/heparan sulphate, hyaluronan, DNA, collagen/gelatin, *Staphylococcus aureus* and cell surfaces, and may act as a substrate for the enzyme **transglutaminase**. The experimental use of controlled proteolytic cleavage of the molecule has led to the recognition of discrete binding domains (Figs 7.18 and 7.19). The fibronectin may be bound to the ligand of interest, for example, collagen or heparin, and then digested with proteases. Alternatively, protease digestion

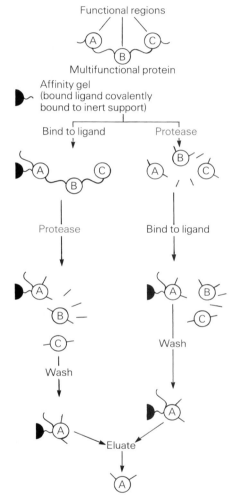

Fig. 7.19 Two approaches for isolating binding domains of multifunctional molecules. The binding molecule has three binding domains, A, B and C. The ligand (molecule bound) is immobilized by covalent coupling to agarose beads. In one approach the binding protein is allowed to adhere to the ligand, via a specific binding site A, and it remains bound after protease treatment. In the alternative strategy, the binding protein is protease digested prior to binding to the immobilized ligand. The protease digestion is controlled so that the binding site, A, remains active. In both approaches, the binding domain is eluted under dissociating conditions.

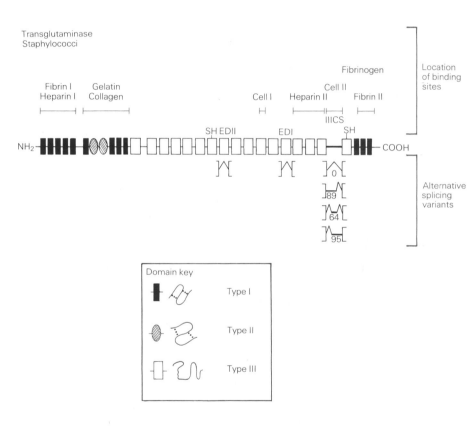

Fig. 7.20 Structure of fibronectin. The distribution of the three types of homologous repeat sequences are indicated, as are the domains involved in the binding to cell surfaces and other extracellular matrix molecules. Three regions of alternative splicing of the fibronectin mRNA (EDI, EDIII, and IIICS) are shown (see text). EDI and EDII segments can be spliced out completely or remain unspliced. In humans, differential splicing of the IIICS domain can generate up to five variants.

may precede affinity chromatography (Fig. 7.19). Such multifunctional molecules can be thought of as consisting of multiple discrete domains each of which perform a specific function. The functional domains are independently folded regions of the protein chain separated by flexible connecting regions, which tend to be more susceptible to cleavage by proteases.

THE FIBRONECTIN POLYPEPTIDE is composed of three types of repeat units, Types I, II and III (Fig. 7.20). The Type I repeat contains 45–50 amino acid residues, held together by two disulphide bonds. Type I repeats are confined to the carboxyl- and amino-terminal regions of the molecule. Type II repeats contain 60 amino acid residues with two internal disulphide bonds. Two of these segments interrupt a sequence of nine Type I segments at the amino-terminus to form the collagen-binding domain. The remaining central region of the molecule is composed of an uninterrupted sequence of 15–17 Type III repeats, each about 90 amino acid residues long and lacking disulphide bonds. Two free sulphydryl groups are present in this region which may be involved in the formation of high M_r multimers. The basic fibronectin molecule is normally composed of two such polypeptide chains

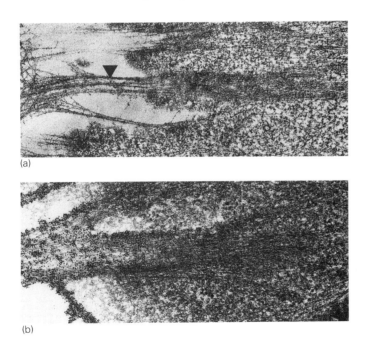

(a)

(b)

Fig. 7.21 Fibronectin in the extracellular matrix. (a) Electron microscopic visualization of the relationship between extracellular fibronectin fibrils (arrowhead) and intracellular microfilament bundles. (b) A similar section stained for fibronectin using an immunoferritin technique. Courtesy of Dr I. Singer, Department of Biochemical and Molecular Pathology, Merck, Sharp & Dohme Research Laboratories, New Jersey, USA.

Reference Ruoslahti, E. (1988) Fibronectin and its receptors. *Annual Reviews in Biochemistry*, **57**, 375–413. Very good, detailed review.

covalently bound through two disulphide bonds, located near the carboxyl-terminus, forming a dimer of M_r 450 000. Two major forms of fibronectin exist: a soluble dimeric form secreted by hepatocytes and endothelial cells, which is found in plasma, and a dimeric or multimeric form, made by many cell types, which is usually deposited in the extracellular matrix as long, insoluble fibrils (cellular fibronectin: see Fig. 7.21).

MULTIPLE FORMS OF FIBRONECTIN have been demonstrated by protein sequencing, and several cDNA clones have been isolated and sequenced. These have shown that structural differences exist not only between plasma and cellular fibronectins but also between the subunits of each type. However, despite the existence of multiple forms of fibronectin there is only one copy of the gene in the human and rat genomes. The human gene contains about 60 exons and the single primary RNA transcript then undergoes a complicated pattern of alternative splicing to produce different fibronectin mRNAs. There are three regions of alternative splicing in the fibronectin transcript (Fig. 7.20). Two of the regions are Type III repeat domains, called ED (Extra Domain) I and EDII in human fibronectin, which can be included or omitted from the mature mRNA. The third region located towards the carboxyl-terminus corresponds to a domain of 120 amino acid residues, termed IIICS. An intricate splicing pattern in this region gives rise to five mRNA variants, some of which may contain up to two extra cell-binding sequences. Alternative splicing of both the EDI and EDII exons is cell-specific. Hepatocytes only synthesize fibronectin lacking both ED domains, while fibroblasts and many other cell types produce all possible variants. The biological significance of alternative splicing is not known, but it could alter the functions of the fibronectin molecule.

EDII contains a potential glycosylation site. Glycosylation of this site may reduce the binding properties of the nearby cell-binding domain. EDII also contains a protease-sensitive site which may give rise to processing differences. This domain is also preferentially expressed in cancer cells which exhibit greatly increased degradation of fibronectin. However, the lack of EDI and EDII domains in plasma fibronectin is not reflected in any gross functional differences when compared to cellular fibronectin.

Fibronectin, as would be expected with a secreted protein, is synthesized with a signal peptide sequence. This is cleaved, together with an associated 5–7 amino acid propeptide, to give the virgin polypeptide. The major post-translational modification of fibronectin is **glycosylation**. Fibronectin contains between 4 and 10% carbohydrate, depending on the source; fibronectins isolated from adult sources have a low carbohydrate content in comparison with fibronectin from embryonal and tumour cell surfaces. Differences in carbohydrate compositions in general reflect differences in glycosyltransferase activities between tissues and cell types. Fibronectin from adult sources contains mostly N-linked carbohydrate with some short O-linked moieties.

Fetal and tumour-derived fibronectins are also substituted with poly-lactosaminoglycan chains and possibly heparan sulphate (Fig. 7.22). The role of glycosylation seems to be to protect the polypeptide against proteolysis, but substitution with large carbohydrate groups, such as polylactosamino-glycan chains, would be expected to interfere with ligand binding. Indeed, placental fibronectin molecules containing polylactosaminoglycan bind less well to gelatin than those without.

FIBRONECTIN HAS BINDING DOMAINS which interact with cell surfaces and several extracellular matrix components. These include two separate binding domains each for fibrin(ogen) and heparin and a single domain each

See *Molecular Biology and Biotechnology*, Chapters 5 and 9

☐ Analysis of the fibronectin gene has shown that each repeat is an individual exon and suggests that the gene has evolved by acquiring pre-existing exons from other genes, with subsequent duplication. Thus, sequences homologous to the Type I, fibrin binding domain, in fibronectin are also present in other fibrin-binding proteins, tissue plasminogen activator and clotting Factor XII. Type II sequences are present in Factor XII, prothrombin, plasminogen and urokinase. Type III sequences have not been detected in other proteins.

Exercise 5

How many possible fibronectin monomers may be produced by alternative splicing in humans?

See Chapter 1

☐ RGD is an example of the single letter nomenclature for designating amino acid sequences, i.e. Ala (A); Arg (R); Asn (N); Asp (D); Asn or Asp (B); Cys (C); Gln (Q); Glu (E); Gln or Glu (Z); Gly (G); His (H); Ile (I); Leu (L); Lys (K); Met (M); Phe (F); Pro (P); Ser (S); Thr (T); Trp (W); Tyr (Y), and Val (V). The letter X is used for an unidentified amino acid. Thus, RGD is the sequence Arg–Gly–Asp.

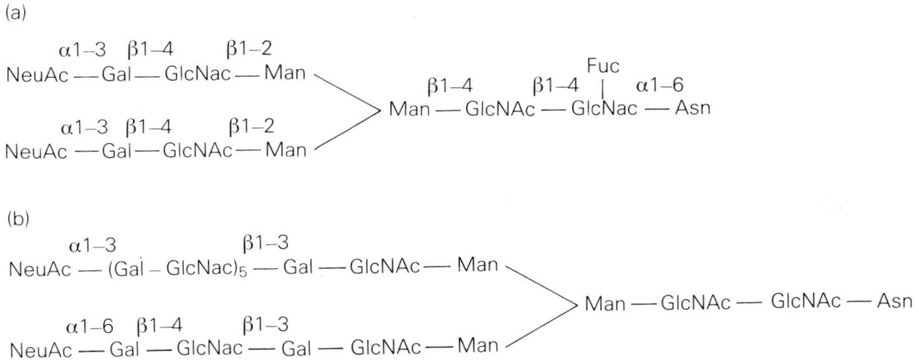

(a)

```
    α1-3    β1-4      β1-2
NeuAc — Gal — GlcNac — Man  ╲
                              ╲              Fuc
                               ╲   β1-4   β1-4  │  α1-6
    α1-3    β1-4      β1-2       ╱  Man — GlcNAc — GlcNAc — Asn
NeuAc — Gal — GlcNAc — Man   ╱
```

(b)

```
     α1-3                    β1-3
NeuAc — (Gal – GlcNac)₅ — Gal — GlcNAc — Man  ╲
                                               ╲
                                                ╲  Man — GlcNAc — GlcNAc — Asn
     α1-6   β1-4      β1-3                      ╱
NeuAc — Gal — GlcNac — Gal — GlcNAc — Man    ╱
```

Fig. 7.22 Examples of *N*-glycans of fibronectin: (a) is an oligosaccharide found on cellular fibronectin and (b) is a fetal lactosaminoglycan found on a number of cells. This is a type of glycosaminoglycan chain containing long side-chains of a repeating *N*-acetylactosaminyl disaccharide (Galβ1-4GlcNAcβ1-3) attached to the normal mannose (Man) and *N*-acetylglucosamine (GlcNAc) core (compare with keratan sulphate, Tables 7.2 and 7.3) *N*-glycans. The chains are much longer (M_r 7–11 000).

for *Staphylococcus* cells, DNA, collagen (gelatin) and cell surfaces (see Fig. 7.20). In addition, two sites for transglutaminase-catalysed cross-linking to other proteins have been identified. Fibronectin may also have specific sites involved in self-association.

The amino-terminal region (Fig. 7.20) contains the fibrin(ogen) I and heparin I binding sites, a *Staphylococcus*-binding site and a transglutaminase cross-linking site (see Table 7.7). In addition, some of the disulphide bonds in this region are involved in fibronectin aggregation via disulphide exchange, i.e. intramolecular disulphide bonds rearrange to form intermolecular bonds (Fig. 7.23). The juxtaposition of the fibrin I site and the transglutaminase cross-linking site (Fig. 7.24) is significant. At physiological temperatures, the fibrin I binding site exhibits only weak binding and trans-glutaminase cross-linking may be essential to stabilize the interaction. Fibronectin–fibrin binding may be important in the clotting process and in wound healing.

Fragmentation studies have localized the collagen (gelatin) binding domain to the two Type II repeat units (Fig. 7.20). Fibronectin binds gelatin more strongly than native collagen, which raises the question of the biological significance of this binding region. However, at 37°C, the collagen triple helix seems to loosen in certain regions. In type I collagen, one such site occurs at the position where it is cleaved by collagenase, and fibronectin also binds to this same region.

The cell attachment site is located in one of the central Type III repeat units which contains the amino acid sequence Arg–Gly–Asp (RGD). This sequence is essential for binding to the cell-surface receptor. The main binding site for heparin is located in the three Type III homology repeats between the two alternatively spliced segments EDI and IIICS. This site has a relatively strong affinity for heparin, and the binding is of an ionic nature. The exact position of the binding site is difficult to locate as fibronectin does not contain a definite cluster of positively charged amino acid residues. This domain also binds the less-sulphated glycosaminoglycans heparan sulphate, dermatan sulphate and chondroitin sulphate, but the interaction is weaker. Thus, fibronectin may interact with the cell surface via its cell-attachment domain and also by binding to cell-surface proteoglycans. The ability of fibronectin to bind proteoglycans and cells suggests that it may play a significant role in ordering matrix deposition, a view given more credence by its formation of multimeric fibrils extending from the cell surface into the extracellular matrix. Another structural glycoprotein, **laminin**, is thought to play a similar role.

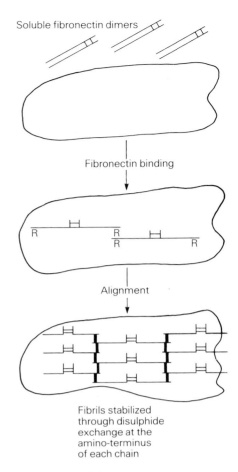

Soluble fibronectin dimers

Fibronectin binding

Alignment

Fibrils stabilized through disulphide exchange at the amino-terminus of each chain

Fig. 7.23 Hypothetical scheme of fibronectin matrix assembly. Solute fibronectin dimers are bound to cell-surface receptors. Subsequent alignment of the receptors may be receptor-mediated and fibril formation occurs by disulphide interchange of amino-terminal sulphydryl groups.

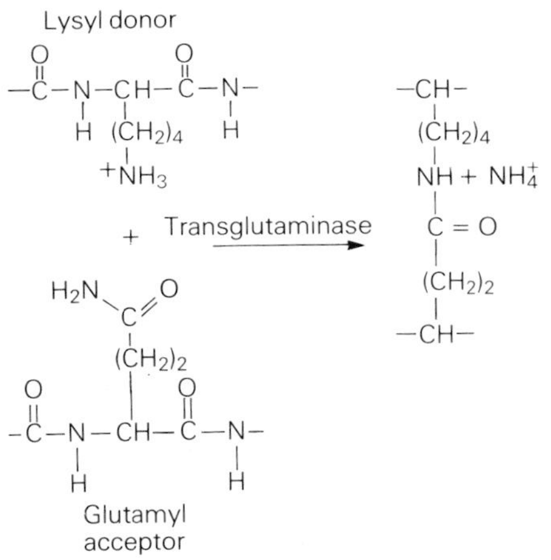

Fig. 7.24 Formation of a covalent intermolecular bond through transglutaminase cross-linking.

LAMININ is a glycoprotein of M_r 900 000, originally isolated from the extracellular matrix of tumours. In normal tissue it is the major non-collagenous glycoprotein of basement membranes, and is located in the *lamina rara* region (Fig. 7.14). Much less is known about the structure–function relationships of laminin than of fibronectin. Laminin consists of two types of disulphide-linked subunits, one of M_r 440 000 called the A chain and two B chains of M_r 205–230 000. Electron microscopic examination of laminin reveals a cross-shaped structure with one long arm and three short arms. Two globular domains are apparent on each of the short arms (Fig. 7.25a), with one globular region at the end of the long arm. The precise location of the chains within this structure is still under investigation, but it is thought that the A chain forms the globular region at the end of the long arm and one of the short arms. Each of the B chains make up the other two short arms and part of the long arm (Fig. 7.25b). Controlled proteolytic digestion of laminin yields a number of fragments able to bind to the cell surface and extracellular matrix components (Fig. 7.25b). The secondary structure of these fragments has been determined.

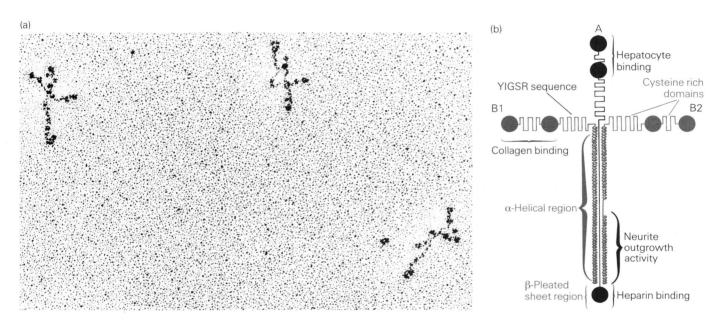

Fig. 7.25 (a) An electron micrograph of laminin, after rotary-shadowing. Courtesy of Prof. J. Engel, Biozentrum der Universität, Basel, Switzerland. (b) Proposed structure of laminin showing the individual A and B chains. Binding domains and area of α-helix or β-pleated sheet are also illustrated.

Immunohistological studies indicate that laminin is a major component of basement membranes in adult tissues and organs such as the kidney, blood vessels, skin, muscle, teeth and others.

Laminin is possibly the earliest adhesion protein to appear during embryonic development. In the mouse embryo the B-chain of laminin has been detected at the two-cell stage. Later, it is found in embryonic membranes such as the yolk sac, chorion and amnion. Its appearance immediately precedes the formation of kidney tubules and may be involved in branching of salivary glands or lung buds.

Its ubiquitous occurrence in the ***lamina rara*** of basement membranes suggests that laminin plays a role in the interaction of epithelial and endothelial cells with other basement membrane constituents. Several lines of study have shown that laminin promotes epithelial and endothelial cell

lamina rara: *microscopically light region of the glomerular basement membrane close to epithelial layer.*

Reference von der Mark, K. and Kuhl, U. (1985) Laminin and its receptor. *Biochimica Biophysica Acta,* **823,** 147–60. Good if slightly dated introduction, gives an alternative mechanism for cellular effects of laminin to that covered in text.

adhesion specifically to Type IV collagen, a major constituent of the **lamina densa** (Fig. 7.14). It was originally thought that laminin specifically enhances the adhesion of epithelial cells, and fibronectin that of mesenchymal cells, but this is not the case. Hepatocytes, fibroblasts, myoblasts and several tumour cell-types adhere to both. However, laminin does seem more active in promoting neurite outgrowth and **myotube** formation than fibronectin.

Systematic examination of the binding of laminin to various glycosaminoglycans indicates it binds most strongly to heparin, with moderate affinity to heparan sulphate, dermatan sulphate and chondroitin 6-sulphate but has little affinity for hyaluronan and chondroitin 4-sulphate. Laminin may also form complexes with another basement membrane glycoprotein, nidogen in tumour and yolk sac basement membranes.

□ As with fibronectin, certain bacteria such as *Escherichia coli* and *Streptococcus pyogenes* have evolved mechanisms to bind to laminin. Presumably this aids adhesion of these parasites to the host basement membranes.

THE MAJOR BINDING DOMAINS OF LAMININ have been determined by controlled protease treatment and binding studies. Treatment with thrombin completely digests the long arm leaving a fragment consisting of the three short arms including globular regions. This fragment binds to tumour cells and also mediates their attachment to type IV collagen. Treatment with most other proteases produces a similar fragment, without the globular domains, which binds to tumour cells but not type IV collagen; that is, it inhibits tumour cell adhesion to a laminin/collagen matrix. These studies indicate that laminin may have three cell attachment domains for neuronal cells, hepatocytes and epithelial cells respectively (Fig. 7.25b). Recently the β-chains have been found to contain the peptide sequence CDPGYIGSR, and a peptide with this sequence inhibits both laminin-mediated attachment and chemotaxis. The cell-surface receptor is not the same as that for fibronectin.

□ CDPGYIGSR is single letter nomenclature for the sequence Cys–Asp–Pro–Gly–Tyr–Ile–Gly–Ser–Arg (see page 192).

Other structural proteins

A number of other glycoprotein molecules which mediate cell–cell and cell–substrate extracellular matrix interactions are known and no doubt many more will be discovered. Several of these are now being sequenced and their structure–function relationships determined.

THROMBOSPONDIN is one such molecule. It is a glycoprotein of M_r 420 000, first identified in human blood platelets. Since then it has been shown to be synthesized and secreted by fibroblasts and endothelial cells, among others, and incorporated into the extracellular matrix. *In vitro* binding studies indicate that thrombospondin can bind to fibrinogen, fibronectin, laminin, heparin, and type V collagen.

The thrombospondin molecule is composed of three identical polypeptide chains (M_r 145 000) which are cross-linked by disulphide bonds. Electron microscopy has shown that, in the presence of Ca^{2+}, thrombospondin possesses four distinct regions: the amino-terminal globular region, and three others ending in a globular carboxyl-terminal region (Fig. 7.26).

The use of monoclonal antibodies has indicated that the globular amino-terminal binds heparin and both the amino- and carboxyl-terminal regions are involved in platelet aggregation. The connecting chains bind laminin, fibronectin, plasminogen and Type V collagen. Cell adhesion appears to be mediated by an RGDA sequence located near the carboxyl-terminal globular regions. The amino-terminal heparin-binding domain may also be important in cell adhesion.

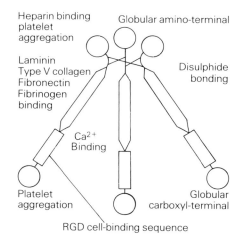

Heparin binding platelet aggregation

Globular amino-terminal

Laminin Type V collagen Fibronectin Fibrinogen binding

Disulphide bonding

Ca^{2+} Binding

Platelet aggregation

RGD cell-binding sequence

Globular carboxyl-terminal

Fig. 7.26 A representation of the electron microscopy image of thrombospondin. The binding domain structure of thrombospondin, derived from proteolytic cleavage and primary sequencing studies is also shown.

---

lamina densa: *microscopically dense region of the glomerular basement membrane.*
myotube: *the primitive muscle forming cells (myoblasts) elongate to become multinucleated myocytes. The elongated tubes are termed myotubes.*

Reference Lawler, J. and Hynes, R.O. (1986) The structure of human thrombospondin. *Journal of Cell Biology*, **103**, 1635. Lawler, J., The structure of human platelet thrombospondin. *Journal of Biological Chemistry*, **260**, 3762-72. A bit advanced but well worth reading.

VON *WILLEBRAND FACTOR* (vWF) is a glycoprotein that mediates the attachment of platelets (and possibly endothelial cells) to the subendothelial vascular basement membrane after injury. It circulates in blood as a heterogeneous series of disulphide-bonded multimers non-covalently complexed with Factor VIII (anti-haemophilic factor). vWF is synthesized by megakaryocytes (platelet precursors) and endothelial cells as a preprovWF of M_r 309 000. This undergoes disulphide-bonding to give a dimeric form, followed by removal of a signal peptide, glycosylation and removal of a M_r 81 000 amino-terminal peptide. The mature protein is composed of two identical subunits of M_r 270 000, connected by a disulphide-bond near the carboxyl-terminal (Fig. 7.27a). Multimeric disulphide-bonded aggregates of M_r 1–20×10^6, are found in serum and intracellular storage granules of platelets and endothelial cells. vWF binds to Type I and Type III collagen, heparin, anti-haemophilic factor and has two cell-binding domains. The multimeric nature of vWF provides a high local density of binding sites and as a consequence a much higher affinity for the substrate.

vWF possesses two cell (or platelet) binding sites (Fig. 7.27b). Site I binds to platelet membrane glycoprotein GpIb by a mechanism which has not been elucidated (probably RGD mediated, see below) and Site II contains an RGD sequence involved in binding to the GpIIb/IIIa membrane complex on platelets and endothelial cells.

(a)

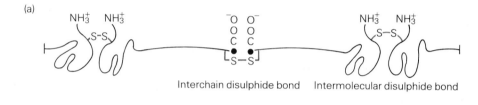

Interchain disulphide bond   Intermolecular disulphide bond

(b)

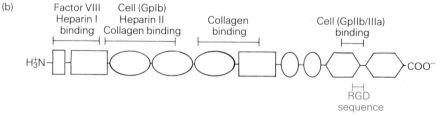

Fig. 7.27 Schematic representation of the multimetric assembly and function domains within vWF. (a) Illustrates intramolecular and intermolecular disulphide bonds involved in dimer formation and the assembly of large aggregates. (b) The mature vWF subunit showing the position of homologous regions (◯, A repeats; Ō, B repeats; ◯, C repeats, and □, D repeats) and binding domains. Both cell-binding domains contain RGD sequences.

CHONDRONECTIN is also a glycoprotein originally isolated from serum, which specifically mediates the attachment of chondrocytes to cartilage Type II collagen. It also resembles fibronectin and laminin in that it binds glycosaminoglycans, specifically chondroitin sulphate. Chondronectin has an M_r of 180 000 and is made up of two disulphide-linked subunits of M_r 70 000. The cell binding domain has not yet been identified.

Several more matrix proteins (for example, **tenascin**) have been described recently which interact with the cell surfaces, probably via the RGD receptor family (see below), but their binding specificity towards the various matrix components is unknown. They appear to form multi-armed oligomeric structures similar to thrombospondin and laminin, and their expression is usually limited to a transient appearance during embryonic development and in the stroma of tumours.

Reference Gehlsen, K.R., Dillner, L., Engvall, E. and Ruoslahti, E. (1988) The human laminin receptor is a member of the integrin family of cell adhesion receptors. *Science*, **241**, 1228–9. Original paper, shows methodology.

Reference Girma, J.-P., Meyer, D., Verweij, C.L. *et al.* (1987) Structure–function relationship of human von Willebrand factor. **70**, 605. Good detailed reviews on structure, function and pathology.

7.8 Membrane receptors for extracellular matrix macromolecules

The structural extracellular glycoproteins described in the previous section interact directly with receptors on the cell surface and through these receptors with the cytoskeleton. Identification of these receptor molecules has involved the use of antibodies which inhibit cell adhesion, affinity chromatography on immobilized adhesion/structural glycoproteins and, more recently, affinity chromatography on immobilized peptides representing the cell binding region of these adhesion proteins. An example of the latter approach is the series of studies carried out to identify the cell binding sequence of fibronectin.

Proteolytic dissection of fibronectin was employed to produce successively smaller peptide fragments which retained the ability to promote cell adhesion. Finally, a series of oligopeptides were synthesized equivalent to overlapping regions of the smallest active proteolytic fragment, and examined for their ability to inhibit cell binding to the original proteolytic fragment. It was deduced that the critical peptide sequence was a tripeptide composed of RGD. This peptide blocked adhesion of cells to fibronectin and promoted cell adhesion when immobilized to solid surfaces.

RGD peptides have been found to prevent cell adhesion to vitronectin, laminin, von Willebrand Factor, thrombospondin and tenascin. Furthermore, RGD sequences have been described in other proteins known to participate in cellular interactions: fibrinogen; type I collagen; II crystallin; the *E. coli* receptor; discoidin I (an adhesive aggregation protein from the slime mould, *Dictyostelium*), and surface proteins of some animal viruses. It is possible that RGD may be a general sequence for mediating cellular interaction, but other cell binding sequences are known, for example, the CDPGYIGSR sequence of laminin. In addition, several receptors have been isolated which appear to interact with extracellular matrix components by sequences other than RGD, since RGD-peptides do not block the binding.

Integrins: a family of RGD receptors

One of the first group of receptors for extracellular matrix molecules identified was a complex termed avian **integrin**, cell-substratum attachment antigen (CSAT) or the 140 000 complex. This complex was isolated using two adhesion-blocking monoclonal antibodies, which also localized the complex to adhesion sites on muscle, fibroblasts, neurons and many other cell types. These antibodies inhibit both migration and adhesion of various cell types to vitronectin, fibronectin, laminin, and Types I and IV collagen.

Integrin isolated by immunoaffinity chromatography consists of three distinct polypeptides. In non-reducing SDS-PAGE these have apparent M_r of 160 000, 120 000 and 110 000 respectively. Under conditions which disrupt disulphide bonds the 160 000 glycoprotein gives two subunits of M_r 140 000 and 25 000, respectively (Table 7.8; Fig. 7.28). The integrin complex has subsequently been shown to bind fibronectin and laminin and their binding is

---

Exercise 6

Many proteins contain the RGD integrin attachment signal, but adhesion receptors can distinguish between these. Suggest possible explanations for this binding specificity.

Box 7.6
RGD and YIGSR peptides and tumours

Both RGD and laminin YIGSR sequences have been shown to inhibit tumour spread via the blood stream (metastasis). This may lead to a use of these in tumour treatment. However, it is unclear if the action of these peptides inhibits tumour cell migration into the vasculature, or adhesion and migration out of the blood vessels at another site.

Reference Hynes, R.O. (1987) Integrins: A family of cell surface receptors. *Cell*, **48**, 549–54. Short advanced review but easy to read.

Reference Juliano, R.L. (1987) Membrane receptors for extracellular matrix macromolecules. *Biochimica Biophysica Acta*, **907**, 261–78. Comprehensive review on integrin structure.

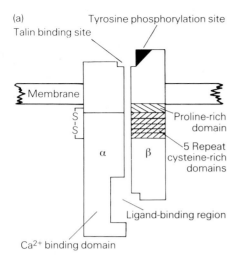

(a)

Tyrosine phosphorylation site

Talin binding site

Membrane

α   β

S
|
S

Proline-rich domain

5 Repeat cysteine-rich domains

Ligand-binding region

Ca^{2+} binding domain

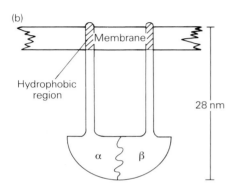

(b)

Membrane

Hydrophobic region

α   β

28 nm

Fig. 7.28 Structure of integrin. (a) Integrin as a transmembrane molecule. The major domains deduced from cDNA cloning and protein sequencing are represented. (b) A representation of the rotary microscopy image of the fibronectin receptor. α- and β-chains are shown, and the hatched regions represent the transmembrane regions. The head region resembles a mushroom or jelly-fish in shape, i.e. a hollow hemisphere.

☐ The integrin family seems to have evolved relatively early. A family of proteins mediating cell aggregation in sea urchin embryos has a polypeptide composition similar to the integrins.

See Chapter 11

See Chapter 12

Table 7.8 Subunit homology in the integrin receptor family

	Subunit $M_r \times 10^{-3}$	β-Subunit form
Chicken integrin complex (140 kDa complex CSAT or JG22 antigen)	140 + 125/130	1
		1
Chick actosialin	170/130	1 (?)
Fibronectin receptor (osteosarcoma) (human)	140 + 25/130	1
Vitronectin receptor (human)	125 + 25/115	3
GpIIb/IIIa (human and others)	120 + 25/110	3
LFA-1 (human)	180/95	2
MAC-1 (OKMI, CRS) (human)	170/95	2
p150, 95 (human)	150/95	2
VLA-1 (human)	210/130	1
VLA-2 (human)	165/130	1
VLA-3 (human)	135/130	1
VLA-4 (human)	150/130	1
VLA-5 (human)	135/130	1
Drosophila (position specific antigens PS-AGS)		
PS-AGS 1	116/110	3 (?)
PS-AGS 2	125/110	3 (?)
PS-AGS 3	120/1120	3 (?)
PS-AGS 4	92/110	3 (?)
Collagen receptor		
(fibrosarcoma) (human)	135/130	1 (?)
(osteosarcoma) (human)	250/70/30	—
(hepatocytes) (rat)	140/120	(?)

inhibited by the RGDS peptide. Integrin has also been demonstrated to interact with talin, but not the vinculin or α-actinin components of intermediate filaments. This binding was inhibited by a peptide corresponding to the carboxyl-terminus of the M_r 110 000 β-chain. Also, at the carboxyl-terminus is a tyrosine residue which can be phosphorylated by the action of **tyrosine kinase**. This has been shown to be highly phosphorylated in cancer cells, and appears to reduce the affinity of integrin for both fibronectin and talin. Individually the α- and β-protein chains show no affinity for extracellular matrix components or talin, confirming the co-operative nature of these interactions.

Independently of the work on avian cells, three other families of integrin-like molecules have been identified, primarily by the use of monoclonal antibodies. One of these is composed of the three leukocyte cell surface adhesive proteins: LFA-1 (involved in the binding of T-lymphocytes to target cells); Mac-1 (the macrophage receptor for complement component C3bi) and p150,95 (important for the binding of leucocytes to endothelial cells in preparation for their passage across the endothelial layer) (Table 7.8).

The second group of proteins are the ***very late antigens*** (VLA-1 to VLA-5), a family of proteins which appear on the surface of stimulated T-lymphocytes. They are also expressed on a variety of cell types other than lymphocytes. Finally, there is the group of *Drosophila* 'position specific' (PS) antigens, which were also originally defined immunologically and whose electrophoretic behaviour is similar to that of integrins. These antigens seem to be involved in the segregation of cell populations during embryonic development.

Whereas adhesion-perturbing monoclonal antibodies were the primary tools used to detect and isolate avian integrin and the other integrin-like receptors mentioned above, a different approach has been employed in isolating RGD binding receptors from mammalian cells. When extracts of human osteosarcoma cells were chromatographed on a column of immobilized fibronectin, vitronectin or type I collagen matrices, the addition of buffers with RGD-containing peptides eluted only a single receptor in each

very late antigens: *group of adhesion receptors found on stimulated lymphocytes that appear at a late period of stimulation.*

case. Each was specific for its respective ligand (that is, fibronectin, vitronectin or Type I collagen) and did not interact with laminin. Thus, these cells contain at least three different receptors that recognize the RGD sequence in one of three different ligands.

Fractionation of platelet extracts on immobilized fibrinogen yielded a fourth RGD binding receptor, which was indistinguishable from the platelet glycoprotein GpIIb/IIIa, and serves as a receptor for four proteins: fibronogen; fibronectin; von Willebrand factor, and vitronectin. It might also bind thrombospondin and type I collagen, each of which also contains RGD sequences. Immunologically-similar proteins have been detected on other cell types including endothelial cells and leukocytes.

HOMOLOGIES BETWEEN INTEGRINS. The general approaches outlined above have resulted in the isolation of a large number of integrins from various cell-types and species. With the possible exception of the avian integrin complex and the osteosarcoma collagen receptor, all consist of two non-covalently linked subunits composed of a large α-chain (M_r 130–210 000) and a smaller β-chain (M_r 95–130 000). In many cases the α-chain is itself composed of two subunits linked by a disulphide bond (Table 7.8; Fig. 7.28).

The cDNA coding for the β-subunit of chick integrin has been cloned and sequenced. The sequences of the β-subunits of LFA-1/Mac-1/p150.95 and human endothelial cell GpIIb/IIIa (that is, GpIIIa) have also been determined and have a high degree of homology with that of chick integrin, largely due to 46 highly conserved cysteine residues.

Immunological comparison between these three homologous (but distinct) β-subunits and several other integrin-like receptors suggests that the vitronectin receptor β-subunit is the same as, or very similar to GpIIIa (β-chain$_3$), and that the β-chains of human VLA, chicken integrin and human fibronectin receptor are all closely related (Table 7.9).

See *Molecular Biology and Biotechnology,* Chapter 9

Table 7.9 *Chain sequence homology. Amino acid homology of amino-terminal sequences in several integrin chains. Conserved residues are shown in colour (conservative aromatic amino acid substitutions are deemed to be homologous).*

Vitronectin receptor	F	N	L	D	V	X	S	P	A	E	Y	S	X	X
GpIIb	L	N	L	D	P	V	Q	L	T	F	Y	A	G	P
Murine LFA-1	Y	N	L	D	T	R	P	T	Q	S	F	L	A	Q
Murine Mac-1	F	N	L	D	T	E	H	P	M	T	F	Q	E	N
Human Mac-1	F	N	L	D	T	E	N	A	V	T	F	Q	E	N
Human p150	F	N	L	D	T	E	E	L	T	A	F	R	V	D
VLA-1	F	(N)	V	D	V	K	D	S	M	T	F	(L)	G	P
VLA-2	(F)	N	L	D	T	X	E	D	N	V	(F)	R	(G)	(P)
VLA-3	F	N	L	D	T	R	F	L	V	V	K	E	A	G
VLA-4	Y	N	V	D	T	E	S	A	L	L	Y	E	G	P
Fibronectin receptor (VLA-5)	F	N	L	D	(T)	E	(E)	P	X	V	L	S	G	P
Position antigen Gp116	F	N	L	E	Q	K	L	P	I	V	K	Y	X	X

Reference Ruoslahti, E. and Pierschbacher, D. (1987) New perspectives in cell adhesion: R-G-D and integrins. *Science*, **238**, 491–7. Extends the Ruoslahti review on fibronectins. Good for integrin specificity.

Homologies also exist between the α-subunits. LFA-1, Mac-1, GpIIb and the vitronectin receptor α-chains show amino-terminal sequence homology (Table 7.9). Also chick integrin α-chain, vitronectin receptor α-chain, fibronectin receptor α-chain and GpIIb show extensive homologies in the carboxyl-terminal region. Immunological cross-reactivity suggests a close relationship between the α-chains of VLA-3 and chick integrin, and VLA-5 and fibronectin receptor. Although there appears to be a close relationship between the α-subunits, current evidence suggests at least ten α-chains exist.

Receptors not belonging to integrin family

It seems that not all cell–matrix interactions are mediated by the integrin family of receptors. Laminin affinity chromatography of membrane extracts of rat, murine and human cultured cells yields a group of surface proteins of M_r 68–70 000 that have a high affinity for laminin ($k_d = 10^{-9}$ mol dm^{-3} compared with 10^{-6} mol dm^{-3} for the integrin family). Antibodies to this group of proteins inhibit the attachment of cells to laminin. Little information is available regarding the structure of these receptors although they can be incorporated into liposomes, suggesting they are hydrophobic integral membrane proteins. They are able to bind to actin and co-localize at cellular adhesion sites suggesting that they interact with the cytoskeleton. The binding of laminin to this receptor is inhibited by the synthetic pentapeptide YIGSR, but not by RGD.

Recently, a laminin receptor has been isolated from human glioblastoma cells by laminin-affinity chromatography. This receptor is a heterodimer with an integrin-like structure and subunits of M_r 120 000 and 30 000/140 000 under conditions which disrupt disulphide bonds. The binding is not inhibited by RGD or YIGSR peptides, but is inhibited by proteolytic fragments of laminin. This may point to the presence of recognition sequences other than RGD and YIGSR.

Many tissues have been reported to possess cell surface or extracellular matrix β-galactoside-binding proteins. Bovine fetal lung, chondroblasts and fibroblasts all have a cell-surface receptor complex which binds both elastin, β-galactosides, laminin and actin. This receptor is a heterotrimer composed of subunits of M_r 67 000, 61 000 and 55 000. The M_r 67 000 subunit is an interesting protein as it binds both elastin and β-galactosides (that is, to their lactosyl residues), and also binds to the M_r 55 000 subunit. Furthermore, the binding of lactose to this subunit releases bound elastin and the M_r 55 000 subunit. This latter subunit appears to be a transmembrane protein which can bind actin, and also associates with the third subunit (Fig. 7.29). Since lactose also blocks elastin fibre formation by tissue-cultured chondroblasts, it appears that the receptor may play an important role in this process. Elastin and laminin peptides block elastin and laminin binding. The latter is not inhibited by the YIGSR peptide, differentiating it from the M_r 67 000 laminin receptor discussed earlier. As laminin contains β-galactoside-rich glyco-conjugates, the binding may also involve that receptor site. However, it appears that at least three laminin receptor complexes exist, none of which appears to be part of the RGD binding integrin family.

Collagen-binding membrane proteins have also been isolated and partially characterized. **Anchorin** is a protein of M_r 31 000 found on the surface of cultured chondrocytes. It binds type II collagen (and to a lesser extent types I and V) in the native, but not denatured, form. Partial sequencing has revealed that it contains a hydrophobic domain, confirming it as an integral membrane protein. **Colligin** (M_r 47 000) type IV collagen-binding protein is found on the surface of *parietal endoderm cells*. A protein with similar

parietal endoderm cells: primitive epithelial cells in the embryonic endoderm (inner layer).

Exercise 7

You have been given a pure preparation of an extracellular matrix glycoprotein which increases the adhesion of human fibroblasts. Suggest how you would identify the 'cell-surface receptor' for this molecule.

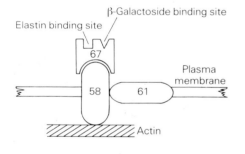

Fig. 7.29 A schematic representation of the probable arrangement of the subunits of the elastin/β-galactoside binding receptor.

characteristics to colligin has been observed in a number of different cell types. The function of these receptors is speculative.

Several apparently different surface receptors for hyaluronan have been detected. The best characterized is the receptor on the surface of cultured baby hamster kidney cells (BHK cells). It is an integral membrane protein of M_r 85 000 which specifically binds hyaluronan and can also interact with actin. Immunological studies have detected its presence on epithelial cells, various tumour cells, fibroblasts and macrophages, but not on endothelial cells. A hyaluronan receptor (M_r 100 000) is present on these cells which binds hyaluronan and chondroitin sulphate, and mediates their endocytosis. From the available information it appears that at least two hyaluronan receptors exist, only one of which is found on endothelial cells.

It is apparent that most cells have membrane receptors able to bind all the major components of the extracellular matrix. Most of these receptors have been shown to bind to components of the cell cytoskeleton. Thus, these receptors probably mediate extracellular matrix–cell regulation and the ordering of the matrix itself.

Exercise 8

An extracellular matrix protein binds to the cell surface of fibroblasts and also to heparin/heparan sulphate. A receptor molecule has not been detected. Suggest further experiments to elucidate the binding mechanism.

7.9 Cell movement and matrix interaction

In many instances, such as embryogenesis and tissue regeneration, specific receptor–matrix interactions appear to direct the movements of specific cell types. Cell migration takes place through membrane recycling. Membrane material is endocytosed via coated pits and transferred to the advancing edge of the cell. This is termed the membrane-flow model of cell movement (Fig. 7.30). How a transient interaction between surface receptor and extracellular matrix components induces their recycling, thus increasing membrane formation in the direction of the binding ligand, is unknown. However, such interactions may direct new membrane formation by cytoskeletal organization. Continued directional movement relies on the transient nature of the receptor binding and diffusion of the receptors to the leading edge. It is noteworthy that many surface receptors for extracellular components are low affinity receptors (k_d 10^{-6}–10^{-7} mol dm^{-3}), thus allowing transient attachment of the cell to the matrix. However, some surface receptors are high affinity receptors (k_d 10^{-9} mol dm^{-3}), notably the M_r 69 000 laminin receptor and the elastin receptor. Binding of the elastin receptor may be regulated by the β-galactoside binding sites, but there is no evidence for such a regulatory mechanism for the laminin receptor. The attached receptor–laminin complex may simply be degraded by cell-surface proteases and newly synthesized receptor secreted at the leading edge of the migrating cell. Laminin, however, is usually associated with the differentiation and organization of epithelial, myoblast and neuronal tissues and its primary function may be tissue organization rather than the direction of cell movement. Laminin is also thought to mediate the binding of circulating tumour cells to endothelial basement membranes during their invasion of capillaries and the homing of pre T-lymphocytes to the embryonic thymus.

See Chapter 12

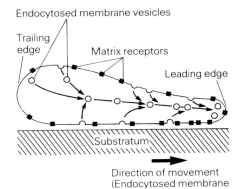

Endocytosed membrane vesicles

Trailing edge

Matrix receptors

Leading edge

Substratum

Direction of movement (Endocytosed membrane + cell)

Fig. 7.30 Membrane recycling and cell movement. Arrows represent the direction of movement of endocytosed membrane and also the cell.

7.10 Regulation of receptor expression and function

Cellular interactions with the surrounding matrix depend on the composition of the matrix and the number and type of matrix receptors present on the cell surface. Little is known about the regulation of adhesion receptors with respect to their expression or function.

Reference Bretscher, M.S. (1987) How animal cells move. *Scientific American*, **257**(6), 44–50. Typically readable overview from *Scientific American*.

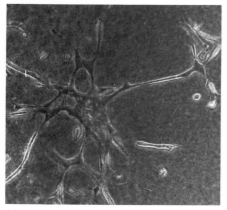

(a)

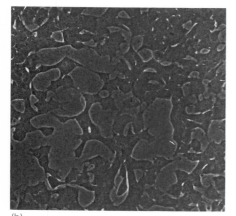

(b)

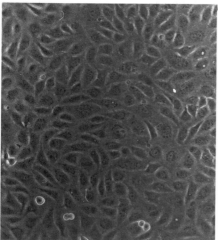

(c)

Fig. 7.31 Cultured cells showing a variety of morphologies depending upon their substratum. (a) Cells in a collagen gel (×259), (b) under a collagen (×306) and (c) on gelatin (×188). Reprinted from Schor, A.M. *et al* (1983) Effects of culture conditions on the proliferation, morphology and migration of bovine aorta endothelial cells. *Journal of Cell Science* **62**, 267–85. Courtesy Dr A.M. Schor, Christie Hospital, Manchester, UK.

Maturation of red blood cell precursors is accompanied by the loss of surface fibronectin receptors. On the other hand, induction of tumour-forming ability in cultured cells does not lead to a loss of receptor but to a decrease in the affinity of the receptor for the extracellular matrix and the cytoskeleton. This seems to be due to increased tyrosine kinase activity and subsequent phosphorylation of the cytoplasmic domain of the receptor or cytoskeleton components, for example, talin.

Studies of cell-surface receptors for growth factors have shown numerous examples where the binding of one growth factor to its specific surface receptor can alter the binding of a second growth factor or hormone to its own receptor. Recent observations suggest that similar effects may occur with matrix receptors. The binding of collagen by BHK cells (Baby Hamster Kidney cells) non-competively inhibits their adhesion to fibronectin, but does not inhibit the binding of fibronectin to its receptor *in vitro*. This suggests that collagen–receptor interaction suppresses the expression of surface fibronectin receptor or greatly reduces its affinity for fibronectin. It has also been observed that the binding of fibronectin to monocytes induces or activates complement Factor C3 receptors, a second integrin.

7.11 Reciprocity, gene expression and cell shape

When cells are isolated from pieces of tissue and cultured *in vitro* they show a different pattern of gene expression than that seen in the original tissue. It has been demonstrated that normal function may be induced by altering the substratum used for culture. For example, the morphology and gene expression of mouse mammary epithelial cells has been compared when cultured on plastic, flat native type I collagen, native basement membranes and floating native type I collagen gel (Fig. 7.31). When cultured on plastic or flat collagen gel, the cells are flat, with few microvilli, and lack a well-developed secretory apparatus. However, when cultured on floating collagen gels or native basement membrane the cells become columnar with many microvilli, a well-developed extracellular matrix and tight junctions, and an extensive directional secretory apparatus.

The secretion of milk proteins, such as caseins, by mouse mammary epithelial cells is increased by culture on basement membrane or floating collagen gels. In cells grown on plastic, the casein polypeptides are retained within the cell and rapidly degraded, suggesting that changes in cell shape can alter the regulation of casein secretion. The individual components of the reconstituted basement membrane are not as effective as the reconstituted supramolecular complex. In addition, the responsiveness of the epithelial cells to prolactin-stimulation of casein secretion is also dependent on the culture substratum. The response is, again, greatest in cells grown on floating collagen gels.

Studies with mammary epithelial and other cell types have demonstrated that the composition of the exogenous extracellular matrix can affect both the amount and composition of the secreted extracellular matrix. The substratum and cell shape can be modulated so that the columnar cells seen on floating gels, or basement membrane, produce mammary basement membrane components.

The relationship between cell shape and phenotypic expression has also been studied using cultured human chondrocytes. Cells that flattened on the culture substratum grew rapidly and secreted little glycosaminoglycan. When the substratum was altered so that the cells were rounded, the cell grew slowly and secreted larger amounts of glycosaminoglycans.

Reference Bissell, M.J., Hall, H.G. and Parry, G. (1982) How does the extracellular matrix direct gene expression? *Journal of Theoretical Biology*, **99**, 31. Detailed account of 'dynamic reciprocity'.

Reference Bissell, M.J. and Barcellos-Hoff, M. (1987) The influence of extracellular matrix on gene expression: is structure the message? *Journal of Cell Science*, suppl. 8, 327–43. Updates previous review.

These two examples clearly show that the composition and shape of the extracellular matrix has profound effects on the cellular phenotype, governing both cell shape and metabolism. As cell shape is regulated by the cytoskeleton, effects of the extracellular matrix on cells are probably mediated by the interaction of the matrix receptors with the cytoskeleton. How this regulates protein expression and secretion is unknown but may be due to changes in the post-transcriptional and post-translational processing of mRNA and secretory protein. Given that cytoskeletal interactions with cell-surface matrix receptors are necessary for ordering of the extracellular matrix, these studies suggest a reciprocal interaction between the cytoskeleton of the cell and the extracellular matrix which is mediated by the cell-surface matrix receptors.

Thus, the extracellular matrix governs the phenotype of the cell while the cell's phenotype determines the extracellular matrix it makes, which in turn governs the cell's phenotype! This has been called **dynamic reciprocity**. It is thought to be regulated by the cytoskeleton–receptor–extracellular matrix continuum.

7.12 Overview

All animal extracellular matrices are composed of the same three groups of molecules: the fibrous proteins, elastin and collagen; the predominantly carbohydrate glycosaminoglycans and proteoglycans, and adhesion/structural glycoproteins. Until recently the extracellular matrix was thought to be an inert scaffolding, composed of fibrous proteins embedded in a highly hydrated amorphous 'ground substance'. This view was based on the structure of the 'classical' load-bearing connective tissues, such as cartilage, which contain fibrous collagens and large aggregating proteoglycans.

More extensive analysis of both these tissues and other extracellular matrices, such as basement membranes, has brought to light the non-fibrous collagens, cell-surface and non-aggregating proteoglycans and an increasing number of structural proteins. The function of the non-fibrous collagens appears to be less mechanical and more for structuring and guiding matrix formation, while the more recently discovered forms of proteoglycan are thought to regulate metabolic processes at the cell surface and control collagen deposition in the matrix. Also, heparan sulphate proteoglycans are essential for the filtering properties of basement membranes.

Structural proteins, by virtue of their ability to bind to both cells and extracellular matrix components and to form fibrillar aggregates, appear to be an essential part of the cell's machinery for interacting with and controlling the extracellular matrix. The proteins bind to the cell through a family of surface receptors (the integrins), which also interact with the cytoskeleton of the cell, and through these possibly its nucleus. Thus, a direct connection appears to exist between the cell nucleus and the extracellular matrix enabling the cell to organize the matrix and the matrix to regulate cell metabolism. This is termed reciprocity.

1. Reinforced concrete, fibreglass.
2. The decrease in both charge density and the smaller, less stable, aggregates would have a decreased water avidity. Thus, the load-absorbing and load-redistributing properties would be less efficient and the cartilage more prone to damage.
3. A pulse–chase experiment is most appropriate. The cells would be incubated with radioactively-labelled precursor for heparan sulphate for a short period. The label precursor is then removed, the incorporated label followed, or chased, over a period of time. If the membrane HS–PG is a precursor of small HS–PG, the label will first appear in the membrane HS–PG and subsequently in the small HS–PG, whilst the label decreases in the membrane HS–PG.
4. As part of their mechanical function most weight-bearing tissues (i.e. articular cartilage and the intervertebral disc) lose and reabsorb interstitial fluid (see text). Waste products and nutrients are interchanged during this process. This seems to be the major source of nutrients for these tissues and immobilized joints quickly show signs of cartilage degeneration. Thus intermittent loading is essential for adequate nutrition of such load-bearing tissues.
5. Each fibronectin monomer contains three regions of alternative splicing; EDI, EDII and IIICS in humans. The regions EDI and EDII undergo alternative splicing to give two possible alternatives for each region, and thus four possible fibronectin monomers. The IIICS can yield five variants in humans. Therefore, the number of monomer variants possible is 20 in humans. Hence, up to 40 possible dimers may be produced. However, only a few are actually produced by each cell type.
6. A number of possibilities exist:

(a) The RGD binding serves only as an initial, possibly allosteric site and the specificity resides in a second binding site unique for each ligand;
(b) The RGD sequence is only part of an extended binding region. However, the amino acid sequences flanking the RGD in the various RGD adhesion proteins show little or no similarity and a number of integrins can bind several of these. Also, peptides with the fibronectin flanking sequences bind to the vitronectin receptor better than to the fibronectin receptor itself.

Recent observations suggest that the specificity does indeed reside in the RGD tripeptide and the residue following it. That the RGD sequence is important would explain why some proteins, such as αII crystallin or the *E. coli* lambda receptor, promote cell attachment *in vitro* for no physiological reason. Three-dimensional modelling of proteins has shown that the RGD sequence can have very different conformations (see Ruoslahti, E. and Pierschbacher, D. (1987) *Science* **328**, 491).
7. The fact that you have a purified ligand would indicate that the affinity chromatography approach would be most fruitful. The extracellular matrix component is immobilized on a chromatography matrix. The fibroblast cell-surface proteins are specifically labelled, usually with ^{125}Iodine, and then solubilized. The solubilized receptor is passed through a column of the immobilized ligand and any proteins bound are eluted with a dissociating medium, normally 4–6 mol dm^{-3} guanidine hydrochloride of 6–8 mol dm^{-3} of urea. The eluted surface proteins are then analysed by electrophoresis. Further characterization of the receptor may include elution with RGD or YIGSR peptides.
8. Most cells have surface heparan sulphate which may be binding to the matrix protein. Two routes of investigation may be employed to determine if heparin sulphate is the 'receptor': (a) enzymically remove the cell surface heparan sulphate and see if the protein still binds, or (b) perform the binding experiments in the presence of various concentrations of extraneous heparan sulphate to see if this inhibits the binding.

QUESTIONS

FILL IN THE BLANKS

1.   Focal contacts, or _____ plaques, are points of close contact between the cell and the _____ _____ . Immunofluorescent studies have shown that _____ proteins, such as _____ , _____ and _____ , together with _____ and _____ _____ proteoglycans co-localize to these sites. More recently, certain _____ have also been localized to these structures suggesting that they form a _____ link between the matrix and the cytoskeleton.

   Fibronectin is an extracellular matrix protein which stimulates cell _____ and _____ . It binds to several integrins, through an _____ sequence, and can also bind _____ , _____ and _____ collagen. Its ability to bind to the cytoskeleton through a _____ receptor, and to _____ _____ forming long fibrils, suggests that it plays an important role in the cellular organization of the _____ _____ .

   Integrins are a _____ of _____ membrane receptors, which bind a number of _____ proteins via an internal RGD sequence. They are composed of _____ large subunits termed the _____ and _____ -subunits. In some cases the _____ -subunit is itself a disulphide-linked heterodimer. Both subunits are transmembrane proteins which co-operate in attaching to the cytoskeleton through _____ . This interaction with the cytoskeleton is regulated by _____ of the _____ domain of the _____ -subunit.

Choose from: α (2 occurrences), actin, adhesion (2 occurrences), β (2 occurrences), cytoplasmic, cytoskeletal, denatured, extracellular matrix (2 occurrences), family, fibrin, fibronectin, heparan sulphate, heparin, integral, integrins, migration, phosphorylation, RGD, self-associate, structural, talin (2 occurrences), transmembrane (2 occurrences), two, vinculin.

MULTIPLE-CHOICE QUESTIONS

State which of the following are true or false:
2.   Hyaluronan is:
A. a glycosaminoglycan
B. a proteoglycan
C. sulphated
D. synthesized on a protein core
E. composed of repeating disaccharides
F. secreted directly through the plasma membrane

3.   Structural proteins are:
A. fibrous proteins
B. multifunctional proteins
C. integral membrane proteins
D. adhesion proteins
E. glycoproteins
F. proteoglycans

4. Fibronectin:
A. has an RGD cell binding domain
B. has interchain disulphide bonds
C. is composed of three subunits
D. is composed of two identical subunits
E. has three types of repeat homologous sequences
F. contains regions formed from variant splicing

5. Fibronectin binds to:
A. all integrins
B. native collagen
C. heparin
D. gelatin
E. fibrin
F. laminin

6. Integrins:
A. bind heparin
B. have two transmembrane subunits
C. are specific for the RGD sequence
D. bind directly to actin
E. bind YIGSR peptides
F. all have the same β-subunit

7. Laminin:
A. has a cross-like structure
B. binds Type I collagen
C. binds Type IV collagen
D. is composed of two disulphide-linked chains
E. has a YIGSR cell-binding domain
F. has an RGD cell-binding domain

SHORT ANSWER QUESTIONS

8. What are the main components of extracellular matrices?

9. Draw a diagram showing the fundamental structural and functional requirements of a structural protein.

10. How are fibronectin fibrils formed?

11. Draw a schematic diagram outlining the basic structure of an integrin.

12. What is the major determinant in the binding of an integrin-like receptor?

13. Which enzyme reaction controls association between the cytoskeleton and an integrin surface receptor?

14. Name two diseases where genetic defects in integrins have been implicated.

ESSAY QUESTIONS

15. Outline the structure and functional properties of proteoglycans.

16. What are the general features of an adhesion protein? How do these properties help them carry out their role as structural proteins?

17. Write an essay on: 'Integrins – a family of surface receptors?'

18. Outline the basis of the reciprocity theory.

Hormones

Objectives

After reading this chapter you should be able to:

☐ explain how the actions of animal hormones are initiated by their combination with receptors;

☐ describe the intracellular signals generated by the binding of a hormone to its receptor;

☐ discuss the mechanisms by which intracellular signals induce changes in cellular function(s);

☐ outline the classification, biosynthesis and action of plant hormones.

8.1 Introduction

Animal hormones are compounds synthesized in endocrine glands and then passed in the blood stream to act on target cells. This chapter will mainly discuss the mechanism of action of some of the hormones which are found in higher animals. Hormones are, however, also present in lower classes of animals: an example is the group of steroidal compounds called **ecdysones** which initiate moulting in insects. Plants also have substances which regulate growth and development, and because their site of action is remote from that of their synthesis they are called **plant hormones**. Some aspects of their action will be discussed.

Animal hormones should be distinguished from **local chemical mediators** which act at a site close to where they are synthesized, and which are not released from glands into the blood. Hormones should also be distinguished from **neurotransmitters** which are released into the space (synapse) between the end of a nerve and a target cell. Nevertheless, there are strong similarities between the mechanisms by which hormones, local chemical mediators and neurotransmitters exert their effects on target tissues.

☐ The term 'hormone' was introduced in 1905 by Starling and is derived from the Greek word meaning to arouse or excite. Some hormones, however, can exert inhibitory effects on cellular activity.

See Chapter 9

8.2 Structure and classification of animal hormones

Some hormones are proteins, others are small peptides, derivatives of amino acids or steroids (Table 8.1). Despite this diversity, hormones have been separated into two classes: those that cannot cross the plasmalemma of their target cells, and those that can. The majority of hormones fall into the first class, but **steroid hormones**, and **thyroxine** and **tri-iodothyronine** are in the second class. *The ability of a hormone to act on a tissue depends upon the presence of a specific receptor protein to which the hormone can bind.* Those hormones which cannot enter the cell bind to receptors incorporated into the cell surface

☐ The thyroid gland secretes two hormones thyroxine and tri-iodothyronine. The only difference between them is that there are four iodine atoms in thyroxine and three in tri-iodothyronine.

Reference Martin, B.R. (1987) *Metabolic Regulation. A Molecular Approach*, Blackwell, Oxford, UK. A lucid and relatively elementary account of the material covered in this chapter. Strongly recommended.

Table 8.1 *Diversity of hormones (those highlighted in colour can cross the cell membrane)*

Name	Site of origin	Chemical nature	Structure
Insulin	Pancreas B cell	Protein	α-chain, 21 amino acids β-chain, 30 amino acids
Luteinizing hormone	Anterior pituitary	Glycoprotein	α-chain, 92 amino acids β-chain, 115 amino acids
Vasopressin	Posterior pituitary	Small peptide	Cys–Tyr–Phe–Gln–Asn–Cys–Pro–Arg–Gly–NH$_2$
Adrenalin	Adrenal medulla	Amino acid derivative	
Thyroxine	Thyroid	Amino acid derivative	
Cortisol	Adrenal cortex	Steroid	
Oestradiol	Ovary, placenta	Steroid	

Exercise 1

Look carefully at the structures of the hormones shown in Table 8.1. Which structural features do you think prevent some hormones from passing through the cell membrane, and what features allow other hormones to pass through freely?

membrane, while hormones which can diffuse freely across the plasmalemma bind to intracellular receptors. The binding of a hormone to its receptor involves fairly weak non-covalent forces, and probably induces a change in the conformation of the receptor.

Hormones which interact with the cell surface

Hormones which interact with receptors in the cell membrane must in some way transmit a signal to the cell interior. They do this by stimulating the production of a **second messenger** (the hormone itself is considered to be the first messenger). Many hormone receptors use a special *transducer protein* to carry information from the activated receptor to the enzyme which catalyses the formation of the second messenger (Fig. 8.1). The second messenger then acts on systems within the cell to produce the effects of the hormone. The effector systems may, by a *negative feedback* loop, inhibit the interaction of the hormone and receptor or reduce the effects of this interaction.

SECOND MESSENGERS increase in concentration in the cell following binding of the hormone to the receptor. The increase in messenger concentration precedes any biological effects of the hormone and when the hormone is removed from the tissue the concentrations of both the second messenger and the biological response decline, in concert. The structures of some second messengers are shown in Fig. 8.2.

transducer: *element which carries a signal from one system to another.*
negative feedback: *the ability of a product of a particular stimulus to inhibit its own further formation.*

Hormones which pass freely through the cell membrane

These hormones have a much simpler pathway of action (Fig. 8.1b); their free passage through the cell membrane eliminates the need for a second messenger.

8.3 Structure of hormone receptors located in the cell membrane

The primary structures of a number of receptor proteins are now known, principally as a result of sequencing the complementary DNA for the receptor. From the amino acid sequence derived from this DNA sequence it is possible to make predictions about the arrangement of the polypeptide chain in the cell membrane. Thus the single polypeptide chain of the β-adrenergic receptor, which binds adrenalin, is thought to pass backwards and forwards through the membrane seven times (Fig. 8.3). There are internal sites for phosphorylation and external sites at which sugar residues may be covalently linked to arginine side-chains. Recent evidence suggests that many different hormone receptors have this general arrangement in the cell membrane.

The **insulin receptor** (Fig. 8.4) belongs to a different family of receptors, all of which when occupied can catalyse the phosphorylation of tyrosine residues of proteins. The insulin receptor consists of two types of subunits.

See *Molecular Biology and Biotechnology*, Chapter 9

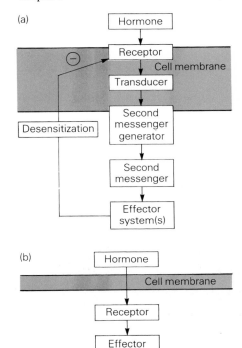

$CH_2OCO(CH_2)_{16}CH_3$
$CHOCO(CH_2)_2(CH_2CH{=}CH)_4CH_2CH_3$
CH_2OH

1-Stearoyl-2-arachidonyl-sn-glycerol
(a diacylglycerol)

Adenosine 3′, 5′ − monophosphate
(cyclic AMP)

Guanosine 3′, 5′ − monophosphate
(cyclic GMP)

Inositol 1, 4, 5 − trisphosphate
(Ins, 1, 4, 5 P₃)

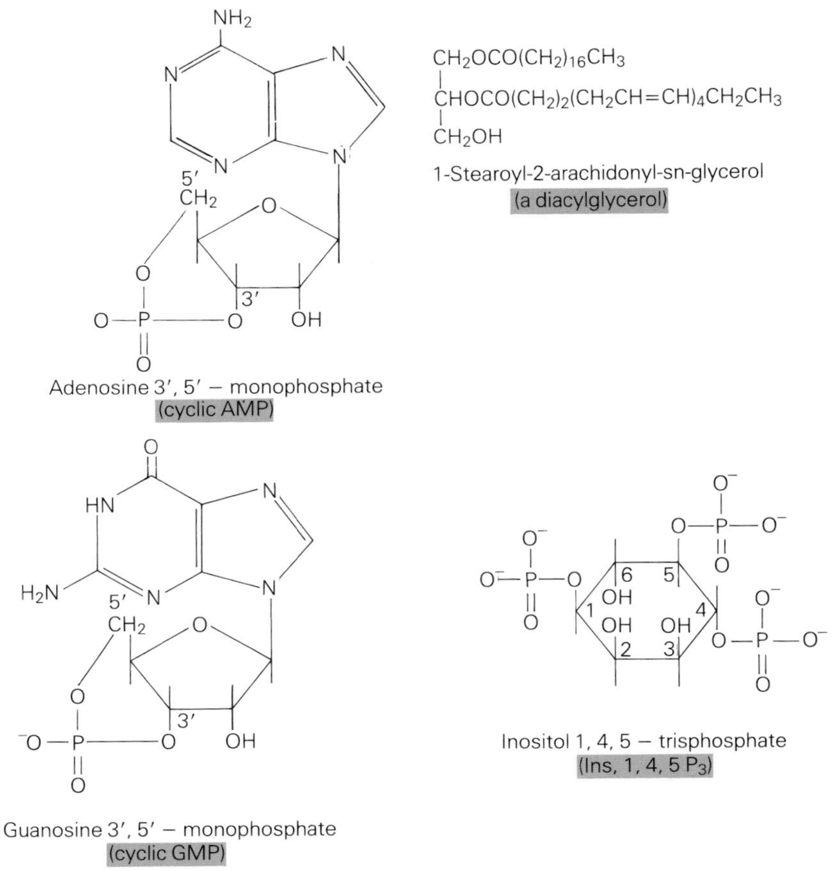

Fig. 8.2 Structures of some second messengers.

Fig. 8.1 (a) Outline of action of hormones that combine with cell membrane receptors. This figure is intended to show the flow of information rather than the physical positioning of the receptor and transducer. (b) Outline of action of hormones that can pass through the cell membrane.

--- Exercise 2 ---

Does the interaction of a hormone with its receptor resemble that of an enzyme with its substrate? What will be the effect of a molecule, analogous to a competitive inhibitor, on the interaction of a hormone with its receptor?

Reference Wallis, M., Howell, S.L. and Taylor, K.W. (1985) *The Biochemistry of the Polypeptide Hormones*, Wiley, Chichester, UK. A comprehensive account of the structure, release and mechanism of action of polypeptide hormones.

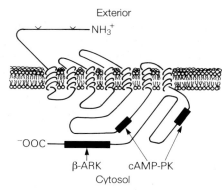

Exterior

—NH$_3^+$

β-ARK   cAMP-PK

Cytosol

Fig. 8.3 Possible arrangement of the polypeptide chain of the adrenergic receptor in the cell membrane. Glycosylation sites are indicated by v Sites for phosphorylation by cyclic AMP-dependent protein kinase and the β adrenergic receptor kinase are denoted by ■.

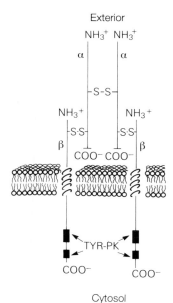

Exterior

NH$_3^+$ NH$_3^+$

α   α

—S–S—

NH$_3^+$ NH$_3^+$

β   —S·S—   —S·S—   β

COO⁻ COO⁻

TYR-PK

COO⁻   COO⁻

Cytosol

Fig. 8.4 Possible arrangement of the insulin receptor in the cell membrane. Sites for phosphorylation on tyrosine residues are shown as ■.

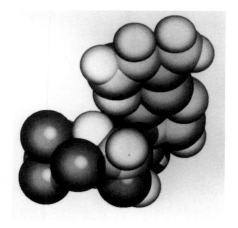

Fig. 8.5 Model of cAMP (see also Fig. 8.2). Courtesy M.J. Hartshorn, Polygen, University of York, UK.

The α subunits have an M_r 130 000 and the β subunits an M_r 90 000. The subunits are held together by disulphide bridges forming a tetrameric complex. The α subunits appear to be outside the cell and together make up the insulin-binding site. The β subunits each cross the membrane only once and have internal amino acid residues which may be phosphorylated.

8.4 Cyclic AMP as a second messenger

The structure of **cyclic AMP** is shown in Figs 8.2 and 8.5. Examples of agents which increase or decrease the cyclic AMP content of tissues are shown in Table 8.2. Adrenalin may either increase or decrease the cyclic AMP content depending upon the receptor with which it combines. Some of the agents listed in Table 8.2 can act through other receptors to generate a different second messenger (see Sections 8.5 and 8.6 and Table 8.6). Thus, a tissue's response to a hormone depends on the kind of receptor it possesses.

Formation of cyclic AMP

Adenylate cyclase catalyses the synthesis of cyclic AMP from ATP with the release of pyrophosphate (Fig. 8.6). The pyrophosphate is subsequently

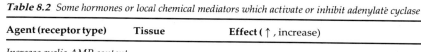

Table 8.2 Some hormones or local chemical mediators which activate or inhibit adenylate cyclase

Agent (receptor type)	Tissue	Effect (↑ , increase)
Increase cyclic AMP content		
Vasopressin (V$_2$)	Kidney	↑ Water reabsorption
Adrenalin (β)	Muscle	↑ Glycogen breakdown
Glucagon (G$_2$?)	Liver	↑ Glycogen breakdown
Thyroid-stimulating hormone	Thyroid	↑ Thyroxine secretion
Corticotropin	Adrenal cortex	Cortisol secretion
Decrease cyclic AMP content		
Adrenalin (α$_2$)	Platelet	Reduction of inhibitory effect on secretion of agents which raise cyclic AMP
Prostaglandin E$_2$	Gastric parietal cell	Inhibition of histamine-induced acid secretion
Adenosine (A$_1$)	White fat cell	Inhibition of triglyceride breakdown

cleaved by a pyrophosphatase to produce two molecules of orthophosphate. This second reaction serves to pull the overall process in the direction of cyclic AMP formation and makes it essentially irreversible. Similar mechanisms are found in fatty acid activation and nucleic acid formation. Adenylate cyclase is located on the inner side of the plasma membrane, where its substrate is a $MgATP^{2-}$ complex. The concentration of the $MgATP^{2-}$ is never rate-limiting and cannot therefore affect the rate of formation of the second messenger.

See *Biological Molecules*, Chapter 7 and *Molecular Biology and Biotechnology*, Chapters 1 and 3

Breakdown of cyclic AMP

Cyclic AMP is converted to 5'-AMP by enzymes called cyclic phosphodiesterases (Fig. 8.6). Some phosphodiesterases are active at quite low concentrations of cyclic AMP such as those found in resting cells. This group of phosphodiesterases maintain a turnover of cyclic AMP, moderate the rise

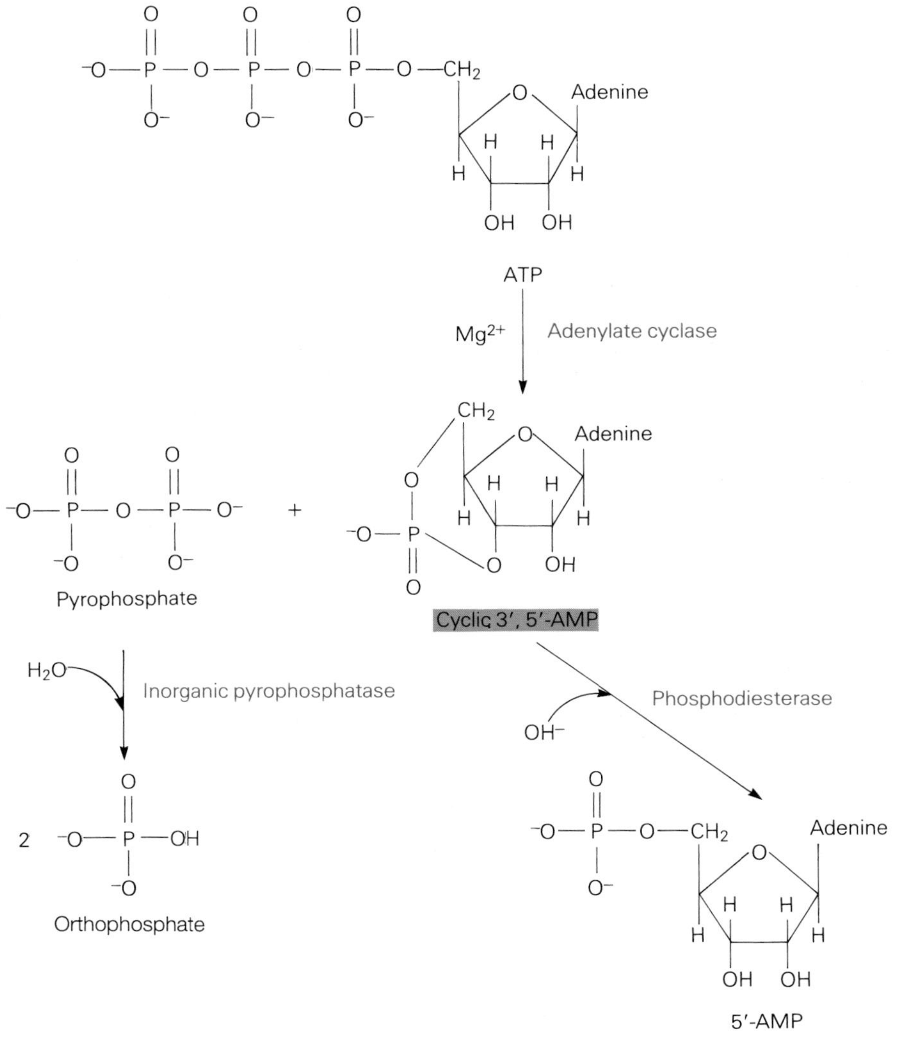

Fig. 8.6 Formation and breakdown of cyclic AMP.

in cyclic AMP that takes place on stimulation with hormone, and return the cyclic AMP concentration to resting levels when the hormone is removed. Other phosphodiesterases are only active at high concentrations of cyclic AMP. This group may set the maximum limit to which cyclic AMP concentration can rise. Phosphodiesterases are found bound to both the plasmalemma and internal membranes, and also in the cytosol. Some phosphodiesterases may be regulated by other hormone or second messenger systems that do not act through cyclic AMP.

Agents which inhibit cyclic AMP phosphodiesterases *potentiate* the action of hormones which stimulate adenylate cyclase activity. Prevention of cyclic AMP breakdown means that the increase in cyclic AMP concentration in response to the hormone will be greater than if the phosphodiesterases were active.

Receptors can move through the membrane to interact with adenylate cyclase

The hormone receptor and adenylate cyclase enzyme are thought to be separate units which do not have a fixed site in the membrane. This can be shown by fusing cells which have an active adenylate cyclase catalytic unit but lack the β receptor for adrenalin, with cells which have β receptors but a chemically-inactivated adenylate cyclase. The fused cells respond to adrenalin by increasing their cyclic AMP content. This means that the receptor must have diffused laterally through the cell membrane to interact with the adenylate cyclase enzyme (Fig. 8.7).

Exercise 3

The plasmalemma of white fat cells has several different types of receptors which can all stimulate adenylate cyclase activity upon binding their respective hormones. If adenylate cyclase is activated to a maximal extent by one highly effective hormone, it cannot be further stimulated by the addition of a second hormone acting through a different receptor. Use this information to show that each receptor molecule cannot be permanently linked to its own pool of adenylate cyclase.

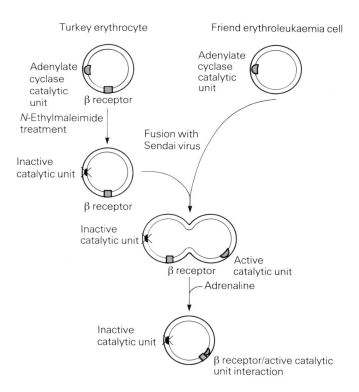

Fig. 8.7 Overview of an experiment showing that the hormone receptor and adenylate cyclase are not permanently coupled.

potentiation: *the response to the two agents together is greater than the sum of the responses to the two agents acting independently.*

Guanine nucleotide-dependent regulatory proteins (G proteins)

Receptors do not interact with adenylate cyclase directly. 'Information' from the activated receptor is transmitted to adenylate cyclase by means of proteins called **G proteins**. The first indication of the importance of G proteins was the demonstration that activation of adenylate cyclase in liver cell plasma membranes by glucagon was dependent upon guanosine triphosphate (GTP). However, the strongest evidence for the involvement of a G protein in the activation of adenylate cyclase was obtained from experiments using the cyc⁻ mutant of the S49 lymphoma cell line. This mutant possesses a *normal* β receptor for adrenalin, and a *normal* adenylate cyclase, but was nevertheless unresponsive to adrenalin. Addition of extracts from normal cell membranes to a preparation of cell membranes from cyc⁻ cells restored the response of the adenylate cyclase in the cyc⁻ membranes to adrenalin. This missing factor was a G protein called G_s.

□ Lymphoma cell lines were originally derived from tumours of a lymph node. S49 indicates the specific cell line.

G PROTEINS are made up of three subunits called α, β and γ associated to form a complex called a **heterotrimer**. The main types of subunit are listed in Table 8.3. Four categories of α subunit have so far been identified, and several of them have subtypes. The β and γ subunits associated with each type of subunit have similar structures to each other.

The G proteins which interact with adenylate cyclase are the α subunits of **G_s** and **G_i**. Each of these subunits has a site for interacting with the occupied hormone receptor; a guanine nucleotide binding site; a site for interacting with $\beta\gamma$ subunits and a site for interacting with adenylate cyclase. G_s also has an arginine side-chain which can be ADP-ribosylated by one of the protein components of cholera toxin, produced by *Vibrio cholerae* (Fig. 8.8). G_i is ADP-ribosylated on a cysteine side-chain by a toxin from *Bordetella pertussis* (the whooping cough bacterium). The effects of these modifications are described below.

Table 8.3 *Simplified list of G protein subunits*

Subunit	Molecular weight range ($M_r \times 10^{-3}$)	ADP-ribosylation catalysed by:	Action
α Subunits			
G_s	45–52	Cholera toxin	Activation of adenylate cyclase
G_i	40–41	Pertussis toxin	Inhibition of adenylate cyclase
G_t (transducin)	39	Cholera toxin Pertussis toxin	Activation of retinal cyclic GMP phosphodiesterase
G_o	39	Pertussis toxin	?
β Subunits	35–36		Binding to α subunits
γ Subunits	5–10		Binding to α subunits

Box 8.1
Cholera

People suffering from cholera may die from dehydration caused by the secretion of large amounts of fluid into the small intestine and the associated diarrhoea. Since cholera toxin causes the irreversible activation of G_s it will elevate the cyclic AMP content of cells. Increases in the cyclic AMP content of intestinal crypt cells causes an increase in Cl⁻ flux across their apical membrane. Water follows the ion and the net result is secretion of fluid into the intestinal lumen. These facts may explain one of the ways by which the bacterium *Vibrio cholerae* exerts its pathological effects.

---

heterotrimer: *protein constructed by the binding together of three different polypeptide chains.*

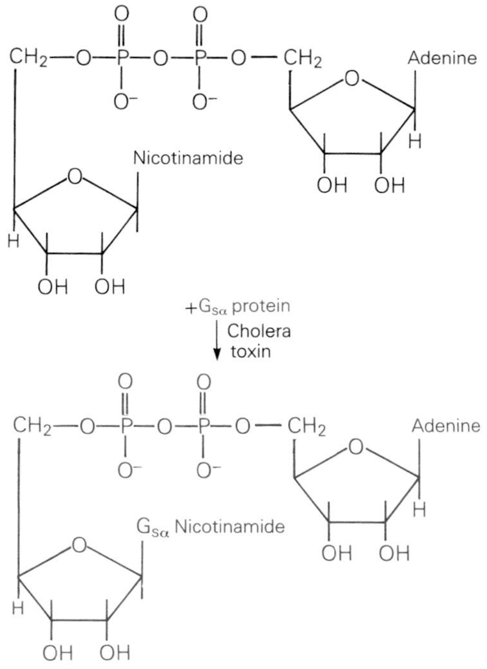

Fig. 8.8 The ADP-ribosylation reaction.

THE MODE OF ACTION OF G PROTEINS involves the dissociation and reassociation of the subunits making up the heterotrimer. In a resting cell the guanine nucleotide binding site on the subunit of the heterotrimer is occupied primarily by guanosine diphosphate (GDP). When a hormone combines with the receptor it probably causes a conformational change in the receptor which enables it to interact with the GDP–G protein complex (Fig. 8.9). The consequence of this interaction is that exchange of GDP for guanosine triphosphate (GTP) is promoted. When GTP is bound, the α subunit becomes activated and dissociates from the β and γ subunits. This form of the α subunit can interact with adenylate cyclase and modify its activity. The α subunit possesses **GTPase** activity, and GTP is converted to GDP:

$$GTP + H_2O \rightarrow GDP + P_i$$

When the guanine nucleotide binding site is reoccupied by GDP, recombination of the α and $\beta\gamma$ subunits can take place and the cycle is complete. Interaction of a hormone with its receptor in an isolated membrane system will therefore stimulate GTPase activity *if* G proteins are involved in the mechanism of action of the hormone. Sometimes this increase in GTPase activity is large enough to be detectable on addition of a hormone to a broken membrane preparation.

ADP-RIBOSYLATION of the α subunit of G_s by cholera toxin inhibits the intrinsic GTPase activity and therefore irreversibly activates this G protein. By contrast, pertussis toxin ADP-ribosylates the α subunit of G_i and prevents the heterotrimer containing G_i and GDP from interacting with the activated receptor. Pertussis toxin, therefore, inhibits the action of any hormones which activate G_i.

Exercise 4

The compound GTPγS has one of the oxygen atoms on the terminal phosphate group replaced by sulphur

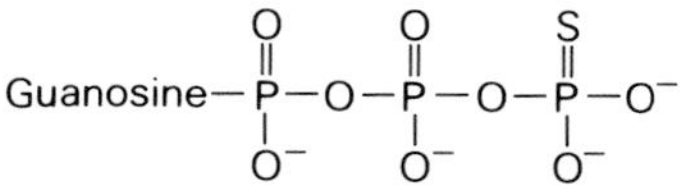

Although this alteration makes it ineffective as a substrate for the GTPase activity of G proteins, it can still bind effectively to the guanine nucleotide binding site. What effect will GTPγS have on the activity of G proteins?

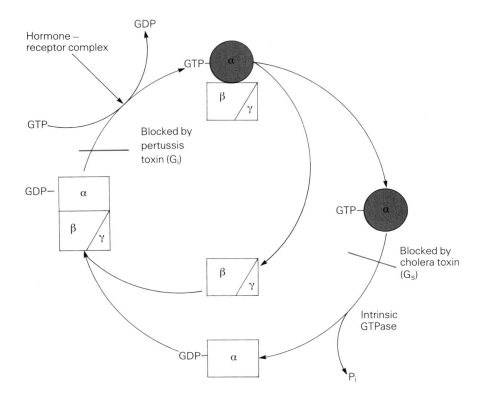

Fig. 8.9 Effect of activated receptors on G proteins (see text for details).

(a)

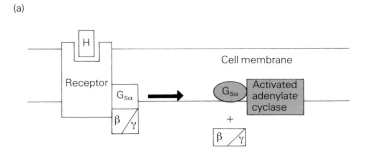

(b)

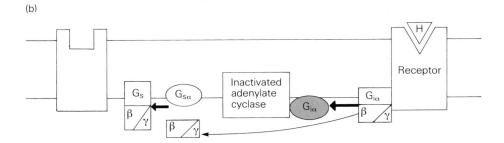

Fig. 8.10 (a) Activation of adenylate cyclase by receptors linked to G_s, (see text). (b) Inhibition of adenylate cyclase by receptors linked to G_i. The α subunit of G_i can either interact with adenylate cyclase directly, or the excess $\beta\gamma$ subunits can reassociate with the α subunit of G_s to form the inactive heterotrimer.

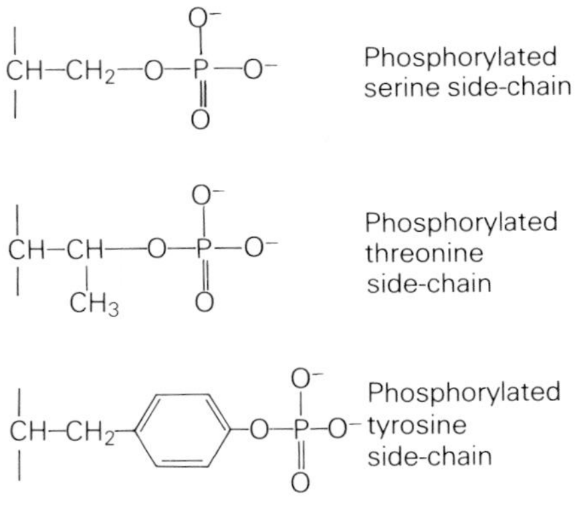

Fig. 8.11 Phosphorylated side-chains of serine, threonine and tyrosine residues.

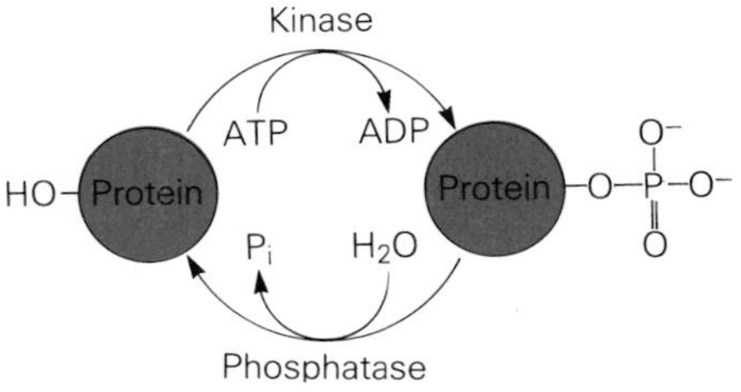

Fig. 8.12 Protein phosphorylation can be altered by either kinase or phosphatase activity.

HORMONES WHICH ACTIVATE ADENYLATE CYCLASE do so by causing the dissociation of the α subunit of G_s from the heterotrimer. The α subunit then activates adenylate cyclase (Fig. 8.10).

HORMONES WHICH DIRECTLY INHIBIT ADENYLATE CYCLASE effect the dissociation of the α subunit of G_i from the G_i complex. G_i may then act directly on adenylate cyclase to inhibit its activity, or the released β and γ subunits may combine with the subunit of G_s forcing the equilibrium back towards the inactive form of G_s (Fig. 8.10).

How does cyclic AMP exert its effects?

Cyclic AMP, in common with some other second messengers, modifies cellular function by causing changes in the phosphorylation of proteins. **Protein phosphorylation** is an important way of controlling cellular activity. Phosphorylation is catalysed by kinase enzymes. Depending upon the enzyme and its substrate, phosphorylation of either a serine, a threonine or a tyrosine residue side-chain (Fig. 8.11) can result from kinase action. The advantage of this method of controlling protein function is that the system can be switched from totally inactive to fully active simply by changing phosphorylation of the protein from zero to 100% or vice versa. If hormone activation is not to be permanent then the phosphate group must be removable. Phosphate removal is catalysed by **phosphatases**. Thus the phosphorylation state of a protein can be changed by altering the activity of a kinase, a phosphatase, or both (Fig. 8.12).

CYCLIC AMP-DEPENDENT PROTEIN KINASE is the enzyme activated by cyclic AMP. This enzyme consists of two catalytic subunits which possess the kinase activity, and two regulatory subunits which always remain attached to one another. Each regulatory subunit has two non-identical sites at which cyclic AMP can bind. The catalytic subunits are inactive when they are bound to the regulatory subunits. However, when cyclic AMP binds to the regulatory subunits the catalytic subunits dissociate from the complex and become active.

$$R_2C_2 + 4 \text{ cAMP} \rightleftharpoons R_2 (\text{cAMP})_4 + 2C$$
$$\text{(inactive)} \qquad\qquad \text{(active)}$$

Two forms of cyclic AMP-dependent protein kinase (Types I and II) exist in most mammalian tissues. The main difference between them is in the nature of the regulatory subunit. In the Type I enzyme the regulatory subunit has an M_r of 49 000 while in the Type II enzyme the M_r is 55 000. The catalytic subunits associated with the different regulatory subunits are similar and have an M_r of 40 000. A further difference between the regulatory subunits is that the Type II regulatory subunits can be **autophosphorylated** by the catalytic subunit. This phosphorylation reduces the rate of reassociation of the Type II regulatory subunit and the catalytic subunits. The type I subunit cannot be autophosphorylated. The proportion of the two forms of cyclic AMP-dependent protein kinase varies from one tissue to another. This suggests that the two forms may exert different effects. Possibilities are that the two forms respond differently to an increase in the concentration of cyclic AMP, or that upon dissociation the regulatory subunits may play different roles in the response of cells to hormones.

Cyclic AMP-dependent protein kinase phosphorylates serine or threonine side-chains at specific sites in proteins. When a serine residue is a substrate there is always an arginine residue between two and five amino acid residues

autophosphorylation: *ability of a protein with kinase activity to phosphorylate itself.*

Table 8.4 *Some effects of phosphorylation of proteins by cyclic AMP-dependent protein kinase*

Enzyme	Effect
Phosphorylase kinase	Phosphorylation of phosphorylase and glycogen breakdown
Glycogen synthetase	Inhibition of glycogen synthesis
Hormone sensitive triglyceride lipase	Stimulation of triglyceride breakdown
Smooth muscle myosin light-chain kinase	Inhibition of phosphorylation of myosin light-chains leading to inhibition of contraction
Phospholamban	Increased calcium uptake into sarcoplasmic reticulum of heart muscle
Troponin I	Sensitivity of troponin to calcium is decreased leading to increased rate of relaxation

on the amino side of the serine which is presumably important in substrate recognition by the enzyme.

Some of the effects of phosphorylation of proteins by cyclic AMP-dependent protein kinase are shown in Table 8.4.

PROTEIN PHOSPHATASES. There appear to be four major classes of **protein phosphatase**, which act on phosphorylated serine or threonine side-chains. They can be distinguished by three major criteria: the effect of the heat-stable inhibitor proteins termed inhibitors 1 and 2; the subunit of phosphorylase kinase which they phosphorylate, and their dependence on divalent cations for activity (Table 8.5). The phosphatases that act on phosphorylated tyrosine are as yet poorly characterized.

Table 8.5 *Classes of protein phosphatase*

Class	Regulation by cations	Inhibition by inhibitor 1 and inhibitor 2	Dephosphorylation of phosphorylase kinase subunits
1	None	Yes	β Subunit
2A	None	No	α Subunit
2B	Ca^{2+} (calmodulin)	No	α Subunit
2C	Mg^{2+}	No	α Subunit

Protein phosphatases 1, 2A and 2C show a broad and overlapping substrate specificity. That is, they will catalyse the dephosphorylation of serine or threonine residues at many sites within isolated proteins. It is therefore difficult to establish which phosphatases are important in controlling a particular hormonal response in an intact cell.

PROTEIN PHOSPHATASE 1 is bound by a specific protein to particles made up of glycogen and other proteins which metabolize glycogen. Cyclic AMP-dependent protein kinase phosphorylates this binding protein, and causes the dissociation of protein phosphatase 1 from the glycogen particles. Cyclic AMP-dependent protein kinase also phosphorylates and activates inhibitor 1, which then binds to, and inactivates, the released phosphatase in the cytosol. The net effect of these changes is that protein phosphatase 1 is prevented from dephosphorylating and inactivating phosphorylase. Phosphorylase has been activated as a consequence of the stimulation of phosphorylase kinase by cyclic AMP-dependent protein kinase.

PROTEIN PHOSPHATASE 2B has a much more restricted substrate specificity and its dependence on Ca^{2+} is due to the presence of calmodulin (see below). Protein phosphatase 2B dephosphorylates and inactivates inhibitor 1.

See *Biological Molecules*, Chapter 6

8.5 Is cyclic GMP a second messenger?

In many cells the content of **cyclic GMP** (Fig. 8.2 and 8.13) may be increased in response to a hormone. This increase is usually secondary to the intracellular actions of the hormone, which are not *caused* by changes in the concentration of cyclic GMP. However, it has recently emerged that **atrial naturetic factor (ANF)**, a group of peptides released from granules in cardiac atrial tissue, may use cyclic GMP as a second messenger. ANF relaxes smooth muscle and increases renal excretion of Na^+ and water. ANF increases the cyclic GMP content in target tissues and can selectively activate a membrane-bound guanylate cyclase. Elevation of cyclic GMP concentration will activate cyclic GMP-dependent protein kinase, which may therefore mediate the actions of ANF.

CYCLIC GMP-DEPENDENT PROTEIN KINASE consists of two identical subunits each of which contain a regulatory and a catalytic site. Activation of the enzyme by cyclic GMP therefore does not involve dissociation of regulatory and catalytic subunits, but instead probably involves a change in shape of each subunit. The enzyme phosphorylates serine and threonine residue side-chains of a wide variety of proteins.

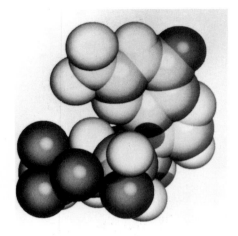

Fig. 8.13 Model of cGMP (see also Fig. 8.2). Courtesy M.J. Hartshorn, Polygen, University of York, UK.

8.6 Hormones which use inositol trisphosphate and diacylglycerol as second messengers

A second major group of hormones (Table 8.6) generates two intracellular messengers by stimulating the hydrolysis of **phosphatidylinositol 4,5-bisphosphate** by a phosphodiesterase (Fig. 8.14). The products are **inositol**

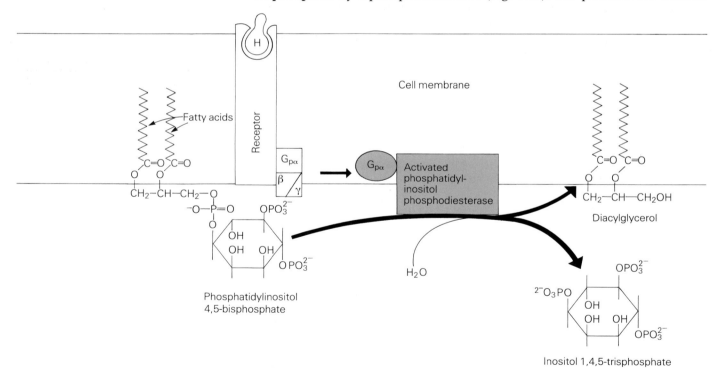

Fig. 8.14 Receptor-mediated breakdown of phosphatidylinositol 4,5-bisphosphate. $Gp\alpha$ is suggested by experiments to be the factor involved in coupling the hormone – receptor complex to the phosphodiesterase, but has not been isolated and purified.

Reference Morgan, N.J. (1989) *Cell Signalling*, Open University Press, Milton Keynes. A good account of signal transduction mechanisms at a fairly elementary level.

1,4,5-trisphosphate and **diacylglycerol**, detailed structures of which are shown in Figs 8.2 and 8.15. Inositol bisphosphate and inositol monophosphate are also formed in hormonally-stimulated tissue, but they accumulate after inositol trisphosphate, and it is the formation of this compound which seems to be the first step.

See *Biological Molecules*, Chapter 7

Table 8.6 *Some hormones which stimulate the breakdown of phosphatidylinositol 4,5 bisphosphate*

Agonist (receptor type)	Tissue	Effect (↑ , increase)
Adrenalin (α_1)	Liver	↑ Glycogen breakdown
Vasopressin (V_1)	Liver	↑ Glycogen breakdown
Angiotensin II	Adrenal cortex	↑ Aldosterone secretion
Cholecystokinin	Exocrine pancreas	↑ Secretion of digestive enzymes

A G PROTEIN (DENOTED G_p) is probably involved in coupling receptors to the phosphodiesterase. Thus the addition of GTP, or of its non-hydrolysable analogues, to cell membrane preparations can stimulate the hormonally-induced formation of inositol phosphates. Furthermore, in some cell-types hormonal induction of phosphatidylinositol 4,5-bisphosphate breakdown is prevented by pretreatment with pertussis toxin. Pertussis toxin is known to ADP-ribosylate certain G proteins and to modify their function (see above). The G proteins involved in the process of generation of inositol trisphosphate have not yet been purified.

Resynthesis of phosphatidylinositol 4,5-bisphosphate

Inositol trisphosphate can be broken down by a series of reactions to give inositol which can then be reincorporated into inositol phospholipids (Fig. 8.16). Diacylglycerol can also be reconverted to phospholipids. Thus a stimulation of phosphatidylinositol breakdown will be followed by a stimulation of resynthesis. Indeed, the action of hormones on inositol phospholipid metabolism was first discovered as an effect on phospholipid turnover.

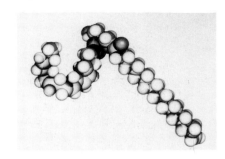

(a)

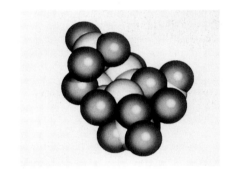

(b)

Fig. 8.15 Molecular models of (a) the diacylglycerol, 1-stearoyl-2-arachidonyl-*sn*-glycerol and (b) inositol 1,4,5-triphosphate (see also Fig. 8.2). Courtesy M.J. Hartshorn, Polygen, University of York, UK.

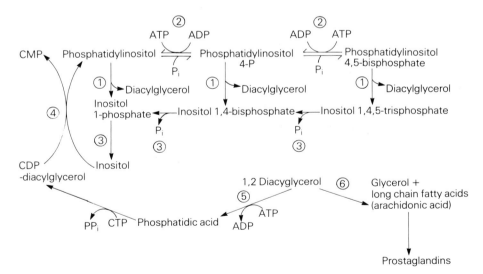

Fig. 8.16 Pathways of inositol and inositol phospholipid metabolism. Metabolism of diacylglycerol. 1, phosphodiesterases; 2, kinase; 3, phosphomonoesterases; 4, cytidine diphosphate–diacylglycerol inositol transferase; 5, diacylglycerol kinase; 6, diacylglycerol and monoacylglycerol lipases.

The breakdown of inositol 1-phosphate and inositol 1,4-bisphosphate is blocked by 10 mmol dm^{-3} Li$^+$ (see Fig. 8.17). Li$^+$ will also block the synthesis of inositol from glucose as this occurs via inositol 1-phosphate. Li$^+$ is used to treat manic depressive mental illness. The brain cannot obtain inositol directly from the blood, so in the presence of Li$^+$, there may be a depletion of inositol phospholipids in brain. Look at Fig. 8.16 to see why. This depletion might partially block the action of neurotransmitters which act at receptors linked to the breakdown of inositol 4,5-bisphosphate, and might be the mechanism by which Li$^+$ exerts its effect on mental illness.

An alternative route for diacylglycerol metabolism

The diacylglycerol released from the breakdown of phosphatidylinositol 4,5-bisphosphate often has arachidonate in the 2 position (Figs 8.2 and 8.15a). Breakdown of the diacylglycerol by lipases will generate free **arachidonate**. This compound can then be converted into **prostaglandins** and related compounds, which may act as further messengers of hormonal action.

ISOMERS OF INOSITOL TRISPHOSPHATE have recently been discovered. Inositol 1,4,5-trisphosphate can be phosphorylated to **inositol 1,3,4,5-tetrakisphosphate**. This compound is then degraded to **inositol 1,3,4-trisphosphate** (Fig. 8.17). The inositol-1,4,5-trisphosphate-3-kinase is activated by Ca^{2+} in the presence of calmodulin (see below). The formation of inositol 1,3,4,5-tetrakisphosphate may provide another messenger concerned with the entry of Ca^{2+} across the cell membrane. Alternatively, it may be a means of regulating the intracellular concentration of the established second messenger inositol 1,4,5-trisphosphate.

How does inositol 1,4,5-trisphosphate act as a second messenger?

Inositol 1,4,5-trisphosphate, but not the 1,3,4 isomer, interacts with a receptor situated on the endoplasmic reticulum. Binding of the messenger to this receptor results in a release of accumulated Ca^{2+} into the cytosol (Fig. 8.18). Thus a rise in intracellular Ca^{2+} concentration can be demonstrated when inositol 1,4,5-trisphosphate is added to cells whose cell membrane has been treated to make it permeable to small molecules. Hormones which initiate a breakdown of phosphatidylinositol 4,5-bisphosphate therefore increase intracellular Ca^{2+} concentration. In many cases this event is prolonged by the opening of Ca^{2+} channels in the cell membrane which allow Ca^{2+} to move into the cell from the extracellular fluid. The mechanism by which these channels are opened has yet to be established. Ca^{2+} can be pumped back into the endoplasmic reticulum, can be taken up by mitochondria, or can be pumped out of the cell (Fig. 8.18).

Many effects of Ca^{2+} are mediated by calmodulin

Calmodulin is a small protein of M_r 16 680 which has four non-identical binding sites for Ca^{2+} (Fig. 8.19). When Ca^{2+} binds to calmodulin it induces a change in its conformation which enables the Ca^{2+}–calmodulin complex to bind to and activate the target proteins. Some targets of calmodulin, which may be important in mediating responses to hormones, are phosphorylase kinase, myosin light-chain kinase and protein phosphatase 2B. In addition, Ca^{2+}–calmodulin can activate the cell membrane Ca^{2+}-ATPase, a Ca^{2+}-pump whose activity will reduce the rise in intracellular Ca^{2+} induced by the

---

Exercise 5

Draw the structures of inositol 1,3,4-trisphosphate and inositol 1,3,4,5-tetrakisphosphate.

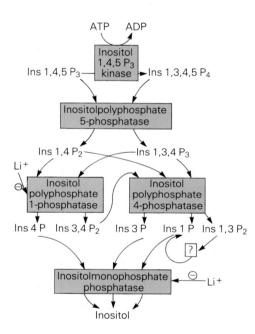

Fig. 8.17 A proposed scheme for the metabolism of isomers of inositol phosphates. Enzymes are in boxes, Ins, inositol; P$_3$, trisphosphate etc. Probable sites for the inhibitory action of Li$^+$ are indicated (see Box 8.2).

☐ In some proteins (for example, phosphorylase kinase) calmodulin is always tightly bound to the enzyme, but otherwise its mechanism of action is similar.

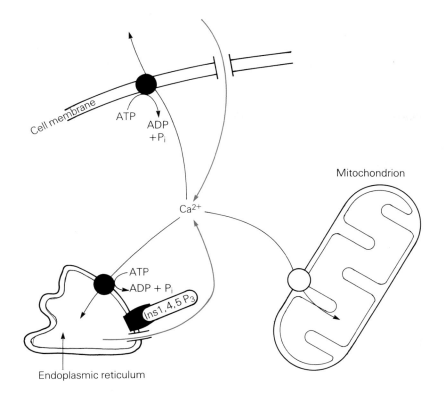

Fig. 8.18 Simplified diagram showing some pathways which can control the rise in Ca^{2+} concentration after hormonal stimulation of a tissue. Pathways in black reduce cytosolic Ca^{2+}, those in colour increase it. ●, Ca^{2+} pump; ||, receptor-operated Ca^{2+} channels; ○, mitochondrial Ca^{2+} uptake system, ◖, the inositol 1,4,5-trisphosphate receptor.

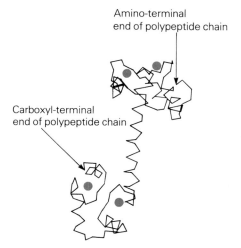

Fig. 8.19 Representation of the three-dimensional arrangement of the polypeptide chain of calmodulin. Positions occupied by Ca^{2+} ions are indicated as ●.

hormone. Also, Ca^{2+}–calmodulin can activate a cAMP phosphodiesterase in several tissues. Hormones acting through receptors linked to the breakdown of phosphatidylinositol 4,5-bisphosphate can thereby affect the cellular content of cyclic AMP. Finally, in brain there is a form of adenylate cyclase which is controlled by Ca^{2+}–calmodulin rather than via G proteins.

Mitochondrial proteins are directly activated by Ca^{2+}

Pyruvate dehydrogenase phosphatase, (NAD^+) isocitrate dehydrogenase and 2-oxoglutarate dehydrogenase are all activated by an increase in mitochondrial Ca^{2+} concentration, which follows from the increase in the concentration of Ca^{2+} in the cytosol. Calmodulin is not involved in this response. This mechanism serves to activate the TCA cycle in heart muscle on contraction, but may also provide an increased supply of ATP in some tissues where hormones raise the intracellular Ca^{2+} concentration.

□ Pyruvate dehydrogenase phosphatase dephosphorylates the α subunit of pyruvate decarboxylase and thus activates the pyruvate dehydrogenase complex. Mitochondrial metabolism of acetyl CoA will thus be enhanced (see *Biological Molecules*, page 50 or Section 2.3).

How does diacylglycerol act as a second messenger?

Diacylglycerol activates a protein kinase which is called protein kinase C because of its dependence on Ca^{2+} for activity. In addition, the purified enzyme requires phospholipid, particularly phosphatidylserine. Diacylglycerol reduces the Ca^{2+} requirement of the isolated enzyme to the concentration found in the resting cell. In an intact cell activation of protein kinase C is thought to occur because diacylglycerol causes the inactive soluble enzyme to bind to the cell membrane thereby bringing it into contact with

Reference Nishizuka, Y. (1986) Studies and perspectives of protein kinase C. *Science*, **233**, 305–12. A review of the field by the discoverer of protein kinase C.

Reference Allan, G.F. *et al.* (1991). Steroid hormone receptors and *in vitro* transcription. *Bioessays*, **13**, 73–8. Fascinating essay on how steroid hormones and their receptors are investigated, and how they act together in cells.

phospholipid resulting in activation of the enzyme. Removal of a portion of the single polypeptide chain (M_r 77 000) of protein kinase C by a proteolytic enzyme gives a product which is active in the absence of Ca^{2+}, phospholipid or diacylglycerol. Activation of the intact enzyme by these agents may therefore involve a loss of the inhibitory effect of that part of the enzyme which is removed by proteolysis (Fig. 8.20).

Protein kinase C catalyses the phosphorylation of a wide range of cellular proteins at threonine or serine residues. Its activation by hormonal induction of the breakdown of phosphatidylinositol 1,4-bisphosphate can be mimicked by the addition of certain synthetic diacylglycerols to cells.

Fig. 8.20 Scheme for the activation of protein kinase C. DAG, diaglyglycerol; PS, phosphatidylserine.

□ An example of a Ca^{2+} ionophore is the compound A23187. Two molecules of A23187 surround one Ca^{2+}. The exterior of the complex is hydrophobic, and can easily be inserted into the lipid bilayer of membranes, which are therefore rendered permeable to Ca^{2+}.

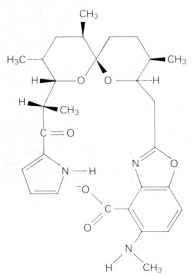

The calcium ionophore A23187.

SEVEN SUBSPECIES OF PROTEIN KINASE C denoted α, β_1, β_2, γ, δ, ε and ζ have been discovered by the screening of cDNA libraries. The β_1 and β_2 forms are derived from alternative splicing of the same gene product. Otherwise each type is coded for by a different gene. The γ enzyme seems to be restricted to the central nervous system. The proportions of the widely distributed α and β forms vary from one tissue to another, and the forms show variations in their activation by Ca^{2+}, diacylglycerol and micromolar concentrations of arachidonate. In contrast, it is now known that the activity of the ε form is completely independent of Ca^{2+}. It seems likely that the different forms of protein kinase C may have different functions.

Why two messengers?

The question arises as to why hormones which stimulate the breakdown of phosphatidylinositol 4,5-bisphosphate generate two second messengers. Interactions between the elevation of intracellular Ca^{2+} concentration and activation of protein kinase C can be investigated by using Ca^{2+} ionophores (Box 4.6) to increase intracellular Ca^{2+}, and by using **TPA** (Box 8.3) or synthetic diacylglycerols to activate protein kinase C. In some cells a strong potentiation can be seen between the Ca^{2+} and protein kinase C signals. *Thus, by having two messengers the response to the hormone is enhanced.* In other cells, Ca^{2+} seems to be responsible for the initial phase of the response, but activation of protein kinase C is required for the response to be sustained.

The compound 12-*O*-tetradecanoylphorbol 13-acetate (TPA) is a potent activator of protein kinase C both in the intact cell and after its purification. This compound is also a tumour promoter. If it is applied, for example, to the skin after a carcinogen then the likelihood of tumour formation is enhanced. Protein kinase C may therefore be involved in the control of cell division.

12-*O*-tetradecanoylphorbol 13-acetate (TPA).

Finally, protein kinase C may phosphorylate the receptor for the hormone and thereby lead to an inhibition of the response. This effect is known as **desensitization** of the receptor and is a common event in many cells stimulated by hormones.

☐ Down-regulation of receptors coupled to adenylate cyclase can also occur. Thus the activated β adrenergic receptor (Fig. 8.3) can be phosphorylated by the β adrenergic receptor kinase, and by cyclic AMP-dependent kinase with a loss of response of the tissue to adrenalin.

8.7 Insulin – a hormone without a messenger?

Insulin (Fig. 8.21) seems to be different from the two main classes of hormones that have been discussed so far. Binding of insulin to its receptor (see Fig. 8.4) activates a **tyrosine kinase** on the β subunit and this kinase phosphorylates tyrosine residue side-chains on the same subunit, that is, an autophosphorylation occurs.

How are the effects of insulin mediated?

Examination of the effects of insulin shows that they involve either phosphorylation or dephosphorylation of threonine or serine residues (Table 8.7), or the translocation of proteins between the plasma membrane and internal membranes. It is possible that the insulin receptor tyrosine kinase can

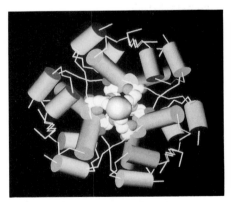

Fig. 8.21 Model of a hexamer of insulin molecules around a central Zn^{2+}. Courtesy Dr J.M. Burridge, IBM, UK.

Table 8.7 *Some effects of insulin*

Phosphorylation of serine/threonine residues	Dephosphorylation of serine/threonine residues	Translocation
Insulin receptor	Glycogen synthetase	Glucose transporter to cell membrane from interior
ATP citrate lyase	Pyruvate dehydrogenase	Insulin receptor from cell membrane to interior
Acetyl CoA carboxylase	Hormone-sensitive triacyl-glyceride lipase	
Ribosomal protein S6	Phosphorylase kinase	

Reference Espinal, J. (1988) What is the role of the insulin receptor tyrosine kinase? *Trends in Biochemical Sciences*, **13**, 367–8. A short interesting article.

See Chapter 4

phosphorylate and activate serine/threonine kinases or phosphatases which then act on the target proteins. After binding insulin, the insulin–receptor complex undergoes internalization. This enables it to interact with proteins present on internal membranes. Such events can result in the translocation of glucose transporter molecules to the plasmalemma and so promote the cellular uptake of glucose associated with insulin. If the receptor is altered by site-specific mutation so that the tyrosine kinase is defective, then translocation of glucose transport proteins to the cell membrane does not occur. Phosphorylation of the insulin receptor may therefore be required for this effect.

In hepatocytes insulin can inhibit adenylate cyclase and activate several cyclic AMP phosphodiesterases. There is evidence that a novel G protein is involved in this mechanism. It is suggested that the insulin–receptor complex may activate this G protein and cause it to activate a phospholipase. This phospholipase then catalyses the enzymic hydrolysis of a glycosyl phosphatidylinositol giving rise to diacylglycerol and an **inositol glycan** second messenger, which includes glucosamine (Fig. 8.22). This proposal remains to be substantiated.

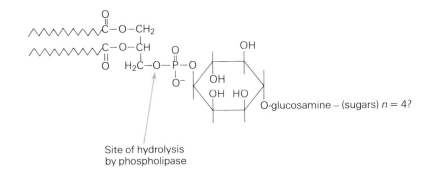

Fig. 8.22 Partial structure of the proposed insulin-sensitive glycosyl-phosphatidylinositol.

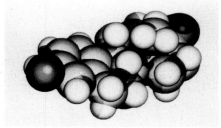

(a)

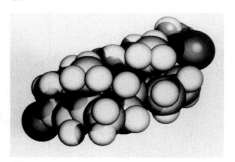

(b)

Fig. 8.23 Models of (a) oestrogen and (b) progesterone. Courtesy M.J. Hartshorn, Polygen, University of York, UK.

8.8 Steroid hormones penetrate the cell membrane

Steroid hormones have a variety of functions. The female sex hormones, the oestrogens and progesterone (Fig. 8.23), are involved in sexual development at puberty, in control of the menstrual cycle and in pregnancy. In males, testosterone is involved, for example, in the development of secondary sexual characteristics at puberty. The mineralocorticoids are a group of steroids involved in the maintenance of salt and water balance, while glucocorticoids are involved in the metabolic responses to starvation.

THE CLASSICAL MODEL OF STEROID HORMONE ACTION is that steroids bind to cytosolic receptors and alter their conformations. The steroid hormone–receptor complexes then migrate to the nucleus where they interact with specific regions of DNA, called **hormone response elements**, to alter the

Reference Saltiel, A.R. and Cuatrecasas, P. (1988) In search of a second messenger for insulin. *American Journal of Physiology*, **255**, C1–C11. Is there an inositol glycan second messenger?

extent of transcription of particular genes (Fig. 8.24). Steroid hormones therefore act by *specifically altering the synthesis of a small number of proteins.* Since they affect protein synthesis, the response to steroid hormones may take several hours to become apparent. By contrast, responses to the hormones that activate adenylate cyclase or induce the breakdown of phosphatidylinositol are extremely rapid. The presence of receptors for a given steroid determines whether or not the tissue will respond to that steroid. Recent work suggests that the presence of unoccupied steroid receptors in cytosolic extracts of cells may be an artifact which results from procedures used for homogenization of the tissue and its subcellular

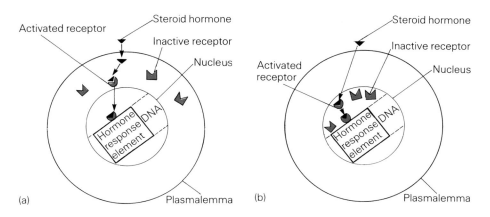

Fig. 8.24 (a) Classical model for steroid hormone action, and (b) model in which receptors are localized to the nucleus.

Box 8.4
Male steroid hormones

In certain tissues of adult males the hormone testosterone is converted by the enzyme testosterone 5 α-reductase to 5α-dihydrotestosterone. In these tissues 5α-dihydrotestosterone rather than testosterone is the active steroid. Activity of the 5α-reductase enzyme increases at puberty. Lack of the enzyme prevents the appearance of secondary sexual characteristics.

Testosterone

Testosterone 5α-reductase

5α-Dihydrotestosterone

Conversion of testosterone to 5α-dihydrotestosterone.

Reference Gehring, U. (1987) Steroid hormone receptors: biochemistry, genetics and molecular biology. *Trends in Biochemical Sciences,* **12**, 399–402. A short, readable review describing steroid hormone receptors.

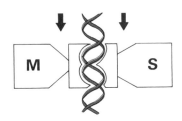

Fig. 8.25 Model for steroid receptor protein showing modulatory (M), DNA-binding and steroid-binding (S) domains. Sites susceptible to proteolytic cleavage are indicated by arrows.

□ Recently it has been found that the nucleotide sequence of part of the c-erb A proto-oncogene (*Biological Molecules*, Chapter 3) is similar to that coding for the DNA binding region of the steroid hormone receptors. The proto-oncogene codes for the thyroxine receptor. The implication of these observations is that the primary action of thyroid hormones is on gene expression.

□ Plant hormones have been named in a variety of non-systematic ways. For example, auxin is derived from the Greek word *auxein*, meaning an increase, and gibberellins from the fungus *Gibberella fujikuroi* from which over 20 hormones have been isolated.

fractionation. Thus, fluorescent monoclonal antibodies to the oestrogen receptor protein only bind to nuclei in tissue sections taken from a number of oestrogen-sensitive tissues (Fig. 8.24).

An example of the action of steroid hormones in inducing the synthesis of particular proteins is provided by the actions of oestrogens and progesterone on the proteins secreted by the chicken oviduct. These proteins are taken up into the developing egg as it passes down the oviduct. Oestrogen induces the formation of mRNA necessary for the synthesis of the proteins ovalbumin, conalbumin, ovomucoid, and lysozyme in the tubular gland cells of the duct. These proteins are packaged into vesicles and secreted into the oviduct. Progesterone induces the synthesis of another protein, avidin, in the goblet cells of the duct.

The current model for steroid hormone receptors suggests that they are made up of three regions, called domains, which are linked by hinge regions which can be cleaved by proteolytic enzymes (Fig. 8.25). There are strong similarities in the amino acid sequences of the DNA binding domains of a variety of steroid hormone receptors. Modification of glucocorticoid receptors by removal of both amino- and carboxyl-terminal amino acid residues gives rise to a truncated receptor which can activate transcription of specific genes in the absence of hormone. This suggests that the hormone-binding domain may prevent the DNA-binding domain from interacting with the DNA in the absence of hormone.

8.9 Plant hormones

Six major classes of substances are known to be produced by plants which regulate their growth and development. These are the auxins, cytokinins, gibberellins, ethylene, abscisic acid and oligosaccharins (Fig. 8.26). Each class contains many active individual hormones and, in most cases, synthetic substances which mimic the action of natural hormones are known and have been used in many studies on plant hormones.

Plant hormones differ from animal hormones in a number of ways. Firstly, although they may be produced in localized areas of the plant, they are made by unspecialized cells. Secondly, the effects of plant hormones are not constant, in the sense that their actions are often concentration-dependent, and are also affected by the presence or absence of other types of hormones. Thirdly, whilst some animal hormones can produce an extremely rapid and localized response, plant hormones generally affect cell division and plant growth, making their effects long-lasting and sustained.

AUXINS stimulate the growth of main shoots and inhibit the growth of lateral shoots. They are synthesized from the amino acid tryptophan in the apical buds of growing shoots (Fig. 8.27).

Auxin-induced growth occurs in two phases. A relatively rapid increase in shoot length, followed by slow changes within the cells. The rapid effects are

---

Box 8.5
Plant cell culture

Growth of plant cells in a test-tube is a useful way of making large numbers of identical and virus-free plants very quickly. A major impetus to the early development of techniques of plant cell culture was the discovery by Van Overbeck and coworkers that coconut milk promoted the growth of isolated embryos (a small part of the seed). There are several cytokinin-like substances in coconut milk which appear to work together to promote growth.

Reference Hill, T.A. (1980) *Endogenous plant growth substances*, Edward Arnold, London, UK. Rather old, but still an excellent account of plant hormones and their actions.

Name	Structure	Molecular models

Ethylene

$$\begin{array}{c} H \\ \end{array} C = C \begin{array}{c} H \\ \end{array}$$

Abscisic acid

CH₃ CH₃ CH₃
OH
O CH₃ COOH

Indole-3-acetic acid – an auxin

CH₂COOH

N
H

Zeatin – a cytokinin

CH₂OH
H N CH₂ CH C CH₃
N N
N N
H

Gibberellic acid – a gibberellin

O
CO
HO OH
CH₃ COOH CH₂

Oligosaccharin – a heptaglucoside

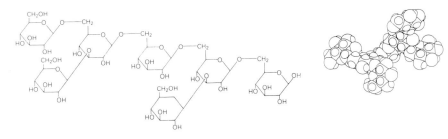

Fig. 8.26 Structures of representative members of the six major classes of plant hormones. Molecular models courtesy of Dr E.E. Eliopoulos, Department of Biochemistry and Molecular Biology, University of Leeds, UK and M.J. Derham, Department of Chemistry, Manchester Polytechnic, UK.

Reference Wilkins, M.B. (ed.) (1984) *Advanced Plant Physiology*, Pitman, London, UK. A rather advanced text, covering virtually all aspects of plant physiology and biochemistry. Chapters 1–6, 19 and 20 are relevant to this chapter.

Reference Roberts, J.A. and Hodey, R. (1988) *Plant Growth Regulators*, Blackie, Glasgow, UK. A short book covering most aspects of plant hormones.

<table>
<tr><td>

Box 8.6
Mammalian growth regulators

</td><td>

Mammals also have growth regulators. These seem to act locally and are therefore not true hormones. The polypeptide epidermal growth factor (EGF) stimulates the increased growth of many cells. The receptor for epidermal growth factor is in the cell membrane. A mutant form of the epidermal growth factor receptor called the v-*erb* B oncogene (*Molecular Biology and Biotechnology*, Box 1.4) is encoded in the genetic material of avian erythroblastosis virus. When this receptor is expressed in cells infected with the virus it exhibits tyrosine kinase activity even in the absence of EGF and the cells grow in an uncontrolled manner.

</td></tr>
</table>

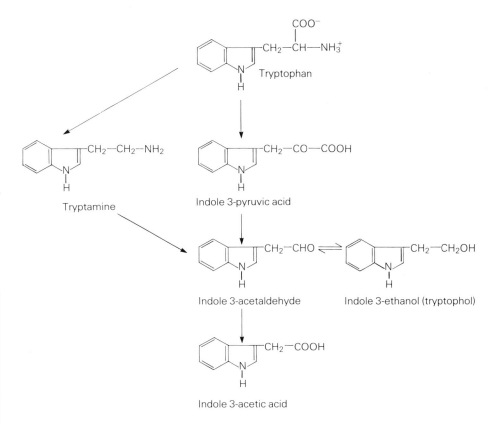

Fig. 8.27 Overview of the synthesis of the auxin, indole acetic acid (IAA) from tryptophan.

Fig. 8.28 Addition of a controlled amount of IAA to a growing coleoptile using a micropipette during investigations of the effects of plant hormones. Courtesy of Dr S. Shaw, Department of Biological Sciences, The Manchester Metropolitan University, UK.

Exercise 6

You think that the curvature of the tip of a grass seedling towards the light is due to the passage of a substance from the tip down into the stem, where it promotes more growth on one side of the stem. Design experiments to test this hypothesis and to investigate the nature of the substance.

---

relatively well understood. Plant cell walls are made up of cellulose fibres embedded in a complex matrix. Their strength enables them to keep the cell in shape despite a considerable internal hydrostatic or turgor pressure. When auxin is applied to the cell (Fig. 8.28), it causes a localized stimulation of a proton pump. The action of this pump causes a drop in pH on the inside of some parts of the wall. This change in pH causes a loosening of the interactions between cellulose fibres, and results in the cell elongating due to internal pressure (Fig. 8.29). The new internal space is occupied not by the cytosol, the volume of which stays fairly constant, but by an enlargement of the vacuole which contains water and a high concentration of dissolved substances.

A sustained (long-term) response to auxin-stimulation is the production of mRNA and proteins which presumably consolidates the earlier enlargement with an increase in biomass.

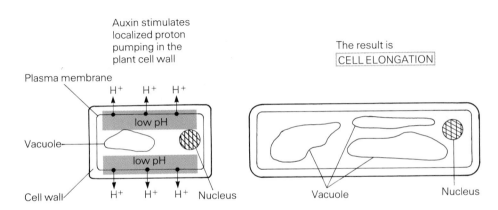

Fig. 8.29 Overview of the effects of auxin in promoting cell elongation.

CYTOKININS promote the division of plant cells and therefore affect the growth and differentiation of plant tissues. They are derivatives of adenine, being synthesized from AMP (Fig. 8.30). Cytokinin-stimulation can overcome the auxin-inhibited growth of lateral shoots. However, in other plant tissues, the two hormones act **synergistically**. Cytokinins and auxins are added to plant tissue cultures to promote growth and differentiation of the cells. The mechanisms by which cytokinins promote these effects are not understood.

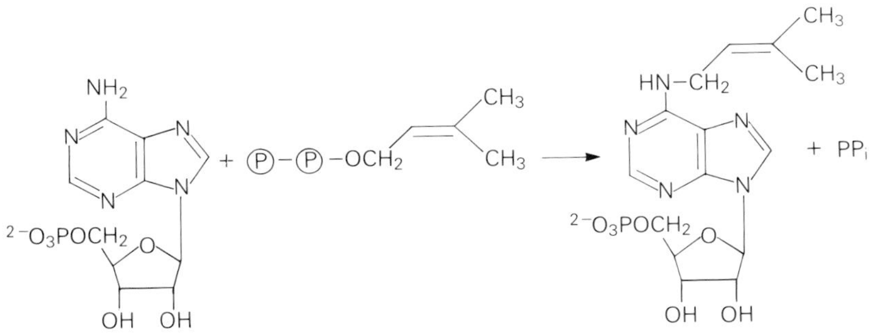

Fig. 8.30 The production of isopentyladenine 5'-monophosphate from AMP and isopentylpyrophosphate, the key step in cytokinin production.

GIBBERELLINS promote shoot elongation (Fig. 8.31) and stimulate the production of specific proteins, for example, amylase, in germinating seeds. Gibberellins act synergistically with auxins. This stimulation appears to occur at the transcriptional level, suggesting that receptors for gibberellins may act at the genetic level, perhaps in a manner similar to vertebrate steroid hormones (Section 8.8).

Box 8.7
Plant tumours

Crown gall tumours are caused by the infection of plant wounds by some strains of *Agrobacterium* bacteria. These bacteria carry the T_i tumour-inducing plasmid. This plasmid encodes several genes involved in plant hormone production. Two gene products control the biosynthesis of auxins and cytokinins respectively. Thus incorporation of the T_i plasmid into the plant genome leads to defective hormone production and the rapid, but undifferentiated growth resulting in crown gall tumour.

---

synergism: *two or more components act together to produce an effect greater than the sum of the individual effects. From the Greek* synergetikos, *to co-operate.*

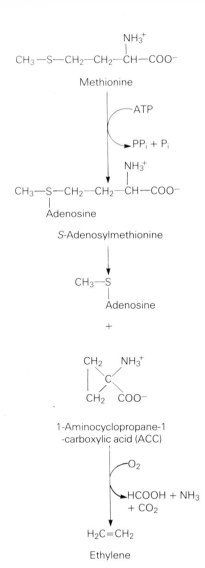

Methionine

S-Adenosylmethionine

1-Aminocyclopropane-1
-carboxylic acid (ACC)

Ethylene

Fig. 8.32 Overview of the biosynthesis of ethylene from methionine.

- ☐ Abcission is the controlled loss of parts of the plant structure. Perhaps the best known example is the shedding of leaves by deciduous trees in autumn. The loss of the plant part is preceded by the formation of a so-called abscission layer across the plant at the point of breakage. This layer consists of several adjacent tiers of extremely thick-walled cells. Abscission usually occurs across the middle lamellae between two sets of cells within the abscission zone (Fig 8.33).

- ☐ The action of oligosaccharins in responsive cells shows some similarity with the release and effects of interferons in viral infected animals (Section 11.3).

Fig. 8.31 The effects of addition of gibberellic acid to the growth of dwarf pea seedlings (*Pisum sativum*). Units are μg cm^{-3}. Courtesy of Dr J. Morris, Department of Biological Sciences, The Manchester Metropolitan University, UK.

ETHYLENE stimulates flower senescence and fruit ripening, in addition to a variety of other effects. For example, it redirects auxin transport in plants leading to transverse rather than longitudinal growth.

Ethylene is produced from the amino acid methionine (Fig. 8.32). The bioconversion of 1-aminocyclopropane-1-carboxylic acid (ACC) to ethylene (Fig. 8.32) is stimulated by auxins and cytokinins and also by mechanical factors such as cell damage.

Ethylene is known to bind to sites associated with the endoplasmic reticulum and Golgi apparatus, possibly to integral membrane proteins. However, as with many plant hormones, the biochemical mechanisms by which ethylene exerts its effects are not understood.

ABSCISSIC ACID (ABA) promotes a variety of physiological responses in plants. Such responses fall into two groups. Slow responses which take hours or days, and fast responses which occur in minutes. Slow responses include the abscission of buds, leaves, flowers and fruits (Fig. 8.33). In contrast, an increase in the concentration of ABA is a stress signal in wilting plants and inhibits K^+ uptake by stomatal guard cells leading to closure of the stomatal opening and reduced moisture loss. It appears that fast responses to ABA are mediated by Ca^{2+} acting as a second messenger (Section 8.6), while slow responses are the effects of changes in the synthesis of nucleic acids and proteins.

OLIGOSACCHARINS are produced as a defence to plant diseases, and they also help control plant growth and differentiation. They are products of the partial hydrolysis of cell walls. For example, bacterial or fungal infection of plants leads to degradation of the cell wall, liberating oligosaccharins which migrate to neighbouring cells. Here, the oligosaccharins stimulate the production of antibiotics which inhibit attack by the pathogens.

Certain plant cells can produce oligosaccharin-releasing enzymes in response to stimulation by other plant hormones. Thus oligosaccharins influence growth and development of the plant. For example, combinations of auxins and cytokinins combined with oligosaccharins has been shown to influence the differentiation of shoots and roots in tissue culture.

Reference Owen, J.H. and Napier, J.A. (1988) *Plants Today*, **1**, 55–9. A short review emphasizing new ideas on the mode of action of abscissic acid at the cellular level.

8.10 Overview

In animals hormones may be divided into two groups. One group of receptors for hormones which cannot cross the cell membrane interacts with adenylate cyclase via either G_s, in which case the cyclic AMP content of the cell is elevated, or via G_i in which case cyclic AMP concentration is reduced. The effects of an increase in the concentration of cyclic AMP are mediated by cyclic AMP-dependent protein kinase, which phosphorylates proteins at specific serine or threonine side-chains. These changes in phosphorylation are countered by protein phosphatases which themselves may be subject to regulation. The second group of hormones combine with receptors on the cell membrane which are coupled by a G protein to the activation of a phosphodiesterase. This converts phosphatidylinositol 4,5-bisphosphate into two second messengers, inositol 1,4,5-trisphosphate and diacylglycerol. Inositol 1,4,5-trisphosphate causes an increase in the intracellular concentration of Ca^{2+} by causing its release from the endoplasmic reticulum. Ca^{2+} binds to calmodulin and, as a consequence, there will be activation of Ca^{2+}–calmodulin-dependent protein kinases and phosphatases. Diacylglycerol activates protein kinase C. Alteration in the phosphorylation of target proteins will then produce the biological effects of the hormone.

The combination of insulin with its receptor stimulates the tyrosine kinase activity of the receptor and the receptor phosphorylates itself. Some of the actions of insulin may be mediated by the activation or inhibition of serine/threonine kinases or phosphatases. These enzymes may be activated directly by phosphorylation of tyrosine residues by the insulin receptor. Alternatively, other actions of insulin may involve a novel inositol glycan second messenger.

Steroid hormones act at receptors within the cell to control the expression of specific genes. The ultimate consequences of this action are that steroid hormones can switch on or off the synthesis of specific proteins. Combination of a hormone with its receptor causes a change in conformation of the receptor which enables it to interact with a specific region of DNA. This interaction between the occupied receptor and DNA alters the transcription of the particular gene.

Plant hormones control the growth and development of plants. Six major classes of plant hormone are known. Plant hormones often act in combination to produce specific effects. In most cases the molecular mechanism of action of plant cell hormones is not known.

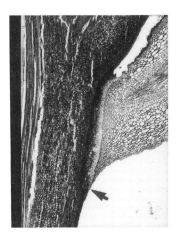

Fig. 8.33 Photomicrograph of a leaf abscission layer (arrowed). Courtesy of M.J. Hoult, Department of Biological Sciences, The Manchester Metropolitan University, UK.

1. Features which prevent passage through the cell membrane are size (e.g. insulin) and the possession of several polar groups (e.g. OH⁻ groups in adrenalin). It is thermodynamically unfavourable to transfer such groups from a water to a lipid environment. By contrast the possession of non-polar ring structures facilitates passage through the lipid bilayer (e.g. steroid hormones).

2. Yes, the types of bonding interaction are similar. Substrates often induce conformational changes in enzymes, just as hormones change the conformation of their receptor. A molecule analagous to a competitive inhibitor would provide a blockade of hormone action which could be surmounted by high concentrations of hormone.

3. If each receptor had its 'personal' adenylate cyclase then the addition of a second hormone would increase adenylate cyclase activity despite maximal stimulation by a first hormone. This does not happen.

4. GTPγS will cause the dissociation of α subunits from the heterotrimer. This activation of α will be irreversible because the GTPγS cannot be hydrolysed. Conversion of GTP to GDP is necessary for reassociation of α with βγ to occur.

5.

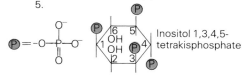

Inositol 1,3,4,5-tetrakisphosphate

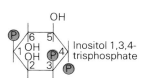

Inositol 1,3,4-trisphosphate

6. Removal of the tip should prevent curvature towards the light. Interposing non-porous plastic between the tip and the stem should prevent curvature, but a porous gel should not. Tips are left in contact with porous gel. Then the gel blocks are placed on one side of the top of a stem from which the tip has been removed. The stem should bend in the dark. The substance could be purified from the gel.

QUESTIONS

FILL IN THE BLANKS

1. Hormones which break down phosphatidylinositol 4,5-bisphosphate yield _____ _____ and _____ as second messengers. Release of intracellular Ca^{2+} from the endoplasmic reticulum is caused by _____ _____ , while _____ causes an activation of protein kinase C. Protein kinase C has to be bound to the _____ to be active. Ca^{2+} binds to a protein called _____ which can then activate several types of protein kinase. Two examples of protein kinases are _____ _____ and _____ _____ _____ _____ .

Choose from: calmodulin, diacylglycerol (2 occurrences), inositol 1,4,5-trisphosphate (2 occurrences), myosin light chain kinase, phosphorylase kinase, plasmalemma.

MULTIPLE CHOICE QUESTIONS

2. State which of the following are true or false:
A. Breakdown of phosphatidylinositol 4,5-bisphosphate can ultimately lead to stimulation of prostaglandin production.
B. There are two routes by which inositol 1,4,5-triphosphate can be metabolized.
C. Pertussis toxin ADP-ribosylates only G_i.
D. Adrenalin can act on three different kinds of receptors.

SHORT ANSWER QUESTIONS

3. Draw two copies of the flow diagram shown in Fig. 8.1a but leave the boxes empty. Now fill in the boxes for (a) the action of adrenalin on muscle glycogen metabolism, and (b) the action of vasopressin (V1) on liver glycogen metabolism. Search through the text for the information that you need.

4. You think that a newly discovered hormone may use cyclic AMP as a second messenger. You are able to measure the response to the hormone continuously in isolated tissue, and have a kit for measuring the cyclic AMP content of the tissue. Design experiments to test your hypothesis.

5. Dibutyryl cyclic AMP:

$$NH-C-(CH_2)_2CH_3$$
(structure of dibutyryl cyclic AMP showing adenine base with $NH-C(=O)-(CH_2)_2CH_3$ substituent, ribose with 5' CH_2, cyclic phosphate, and 3' $O-C(=O)-(CH_2)_2CH_3$ group)

crosses cell membranes more readily than cyclic AMP itself. Why is this? It is also a poorer substrate than cyclic AMP for phosphodiesterases. What might be the effect of adding some of this compound to intact cells which responded to a particular hormone by increasing cyclic AMP content? Why might dibutyryl cyclic AMP be more effective than cyclic AMP itself in producing effects in this experiment?

6. How could you use an inhibitor of cyclic AMP phosphodiesterase to provide evidence that a particular hormone exerted its effect on a tissue by increasing cyclic AMP content?

7. Make a table comparing cyclic AMP-dependent protein kinase, cyclic GMP-dependent protein kinase and protein kinase C.

8. Answer the following questions:
(a) How many subunits has the insulin receptor? Are they all of the same type?
(b) The insulin receptor possesses enzyme activity. What can this enzyme activity do?
(c) Explain how the action of insulin on a tissue could result in both increases and decreases in the phosphorylation of serine and threonine residue side-chains in proteins.

9. Using genetic engineering methods, the coding region for the DNA binding site in the gene for the oestrogen receptor was replaced by the DNA binding region for the glucocorticoid receptor. What will be the effect of oestrogen in cells expressing this new gene?

10. Given that auxins promote shoot elongation and cytokinins stimulate cell division, what would be the effects of adding a low concentration of auxins and relatively high concentrations of cytokinins to undifferentiated plant cells growing in tissue culture?

ESSAY QUESTION

11. Using the headings:
Site of action, Mechanism of action and Speed of action, write a short essay comparing steroid hormones with those which stimulate adenylate cyclase.

9

Nerves, neuro-transmitters and their receptors

Objectives

After reading this chapter you should be able to:

☐ describe the molecular basis of the resting potential and the action potential;

☐ explain the action of neurotransmitters in transmitting a nerve impulse from one cell to another;

☐ relate the structure of nerve receptors to their function;

☐ explain how action potentials are generated as a response to external stimuli, in particular light and sound.

9.1 Introduction

Cellular metabolism is under precise control at the enzymic and genetic levels. In addition, multicellular organisms, the so-called 'higher' animals and plants, require a higher degree of co-ordination, because their cells need to communicate with each other. The various cellular functions must be integrated so that cells, organs and systems work together to satisfy the complex requirements of these organisms.

Two main control systems operate at the supracellular level, namely the endocrine and the nervous systems. Endocrine control operates through hormones (Chapter 8). Both animals and plants possess hormones, but a nervous system is only found in animals. Although this chapter is devoted to the nervous system as a typically animal control mechanism, it is possible to start by considering a phenomenon common to both animal and plant cells.

A **difference** in electrical potential exists in all cells between the inner and outer faces of the plasma membrane. This potential difference is called the **resting potential**, and in most cells it has a value between -70 and $-90\,\text{mV}$ (with the inside being negative). The resting potential results from the unequal distribution of ions on the two sides of the membrane: *there is always a minute excess of cations outside the cell*.

In a special class of animal cells, the so-called **excitable cells**, the membrane potential may temporarily depart from its equilibrium value as a response to a given stimulus. A localized **action potential** is then produced, with a value around $+30\,\text{mV}$, the interior of the cell being positive with respect to the exterior. This localized action potential may be transmitted along the plasma membrane and even pass from one cell to another. The transmission of the action potential is the basis of the **nerve impulse** which is transmitted along the **axon** which is an extended nerve-cell process. Outside the central nervous system, axons are organized in bundles called **nerves**. Sometimes, axons are surrounded by a complex membranous structure, the myelin sheath (Fig. 9.1).

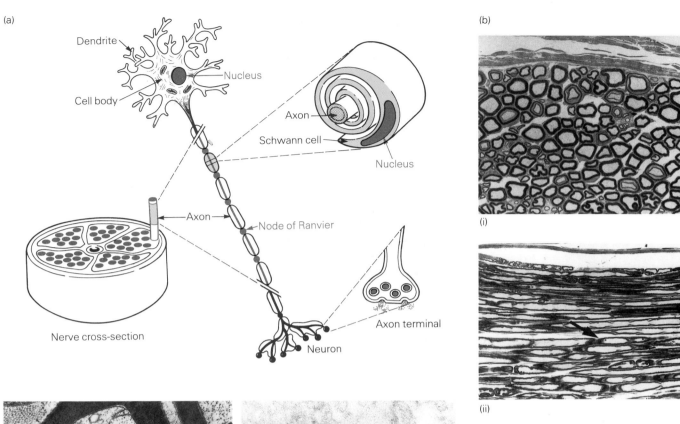

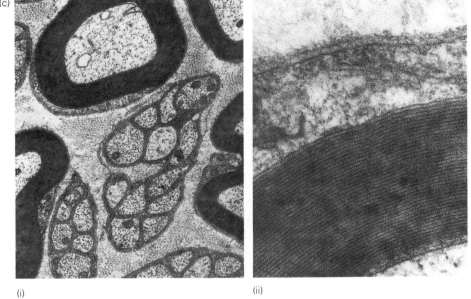

Fig. 9.1 (a) Schematic representation of a nerve, a myelinated nerve fibre, and a neuron. (b) Microphotography of a peripheral rat nerve, fixed with glutaraldehyde and osmium tetroxide. (i) Cross-section: myelin sheaths appear as dark rings surrounding the axons (×400). (ii) Longitudinal section: the arrow points to a Ranvier node (×160). Stain: toluidine blue. (c) Cross-section of nerve fibres from rat brachial nerve. (i) Myelinated and non-myelinated fibres; the former are surrounded by a thick electron-dense sheath (×7000). (ii) Detail of the multilamellar organization of the myelin sheath (×14 000). Courtesy of Drs S.F. Aliño and E. Hilario, Departamento de Biologia Celulary, Ciencias Morfologicas, Universidad del Pais Vasco, Bilbao, Spain.

9.2 Resting potential

The resting potential is due to a slight excess of cations, not compensated by the corresponding anions on the outside of the cell (Fig. 9.2). The cation excess is not detectable by chemical analysis, being about 1 000 000 cations per 999 999 anions, on average. This small ion imbalance is enough to give rise to the resting potential, because the **electrical capacity** of the membrane is low.

Various metabolic processes maintain the cation excess outside the cell. Some of these processes are antagonistic; that is, they work against each other. However, their *total* activities result in a steady-state equilibrium (Fig. 9.2). Three main processes are involved:

1. The '**sodium pump**', which is an ATPase that actively co-transports Na^+ out of and K^+ into the cell, whilst hydrolysing ATP. All cells possess a sodium pump, which is a complex of intrinsic proteins in the plasma membrane (Section 4.6).
2. The **passive diffusion of Na^+ and K^+ ions**, opposing the sodium pump and discharging the gradients generated by it.
3. The **passive diffusion of other ions**, mainly Cl^-, as a consequence of electrochemical potential gradients generated by a variety of mechanisms.

The maintenance of the steady state resting potential relies, therefore, on specific ion gradients and the selective permeability of the cell membrane to the various ions.

In physical terms, the resting potential can be assimilated into a **diffusion potential** that can be estimated from the equilibrium concentrations of ions using the Goldman–Hodgkin–Katz equation:

$$V = \frac{RT}{F} \ln \frac{(Na_i^+)(P_{Na^+}) + (K_i^+)(P_{K^+}) + (Cl_o^-)(P_{Cl^-})}{(Na_o^+)(P_{Na^+}) + (K_o^+)(P_{K^+}) + (Cl_i^-)(P_{Cl^-})}$$

where V is the resting potential, P is the permeability coefficient of a particular ion, and the subscripts i and o correspond, respectively, to the ion concentrations inside and outside the cell. R, T and F have their usual meanings: the gas constant, absolute temperature and Faraday respectively.

Strictly speaking, all anions and cations present should be taken into account in this equation. In practice, however, only Na^+, K^+ and Cl^- are in sufficiently high concentrations, or have permeability coefficients high enough, to have a significant influence on the final value of the resting potential.

□ The electrical capacity of a capacitor (usually called a 'condenser' in electronics) is defined as the ratio of electrical charge to potential difference between the plates. In the case of cell membranes, the lipid bilayer acts as a dielectric, separating the intra- and extracellular conducting media. The capacity (per cm^2) of biological membranes is estimated at about 1 pF, or 10^{-12} F (1 farad (F) = 1 coulomb/1 volt). Condensers used in electronic circuits have capacities ranging from 1 pF to 0.1 F.

□ Absolute values of permeability coefficients P_j are seldom used. **Relative permeability coefficients**, usually with respect to K^+, are used instead. Typical values are $P_{K^+} = 1$, $P_{Na^+} = 0.04$, $P_{Cl^-} = 0.45$.

Exercise 1

Estimate the resting potential of the cell depicted in Fig. 9.2b, at 37°C, with the data included in Marginal Note above. Gas constant $R \approx 8\ JK^{-1}\ mol^{-1}$.

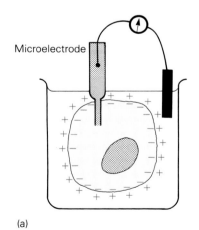

(a)

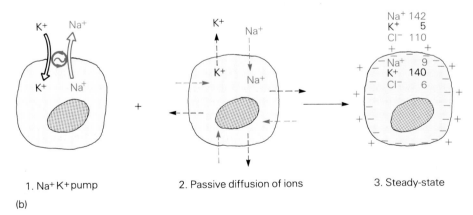

1. Na⁺ K⁺ pump   2. Passive diffusion of ions   3. Steady-state

(b)

Fig. 9.2 (a) Measurement of the cell resting potential. The glass microelectrode filled with a salt solution can penetrate the cell. The resting potential is the potential difference between the inside and the outside of the cell. The outside potential is detected with a reference electrode (e.g. calomel or platinum). (b) The origin of the cell resting potential. The figures in diagram 3 are millimolar (mmol dm⁻³) concentrations of ions.

Reference Chang, R. (1981) *Physical Chemistry with Applications to Biological Systems*, 2nd edn, Macmillan, New York, USA. Good discussion of the physico-chemical principles underlying bioelectric phenomena.

Reference Nobel, P.S. (1983) *Biophysical Plant Physiology and Ecology*, W.H. Freeman, San Francisco, USA. Contains a very readable derivation of the Goldman–Hodgkin–Katz equation, together with good coverage of other aspects of diffusion.

9.3 Action potential

The action potential is produced when, after an appropriate stimulus, the ionic permeability of an excitable cell membrane is modified. In particular, **specific Na$^+$ channels** are opened in the membrane. Only excitable cells possess these kinds of channels. Na$^+$-channels are intrinsic membrane proteins that may exist in various conformations, making the membrane permeable or impermeable to Na$^+$. When the Na$^+$ channels are open, a rapid Na$^+$ influx occurs, down the gradient generated by the sodium pump. The entry of Na$^+$ produces a local excess of positive charge inside the cell (that is, opposite to the resting situation). As a consequence, the potential difference between both sides of the plasma membrane inverts going from, for example, -70 mV (inside negative) to $+30$ mV (inside positive) (Fig. 9.3). In physiological terms, the resting membrane is regarded as polarized, and becomes depolarized when an action potential occurs.

—— *Exercise 2* ——

Point out the main differences, from the biochemical point of view, between an excitable and a non-excitable cell.

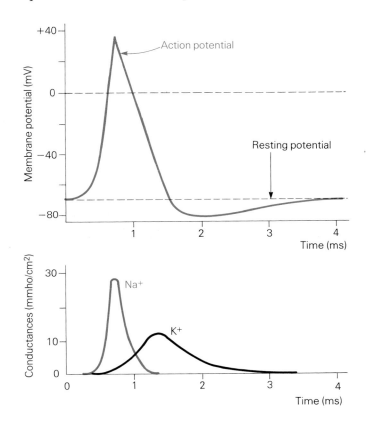

Fig. 9.3 The resting and action potentials as measured across a neuron plasma membrane. (The unit of electric conductance is the mho.)

The action potential generated at a single point in the membrane is rapidly transmitted in all directions. This is due to a **positive feedback** effect, according to which depolarization causes opening of the neighbouring sodium channels. Since neurons are fibre-like structures (see Fig. 9.1), depolarization propagates along the fibre. This propagation is termed the **nerve impulse.**

Na$^+$-channels stay open for only about 1 ms. At the same time, specific K$^+$-channels, which are also characteristic of excitable cells, open. Then, because of the K$^+$ gradient generated by the sodium pump (remember that the pump

Reference Siegel, G.J. (ed.) (1981) *Basic Neurochemistry*, 3rd edn, Little, Brown Ltd, London, UK. Rather old, but a good textbook of general interest for the study of excitable cells.

transports Na^+ to the outside and K^+ to the inside of the cell), there is an outflow of K^+ that restores the resting potential in 2–3 ms (Fig. 9.3). At the end of this process the K^+-channels also close down. It must be stressed that very small changes in the local ion concentrations are enough to produce the observed variations in membrane potential; the overall ionic concentrations remain almost unchanged. Although the action potential is in principle transmitted in all directions, the observed macroscopic effect is that the nerve impulse moves in a definite direction. This occurs because the ion channels that have just closed cannot open up again for a length of time known as the **refractory period**. This refractory period allows the propagation of different nerve impulses along the same nerve fibre, at intervals of only a few milliseconds (Fig. 9.4).

The ion channels

The 'voltage clamp' technique allowed Hodgkin and Huxley in 1952 to show the existence of separate channels for Na^+ and K^+. Briefly, this technique allows the experimenter to select a membrane potential in a large axon from squid and to keep it constant (Fig. 9.5). In addition, ion fluxes to or from the axon can be monitored under these conditions. By inducing depolarization (for example, from -65 mV to -9 mV) and then 'clamping' that is, fixing

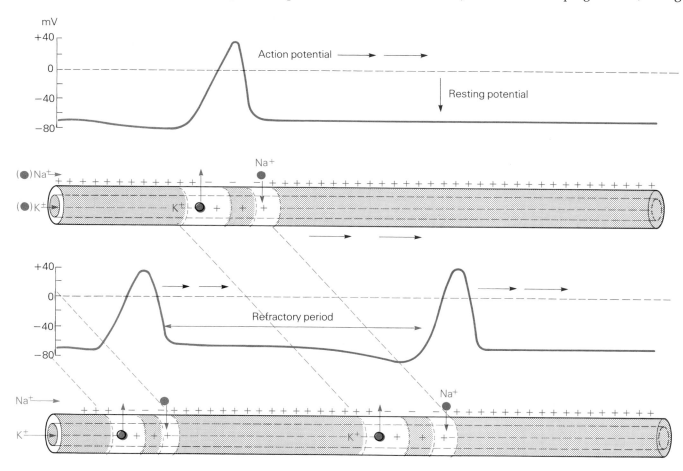

Fig. 9.4 Propagation of a nerve impulse along an axonic fibre. The action potential 'travels' along the fibre from left to right. After some time, longer than the refractory period, another nerve impulse may travel along the same fibre. Membrane regions permeable to Na^+ or K^+ ions are shown in red.

externally the voltage at $-9\,mV$, results in a rapid cation influx being observed, followed by a slower outflow, as described previously. If Na^+ in the extracellular fluid is replaced by **choline** (another cation, Fig. 9.6), a rapid cation influx is not observed, although the slower outflow does occur. There are substances such as the tetraethylammonium ion, TEA (Fig. 9.7), that selectively inhibit the cation outflow.

The above experiments suggested the existence of separate channels for Na^+ and K^+. In the absence of Na^+ a fast entry of positively charged species does not occur, while TEA appears to block the channel for K^+ outflow. In the same way, certain substances containing charged guanidinium groups, such as tetrodotoxin and saxitoxin (Fig. 9.7), specifically inhibit the membrane conductance to Na^+, without affecting K^+ movements, perhaps through binding to a carboxylate group in the Na^+ channel. In contrast, the alkaloid veratridin (Fig. 9.7) binds to a different region of the Na^+ channel, fixing it in the open conformation resulting in a permanent depolarization of the axon.

More recent studies have shown that these channels are **voltage-sensitive**. In particular, the Na^+-channels appear to contain some kind of 'voltage sensor' that is able to detect the voltage change that accompanies depolarization. The protein, after 'sensing' the depolarization, would experience a conformational change, and the Na^+-duct would open. Some neurotoxins, for example that in scorpion venom, are basic polypeptides that bind to the Na^+-channel in a region different from either that of tetrodotoxin

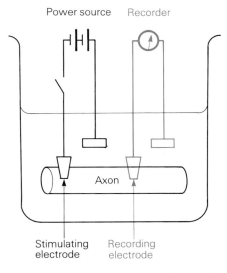

Fig. 9.5 The 'voltage clamp' experiment (see text for details).

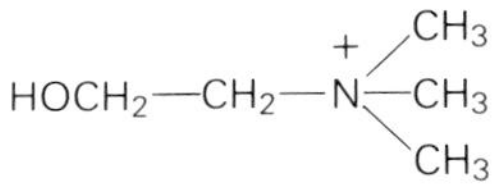

Fig. 9.6 Choline.

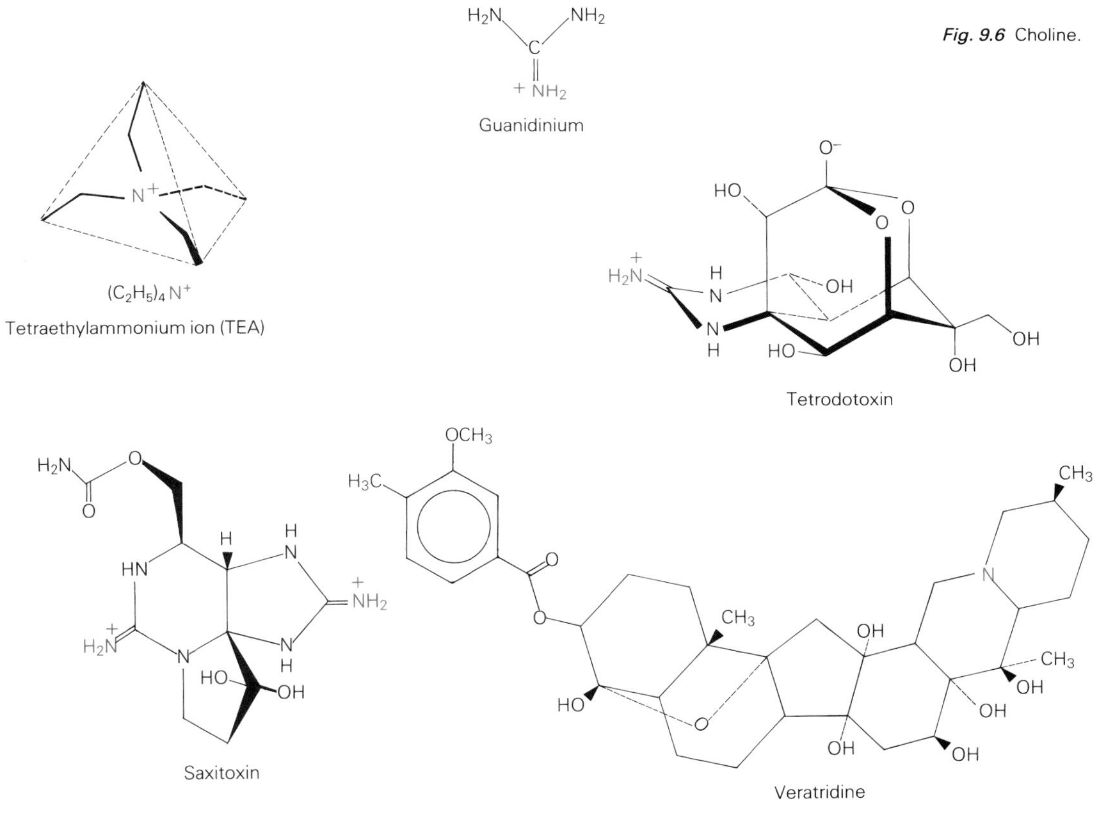

Guanidinium

Tetraethylammonium ion (TEA)

Tetrodotoxin

Saxitoxin

Veratridine

Fig. 9.7 Some drugs which act on the ion channels. Guanidinium is shown for comparison.

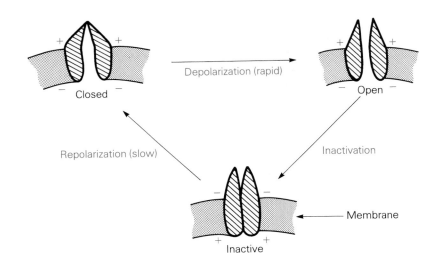

Fig. 9.8 A simplified scheme of the proposed three extreme conformations of the Na$^+$ channel.

or veratridin. The binding of these toxins has the effect of slowing down the closure (inactivation) of the channel after activation during the formation of an action potential. The binding of this kind of toxin is voltage-dependent, suggesting that a channel conformation is recognized that is also sensitive to the membrane potential. Consequently, it has been proposed that the Na$^+$ channel can exist in at least *three* different conformations (Fig. 9.8): **closed** (resting potential), **open** (increased Na$^+$ permeability) and **inactive** (decreased Na$^+$ permeability during the action potential).

The Na$^+$-channel from cells of the electric organ (electroplaxes) of the electric eel, *Electrophorus*, has been purified and reconstituted. It is an intrinsic membrane protein, consisting of at least one major polypeptide (M_r 270 000), and perhaps some smaller ones. The cDNA of various Na$^+$-channels have been cloned and sequenced and found to show a high degree of homology. The data suggest that the amino acid sequence of the main Na$^+$-channel polypeptide has been conserved over long periods in evolution. This polypeptide consists of four transmembrane domains, each about 300 residues long, interconnected by shorter sequences (Fig. 9.9). The Na$^+$-channel has been reconstituted into liposomes (see Fig. 4.3), and shows the properties predicted by physiological and pharmacological studies. The channel is voltage-dependent, adopts the open conformation in the presence of veratridine and is blocked by tetrodotoxin or saxitoxin.

THE PATCH–CLAMP TECHNIQUE allows the observation of ionic movements through a single channel. For this purpose, a micropipette about 0.5 μm in diameter at its narrow end filled with an electrolyte solution is used (Fig. 9.10). By applying gentle suction, a small patch of the cell plasma membrane can be isolated and adhered to the micropipette tip. The patch size is such that it is highly probable that it contains only a single ion channel. The electrical potential across the patch may be clamped at any desired value, and the ionic compositions at both sides of the membrane can also be varied at will in order to study the ion flux through a single ion channel.

By means of this kind of experiment it has been established that each Na$^+$-channel remains open for an average time of 0.7 ms, and that the average current across each channel is 1.6×10^{-12} amps (or Cs^{-1}). It is possible to use

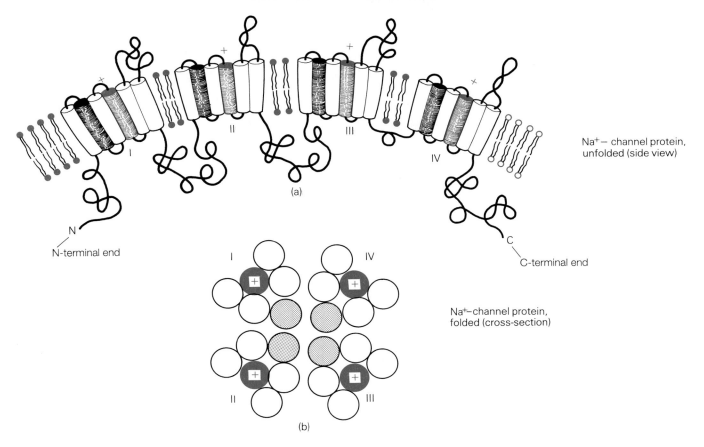

Na$^+$— channel protein, unfolded (side view)

N
N-terminal end

C
C-terminal end

Na$^+$—channel protein, folded (cross-section)

(b)

Fig. 9.9 A schematic representation of (a) the hypothetical secondary and (b) tertiary structures of the Na$^+$-channel protein, as deduced from analysis of its primary structure.

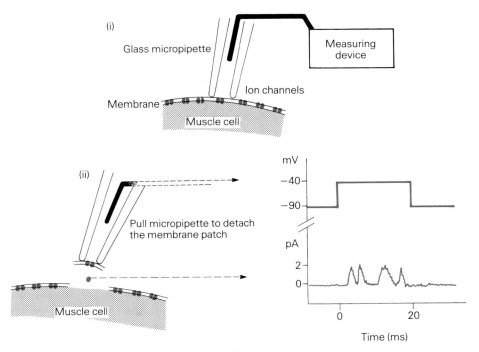

Fig. 9.10 The 'patch—clamp' technique (see text for details).

these data to estimate the number of Na$^+$ passing through a single channel in a depolarization cycle using the expression:

$$\text{No. of Na}^+/\text{channel} = \frac{\text{current (C s}^{-1}) \times \text{time(s)} \times \text{Avogadro's number}}{F \text{(C mol}^{-1})}$$

$$\text{Hence, No. of Na}^+/\text{channel} = \frac{1.6 \times 10^{-12} \times 0.7 \times 10^{-3} \times 6 \times 10^{23}}{96\,500}$$

$$= 7000 \text{ Na}^+/\text{channel}$$

9.4 Synaptic transmission, neurotransmitters and receptors

The nerve impulse (action potential) passes from one cell to another in the process known as **synaptic transmission**. Synaptic transmission is a chemically-mediated phenomenon, the mediators being small, specialized molecules called **neurotransmitters**.

The synapse

The structure of the connections between excitable cells, that is, between a sensory cell and a neuron, or between two neurons, or between a neuron and

□ A great deal of our knowledge on neurotransmitters has been obtained experimentally through the use of **synaptosomes**. It should be noted that this name does not correspond to any structure visible in the intact cell, but rather refers to a subcellular preparation consisting of the terminal ends of neuronal processes, containing mainly synaptic vesicles and mitochondria.

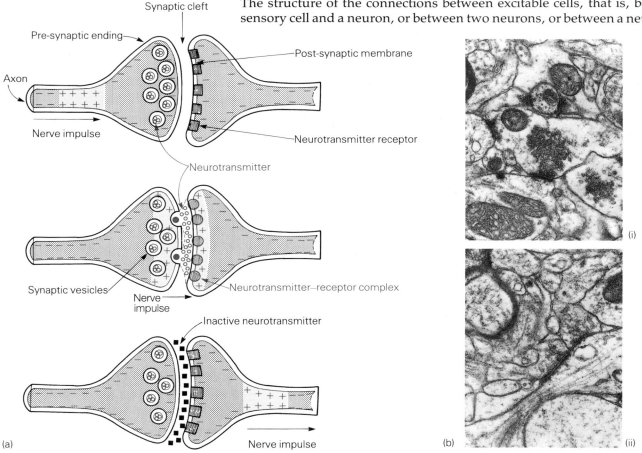

Fig. 9.11 The synaptic transmission of a nerve impulse from cell to cell. (a) Schematic representation. (b) Electron micrographs. (i) Nerve endings with synaptic vesicles and mitochondria (×14 000). (ii) A detail of synaptic structures, showing the synaptic vesicles inside the pre-synaptic ending, the synaptic cleft (clear line) and the electron-dense post-synaptic membrane (×14 000). Courtesy of Drs S.F. Aliño and E. Hilario, Departamento de Biologia Celular y Ciencias Morfologicas, Universidad del Pais Vasco, Bilbao, Spain.

a muscle cell, has been studied in detail by electron microscopy. In all these connections or *synapses*, three components can be distinguished (Fig. 9.11):

1. The **pre-synaptic ending**, which is limited by the presynaptic membrane, and contains the synaptic vesicles, loaded with neurotransmitter;
2. The **synaptic cleft**, or intersynaptic (intercellular) space;
3. the **post-synaptic membrane**, where the neurotransmitter receptors are located.

The molecular mechanism of synaptic transmission is as follows. Upon arrival of the nerve impulse, the pre-synaptic membrane is depolarized. In turn, depolarization determines the opening of voltage-sensitive Ca^{2+}-channels and subsequent entry of Ca^{2+} into the cytosol. These Ca^{2+}-channels have been characterized using the same techniques as described for the sodium channels. The increase in cytosolic Ca^{2+} concentration induces in turn the fusion of the membranes of some synaptic vesicles with the cell plasma membrane (Fig. 9.11). This Ca^{2+}-induced fusion process is probably mediated by a protein, synapsin I. Vesicle fusion causes the liberation of neurotransmitter into the synaptic cleft in a process of **exocytosis**. The neurotransmitter molecules diffuse across the intersynaptic space to the post-synaptic membrane where they bind to specific receptors. Neurotransmitter binding to a receptor molecule stimulates the formation of an action potential in the post-synaptic membrane. This results in the depolarization being immediately propagated along the plasma membrane of the second cell (Fig. 9.11).

The depolarization signal is destroyed by one or more of the following mechanisms: (a) diffusion of neurotransmitter outside the synaptic cleft, (b) uptake of neurotransmitter by the pre-synaptic membrane, or (c) enzymic breakdown of neurotransmitter.

It should be noted that the same neuron, because of its axon and dendrite branches (Fig. 9.12), may establish synapses with many other neurons and in turn receive connections from a large number of cells.

☐ Membrane fusion occurs in many important events at the cellular level: exocytosis (of which neurotransmitter release is but one example), egg fertilization, cell infection by membrane-coated viruses (Chapter 4).

See Chapter 1

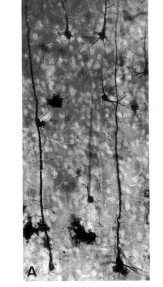

Fig. 9.12 Micrographs of a thin section of rat brain cortex, Golgi stain. A–C are of different regions of the brain cortex illustrating neuronal polymorphism (×73).

Exercise 3

List the main morphological elements of a synapse.

---

synapse: the extremely narrow gap separating two excitable cells. From the Greek synapsis meaning conjunction or union.

Neurotransmitters

□ The so-called **autonomous nervous system** regulates a large number of visceral functions without the control of consciousness. Two sections, **sympathetic** and **parasympathetic**, may be distinguished in the autonomous system. They differ in morphology and function; they also use different neurotransmitters.

□ Catecholamine synthesis is another example of a metabolic pathway in which the key regulatory enzyme catalyses the first, or one of the first, steps in the pathway. The presence of regulatory enzymes at or near the beginning of a multienzyme sequence is one of the main general mechanisms regulating metabolic sequences.

Many different neurotransmitters are known and some examples are shown in Fig. 9.13. Different kinds of neurons synthesize different neurotransmitters. For example, the main neurotransmitters of the sympathetic nervous system are the catecholamines, *adrenalin* (epinephrin) and **noradrenalin** (norepinephrin), while the transmitter of the parasympathetic system is **acetylcholine** (Fig. 9.14).

THE CATECHOLAMINES, adrenalin and noradrenalin, are synthesized from tyrosine (Fig. 9.13). The limiting step of this metabolic pathway is the first reaction: the hydroxylation of the tyrosine aromatic ring to give 3,4-dihydroxyphenylalanine (DOPA). Catecholamine biosynthesis occurs in the cytosol of specific cells (**adrenergic** neurons and adrenal cells), and the products are actively transported into the synaptic vesicles.

The effects of catecholamines on the post-synaptic membrane are poorly understood at present. Various kinds of adrenergic receptors have been distinguished mainly on the basis of their pharmacological properties. The

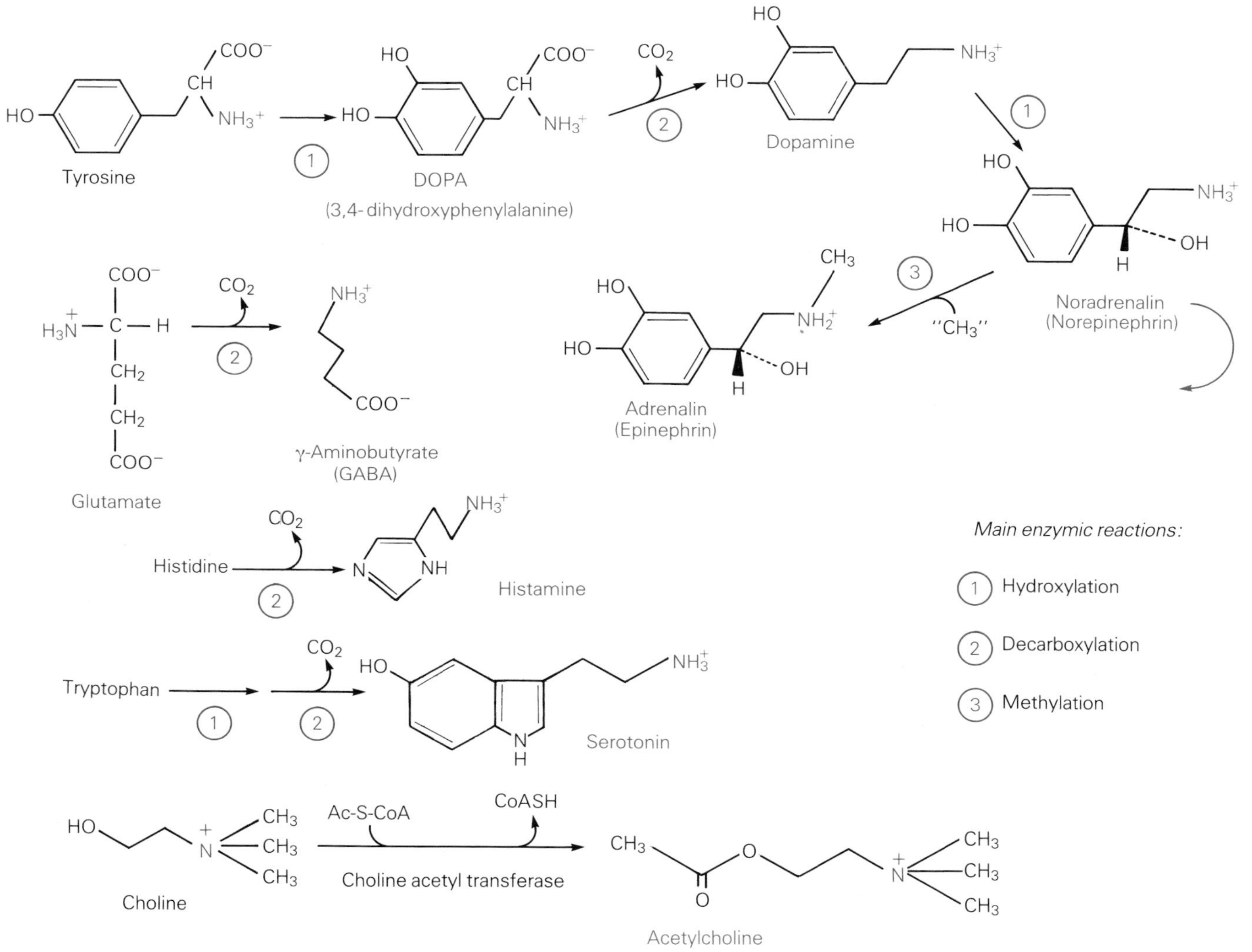

Fig. 9.13 Some neurotransmitters and their biosynthetic pathways.

adrenalin (epinephrine): *meaning roughly 'on top of the kidney'. Refers to the location of the adrenal glands, above each kidney – Latin, renes, kidney – from which adrenalin was first extracted.*

catecholamines: *adrenalin and noradrenalin are jointly called catecholamines, since they contain both an amino and a catechol (1,2-diphenol) group. Catecholamines are important mediators in the nervous (sympathetic) and hormonal sytems.*

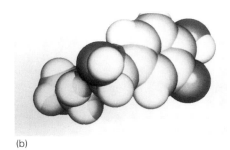
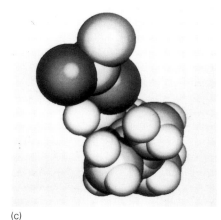

(a)

(b)

(c)

Fig. 9.14 Computer-drawn models of (a) adrenalin, (b) noradrenalin and (c) acetylcholine. Courtesy M.J. Hartshorn, Polygen, University of York, UK.

Reference Goodman, L. and Gilman, A. (editors) (1990) *The Pharmacological Basis of Therapeutics* (2nd edn), Pergamon, Oxford, UK. This standard text on pharmacology is regularly updated. The current edition is always an excellent source of information on neurotransmitters and drugs affecting neurotransmission.

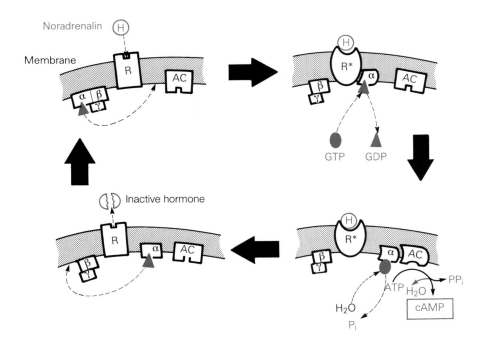

Fig. 9.15 A simplified scheme of the cAMP-mediated action of noradrenalin. R, receptor; R*, activated receptor: AC, adenylate cyclase; α, β and γ are subunits of the G protein (see Section 8.7).

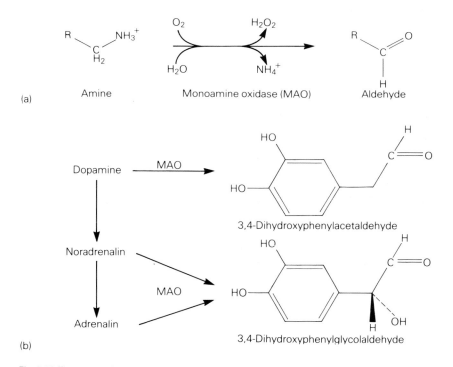

Fig. 9.16 The monoamine oxidase (MAO) reaction: (a) general reaction; (b) MAO in catecholamine metabolism.

binding of noradrenalin to the so-called β-adrenergic receptors in neurons produces activation of the membrane-bound adenylate cyclase, the enzyme responsible for cyclic AMP synthesis (Fig. 9.15). Catecholamine breakdown occurs at least partly through the action of **monoamine oxidase** (Fig. 9.16).

See Chapter 8

ACETYLCHOLINE is synthesized from acetyl CoA and choline, in a reaction catalysed by choline acetyl transferase (Fig. 9.13), and is also stored in vesicles. As in the case of catecholamines, acetylcholine may elicit different physiological responses in the various types of postsynaptic membranes. Two main kinds of **cholinergic** receptors are known, distinguishable according to their sensitivity to various drugs called **nicotinic** and **muscarinic** acetylcholine receptors, respectively. Acetylcholine binds reversibly to cholinergic receptors. Upon dissociation, the neurotransmitter is hydrolysed,

---

Box 9.2
Drugs affecting cholinergic transmission

Drugs acting on the cholinergic synapse have found extensive clinical use. Most of them act either on the cholinergic receptor or on cholinesterase.

1. *Drugs acting on the cholinergic receptor(s).*
   Curare alkaloids (for example, tubocurarine) or cobra toxin bind to the acetylcholine receptor at the neuromuscular junction, competing with this neurotransmitter, and thus inhibiting post-synaptic depolarization. As a consequence, muscle paralysis occurs. Succinylcholine, and related drugs bind the receptor in a similar way to acetylcholine, but for a longer time. In this case Na^+-permeability in the post-synaptic membrane remains high, and permanent depolarization ensues. Succinylcholine has a clinical application as a muscle relaxing drug used prior to anaesthesia.

2. *Drugs acting on cholinesterase.*
   Physostygmine and neostygmine are drugs which inhibit the hydrolytic action of cholinesterase, thus potentiating the cholinergic transmission. They are used in treating, for example, myasthenia gravis, an autoimmune disease characterized by low levels of functional receptors. Other drugs act as irreversible inhibitors of cholinesterase. This is the case of the organophosphorus compounds used as 'nerve gas', for example, sarin (in chemical warfare) or as insecticides (parathion) (see also *Biological Molecules*, Chapters 4 and 5).

Computer-drawn model of the nicotinic acetylcholine receptor embedded in a lipid bilayer. α-Bungarotoxin (1) is also shown, joined to the receptor. Courtesy Dr D. Osguthorpe, Molecular Graphics Unit, University of Bath, UK.

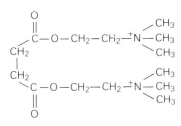

Succinylcholine

Parathion

Neostygmine

Sarin

Reference Conti-Tronconi, B.M. (1982) The nicotinic cholinergic receptor: correlation of molecular structure with functional properties. *Annual Review of Biochemistry*, **51**, 491–530. Rather an old article and written at a rather advanced level, but a reasonable place to start reading about the receptor.

and thus inactivated. The hydrolysis is catalysed by acetylcholinesterase, an enzyme also present in the postsynaptic membrane:

$$CH_3-\overset{\overset{O}{\|}}{C}-O-CH_2CH_2N^+\overset{CH_3}{\underset{CH_3}{-CH_3}} \xrightarrow{H_2O} CH_3COO^- + HOCH_2CH_2N^+\overset{CH_3}{\underset{CH_3}{-CH_3}}$$

acetylcholine   acetate   choline

A great deal of our knowledge about acetylcholine (and other neurotransmitters) is based on studies using various drugs. Some of these are described in Box 9.2.

γ-AMINOBUTYRIC ACID (GABA) is produced by decarboxylation of glutamate (Fig. 9.13). Its main effect on the post-synaptic membrane consists of increasing permeability to chloride ions. This causes the entry of Cl^- and the **hyperpolarization** of the membrane, i.e. the potential becomes more negative inside. As a result, depolarization is made more difficult. GABA-mediated synapses are **inhibitory**. Other neurotransmitters, such as **histamine** or **serotonin** (Fig. 9.13) participate in specific neuronal circuits.

The nicotinic acetylcholine receptor

Neurotransmitters, once they are released into the synaptic cleft, diffuse and bind to their corresponding receptors in the post-synaptic membrane. In general, receptors are molecules, usually proteins, that selectively bind a given neurotransmitter, hormone, or other effector of signal transduction. All nerve receptors characterized so far are intrinsic membrane proteins. Each of them is able to bind one particular neurotransmitter with a high degree of specificity.

Exercise 5

Sarin is a nerve gas that blocks cholinesterase in cholinergic synapses. What will be the effect of Sarin on the acetylcholine levels in these synapses?

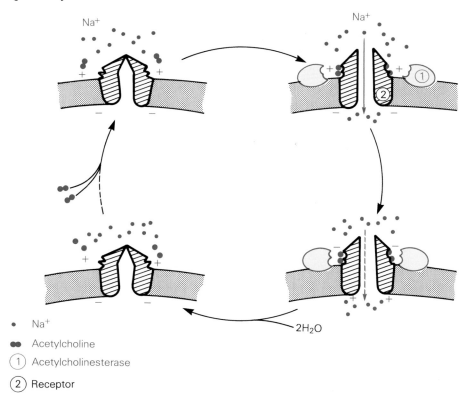

-   Na^+
- ●●   Acetylcholine
- ①   Acetylcholinesterase
- ②   Receptor

Fig. 9.17 The effect of acetylcholine on the post-synaptic membrane, including neurotransmitter breakdown.

Reference Changeux, J.-P. *et al.* (1984) Acetylcholine receptor: an allosteric protein. *Science*, **225**, 1335–45. Excellent article covering the known morphology and action of the nicotinic receptor for acetylcholine up to that date.

The nicotinic acetylcholine receptor is the best characterized of all nerve receptors. It is an intrinsic membrane protein that, upon binding of acetylcholine, allows the entry of cations into the cell (Fig. 9.17). Since the resting potential favours the influx of Na^+ the membrane is depolarized.

The receptor protein has been isolated from the electroplax membranes of the electric ray *Torpedo* and electric eel *Electrophorus*. Purification procedures involve solubilizing the membrane proteins with detergents and purifying the protein by affinity chromatography using cobra toxin, to which the receptor binds specifically.

The M_r of the receptor is 268 000. It is formed by the aggregate of four types of polypeptides: two α subunits (M_r 40 000), a β subunit (M_r 49 000), a γ subunit (M_r 57 000), and a δ subunit (M_r 65 000) giving a molecular composition $\alpha_2\beta\gamma\delta$. Subunit α is known to bind acetylcholine and also α-bungarotoxin, a specific inhibitor. All subunits show a high degree of sequence homology (Fig. 9.18) suggesting a common ancestral gene. All contain five transmembrane α-helices, one of which is amphipathic; that is, it has one hydrophobic and one hydrophilic face. The latter is believed to be part of the ion channel (Fig. 9.19).

Electron diffraction and neutron scattering studies together with structure prediction techniques have allowed a model to be proposed for this receptor (Fig. 9.19 and Box 9.2). The molecule spans the membrane and has an internal ionic channel, 0.7–0.8 nm in diameter, which allows the passage of both Na^+

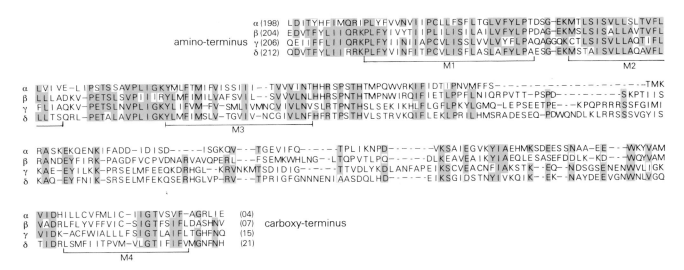

Fig. 9.18 Comparison of partial amino acid sequences, including the transmembrane portions, of the α, β, γ and δ subunits of the acetylcholine receptor. The α, β, γ and δ subunit are aligned. The one-letter amino acid notation is used. Gaps (—) have been inserted to achieve maximum homology, indicated by blocks of colour. M1–M4 indicate probable transmembrane helices. The values in parentheses indicate the numbers of amino acid residues which are found either side of sequences given. Data from Noda *et al.* (1993) *Nature* **302**, 528–32.

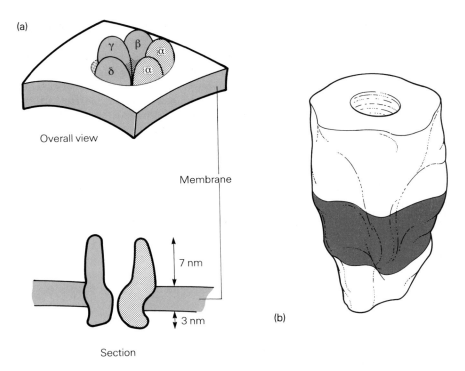

(a)

γ β α
δ α α

Overall view

Membrane

7 nm

3 nm

(b)

Section

Fig. 9.19 (a) A schematic representation of the acetylcholine receptor. (b) Model of the acetylcholine receptor determined by an electron microscopy study of the receptor from post-synaptic membranes from *Torpedo marmorata*. The coloured area represents the portion of the protein which is embedded in the membrane. Redrawn from Unwin, N. *et al.* (1988). Arrangement of the acetylcholine receptor subunits in the resting and desensitized *Torpedo* postsynaptic membranes. *Journal of Cell Biology*, **107**, 1123–38.

and K^+. The nicotinic acetylcholine receptor has been reconstituted into liposomes where it is sensitive to α-bungarotoxin and other drugs. Under these conditions, addition of acetylcholine induces the outflow of Na^+ and K^+ previously trapped inside the liposome.

9.5 *The generation of action potentials by sensory stimuli*

All sensory receptors contain cells specialized in the detection of specific stimuli (pressure, light, sound, chemicals, etc) and their conversion (**transduction**) into nerve impulses. These impulses are transmitted to the central nervous system via the corresponding nerve fibres, and are interpreted as tactile, visual, auditory, gustatory, etc. sensations. From the biochemical point of view, the molecular mechanisms of hearing and vision are the best understood.

Hearing

The transduction of auditory stimuli into nerve impulses takes place in **ciliated cells** covering the cavities of the internal ear (Fig. 9.20). Sound waves are transmitted through the tympanum and are amplified by a chain of small bones, the ossicles, before being transmitted to the endolymph of the internal ear. Sounds are thus converted into vibrations of this fluid. As a result, the basilar membrane of the cochlea vibrates and the crest of the ciliated cells is displaced (Fig. 9.21). Such displacement is accompanied by the opening of

☐ The middle ear (between the tympanic membrane and the labyrinth) contains three small bones, the **ossicles**, consisting of the **malleus**, **incus**, and **stapes**. The ossicles form a chain-like structure and transmit the acoustic waves from the external ear to the fluid in the cochlea.

The inner ear consists of a complex ensemble of membranous cavities, excavated in the petrous part of the temporal bone. The so-called membranous labyrinth (utricle, saccule, semicircular canals, cochlea) is filled with a clear fluid, the **endolymph**. Endolymph has a peculiar ionic composition: it contains high K^+ and low Na^+ concentrations.

specific K⁺-channels. K^+ is the most abundant cation in endolymph, and its entry produces the depolarization of the ciliated cell. In these cells, depolarization occurs through entry of K^+, and not of Na^+, as is usually the case. In turn, depolarization causes the opening of Ca^{2+}-channels located in the basal part of the cell. Ca^{2+} entry triggers the fusion of synaptic vesicles with a domain of the plasma membrane of the cell, close to the neuron. In this way, the sound vibrations give rise to an action potential in a neuron that is transmitted to the brain where it is interpreted as sound.

Vision

The retinal rods and cones are cells specialized in the transduction of light stimuli into nerve impulses. Cones detect bright light and colour; rods perceive dim light but do not detect colour. The retinal rods consist of an outer

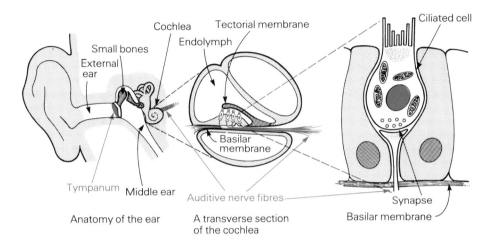

Fig. 9.20 Structure of the mammalian ear and location of the ciliated cells.

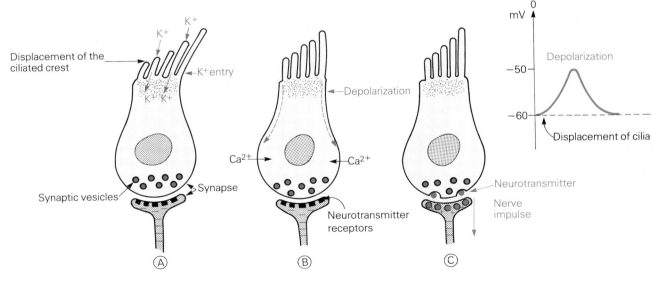

Fig. 9.21 Ciliated cells of the inner ear. Transduction of movements of the ciliated crest into nerve impulses.

Reference Neely, J.G. (1985) Mechanisms of hearing: cochlear physiology. *Ear, Nose and Throat Journal*, **64**, 292–307. Excellent account of how sound waves are turned into action potentials.

Reference Mashland, R.H. (1986) The functional anatomy of the retina. *Scientific American*, **255**(6), 90–7. A sharply focussed review on the different types of cells in the retina and their roles in vision. Excellent diagrams and some beautiful photographs.

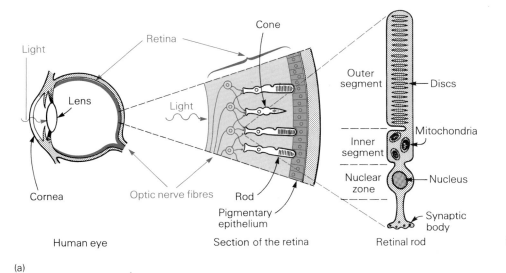

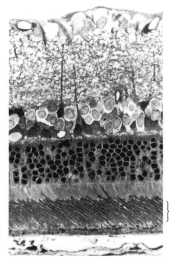

Fig. 9.22 The vertebrate eye, and the location of the retinal rods. (a) Schematic representation. (b) Micrography of the rat retina. The brackets indicate the limits of the layers of rods and cones. Richardson stain. ×320. Courtesy of Drs S.F. Aliño and E. Hilario, Departmento de Biologia Celular y Ciencias Morfologicas, Universidad del Pais Vasco, Bilbao, Spain.

segment, an inner segment, a nuclear region and a synaptic body (Fig. 9.22). The latter is in contact with a sensory neuron of the optic nerve.

The outer segment of the rod gives the cell its characteristic shape, and contains as many as one thousand **discs**, or flattened bags. The disc membrane is rich in **rhodopsin**, an integral protein containing **11-*cis*-retinal** as a prosthetic group (Fig. 9.23). Retinal, the aldehyde of vitamin A, is the light-sensitive pigment. In addition, the discs are loaded with 3′,5′-GMP (cyclic GMP, or cGMP, Fig. 9.23). The rod inner segment contains the Na$^+$-pump that is essential in the generation of the resting potential.

11-*cis*-Retinal absorbs light in the visible range (400–700 nm). When one photon of the appropriate energy is captured by 11-*cis*-retinal, its configuration is changed to **11-*trans*-retinal** (Fig. 9.24). This conversion occurs in a few picoseconds. 11-*trans*-Retinal cannot form a stable complex with rhodopsin and, as a result, free 11-*trans*-retinal and **opsin** (that is,

---

Exercise 7

The retina contains three types of conical cells whose different *opsins* (proteins) determine whether their absorption maxima are in the red, green or blue regions of the visible spectrum. Most patients suffering from *daltonism* cannot perceive the green colour. Comment on the molecular basis of daltonism.

☐ Electromagnetic radiation, of which light is an example, can be characterized either by its frequency (ν) or its wavelength (λ). Both are related by $c = \lambda\nu$, c being the velocity of light. The energy E of a photon is thus related to the frequency (or wavelength) of its associated radiation: $E = h\nu = hc/\lambda$, where h is Planck's constant (6.62×10^{-34} Js). In order to excite the visual pigments, electro-magnetic radiation must have a wavelength in the range 400–700 nm; that is, it must be in the so-called 'visible range' of the spectrum.

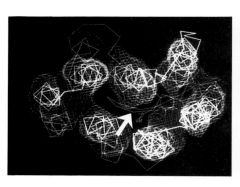

(a)

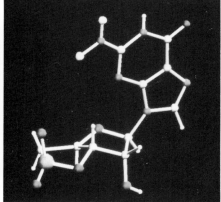

(b)

Fig. 9.23 (a) Computer-drawn representation through a rhodopsin molecule showing the bound 11-*cis*-retinal (arrowed). Courtesy Dr J.B.C. Findlay, Department of Biochemistry and Molecular Biology, University of Leeds, UK. (b) Model of cGMP. Courtesy M.J. Hartshorn, Polygen, University of York, UK.

Reference Schnapf, J.L. (1987) How photoreceptors respond to light. *Scientific American*, **256**(4), 32–40. A superbly illustrated account of the electrophysiology of retinal rods and cones.

Reference Nathans, J. (1989) The genes for color vision. *Scientific American*, **260**(2), 42–9. A colourful account of the isolation of the genes encoding the colour-detecting proteins of the human eye.

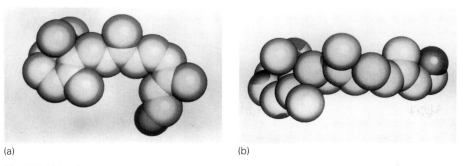

(a)   (b)

Fig. 9.24 Molecular models representing the isomerization of (a) 11-*cis*-retinal into (b) 11-*trans*-retinal. Courtesy M.J. Hartshorn, Polygen, University of York, UK.

rhodopsin minus retinal) are formed. Opsin then activates the guanosine nucleotide-bound protein **transducin**, at the expense of the hydrolysis of one molecule of GTP. GTP–transducin activates a phosphodiesterase, and this in turn hydrolyses cGMP to 5′-GMP (Fig. 9.25). According to results obtained using the patch–clamp technique (see above), the role of cGMP is to act as an allosteric effector to the Na$^+$-channels, favouring the depolarization of the rod plasma membrane. The ultimate effect of the photon is to decrease the cGMP concentration in the cell and, consequently, to produce a **hyperpolarization** of the cell. The hyperpolarization is transmitted to the synaptic body and initiates a nerve impulse in a visual neuron (Fig. 9.26). As can be seen in Figs 9.25 and 9.26, a single photon initiates a mechanism involving increasing numbers of molecules, which eventually elicits a macroscopically-observable physiological effect (that is, a nerve impulse). Similar cascade processes are found in other aspects of metabolism, such as activation by hormones (Chapter 8).

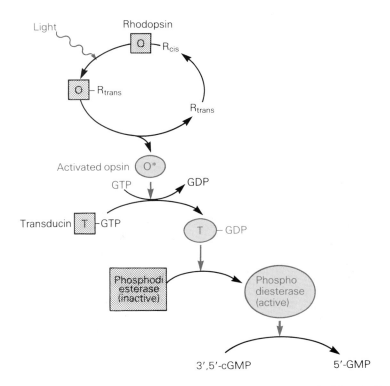

Fig. 9.25 The light-triggered cascade activation of cGMP phosphodiesterase in the retinal rod. 0, opsin; R, retinal.

References Anholt, R.R.H. (1987) Primary events in olfactory reception. *Trends in Biochemical Sciences*, **12**, 58–62; Lancet, D. and Pace, U. (1987) The molecular basis of odor recognition. *Trends in Biochemical Sciences*, **12**, 63–6; Reed, R.R. (1990) How does the nose know? *Cell*, **60**, 1–2. A detailed discussion of olfaction is outside the scope of this chapter. However, these three short, readable articles are an excellent starting place for those wishing to learn about the topic.

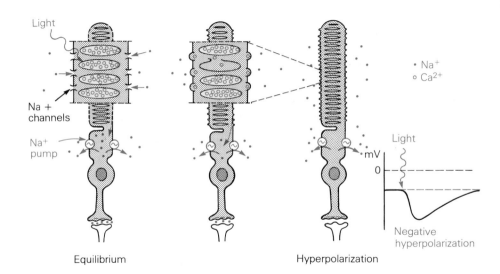

Light

Na + channels

Na$^+$ pump

• Na$^+$
∘ Ca^{2+}

Light

mV

0

Negative hyperpolarization

Equilibrium   Hyperpolarization

Fig. 9.26 The role of retinal rod cells in the transduction of light energy into nerve impulses.

9.6 Overview

Action potentials can only arise in excitable cells. The action potential is produced when the ionic permeability of an excitable cell membrane is modified through opening of specific Na$^+$ channels. When the Na$^+$ channels open, a rapid Na$^+$ influx occurs, producing a local excess of positive charge inside the cell: the membrane is depolarized. The action potential is rapidly transmitted along the nerve fibres. Na$^+$ channels stay open for about 1 ms, then close when, simultaneously, K$^+$ channels open. The outflow of K$^+$ from the cell restores the original (resting) potential in 2–3 ms. The K$^+$ and Na$^+$ channels are intrinsic membrane proteins.

The action potential is transmitted from one cell to another by synaptic transmission which is a chemically-mediated phenomenon. The mediators are specialized molecules called neurotransmitters. Many different neurotransmitters are known. Catecholamines (adrenalin, noradrenalin) and acetylcholine are the most common. Once released into the synaptic cleft, neurotransmitters diffuse and bind to their corresponding receptors in the post-synaptic membrane. Receptors are intrinsic membrane proteins.

Sensory cells are specialized in the detection of specific stimuli (for example, light) and their transduction into nerve impulses. The molecular mechanisms of hearing and vision are well understood. In hearing, sound vibrations lead to the depolarization of the ciliated cells of the internal ear through the entry of K$^+$ not Na$^+$. Visual cells are hyperpolarized upon the arrival of a photon. Depolarization and hyperpolarization are transmitted to the central nervous system via specific nerve fibres and interpreted as auditory or visual sensations.

Answers to Exercises

1. −70 mV
2. The excitable cell contains specific Na$^+$ and K$^+$ channels, which allow depolarization to occur.

3. Pre-synaptic ending, synaptic membrane, synaptic cleft and post-synaptic membrane.
4. GABA; Glu, Gly; noradrenalin, serotonin, histamine.

5. Increased neurotransmitter levels.
6. Resonance.
7. Opsin in cones sensitive to green light is absent, or is not functional.

FILL IN THE BLANKS

1. Although all cells exhibit a _____ potential, only _____ cells can elicit _____

potentials. Polarized membranes have an electric potential _____ inside;

depolarization occurs most commonly through an entry of _____ . Depolarization of

the pre-synaptic ending produces a release of _____ . The main _____ in para-

sympathetic neurons is _____ , while the _____ system uses catecholamines.

All receptors are _____ _____ proteins. Some of them have been purified and

_____ into artificial lipid bilayers.

Sound waves are transmitted to the _____ that fills the _____ _____ . Vibrations

of this fluid produce _____ of the ciliated cell.

In retinal rod outer segments, the disc membrane contains _____ , a protein

possessing _____ as a prosthetic group.

Choose from: action, acetylcholine, 11-*cis*-retinal, depolarization, endolymph, excitable, inner ear, intrinsic membrane, Na^+, negative, neurotransmitter (2 occurrences), reconstituted, resting, rhodopsin, sympathetic.

MULTIPLE-CHOICE QUESTIONS

Identify the correct statements. Note that more than one statement might be correct in each question.

2. In a neuron, the resting potential is due, at least in part, to:
A. the Na^+-K^+-ATPase of the plasma membrane
B. the Ca^{2+}-ATPase of the sarcoplasmic reticulum
C. the volume flux of cations outside the cell
D. the excess of cations inside the cell

3. Repolarization after an action potential occurs:
A. through opening of Na^+ channels
B. through closure of K^+ channels
C. through opening of Na^+ and closure of K^+ channels
D. through closure of Na^+ and opening of K^+ channels

4. The patch–clamp technique:
A. was developed in the 1970s
B. allows the separate observation of Na^+ and K^+ channels
C. allows the observation of individual Na^+ or K^+ channels
D. requires the use of micropipettes with a diameter of about 0.5 μm.

5. The nicotinic acetylcholine receptor:
A. is an integral membrane protein
B. is usually isolated from electroplax membranes
C. consists of three polypeptides
D. consists of two different types of subunits

6. Which one of the following is not essential in the molecular mechanism of light transduction (vision):
A. rhodopsin
B. 11-*cis*-retinal
C. transducin
D. actinin

7. Select the function (A–E) for each of the substances (i)–(v).

A. neurotransmitter   (i)   tetrodotoxin
B. neurotransmitter analogue   (ii) histamine
C. inhibitor of ion channel   (iii) α-bungarotoxin
D. inhibitor of cholinesterase   (iv) parathion
E. inhibitor of receptor   (v)   succinylcholine

SHORT-ANSWER QUESTIONS

8. The ionic concentrations ($mmol\,dm^{-3}$) inside and outside squid giant axon are: Na^+, 50 and 440; K^+, 400 and 20; Cl^-, 50 and 560, respectively. Together with the data in the marginal note on p. 236 estimate the resting potential of the squid giant axon.

9. Name three substances which interfere with cation fluxes in excitable cells, and explain their mechanism of action.

10.   What, if anything, does acetylcholine have in common with most drugs acting on the cholinergic synapse?

11.   Suggest a possible pharmacological action for decamethonium, $(CH_3)_3N^+-(CH_2)_{10}-N^+(CH_3)_3$?

12.   What is the energy associated with one mole of photons (an Einstein) of red light of wavelength 600 nm?

Muscle contraction

Objectives

After reading this chapter you should be able to:

☐ describe the structure of muscle tissues in terms of their molecular components and their organization at the cellular level;

☐ explain how the movement of protein filaments relative to one another is the basis of the contractile mechanism;

☐ describe how the energy of ATP hydrolysis is used to bring about a change in the protein myosin which generates movement;

☐ relate the ways in which the contractile events are initiated and integrated with metabolic aspects of muscle physiology.

10.1 Introduction

Many forms of life have evolved means of moving parts of their structures or of moving from one location to another. Some bacteria have an erratic tumbling motion using flagella (see Fig. 2.19). In other unicellular organisms the movement is directed and made purposeful by an exaggeration of cyto-

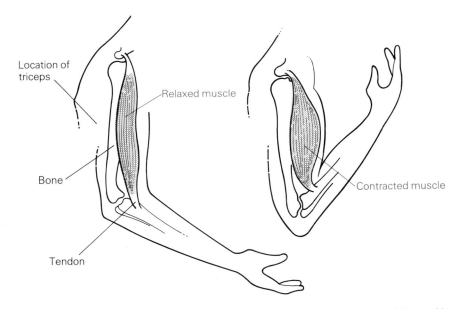

Fig. 10.1 Muscle contraction and the movement of limbs. Muscles in vertebrates tend to work in opposition. That shown is the biceps. The triceps, behind the bone, does roughly speaking the opposite of the biceps, giving control of movement.

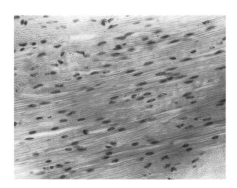

(a)

(b)

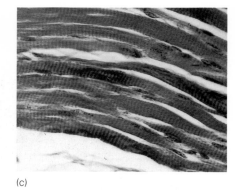

(c)

Fig. 10.2 Photomicrographs of (a) smooth (×210), (b) cardiac (×210) and (c) skeletal muscle tissues. (×500). Courtesy M.J. Hoult, Dept of Biological Sciences, The Manchester Metropolitan University, UK.

plasmic streaming (Chapter 6). Multicellular animals have developed the most highly specialized tissues for producing movement. In such tissues, large numbers of cells co-operate to bring about a reversible shortening, a process known as **muscle contraction**. This term immediately brings to mind the interplay of large muscles and bony skeleton which controls the movement of limbs and locomotion (Fig. 10.1). However, there are many different types of muscle such as **cardiac** and **smooth** muscle which operate pumping or peristaltic movements in the internal organs (Fig. 10.2).

Muscle studies have long been popular with biophysicists, physiologists and biochemists because of the unusually high degree of structural organization in the tissue and its unique mechanical response to external stimulation.

The appeal of the sliding filament model, although now over 30 years old, is so great that unfortunately emphasis is often almost solely presented in terms of the sarcomeric machinery. In many respects, however, the fibre can be considered a 'soft' or plastic machine whose molecular components are continually being removed and replaced without disturbance of function.

The immense sensitivity and sophistication of modern immunochemical and molecular biology techniques will increasingly be called upon to discover subtleties of structural detail, and the response of the muscle protein genes to the external prompting of motor nerves and hormones.

10.2 The cell biology of skeletal muscle

A superficial glance at the skeletal musculature of different species shows that muscles vary in size, shape and colour. In simple terms, a deep red colour indicates the presence of a rich blood supply, a high content of the oxygen store, myoglobin, an **aerobic** metabolism and often a capacity to contract and maintain tension for long periods without tiring. Paler muscles contract more rapidly and **anaerobically** but quickly become exhausted due to depletion of energy reserves and the accumulation of waste products. Irrespective of colour, examination by microscopy shows the presence of **fibres** lying parallel to one another.

Muscle fibres

Muscle fibres are about 100 μm in diameter and may run the full length of the muscle. They are attached to tendons at their extremities (Fig. 10.1). Muscle fibres arise from a fusion of many cells, and therefore form a *syncytium* (Fig. 10.4). The fused plasma membranes now become the **sarcolemma** which

---

Exercise 1

Muscle has been called an 'energy-transduction' device. What forms of energy are interconverted during a muscle contraction?

☐ Chicken and pork are whiter meats than beef and lamb, reflecting differences in fibre types and metabolism.

syncytium: a multinucleated cell. The term is derived from a Greek word meaning together.

Reference Jones, D. (1990) The fast whites and the slow reds. *Biological Sciences Review*, **2**, 2–5. A witty, readable and short review of muscle physiology.

Fig. 10.3 Superficial muscle in the trunk of skipjack tuna (*Euthynnus pelamis*).
Courtesy Professor I.A. Johnston, Dept of Biology and Preclinical Medicine, University of St Andrews, UK.

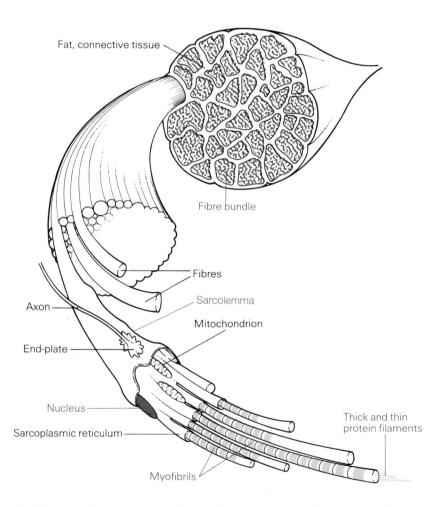

Fat, connective tissue

Fibre bundle

Fibres

Sarcolemma

Axon

Mitochondrion

End-plate

Nucleus

Thick and thin
protein filaments

Sarcoplasmic reticulum

Myofibrils

Fig. 10.4 The relationship between muscle fibres (multi-nucleated cells), myofibrils and protein filaments.

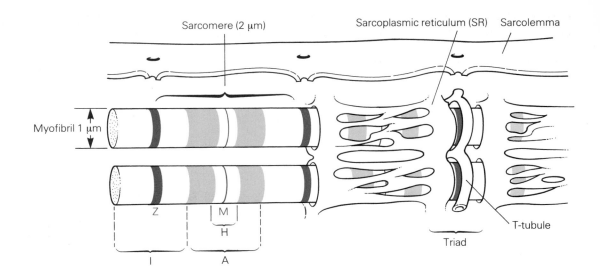

Fig. 10.5 Cutaway drawing of fibre showing relationship between membranes and striated myofibrils.

sheaths the whole fibre. The **nuclei** of the fused cells are found lying just below the sarcolemma often near the **end-plates** of nerve axons and are more elongated than in most cells. Nuclei are forced to lie at the periphery of the fibre because bundles of structural protein (**myofibrils**) occupy 80% of the fibre volume; there is little space either for the cytosolic matrix called **sarcosol** in the muscle syncytia or for other organelles such as **mitochondria**. The endoplasmic reticulum in muscle is called the **sarcoplasmic reticulum** (SR) (Fig. 10.5). It surrounds the myofibrils and acts as a store for Ca^{2+}. It releases Ca^{2+} to initiate contractile activity and must reabsorb Ca^{2+} rapidly when stimulation ceases.

A fibre contains about a thousand myofibrils which, when observed by phase contrast microscopy, have alternating light and dark regions (Fig. 10.5) referred to as **isotropic (I)** and **anisotropic (A)** respectively, based on their refraction characteristics. Adjacent myofibrils are precisely aligned, sarcomere to sarcomere, so that the striation pattern runs continuously across the muscle fibre. The striped pattern gives the **striated** fibre its name. The light (I) bands are bisected by a narrow **Z-line**, while an **H-zone** surrounds the **M-line** which occurs at the centre of the darker A-band (Figs 10.5 and 10.6). The distance between two Z-lines constitutes a **sarcomere**. When a myofibril contracts the sarcomere shortens from about 1–2 μm. The pale I-band is reduced on shortening, the A-band remaining constant.

At the A–I boundary, the sarcolemma folds inwards forming a tubular network, the **T-tubule system**, which deeply penetrates the fibre, surrounding each Z-line with a hollow collar. Its hollow interior corresponds to the extracellular space. This T-system transmits nerve impulses in the form of membrane depolarizations throughout the fibre. Where the T-system approaches two distended SR sacs the region is termed a **triad** unit which can be isolated experimentally. Communication across the 50 nm gap between T-tubule and SR leads to the initiation of contraction.

Fibre types

Both deep red and paler muscle types contain fibres of similar dimensions and striation pattern. However, the former tends to contain slow-contracting, **Type I** fibres, characterized by the presence of many mitochondria but a poorly

□ Mitochondria and particles composed of glycogen and enzymes are squeezed between myofibrils.

□ Anisotropic and isotropic refer to those structures which have physical properties which **differ** in different directions and those which have the **same** physical properties in every direction respectively. A rope is anisotropic whereas a piece of concrete is isotropic.

□ The inside of the T-tubule is equivalent to the outside of the cell. Compare with, for example, the endomembranes of 'typical' cells (Section 1.6).

Exercise 2

Name four kinds of membranes likely to be found in the muscle fibre.

---

sarco-: derived from the Greek sarkos, meaning flesh. Thus sarcoplasm, sarcolemma, sarcoplasmic reticulum and sarcosol are all applied to the muscle syncytium.

Reference Bourne, G.H. (1972) *The Structure and Function of Muscle*, vols I–IV, 2nd edn, Academic Press, New York, USA, and London, UK. A broad and detailed coverage of all types of muscle across the animal kingdom. A physiological and medical approach.

developed SR. In paler muscle, **Type II** or fast-twitch fibres predominate, with fewer mitochondria but an extensive SR. Most muscles contain a mixture of these fibre types, and others which seem to be intermediate in character. The proportions of these fibres determine the speed with which the muscle can contract and also its resistance to fatigue.

Fibre types differ also in the way in which they are **innervated** and probably also in the number and type of hormone receptors which are found in the sarcolemma.

Protein filaments

Huxley and colleagues in the 1950s discovered that the striped appearance of the myofibril could be explained in terms of two sets of **protein filaments** (Fig. 10.7). **Thick filaments** of approximately 15 nm diameter were observed predominantly in the A-band on electron microscopy of transverse sections through the fibre. These are mainly comprised of aggregates of the protein **myosin**. **Thin filaments** of diameter 7 nm dominate corresponding sections through the I-band. Thin filaments are comprised of aggregates of the protein **actin**. Depending on the degree of extension of the muscle fibre, the thick and thin filaments overlap at the interface between A and I bands. In overlap regions each thick filament is surrounded by six thin filaments in a hexagonal pattern (Fig. 10.7). Projections from the thick to the thin filament, called **cross-bridges**, are known to mediate the sliding of thick and thin filaments relative

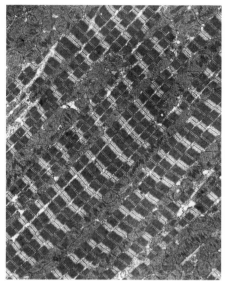

Fig. 10.6 Electron micrograph of striated muscle of anchovy (*Engraulis encrasicolus*) (×3750). Courtesy Professor I.A. Johnston, Department of Biology and Preclinical Medicine, University of St Andrews, UK.

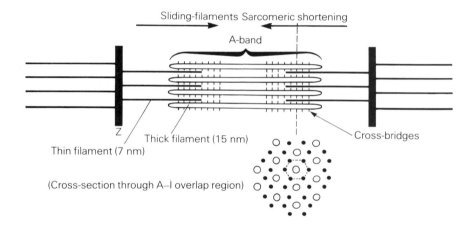

Fig. 10.7 Protein filaments of the sarcomere.

Exercise 3

Examine Fig. 10.7. If a thick filament is 1.5 μm long, what is the minimum length of a fully contracted sarcomere? What is the maximum extended length possible if thick and thin filaments are to remain just in contact?

Box 10.1
Exercise, training and muscle growth

Newsholme, E.A. and Leach, A.R. (1983) *Biochemistry for the Medical Sciences*, John Wiley & Sons, Chichester.

People are born with muscles which will have a fixed number of fibres throughout life and a unique distribution of 'fast' and 'slow' fibre types. In terms of athletic ability, this distribution may predispose an individual either for **endurance** events or those involving sudden bursts of **power**.

Early in exercise, glycolytic fibres are used to provide contraction at the expense of stored glycogen. Progressively, after about 30 minutes, more oxidative fibres are recruited, the exercise becomes 'aerobic' and fat becomes the major fuel.

Training muscles stimulates the production of myofibrillar proteins, especially in the fast fibres used in power events such as sprinting and weight-lifting. Endurance training, however, tends to increase oxidative capacity via increased mitochondrial numbers with less emphasis on fibre bulk.

to each other to bring about shortening of the sarcomere and contraction of the fibre. Since each half of the thick filament draws thin filaments in opposite directions, there must be an opposite **polarity** in its two halves. Similarly, the thin filaments at each end of the sarcomere 'point' in opposite directions. The **sliding filament theory** has become the accepted view of myofibrillar contractions.

Metabolism of fibres

The metabolic energy required to bring about contraction is generated by reactions occurring in the sarcosol or in mitochondria squeezed between the packed myofibrillar arrays. Type II fibres rely mainly on the **glycolytic** pathways and energy stored in the form of **glycogen**. Type I fibres, which contain more mitochondria, oxidize **fatty acids**, hence their greater requirement for oxygen. However, within the fibre, as in all cells, other processes such as **transcription** and **translation** must be carried out. Newly-formed myofibrillar proteins must then be assembled into filaments before they can perform their proper role in the cell.

The mass of myofibrillar protein, in addition to its contractile role, acts as a **protein reservoir** which can be utilized in times of nutritional deprivation. However, under such circumstances the breakdown of the myofibrillar assemblies must presumably be carried out in a tightly-controlled fashion to minimize the disruption to the contractile process.

See Energy in Biological Systems and *Molecular Biology and Biotechnology*

Satellite cells

The mature muscle syncytium develops from mononucleated cells called **myoblasts**. In adults, a few myoblasts persist as small mononucleated, flattened, inactive cells called **satellite cells**. These are found in close contact with the muscle fibre. **Satellite cells** have the ability to multiply and fuse with fibres in regenerating muscle damaged by injury, for example. However, the overall number and type of fibres in a particular muscle seem to be **genetically programmed** and do not alter appreciably during life.

See Chapter 12

10.3 Other muscle types

Cardiac muscle

Heart muscle resembles skeletal muscle in that fibres are present and have similar striations to those already described (Fig. 10.8). However, the fibres are narrower, about 15 μm in diameter, and are made up of mononucleated cells joined end to end, rather than syncytia as in skeletal muscle fibres. **Intercalated discs** are found where the cells join. These are a reinforced double sarcolemma and contribute to the structural integrity of the fibre. The cells of the fibre may also form branches which fuse with similar branches on adjacent fibres. The nuclei are central rather than peripheral and myofibrils are present but are less densely packed than in skeletal muscle. Accordingly there is relatively more sarcosolic space, especially adjacent to the nuclei of the cell where there are myofibril-free cones, for organelles such as mitochondria and fuel such as glycogen particles and fat droplets. Despite the non-continuity of the cells, contractions in one cell are passed on to the neighbouring cells which thus co-operate to give the rhythmic beating characteristic of cardiac muscle.

As in skeletal muscle, the cardiac cell sarcolemma forms deep invaginations into the cell at the level of the Z-line on the myofibrils. However, this T-tubule system branches *longitudinally* in order to spread the nerve impulse to the

(a)

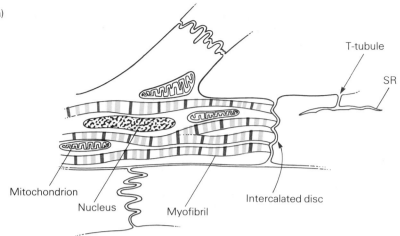

T-tubule

SR

Mitochondrion

Nucleus

Myofibril

Intercalated disc

(b)

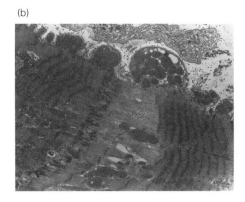

Fig. 10.8 (a) Cell structure of cardiac muscle. (b) Electron micrograph of human cardiac muscle (×6250). Courtesy Ms E. Torgenson, Dept of Pathology, University of Manchester, UK.

underlying contractile elements. There is thus no separate sarcoplasmic reticulum around the myofibrils as in skeletal fibres. A similar modification to the T-tubule system is seen in other muscle which beats rapidly and rhythmically, such as insect flight muscle.

Smooth muscle

Smooth muscle tissue is found in arterial walls and internal organs in the body, where its ability to induce waves of contractile activity or maintain tension is more appropriate than the rapid shortening encountered in skeletal muscle. Fibres are *not* easily seen in the tissue. Smooth muscle consists of sheets or bundles of elongated, tapering, mononucleated cells (Fig. 10.9) lying parallel to each other and embedded in **connective tissue**. Small projections are present on the surface of smooth muscle cells which may mesh with similar spikes on neighbouring cells so that waves of contraction develop in the tissue.

Smooth muscle cells also contain mitochondria and a sarcoplasmic reticulum, in addition to the nucleus and structures which are darkly stained when examined by electron microscopy (**dense bodies**) and are perhaps analogous to the Z-line in striated tissue. Myofibrils of the types described for

(b)

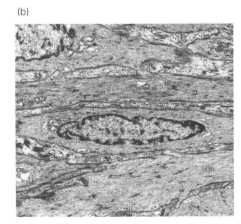

(a)

Membrane spikes

Dense bodies

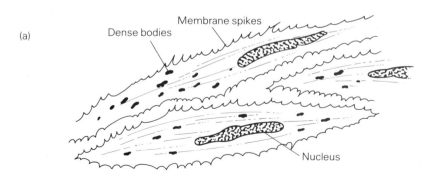

Nucleus

Fig. 10.9 (a) Cell structure of smooth muscle. (b) Electron micrograph of smooth muscle from human renal pelvis (×4450). Courtesy of Dr J. Gilpin, Dept of Cell and Structural Biology, University of Manchester, UK.

skeletal and cardiac tissue are absent. Filaments of diameters and lengths roughly corresponding to the thick and thin filaments of the myofibril can be demonstrated under certain conditions of sample fixation, but are not always observed. Thick filaments lying approximately parallel to the long axis of the cell may form *transiently*, and interact in a fashion analogous to that of the sarcomere before the protein components are re-dispersed. Despite the instability and less regular organization of the filaments, their sliding relative to each other is the basis of the contractile process as in the other muscle types.

Filaments of **intermediate** diameter (10 nm) link the dense bodies forming a cytoskeletal framework.

Non-muscle cells

Many, and probably most, animal cells contain unstable protein filaments exhibiting similar properties to those of smooth muscle. Increasingly, parallels are being found between processes underlying cytoplasmic streaming and mobility in many types of cells and the highly specialized systems developed for muscle tissue.

10.4 Structural proteins of muscle

The proteins of the striated muscle fibre can be subdivided into those associated with the myofibrils, comprising about 60% of the total, those of organelles such as the mitochondria and SR, and the soluble enzymes and proteins in the sarcosol associated with general metabolic activity.

The myofibrillar proteins are primarily those constituting the thick and thin filaments and are directly involved in contractile activity. The analogous filamentous proteins of the smooth muscle cell will also be discussed. Recently a number of ancillary and cytoskeletal proteins have been discovered and the identification of their precise location and function is an exciting area of muscle research.

Proteins of the thick filament

Thick filaments (15 nm diameter) contain mainly **myosin**. Striated and cardiac muscle myosins dissolve readily at an ionic strength of approximately 0.5 mol dm^{-3}, especially in the presence of ATP. This allows their easy extraction from tissues. Lowering the **ionic strength** to 0.1 mol dm^{-3} leads to reassociation of the protein into aggregates which resemble native filaments (Fig. 10.10). Repeated cycles of solubilization and precipitation give fairly pure myosin preparations. The extraction of smooth muscle myosin is less easy, generally requiring low ionic strength and high ATP conditions, and careful attention to homogenization and centrifugation protocols.

The organization of various types of myosin monomers into filaments is an interesting structural problem. In addition to its structural and contractile roles, myosin is also an **enzyme** which catalyses the hydrolysis of ATP, although with rather low efficiency *in vitro*. These two aspects of the myosin molecule, its capacity for self-association and its enzymic activity, reflect the presence of two distinctive **domains** in the proteins. Self-association involves interactions between long **helical tails (or rods)** whereas ATPase activity resides in globular **heads** (Fig. 10.10). The ATPase heads are closely identified with the **cross-bridges** and the **sliding filament** mechanism outlined earlier.

All types of myosin so far studied have an M_r of about 520 000 and are composed of **six subunits**. There are two **heavy** chains (each M_r 220 000) and four **light** chains of M_r ranging from 17 000–25 000, depending on myosin

See Chapter 6

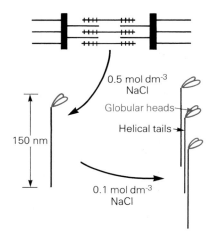

0.5 mol dm^{-3} NaCl

Globular heads

Helical tails

150 nm

0.1 mol dm^{-3} NaCl

Fig. 10.10 Effects on myosin of varying ionic strengths of NaCl.

☐ The self-association of myosin can be followed by observing changes in turbidity since large aggregates scatter more light.

Exercise 4

Suggest a plausible reason why NaCl dissolves myosin aggregates.

Reference Bagshaw, C.R. (1982) *Muscle Contraction*, Outline Studies Series, Chapman and Hall, London, UK. A very readable text, particularly strong on structural details.

type. Two dissimilar light chains are found associated with the amino-terminal region of each heavy chain to form one head of the two-headed myosin molecule (Fig. 10.11).

MYOSIN HEAVY CHAINS. The lower carboxy-terminal half of each heavy chain takes the form of an α-helix. These helices further twist around each other to form a **coiled-coil** structure (Fig. 10.12). Recent studies of myosin **genes** by DNA sequencing procedures have allowed its primary structure to be determined. A **repeating pattern** of seven amino acid residues is observed, which satisfies the α-helical requirement. A hydrophobic residue is found every three or four residues, while a longer repeat sequence every 28 amino acids contains six charged regions. Hydrophobic interactions bind the two helices together in a coiled-coil conformation. The bands of charged residues are found mainly on the surface and are involved in the **aggregation** of myosin monomers into filaments.

Moving 'up' the heavy chain, away from the carboxy-terminus, the strict rules of **periodicity** begin to be relaxed about a third of the distance along, so that there is a weak spot in the coiled-coil forming a **hinge region** which allows the stiff rod-like structure to bend (Fig. 10.13). Moving along further, first the coiled-coil and eventually the α-helices disappear, as the heavy chains become closely associated with the light chains in the head region.

Relatively new data on the **primary structure** of the heavy chain explain earlier studies which involved the use of proteolytic enzymes to 'dissect' the myosin molecule (Fig. 10.13). Cleavage of both chains at the weaker and relatively more open hinge region by trypsin or chymotrypsin generates **light meromyosin (LMM)**, which is insoluble at low ionic strength, and double-headed **heavy meromyosin (HMM)** which retains ATPase activity. Alternatively, papain cleaves in the neck region to form myosin **rod** and single-headed **subfragment-1 (S-1)**. These procedures have demonstrated that the coiled-coil regions confer insolubility on the myosin molecule and are

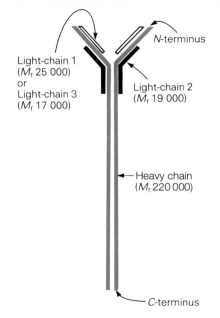

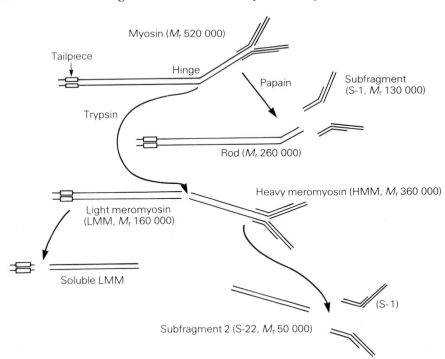

Fig. 10.13 Proteolytic fragmentation of myosin; removal of tailpiece inhibits aggregation of LMM.

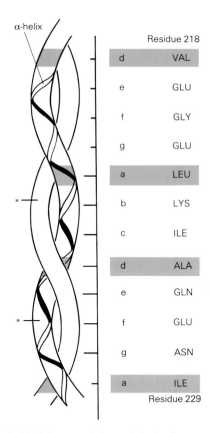

Fig. 10.12 Two myosin heavy chains (each an α-helix) forming a coiled-coil: a and d are hydrophobic residues; * indicates charged residues on outer surface of coiled-coil (myosin sequence from *Caenorhabditis elegans*, a soil nematode, see Fig. 12.8(c)).

therefore involved in self-association. They have also yielded 'active heads' or 'isolated crossbridges' for ATPase studies free from insolubility considerations. Proteolytic dissection is still valuable for probing the myosin molecule. Near the carboxy-terminus of the heavy chain for example, trimming away of a peptide segment (M_r 5000) destroys the self-associating character of LMM. It is interesting to speculate that this **tail-piece**, presumably the last part of the heavy chain to be completed during translation, regulates filament formation of myosin *in vivo*.

The heavy chain polypeptide linking LMM and the myosin heads is termed **subfragment-2 (S-2)**. Thus the myosin rod contains both LMM and S-2 domains. The S-2 domain is less rigid than LMM, and some studies have suggested that it undergoes a **helix–coil** transition as part of the contractile process. In the 'top half' of the myosin molecule, protection against proteolytic cleavage of the heavy chain by binding of light chains, ATP or the thin-filament protein, **actin**, has enabled the **topography** of the head region to be mapped. Such studies have indicated that the ATP and actin-binding sites are located on the heavy-chain polypeptide (Fig. 10.14). The role of the associated light chains has attracted much interest.

MYOSIN LIGHT CHAINS. Each head of HMM contains an identical light chain, LC-2 of M_r 20 000. This has also been termed the **regulatory** or **phosphorylatable** light chain. It is sometimes possible to exchange LC-2 between myosins of different type, or species, or even replace it with versions produced by genetic engineering. LC-2 contains a metal ion binding site and a

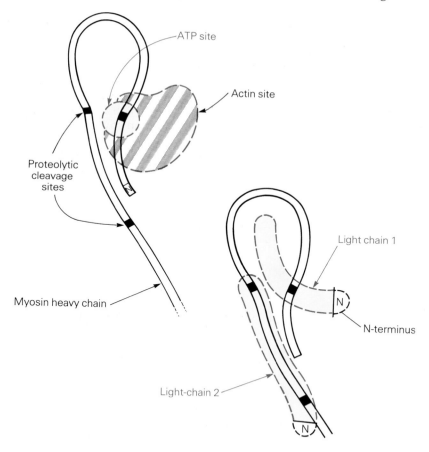

Fig. 10.14 Binding sites of ATP, actin and light-chains to heavy chain of myosin.

phosphorylation site which is the substrate for a specific **protein kinase** (myosin light chain kinase, MLCK). In smooth muscle, Ca^{2+}-binding by LC-2 and its phosphorylation by MLCK are directly involved in the initiation of contraction. The role of LC-2 in striated muscle myosin is less well understood.

The other light-chain (LC-1, M_r 25 000) possessed by each head has been termed the **alkali** or **essential** light chain and was difficult to remove without loss of myosin ATPase activity. The LC-3 light chain (M_r 17 000) is a shortened version of LC-1 produced by alternative **splicing** of the same gene (Fig. 10.15) and is usually found in equal amounts in a preparation of myosin molecules. However, it is not thought that LC-1 and LC-3 are present on adjacent heads in the same myosin monomer. The role of the alkali light chains is probably to modify ATPase and actin-binding characteristics. The LC-2 light chain lies alongside the **neck** region, while the alkali light chain covers the ATP binding site (Fig. 10.14).

Light chain **variants** are found in different fibre types. Similarly, numerous heavy chains variants have been identified. These are the products of a

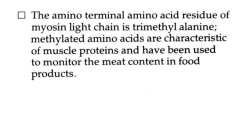

See *Molecular Biology and Biotechnology,* Chapter 5

☐ The amino terminal amino acid residue of myosin light chain is trimethyl alanine; methylated amino acids are characteristic of muscle proteins and have been used to monitor the meat content in food products.

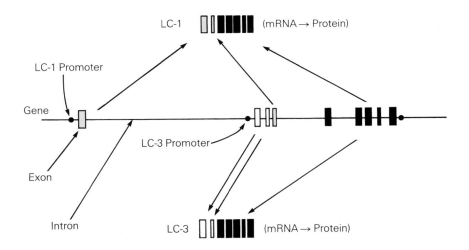

Fig. 10.15 Alternative splicing of light-chain gene to produce LC-1 and LC-3 (chicken myosin).

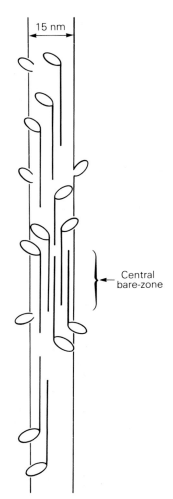

Fig. 10.18 Antiparallel packing of myosin molecules of skeletal muscle to create a bare-zone in the thick filament.

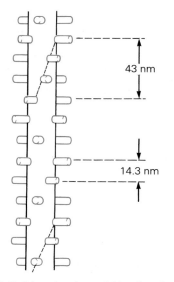

Fig. 10.19 Orientation of cross-bridges (in red) based on X-ray diffraction studies.

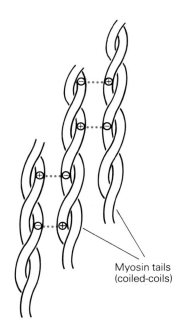

Fig. 10.16 Myosin tails binding together by charge–charge interactions.

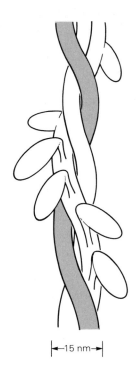

Fig. 10.17 Three 'ropes' of myosin forming a thick filament (heads shown only on one rope).

multigene family with 7 to 13 members in vertebrates. Factors co-ordinating myosin heavy- and light-chain gene expression and the assembly of the newly-synthesized polypeptides into functional myosin remain largely unknown.

MYOSIN FILAMENTS. The twin α-helices of the myosin rod are stabilized by hydrophobic interactions. Charged patches on the rod surface enable neighbouring myosins to bind together (Fig. 10.16) in stable aggregates, perhaps initiated by the tail-piece association described earlier.

In striated muscle some 300 myosin monomers form each thick filament, a structure 15 nm in diameter and 1500 nm long with tapering ends. Since each myosin monomer is 150 nm long and 2 nm thick, the packing of 300 monomers must be very dense. Currently it is proposed that three 'ropes' or subfilaments of myosin twist around each other in such a way that the heads or cross-bridges point to the exterior of the filament, forming a regular repeating pattern (Fig. 10.17).

Perhaps most remarkably, the filament is in two halves, with the myosin molecules lying in opposite directions or **polarities** in each half. This creates a central **bare-zone** having no cross-bridges. The myosins seem to associate in an **antiparallel** (tail-to-tail) manner in the bare-zone, but in a **parallel** sense (head-to-tail) in the rest of the filament (Fig. 10.18). Whatever the full detail of substructure, evidence from X-ray diffraction studies has established that individual myosins are **staggered** by 14.3 nm (about 10% of their length) in binding to adjacent molecules and that this leads to a helical arrangement of cross-bridges around the surface of the filament. Along one 'edge' of the filament cross-bridges are repeated every 43 nm (Fig. 10.19). The cross-bridges are 20 nm long, the length of the myosin head, enough to span the

Reference Squire, J. (1981) *The Structural Basis of Muscular Contraction*, Plenum Press, New York, USA. An advanced text written from a biophysical standpoint.

distance to the thin filament. However, there is evidence that the S-2 region of the heavy-chain may 'swing-out' above the hinge as part of the contractile process.

Filament assembly is dependent on pH and ionic strength. Under carefully controlled conditions, synthetic myosin filaments can be grown which closely resemble the native filament by X-ray diffraction and electron microscope criteria. The forces involved in binding myosin monomers together gradually weaken towards the end of the assembly process, leading to the formation of tapering ends. The number of myosins in a filament is remarkably constant as a result of subtle changes in packing geometry along the length of the filament.

Native thick filaments can be prepared by gentle dispersion of myofibrils and careful centrifugation, and seem to have a greater sensitivity to regulatory factors than synthetic filaments, probably because ancillary proteins are present. Isolated bare-zones or a synthetic equivalent, **minifilaments**, may be used to nucleate the filament growth process.

SMOOTH MUSCLE MYOSIN FILAMENTS. Using smooth muscle and non-muscle myosins, filaments can also be grown *in vitro* and examined by electron microscopy. A central bare-zone is not observable and length regulation is much less controlled. Despite the similarities in myosin structure and filament diameter between smooth and skeletal versions of the protein, the mode of packing may be completely different (Fig. 10.20). Smooth myosin forms tail-to-tail or anti-parallel dimers which subsequently aggregate side-to-side producing long **ribbons** with heads or cross-bridges along the edges. Coiling of the ribbon gives the overall filament its superficial resemblance to the thick filaments of the sarcomere.

Smooth muscle myosin is highly flexible and the rod portion can loop around to bind the neck region (Fig. 10.20). In this form, called 10S, the myosin is unable to polymerize into filaments and may represent a storage form of the myosin in resting smooth muscle. Straightening of the myosin and subsequent aggregation into filaments is brought about by LC-2 **phosphorylation** by a specific light-chain kinase.

The ability of essentially similar myosin molecules to form either stable filaments in the myofibril or unstable filaments in smooth or non-muscle cells underlines the enormous versatility of the myosin structure and perhaps explains the need for large numbers of closely-related genes in vertebrates.

OTHER THICK-FILAMENT PROTEINS. The bare-zones of the thick filaments are closely aligned in the myofibril to form the **M-line** (Fig. 10.21). Here, the three myosin subfilaments change from head-to-tail binding to tail-to-tail antiparallel packing. Three filaments of **M-protein** (M_r 165 000) are linked to the myosin strands by protein cross-bridges including the enzyme **creatine kinase**. This enzyme replenishes stocks of ATP depleted by contractile activity, using **creatine phosphate** as its substrate (see later and Fig. 10.34).

In each half-filament are located seven bands of **C-protein** of M_r 150 000 (Fig. 10.22) which can be visualized by **immunochemical** techniques (Box 6.2). C-protein is a V-shaped protein with an estimated three molecules per complex. It has been proposed that the three Vs form a ring, encircling the thick filament, and act as a molecular clamp. Alternatively, since the arms of the V are of equal length (20 nm) they could project away from the thick towards the thin filament in a similar fashion to the cross-bridges. Other proteins can be detected as distinctive stripes after immunochemical labelling of the A-band. These include H, I, F and X proteins of largely unknown function! They may act as filament length regulators.

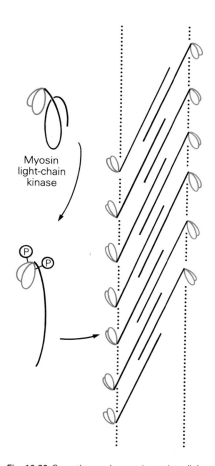

Myosin light-chain kinase

Fig. 10.20 Smooth muscle myosin: antiparallel packing in 'ribbons'. Compare with Fig. 10.18. Packing requires prior phosphorylation of the myosin (see text).

☐ Myosin filaments grown experimentally from pure protein are usually longer than the native thick filaments of the myofibril.

☐ The appearance of enzymes such as creatine phosphokinase in serum indicates leaky muscle membranes, and muscle damage or disease, for example in heart conditions.

Exercise 7

How many molecules of C-protein would be associated with one native myosin thick filament?

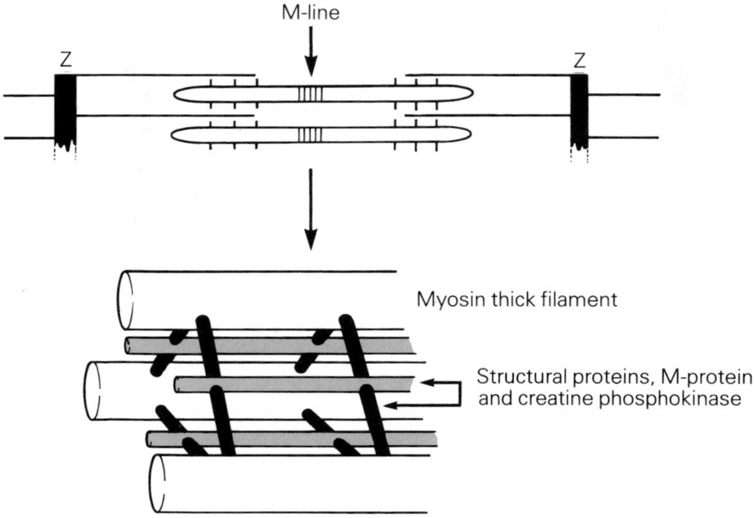

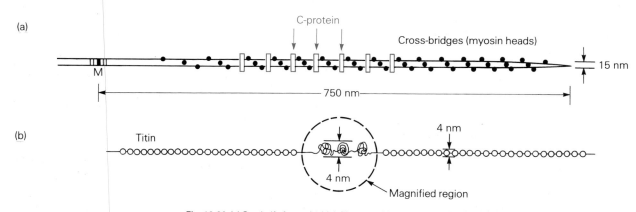

Fig. 10.21 Simplified structure of the M-line.

Myosin thick filament

Structural proteins, M-protein and creatine phosphokinase

□ Despite its abundance, titin remained undiscovered because of its inability to penetrate the commonly-used concentrations of polyacrylamide used during SDS-PAGE (Box 4.1).

□ Myosin can be dissolved out of skinned fibres and then allowed to reform correctly, suggesting the existence of an insoluble matrix which perhaps consists of the elastic filaments titin and nebulin.

Table 10.1 *Comparison of actin primary sequence in different muscle types. Numbers of amino acid substitutions out of a total of 375*

	Smooth	Cardiac	Non-muscle
Striated	6	4	24
Smooth	—	—	22

Much interest has been aroused by the discovery of two enormous proteins, **titin** (sometimes called **connectin**) and **nebulin**, (M_r 3 000 000 and 500 000 respectively) which also seem to be located at the end of the thick filament. Titin is among the largest proteins known. It is a string-like, elastic molecule. Electron micrographs of the purified protein suggest a repeating domain structure giving a beads-on-a-string appearance (Fig. 10.22b). The titin molecule has a diameter of 4 nm and may be equivalent to **gap filaments** seen on some micrographs, which link the tips of myosin thick filaments to Z-lines. This network would prevent over-extension of the sarcomere and maintain the thick filaments in an appropriate array (Fig. 10.23). However, other studies suggest that titin lies alongside the thick filament acting as a **length regulator** and that it is nebulin which forms the elastic network. Titin is thought to be absent from smooth and non-muscle cells.

Proteins of the thin filament

The major protein of the thin filament is **actin** (M_r 42 000). The primary sequence of the protein has been highly conserved throughout evolution

(a)

C-protein

Cross-bridges (myosin heads)

15 nm

M

750 nm

(b)

Titin

4 nm

4 nm

Magnified region

Fig. 10.22 (a) One half of myosin thick filament with seven bands (red) of C-protein. (b) A titin molecule, see text, is shown for comparison.

Reference Fulton, A. B. and Isaacs, W. B. (1991) Titin, a huge, elastic sarcomeric protein with a probable role in morphogenesis. *BioEssays*, **13**, 157–61. The latest on the largest.

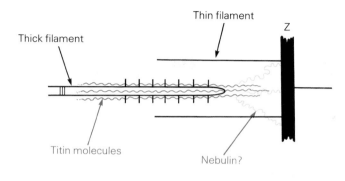

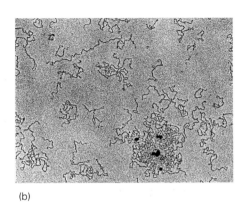

(a)

(b)

Fig. 10.23 (a) Probable location of the giant elastic proteins titin (red) and nebulin (pink) in the sarcomere. (b) Electron micrograph of titin showing molecules upto 800 nm long. Courtesy of Dr J. Trinick, Department of Veterinary Medicine, University of Bristol, UK.

Fig. 10.24 The ATP-dependent G–F transformation of actin.

(Table 10.1). This suggests that even minor variations would interfere with one or more of the functions of this versatile protein and therefore cannot be tolerated. Monomeric or **G-actin** polymerizes *in vitro* with the stoichiometric hydrolysis of one ATP molecule per monomer as it becomes incorporated into fibrous **F-actin** (Fig. 10.24). Actin is found extensively in cardiac and smooth muscle and probably in all cells. The reversible G–F transformation is manipulated by an ever-increasing list of actin binding, bundling, cross-linking and capping proteins (Fig. 10.25). Transiently-formed actin filaments probably have a role in cytoplasmic streaming.

The actin in sarcomeres forms permanent **thin filaments** 7 nm wide anchored in the Z-line. The thin filament consists of two strands of end-to-end linked actin monomers wound around each other to form a double helix (Fig. 10.26). The monomers are not spherical, as usually depicted, but have a dumbbell shape so that the linear polymers have a directionality, or polarity, creating pointed and barbed ends. This can be revealed by decorating F-actin with S-1 heads of myosin (Fig. 10.26). The polarity of the actin filaments is reversed in each of the I bands of the sarcomere, an essential feature of the sliding filament mechanism. A full turn of the actin double-helix corresponds to about 76 nm, compared to the myosin cross-bridge spacing of 43 nm described earlier (see Fig. 10.19).

Exercise 8

From Table 10.1 what is the percentage homology between striated and cardiac actins?

See Chapter 6

□ Actin has been detected in all cells so far examined. Two major types exist. The α-form in skeletal muscle and the β-form in non-muscle cells.

See Chapter 6

Reference Ohtsuki, I., Maruyama, K. and Ebashi, S. (1986) Regulatory and cytoskeletal proteins of vertebrate skeletal muscle, in *Advances in Protein Chemistry*, vol. 38, (eds C.B. Anfinsen, J.T. Edsall and F.M. Richards), Academic Press, London, UK. An advanced review which concentrates on thin filaments.

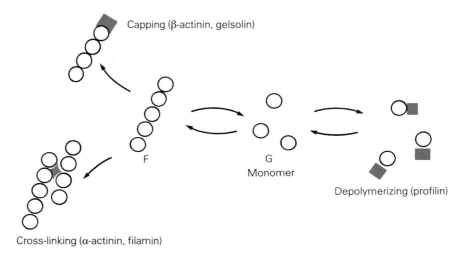

Capping (β-actinin, gelsolin)

F

G
Monomer

Depolymerizing (profilin)

Cross-linking (α-actinin, filamin)

Fig. 10.25 Overview of the interrelationships between the different forms of actin.

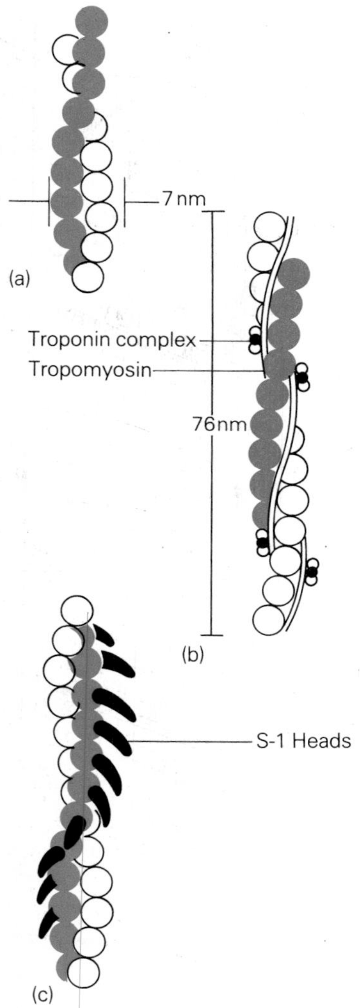

REGULATORY PROTEINS. In the groove between the two strands of the F-actin helix lies a long rod-like protein, **tropomyosin**, a dimer (monomer M_r 35 000) which interacts with seven actin monomers. Tropomyosin has a coiled-coil structure similar to that of the rod portion of myosin heavy chain. At the end of the tropomyosin molecule is found a multi-subunit protein **troponin** (Fig. 10.26). The three components of this complex have the ability to respond to fluctuating Ca^{2+} concentration by repositioning the tropomyosin slightly so as to allow the F-actin monomers to interact with the myosin cross-bridges and initiate the sliding process (Fig. 10.27).

The **troponin : tropomyosin complex** thus acts as a switch responding to the Ca^{2+} signal, which is itself the result of nervous stimulation, to bring about the mechanical events of the contractile process. **Troponin C** (M_r 18 000) binds Ca^{2+}. It has sequences in common with other Ca^{2+}-binding proteins such as myosin LC-2, **calmodulin** and the Ca^{2+}-dependent proteinases, or **calpains**. All such proteins contain the E–F hand structures comprised of α-helical domains which constitute Ca^{2+}-binding sites (Fig. 10.28). On binding Ca^{2+}, troponin C changes its conformation and enables **troponin T** (M_r 37 000) to interact with tropomyosin which shifts its alignment slightly as

7 nm

(a)

Troponin complex

Tropomyosin

76nm

(b)

S-1 Heads

(c)

Fig. 10.26 Actin thin filament. (a) F-actin double helix. (b) Alignment of tropomyosin and troponin complex. (c) F-actin 'decorated' with S-1 heads showing polarity of thin filament.

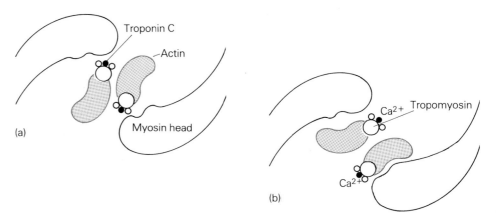

Troponin C

Actin

Myosin head

(a)

Ca^{2+} Tropomyosin

Ca^{2+}

(b)

Fig. 10.27 Cross-section through a thin filament at level of troponin complex: (a) in the absence of Ca^{2+} myosin is unable to bind to actin; (b) Ca^{2+} induces realignment so that myosin may interact with actin.

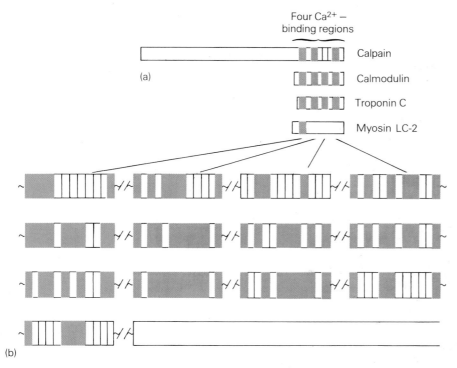

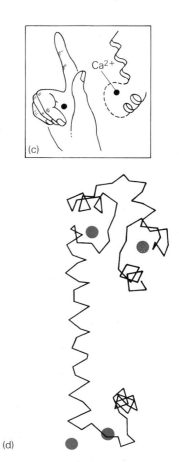

Fig. 10.28 Ca²⁺-binding sites in four muscle proteins: (a) location of Ca²⁺-binding sequences twelve amino acid residues long; (b) amino acid homologies between sequences, common residues are shaded; (c) E-F 'hand' structure; (d) molecular model of calmodulin.

described above. **Troponin I** (M_r 24 000) inhibits the tropomyosin shift by competing with troponin T for tropomyosin binding sites in the absence of Ca²⁺. Thus, regulation of the actin–myosin interaction in the striated myofibril is a **thin filament-mediated** process initiated by Ca²⁺ release from the SR.

SMOOTH MUSCLE ACTIN FILAMENTS do not contain troponin and tropomyosin as such, although it has been proposed that the protein **leiotonin** performs a similar regulatory function. More probably, the Ca²⁺ signal in smooth muscle brings about phosphorylation of myosin LC-2 which transforms the smooth myosin into a form able to interact with actin (see Fig. 10.20). Thus Ca²⁺ signals in smooth muscle operate by a thick filament regulatory mechanism.

Other myofibrillar and cytoskeletal proteins

Z-LINE PROTEINS. The major protein of the Z-line is α-**actinin** (M_r 95 000), a rod-like dimer 4 nm thick and 40 nm long. In this protein the rod-like region is formed by the α-helical polypeptide folding back upon itself to produce three **triple-helical** domains (Fig. 10.29). α-Actinin is an actin-binding protein which cross-links the thin filaments of adjacent sarcomeres in a regular array forming the characteristic Z-line in the centre of the I band (Fig. 10.29).

Recently it has been possible to prepare *in vitro* Z-line sheets as side-by-side aggregates of Z-discs from adjacent myofibrils, free of contractile proteins. Labelled-antibody staining of such structures, which are analogous to transverse sections through the fibre at the level of the Z-line, reveals that the central portion of the disc is α-actinin, surrounded by a collar of the cytoskeletal protein **desmin** (Section 6.5).

□ Fig. 10.28c shows a Ca²⁺-binding site of calmodulin as a hand. The index finger and thumb represent the E and F helices of the molecule respectively, hence the E–F hand.

Exercise 9

From Fig. 10.28b determine the overall homology between the four Ca²⁺-binding regions (each 12 amino acid residues) of calmodulin and troponin C.

□ Calpain (or Ca²⁺-dependent proteinases) have the remarkable property of promoting the solubilization of Z-lines in myofibrillar preparations.

Reference Korn, E.D., Carlier, M.F. and Pantaloni, D. (1987) Actin polymerization and ATP hydrolysis. *Science*, **238**, 638–44. Background to actin and how ATP hydrolysis is connected with monomer–polymer transition.

In certain muscle wasting diseases such as **Duchenne muscular dystrophy**, excessive catabolism of myofibrillar proteins leads to progressive weakening of the skeletal musculature from the age of 4 or 5 onwards in boys. It has proved difficult to identify the primary defect in this process, which has a strong genetic basis, since most of the major muscle proteins seem entirely normal.

Recently the defective gene was identified by molecular genetics techniques and found to code for a large (and scarce) polypeptide of approximate M_r 400 000. (See *Molecular Biology and Biotechnology*, Chapter 9). Although only titin and nebulin among muscle proteins were known to be in this size range, both are present normally in dystrophic muscle.

From the DNA sequence, the amino acid sequence of the defective protein, **dystrophin**, has now been predicted. The sequence is compatible with a rod-like protein with features like α-**actinin**, which binds to actin and membranes. Using immunolocalization techniques, it seems likely that dystrophin is found in minute quantities in the T-tubule/SR junction where it may align these membranes in precise relation both to the myofibrillar Z-lines and sites in the sarcolemma.

In dystrophic muscle fibre, regeneration seems to fail largely because of an inability to organize newly-synthesized proteins into functional myofibrils.

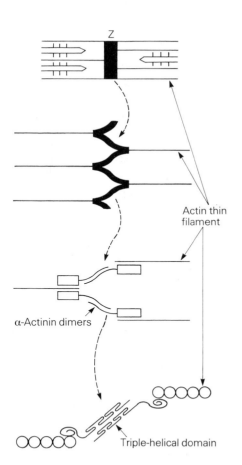

Fig. 10.29 A speculative representation of a Z-line showing the role of α-actinin.

INTERMEDIATE FILAMENTS. Desmin (M_r 55 000) forms filaments 10 nm in diameter, that is, intermediate in size between thick myosin and thin filaments. The protein has a largely α-helical central rod portion with head and tail pieces. Intermediate filament formation seems to involve the association of desmin rods in a coiled-coil fashion. However, the conformation of the desmin at the Z-line may not be filamentous, and the protein seems to be involved in maintaining sarcomeres in register via Z-line alignment (Fig. 10.30). More recently, it has been proposed that longitudinal strands of fibrous desmin link Z-lines in the same myofibril; these presumably lie below the SR membranes which sheath each myofibril. If this is the case, these filaments must be elastic to allow for sarcomeric extension, and must not enter the thick and thin filament arrays where they would interfere with the sliding filament mechanism.

Like many cytoskeletal proteins, desmin can be phosphorylated. It is also a substrate for limited proteolysis by the Ca^{2+}-dependent neutral proteinase which has a remarkable property of selectively removing Z-lines from myofibrillar preparations. Whatever its role, skeletal muscle desmin can be considered a minor myofibrillar component. In contrast, in smooth muscle cells desmin is exceedingly abundant, and the 10 nm filaments are very prominent. Desmin and actin filaments probably connect the **dense bodies** of the smooth muscle cell, which are thought to be analogous to myofibrillar Z-lines and similarly are rich in α-actinin. Certain of the filaments bind to the plasma membrane by **attachment plaques** which contain proteins such as **vinculin** (M_r 117 000). In striated muscle fibres too, vinculin is found at the level of each Z-line, probably acting as a bridge between the Z-line and the sarcolemma (Fig. 10.30).

Maintenance of muscle structure

Figure 10.31 and Table 10.2 emphasize that far from being a simple array of actin and myosin filaments, the muscle fibre and smooth muscle cell seem packed with additional structures and proteins whose functions are poorly understood, together with mitochondria, membranous reticular systems, glycogen particles and nuclei. In the living state, the continual renewal and replacement of the protein structures in response to exercise, hormones and

Reference Maruyama, K. (1985) Myofibrillar cytoskeletal proteins of vertebrate skeletal muscle, in *Developments in Meat Science*, vol. 3 (ed. R.A. Lawrie), Elsevier Applied Science, London, UK, and New York, USA. A useful account of recent discoveries about sarcomeric proteins.

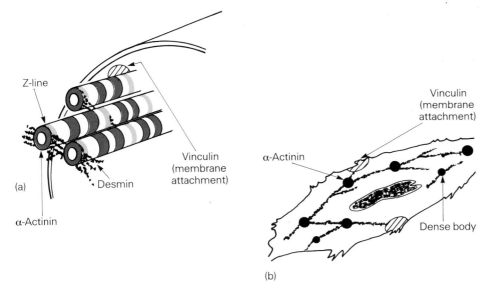

Fig. 10.30 Probable location of α-actinin, desmin and vinculin in (a) striated fibres and (b) smooth muscle cells.

Exercise 10

From Table 10.2 calculate the molar ratio of titin:myosin in the myofibril. Since there are approximately 300 myosin molecules in a thick filament, how many titin molecules could be associated with each filament?

Table 10.2 Major myofibrillar proteins of vertebrate skeletal muscle

Protein	Location	M_r ($\times 10^{-3}$)	Content (%)
Myosin	Thick filament	520	43
Titin	Thick filament	3000	10
C-protein	Thick filament	135	2
M-protein	Thick filament	165	2
Actin	Thin filament	43	22
Troponin complex (C + I + T)	Thin filament	70	5
Tropomyosin (dimer)	Thin filament	66	5
α-Actinin (dimer)	Z-line	190	2
Desmin	Z-line	53	1
Nebulin	I-band	800	3

nutritional factors must require a highly co-ordinated pattern of synthetic and degradative activities. The maintenance and turnover of muscle fibre constituents needs to occur continually without disrupting normal contractile activity.

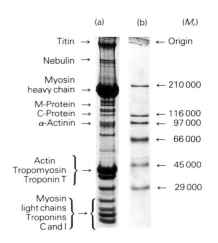

Fig. 10.31 Sodium dodecyl sulphate polyacrylamide gel electrophoresis (SDS-PAGE) of (a) trout myofibrillar proteins, (b) molecular weight calibration proteins. Photograph courtesy J.Q. Zhang, Dept of Applied Biochemistry and Food Science, University of Nottingham, UK.

10.5 Energetics of muscle contraction

In all cells the biosynthesis and turnover of cellular constituents requires the expenditure of metabolic energy in the form of ATP. In muscle fibres, however, most of the ATP is consumed by the sliding of the thick and thin protein filaments which underlies sarcomere shortening and myofibrillar contraction. In addition, approximately one third of muscle ATP is used to return the sarcosolic Ca^{2+} to its storage site in the sarcoplasmic reticulum via a Ca^{2+} **pump** (Section 4.6).

Energy, metabolism and origins of ATP

ATP (Fig. 10.32) is obtained from the partial oxidation of carbohydrate or lipid fuels by the processes of glycolysis and β-oxidation respectively, followed by

---

Exercise 11

Using the marker proteins shown in Fig. 10.31b plot a calibration curve for this SDS-PAGE experiment (plot log molecular weight versus the distance moved by the protein bands). In Figure 10.31a, two major bands are found between those marked α-actinin and actin. From the calibration curve estimate the molecular weight of the lower band and suggest its possible identity.

Reference Buckingham, M.E. and Minty, A.J. (1983) Contractile protein genes, in *Eukaryotic Genes: Their Structure, Activity and Regulation* (eds N. Mclean, S.P. Gregory and R.A. Flavell), Butterworths, London, UK. A quite different, molecular biological approach.

The biosynthetic pathways that supply metabolic energy in the form of ATP for contraction are inevitably accompanied by heat production, as is the hydrolysis of ATP itself.

During exercise this heat must be dissipated by sweating. However, even in resting muscles the capacity of the fibres to generate heat can be utilized by shivering, i.e. the uncoordinated contraction of fibres, so that ATP hydrolysis is effectively **uncoupled** from mechanical work. In cold environments this mechanism is important for **thermoregulation**.

In some animals and a small number of susceptible people, stress, or exposure to some of the gases used as general anaesthetics, can trigger off uncontrolled heat production by a similar process of uncoupling by making muscle membranes leaky to Ca^{2+}. In severe cases the resulting **hyperthermia** can prove fatal.

See *Energy in Biological Systems*

the activities of the TCA cycle and electron transport chain in mitochondria (Fig. 10.33). However, the concentration of ATP in the muscle has been estimated at only $5\,mmol\,kg^{-1}$, enough for only one or two contractions and, surprisingly, it remains fairly constant throughout contractile activity. The explanation is that ADP can be rephosphorylated from a larger pool of **creatine phosphate** via the action of **creatine phosphokinase** (Fig. 10.34) which was mentioned earlier (Fig. 10.21) with reference to M-line structure in thick filaments.

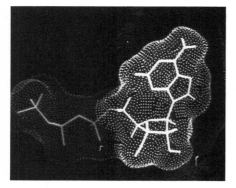

Fig. 10.32 Molecular model of ATP. Courtesy Dr C. Freeman, Polygen, University of York, UK.

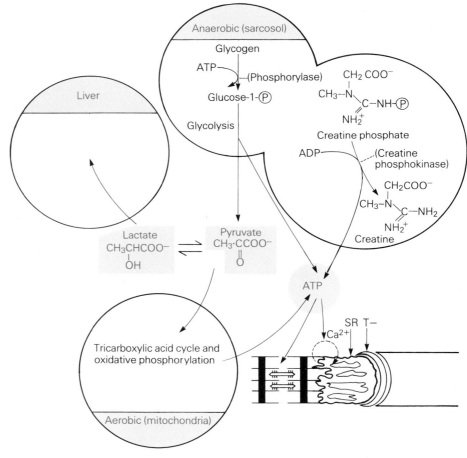

Fig. 10.33 Overview of the origins of ATP required for sarcomeric contraction and re-uptake of Ca^{2+} into SR.

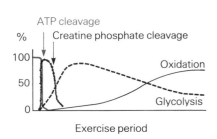

Fig. 10.34 Summary of metabolic energy supplies during exercise.

The type of muscle fibre determines whether carbohydrate or lipid is the major fuel for oxidation. Type I fibres of slow-twitch and cardiac muscle have large numbers of mitochondria and access to constant, high concentrations of oxygen via a rich blood supply allowing fatty acid oxidation. Type II fibres are essentially anaerobic in function and rely on a large glycogen reservoir and much less efficient generation of ATP by glycolysis. The enzymes of glycolysis generate **pyruvate** but this is rapidly equilibrated with lactate by **lactate dehydrogenase**, which is present in high concentrations in muscle tissue. Lactate is transported in the blood to the liver for reprocessing into glycogen during recovery from exercise (Fig. 10.33).

Few muscles are purely of one type. An advantage of the Type II glycolytic fibre is its ability to respond rapidly to external stimuli such as the hormone **adrenalin** in times of great demand for flight or fight. The mechanism for this response involves release of **cAMP** and a **cascade** of protein phosphorylation reactions leading to the rapid activation of the enzyme **phosphorylase** which initiates glycogen utilization (see Chapter 8 and Fig. 10.44).

Exercise 12

Muscle fibres can be classified into Types I or II using histochemical techniques based on enzyme activities. What enzymes might be useful in this context?

Enzymology of the actin–myosin interaction

The contractile process at its simplest involves movement of filaments of myosin relative to those of actin (see Fig. 10.7). This movement is generally believed to be generated by **conformational changes** in the myosin cross-bridges brought about by hydrolysis of ATP. The **ATPase site** of myosin is localized in the head region, or S-1, of the protein (see Fig. 10.14).

A prerequisite for normal enzyme kinetic studies is usually the availability of soluble enzymes and substrates. Unfortunately, myosin and F-actin preparations form turbid or gelatinous suspensions, especially at low ionic strengths. ATP hydrolysis cannot therefore be followed directly by spectro-photometric procedures. These problems were circumvented using soluble **HMM** or **S-1** preparations (see Fig. 10.13). Although these proteins were probably damaged by proteolytic nicking, especially in the light chains, subsequent studies with improved preparations have largely borne out the original measurements.

See *Biological Molecules*, Chapters 4 and 5

Studies of the **actomyosin ATPase** have been of two types, using steady-state conditions or transient kinetic techniques. In the former, using low concentrations of HMM, S-1, or acto-HMM in the presence of excess ATP, phosphate release over several minutes is followed chemically. In the latter, generally requiring higher protein concentrations, reactions can be followed in the millisecond time-range by special rapid-reaction absorbance or fluorescence techniques.

CROSS-BRIDGE CYCLING. An important observation was that the addition of ATP to acto-HMM led to a rapid decrease in turbidity, followed by a slow recovery back to the original turbidity after exhaustion of ATP (Fig. 10.35).

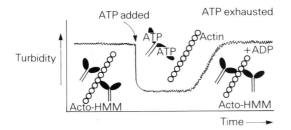

Fig. 10.35 Changes in turbidity (light scattering) associated with acto-HMM ATPase interactions.

□ The Lymn–Taylor model is named after its first proposers (1970), following their studies at the Department of Biophysics, University of Chicago.

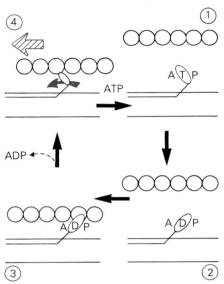

Fig. 10.36 Lymn–Taylor cyclic scheme for cross-bridge mechanism (see text for details).

□ Small molecules such as ATP enter their binding sites on protein molecules at very rapid, diffusion-controlled rates.

□ The binding of actin to HMM reduces the affinity for ATP, and, conversely the binding of ATP to HMM reduces its affinity for actin!

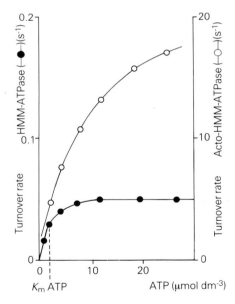

Fig. 10.37 The effect of actin activation on the activity of HMM-ATPase at varying concentrations of ATP.

It was reasoned that ATP caused dissociation of actin from HMM and that ATP hydrolysis took place *only* on the HMM protein. This dissociation–reassociation reaction is a useful model for the cross-bridge cycle occurring during the sliding filament procedure between thick and thin filaments (see Fig. 10.7). An essential feature of the cycle, often referred to as the **Lymn–Taylor scheme**, was that during at least two stages myosin must undergo conformational changes which could generate shifts in the relative positions of the cross-bridges or myosin heads and the F-actin strands. Presumably one such conformational change occurs when myosin is unattached to actin (Fig. 10.36, $1 \rightarrow 2$), and a second when myosin is bound to actin to provide the drive-stroke for the sliding procedure (Fig. 10.36, $3 \rightarrow 4$).

It is possible to treat actin as a second substrate and measure the V_{max} and K_m at high ATP or actin concentrations (Fig. 10.37). The overall rate constant, or **turnover number** for myosin or HMM-catalysed hydrolysis of ATP was found to be close to $0.05 \, s^{-1}$, making it one of the slowest enzymes known (compared for example to the equivalent acetylcholinesterase value of $2 \times 10^4 \, s^{-1}$). However, the addition of F-actin increased this rate by several hundredfold to a maximum of $20 \, s^{-1}$. It was important to identify the rate-limiting step(s) in the ATPase reaction and the way in which actin can apparently speed up or circumvent this step.

INTERMEDIATES DURING THE CROSS-BRIDGE CYCLE. Specialized kinetic techniques established that the rates of (i) ATP binding to acto-HMM, (ii) actin release, and (iii) hydrolysis of the terminal phosphate of ATP ($150 \, s^{-1}$) were rapid (Fig. 10.38) suggesting that ADP release limited the rate with which the drive-stroke could occur. In fact it has been possible to identify many other intermediates, their rates of interconversion and whether they involve covalent changes in substrate or conformational changes in protein (Fig. 10.38). Several features emerge from this complex scheme. The **free energy** normally released by the hydrolysis of ATP is transferred to the myosin protein allowing it to adopt an energized conformation (M^{**}). In the absence of F-actin, M^{**} returns to a less-stressed position (M^{*}) very slowly, accompanied by release of phosphate. This step is probably equivalent to the rate limiting step ($0.05 \, s^{-1}$) described earlier.

In the presence of actin, it is believed that actin rapidly re-attaches to $M^{**}ADP..\circled{P}$ and facilitates the release of products and a return to the normal conformational state. In this model the $AM^{**} \rightarrow AM$ transition represents the drive-stroke in contraction with a rate constant $\sim 20 \, s^{-1}$. In the absence of sufficient ATP, the AM complex would be permanent and represents the state of the cross-bridge–actin interaction during ATP exhaustion in *rigor mortis*. However, during normal contraction, the cross-bridge would dissociate on binding further ATP and the cycle would be repeated.

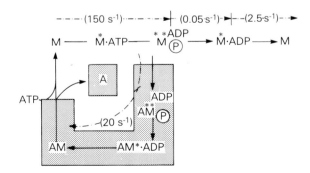

Fig. 10.38 Scheme showing rates (i.e. rate constants) of interconversion of the intermediates in the HMN-ATPase reaction (in the box; — · — is rate acceleration in the presence of actin).

Reference Woledge, R.C., Curtin, N.A. and Homsher, E. (1985) *Energetic Aspects of Muscle Contraction*, Academic Press, London, UK. A fairly advanced text, unusual in that it relates physiological and biochemical studies of the contractile process.

Box 10.5
Transformation of muscle to meat

Immediately following the death of an animal, the leakage of Ca^{2+} from the SR can continue to initiate the **cross-bridge cycle**, exhausting ATP and increasing phosphate until eventually thick and thin filaments become irreversibly cross-linked in rigor mortis. Ca^{2+} can also continue to stimulate glycolysis and the production of lactate. Depending on the amount of stored glycogen, the pH of muscle falls slowly to below 6.0.

In meat animals this process is encouraged by holding the carcass at ambient temperature, since low pH discourages cross-bridge cycling and fibre shortening. Too rapid chilling or freezing causes extensive sarcomeric contraction or cold-shortening, which leads to loss of water-holding capacity and a tough, fibrous product. Low pH during storage also encourages muscle proteolytic enzymes such as cathepsins to cleave structural proteins and connective tissue and so carry out a natural **tenderization** process.

In fish muscle the fall in pH is less marked, and autolytic tenderization is generally unnecessary, since fibres are shorter and there is less collagen and connective tissue.

The Lymn–Taylor scheme forms a satisfying basis for understanding the energetics of the sliding filament mechanism. It is important to realize that it has been proposed on the basis of studies using purified proteins in solution, whereas in intact muscle the proteins are in highly structured arrays, with closely associated ancillary proteins.

Exercise 13

The half-life ($t_{1/2}$) of an approximately first-order reaction can be related to the rate constant (k) by the expression $k = \ln 2/t_{1/2}$. Assuming that the maximum hydrolysis of ATP by actin-activated HMM has $k = 200$ s, what would be the half-life of one cross-bridge cycle?

An elastic or spring-like component in cycling?

Are conformational changes in the S-1 region all that is required to explain the mechanics of cross-bridge function inferred from physiological measurements of tension development and shortening in intact tissue? Most such studies postulate the presence of some elastic or spring-like component in the cross-bridge, physically distinct from the ATPase or actin-binding sites on the myosin heavy chain (see Fig. 10.14). As described earlier, the neck region (S–2) of the myosin heavy-chain contains a flexible hinge which could allow the swing-out of the head (S-1) towards the thin filament as part of the normal cross-bridge cycle (Fig. 10.39). Furthermore, S-2 in some myosins has been found to be **thermally unstable**, suggesting that a reversible helix-to-coil transition may be an additional component of the mechanism (Fig. 10.39).

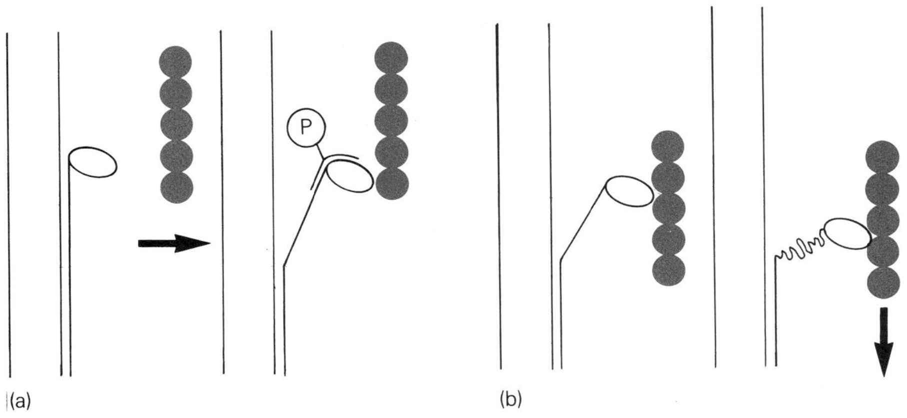

Fig. 10.39 (a) A possible role of LC-2 phosphorylation in 'swing-out' of cross-bridge and (b) the **collapsing cross-bridge** theory of contraction.

Fig. 10.40 Photomicrograph of a motor end plate (×210). Courtesy M.J. Hoult, Dept of Biological Sciences, The Manchester Metropolitan University, UK.

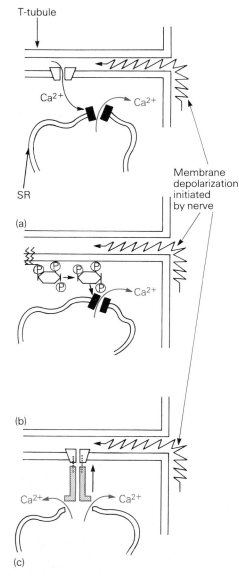

Fig. 10.41 Alternative theories for the coupling of excitation and contraction. (a) Ca^{2+} signal; (b) inositol trisphosphate signal; (c) 'Spanning' or 'foot' protein.

It is interesting that LC-2, the regulatory light-chain, spans the S-1/S-2 junction (see Fig. 10.14). Phosphorylation of this light chain is able to increase efficiency of the contractile process in intact skeletal muscle. However, this phosphorylation does not seem to affect the ATPase of isolated myosin or its subfragments.

There still are further aspects of the actin–myosin interaction which need to be understood. For example, the necessity for a *double-headed* myosin molecule, which could perhaps interact with two thin filaments at the same time is not understood.

It is generally assumed that smooth-, cardiac- and non-muscle myosins interact with actin to produce relative movement in much the same way as the striated muscle myosin whose kinetics have predominantly been studied. Smooth-muscle myosins have lower ATPase rates and form filaments *in vitro* only in response to light chain phosphorylation. ATP can be trapped at the ATPase site by formation of the looped conformation (see Fig. 10.14), but the relevance of this to physiological events remains to be established.

10.6 The role of Ca^{2+} in the regulation of muscle contraction and metabolism

In skeletal and cardiac fibres, in simple terms, the arrival of a nerve impulse at the **motor end plate** (Fig. 10.40) leads to Ca^{2+} release from stores in the SR, such that sarcoplasmic concentration rises from 10^{-7} to 10^{-5} mol dm^{-3}.

Communication at triad junctions

The precise way in which **depolarization** (Section 9.4) of the sarcolemma is transformed, via the T-tubule system, into Ca^{2+} release by the SR is not fully understood in biochemical terms. Since a 50 nm gap exists between the two membranes, communication could be chemical, or involve channels and protein structures (Fig. 10.41).

In the former case, the diffusion of Ca^{2+} from the T-system to receptors in the SR could in turn stimulate release of Ca^{2+} from the SR (Fig. 10.41a). Alternatively, in some cells **phospholipase C**-mediated release of **inositol phosphates** from membrane phospholipids is known to be a powerful chemical signal for the efflux of Ca^{2+} from internal stores such as SR (Fig. 10.41b). Other observations suggest a direct protein link in the form of a **spanning protein** between the T-system and SR, linking nerve impulse to SR via protein conformational changes (Fig. 10.41c). Whatever the mechanism, the overall outcome is the release of Ca^{2+} from the SR lumen where it is normally bound to the storage protein **calsequestrin** (M_r 45 000).

Ca^{2+} stimulation of striated and smooth muscle

The earliest detectable event in the interaction of Ca^{2+} with the myofibrillar apparatus is observed by X-ray diffraction measurements to be a slight shift in the alignment of tropomyosin along the F-actin groove brought about by interactions within the troponin complex. The F-actin monomers thus intervene in the 'idling' myosin ATPase cycling mechanism and bind to the M** form of myosin prior to the drive-stroke (see Fig. 10.38).

In typical vertebrate fibres, *one* nerve impulse brings about one round of accelerated ATPase activity and a sarcomeric shortening. In other fibres, notably the rapid flight muscle of insects (Fig. 10.42), a single impulse may lead to a *repetitive*, rhythmic series of contractions, perhaps reflecting an even higher degree of filament organization in these muscles.

Reference Cross, R.A. (1989) Smooth operators. The molecular mechanics of smooth muscle contraction. *BioEssays*, **11**, 18–21. A short, readable account of the biomolecular events involved in the contraction of smooth muscle.

Reference Taylor, K.A., Reedy, M.C., Cordova, L. and Reedy, M.K. (1984) Three-dimensional reconstruction of rigor insect flight muscle from tilted thin sections. *Nature*, **310**, 285–91. An account of computer three-dimensional reconstruction applied to insect flight muscle.

Damage or demyelinating diseases such as multiple sclerosis leading to motor neurone malfunction are usually accompanied by paralysis and atrophy of the recipient muscle.

At the skeletal motor end plate, the autoimmune condition myasthenia gravis results in impaired activity of the acetylcholine receptors (see Section 9.4) on the post-synaptic region of the sarcolemma. Accordingly the nerve signal is not transduced leading to the normal depolarization of the sarcolemma and T-tubules and activation of the Ca^{2+} channel. The muscle fatigue or paralysis is similar to that brought about by curare and some snake venoms which also act on acetylcholine receptors.

In cardiac muscle, more common medical problems are associated with damaged membranes, such as the sarcolemma, which produce abnormal Ca^{2+} influx via the channels. This in turn leads to uncontrolled release of Ca^{2+} from the sarcoplasmic reticulum. Uncoordinated contractions, fibrillations, arrythmias and spasm are all features of coronary heart disease. Drugs such as the dihydropyridines can be useful modifiers of Ca^{2+}-channel function.

Ca^{2+} influx is also stimulated by catecholamines such as adrenalin acting via β-receptors and cAMP (see Sections 8.4 and 9.4) as part of the normal fight or flight response. In damaged heart tissue, such stress could lead to widespread spasm and heart failure and may prove fatal. Treatment is by drugs known as β-blockers which suppress overstimulation of the Ca^{2+} channels.

The release of Ca^{2+} from stores in smooth muscle cells is almost certainly initiated by the inositol phosphate pathway mentioned above (see Fig. 10.41), in response to external signals such as the interaction of **adrenalin** with plasma membrane receptors. The myosin light chain kinase of smooth muscle responds to Ca^{2+} by phosphorylating myosin LC-2, initiating the thick–thin filament interaction and the contractile process. Relaxation requires the reabsorption of Ca^{2+} to the SR lumen, a process dependent upon ATP and the **Ca^{2+}-ATPase pump** in the SR membrane. The pump protein (M_r 110 000) is abundant in the membrane. It becomes phosphorylated by ATP in the process of transporting two Ca^{2+} against a concentration gradient (Fig. 10.43). An ancillary protein component, **phospholamban**, facilitates this uptake.

Ca^{2+} AND PROTEIN PHOSPHORYLATION. In both types of muscle cell (and indeed in all cells) Ca^{2+} is a **second messenger**, analogous to **cyclic nucleotides**, affecting a variety of processes and metabolic pathways. Frequently these effects require the involvement of the Ca^{2+}-binding protein **calmodulin** (see Fig. 10.28) and the activation of protein kinases. The myosin light chain kinases already described incorporate a calmodulin subunit into their structure.

In fast Type II fibres especially, an immediate response to increased Ca^{2+} would be the stimulation of **phosphorylase kinase** and the glycogen breakdown cascade (Fig. 10.33) which provides the ATP necessary for contraction. Phosphorylase kinase can be further phosphorylated and activated by a **cAMP-dependent kinase** responding to the stimulation of receptors in the sarcolemma by agents such as adrenalin. Thus energy metabolism is responsive to neuronal and hormonal signals via the phosphorylase kinase reaction (Fig. 10.44). A similar interplay of Ca^{2+}- and cAMP-dependent protein kinases can bring about the interconversion and regulation of numerous other myofibrillar proteins such as troponins I, T and C, cytoskeletal proteins such as desmin and the Ca^{2+}-ATPase pump of SR. Phosphorylation of smooth myosin light chain kinase by cAMP-dependent protein kinase inhibits myosin phosphorylation and hence contractile activity.

See Chapter 8

☐ In fatigued muscle the fibres become progressively unable to restore Ca^{2+} to the SR, so that the fibres cannot respond to incoming nerve impulses.

Fig. 10.42 Bumble bee in flight. Courtesy M.J. Hoult, Dept of Biological Sciences, The Manchester Metropolitan University, UK.

The protein content of a muscle is determined by a balance between protein synthesis and breakdown, a process referred to as **turnover**.

Stretching a muscle stimulates synthesis by a feedback loop involving stretch receptors in the muscle and the nervous system. Immobilization of a muscle, in a plaster cast for example, in contrast leads to rapid loss of protein by accelerated breakdown.

A wide range of hormones or their analogues influence turnover. Some steroid hormones such as testosterone and its derivatives are **anabolic** (stimulate synthesis) whereas others such as cortisol are **catabolic** (stimulate breakdown). There are complex interactions with other endocrine systems such as those involving the thyroid and polypeptide hormones, for example thyroxine and insulin respectively, which also stimulate anabolic processes.

Analogues of **adrenalin** stimulate protein synthesis and fat breakdown, and seem able to reduce the rates of muscle wasting in the muscle of immobilized limbs.

It is hardly surprising that hormonal control of muscle growth interests doctors, unscrupulous athletes (see *Biosynthesis*, Box 1.1) and farmers who are producing animals for meat.

The function of many of these phosphorylation signals, especially those involving structural proteins, remains obscure. Further complexity has been given to the field by the discovery of **protein kinase C** and the phospho-inositide **signal transduction** pathway concerned with Ca^{2+} fluxes in cell membranes. In this system, receptor stimulation leads to conversion of **phosphatidyl inositol** (PI) to **inositol trisphosphate** (IPs) and **diacyl glycerol** (DAG). This last compound stimulates protein kinase C and Ca^{2+} flux (Fig. 10.44).

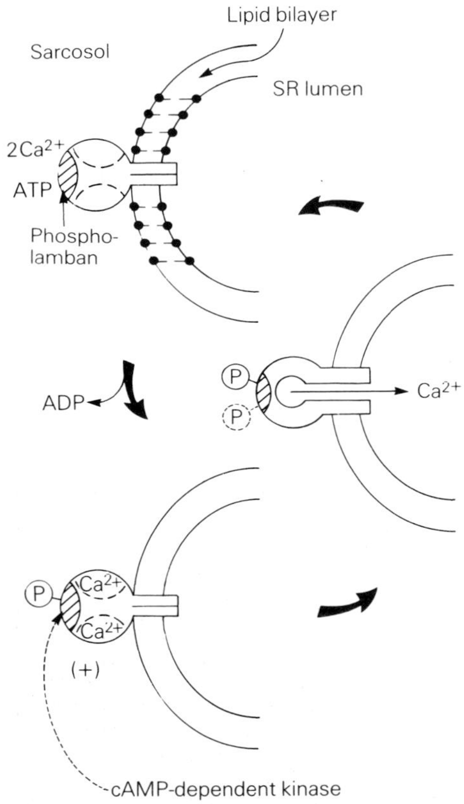

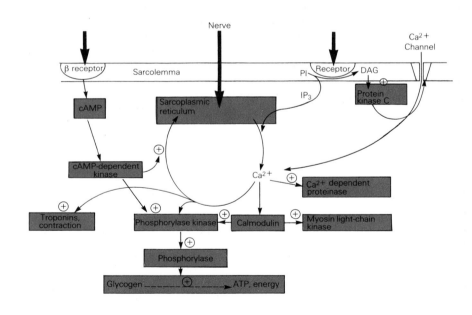

Fig. 10.44 Effect of external stimuli (large arrows) on the release of Ca^{2+} and the interplay of protein kinases in muscle. PI, phosphatidyl inositol, a regulatory membrane lipid; DAG, diacylglycerol, released from PI and stimulating protein kinase C; IP₃, inositol trisphosphate; ⊕, stimulation of activity (see also Chapter 8).

Fig. 10.43 Re-uptake of Ca^{2+} into SR via the Ca^{2+}-pump. A secondary phosphorylation by cAMP-dependent protein kinase increases the rate of transport.

Ca²⁺-DEPENDENT PROTEOLYSIS. The binding of Ca^{2+} to regulatory proteins such as troponin and protein kinases brings about reversible changes in contractile activity and metabolic processes, respectively. It has been suggested that Ca^{2+} has irreversible effects in muscle structure and function by stimulation of **Ca^{2+}-dependent neutral proteinase** (or calpains, see Fig. 10.28). Susceptible proteins include components of myofibrils and Z-lines, for example, titin and desmin, cytosolic steroid hormone receptors and membrane proteins such as vinculin and protein kinase C. Thus Ca^{2+} may have additional catabolic or regulatory functions in the muscle fibre more related to **growth** and **protein turnover** than to contractile activity.

10.7 Overview

All muscle types ultimately rely on interactions between **protein molecules** to generate movement, a process driven by chemical energy released from compounds such as ATP. Cytoplasmic streaming probably involves the same or similar proteins but operates in a less co-ordinated manner. In muscle cells the ways in which the various proteins are organized into complexes and filaments are truly remarkable and have been studied by physiologists, microscopists, crystallographers and protein chemists, all sharing a common fascination with the molecular architecture of muscle tissues.

This chapter dealt firstly with some aspects of the cell biology of different muscle types, concentrating on the best-studied example, **striated or skeletal muscle**. Secondly, the proteins of the **contractile machinery** were described in biochemical detail. The next section showed **free energy** generated by the hydrolysis of ATP is utilized to produce movement of the **protein filaments** relative to each other. These sequences of molecular events are initiated by the **neuronal stimulation** of the muscle cell. These are intimately linked with Ca^{2+} fluxes, which produce many additional effects in muscle metabolism.

This molecular machinery is in a **dynamic** state, with components continually being synthesized and degraded in response to a number of physiological factors such as exercise, hormones and nutrition; it is also self-repairing. Knowledge about muscle regeneration and growth is required particularly by medical researchers dealing with muscle diseases and injuries, those interested in sports training and farmers producing animals for meat.

Answers to Exercises

1. The chemical energy of ATP is converted to the kinetic energy of the thick and thin filaments.
2. Sarcolemma, T-tubule, sarcoplasmic reticulum, mitochondrial and nuclear membranes for example.
3. Contracted, 1.5 μm; extended, 3.0 μm.
4. The high concentrations of ions mask the charge–charge interactions between myosin rods.
5. One molecule of myosin (M_r 520 000) gives one molecule of LMM (M_r 160 000). The maximum yield is therefore 160 000/520 000 g, that is 0.31 g.

6. From Fig. 10.13 it can be determined that the M_r of S-1 is 120 000, and from the text the M_r of actin is 42 000.
7. Seven stripes per half filament means 14 stripes in total; with three molecules per complex the total would be 42.
8. Of 375 amino acids, four are different. Therefore 371 (99%) are homologous.
9. From Fig. 10.28 calmodulin and troponin C have 26 'shared' amino acids (shaded) out of 48, corresponding to 54% homology.
10. There is approximately 43 g myosin per 10 g titin: thus there is 43/520 000 mol of myosin per 10/3 000 000 mol titin. This

represents 83 μmol myosin per 3.3 μmol titin a ratio of 25 : 1. However, in terms of myosin thick filaments there is 83/300 μmol filaments, i.e. 0.28 μmol filament per 3.3 μmol titin. Each thick filament could therefore be associated with 12 titin molecules.
11. A muscle protein of M_r 50 000–60 000 could be desmin as described in the text.
12. Mitochondrial enzymes such as succinate dehydrogenase are more prevalent in Type I, and glycogen phosphorylase activity would be greater in Type II, i.e. anaerobic fibres.
13. $t_{1/2} = \ln 2/k = 3.5$ ms.

FILL IN THE BLANKS

1.   Muscle is a specialized tissue designed to produce _____ in animals. The basic unit in skeletal and cardiac muscle is the _____ . Alternating light and dark bands give rise to _____ , the distance between two _____ constituting one _____ . The contractile unit is built up out or arrays of _____ and _____ filaments. These are largely constructed from the proteins _____ and _____ . The filaments interact via _____ and use the energy source, _____ to drive the process.

The major thick filament protein _____ is made up of _____ subunits. The use of _____ enzymes to dissect the molecule leads to subfragments such as _____ and _____ . ATPase activity is located in the _____ region of the molecule. Repeating sequences are found in the heavy chain polypeptide which adopts a _____ conformation.

Thin filaments are constructed of the globular protein _____ . Regulatory proteins include _____ and _____ . _____ are released by the _____ _____ in response to the arrival of a _____ _____ . A cyclic series of interactions between thick and thin filaments, known as the _____ cycle, drives shortening of the _____ .

Choose from: actin (2 occurrences), ATP, Ca^{2+}, coiled-coil, cross-bridges, fibre, HMM, LMM, Lymn–Taylor, movement, myosin (2 occurrences), nerve impulse, proteolytic, S-1, sarcomere (2 occurrences), sarcoplasmic reticulum, six, striations, thick, thin, tropomyosin, troponins, Z-lines.

MULTIPLE-CHOICE QUESTION

2.   State whether the following are true or false:
A. In skeletal muscle, Z-bands are bisected by I-bands and the H-zone surrounds the M line.
B. The T-tubule system is derived from the sarcoplasmic reticulum.
C. Satellite cells may be induced to form new muscle fibres.
D. Myosin filaments consist of two α-helices stabilized by hydrophobic interactions.
E. Actin in sarcomeres consists of filaments 7 nm wide.
F. The extraction of myosin from all muscle tissues requires the use of media of high ionic strength.
G. HMM can catalyse the hydrolysis of ATP at a high rate.
H. The addition of F-actin decreases the turnover number of myosin-catalysed hydrolysis of ATP.
I. Creatine phosphokinase can donate its P$_i$ to ADP to replenish muscle ATP.
J. Calsequestrin is a Ca^{2+}-storage protein in the sarcolemma of skeletal muscle.

SHORT-ANSWER QUESTIONS

3.   List at least four different kinds of helical structures found in thick and thin filaments.

4.   What characteristic do C-protein and M-protein have in common?

5.   How do the thick filaments differ in striated and smooth muscle? Use a diagram to illustrate your answer.

6.   In what type of fibre would you expect to find high concentrations of the oxygen-storage protein myoglobin and why?

7.   Thin filaments decorated with S-1 (myosin heads) can be formed only in the presence of ATP. True or false?

8.   Compare the structures of the myofibrillar protein filaments with the fibrous proteins of the cytoskeleton (Chapter 6) and those of the extra cellular matrix (Chapter 7).

ESSAY QUESTIONS

9.   Muscle protein breakdown occurs on denervation or immobilization of muscle in wasting diseases such as dystrophy and in post-mortem tenderization of meat. Describe the probable sequence of events in such 'breakdown'.

10.   Use the text and the bibliography to speculate on how exercise and training can influence the structure and activity of different muscles.

11.   Several types of protein kinases are found in the muscle fibre; how do they affect contractile function and fibre metabolism?

12.   Where are the muscle genes situated in the fibre? What factors are known to influence their expression? How many different genes are transcribed to make a myosin molecule?

13.   Find out which smooth muscles are used as sources of muscle proteins. It is thought that prostaglandins have a role in smooth muscle contraction; what is a likely source of these compounds? Do prostaglandins also have a role in skeletal muscle?

11

The immune response

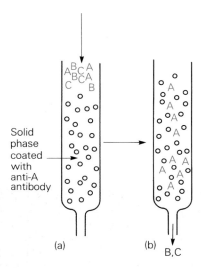

Fig. 11.1 Purification of proteins using affinity chromatography. (a) The separation column consists of Sepharose particles coated with antibodies to one of the proteins to be purified, in this case anti-A. The protein mixture, containing, say, proteins A, B and C is eluted through the column at neutral pH. (b) The protein, A, binds to the specific antibody on the solid phase, while the remaining proteins pass straight through. Pure protein A can now be washed off the column using a buffer of low pH.

Exercise 1

Look at Table 11.1. Construct a similar table giving *in each case* two diseases caused by viruses, bacteria, fungi, protozoa and helminths. List the source of the pathogen as shown in this table but do not use the same organisms.

Objectives

After reading this chapter you should be able to:

☐ discuss the nature of non-specific immune responses;

☐ describe the tissues, cells and macromolecules involved in specific immunity;

☐ discuss the interaction of components of specific immunity with the non-specific mechanisms;

☐ review the role of antibodies in the elimination of microorganisms from the body.

11.1 Introduction

Animals and plants are constantly exposed to a range of microorganisms which have the potential to cause disease. This microbial onslaught comes from many sources, including air, water, soil, food, and the body fluids of infected animals. Viruses and bacteria are the most frequent *pathogens* but many fungi, protozoa and some multicellular animals can also cause disease (Table 11.1). Most animals, however, are not constantly ill and they usually recover from infection. The ability to survive in the face of this 'mass attack' can be attributed to the combined efforts of the cells and tissues of the **immune system**. All multicellular animals have such a system which serves both to prevent the entry of microorganisms and to remove those which breach these defences.

The immune response is not restricted to reactions against pathogenic microorganisms. It is shown towards all material which is 'foreign' to the animal. The most highly developed immune systems are found in the 'higher' vertebrates, such as birds and mammals. This account is limited to the immune responses of those animals.

Immunology, the study of the immune response, has made tremendous advances in the last 30 years and several immunologists have received Nobel prizes for their studies (Table 11.2). Immunological research has revealed the complexity of the cellular interactions which take place during the production of an immune response. This knowledge has been used to understand and devise treatments for a variety of conditions in which the immune response fails or works inappropriately. In addition, immunological techniques have been applied with great success in almost every area of biology. In biochemistry, antibodies are used to purify proteins from complex mixtures (Fig. 11.1); in cell biology, 'fluorescent' antibodies are used to locate molecules in cells (Fig. 11.2 and Chapter 6); in pharmacology, radiolabelled antibodies are used to assay drug levels in body fluids.

---

pathogen: an organism which causes disease.

Plants, like animals, are susceptible to pathogenic microorganisms including viruses, bacteria and multicellular parasites (see Box 2.6). Plants also have defence mechanisms which resist infection and these defences are more akin to those of the non-specific mechanisms in animals. For example, plants have effective barriers such as the cuticle which prevent entry of organisms. Plants also extrude chemicals which repel pathogens. For example, the roots of peaches have glucosides which, on cleavage by glucosidases of invading pathogens, release hydrogen cyanide and poison the pathogen. Antibiotic substances frequently found in plants include caffeic and chlorogenic acid and dihydroxyphenylalanine.

Plants may also produce **phytoalexins**. These are toxins which the plant synthesizes following invasion tissue damage. They include substances such as phaseolin and rishitin. Plants may also have mechanisms for detoxifying poisons produced by the pathogen itself.

Plants can display a hypersensitivity reaction. The hypersensitive reaction in plants is a response to large numbers of invading 'incompatible' pathogens. At the site of invasion there are changes in plant cell membrane permeability, tissue collapse, and necrosis. This response can be found within hours of bacterial invasion. The hypersensitive necrosis is also associated with a build-up of phytoalexins in the necrotic area.

(a) Rishitin

(b) Phaseolin

Table 11.1 *Pathogenic microorganisms and disease*

Disease	Type of organism	Source of pathogen
Measles	Virus	Mucus of respiratory tract (coughs and sneezes)
Rabies	Virus	Saliva of infected animals
Whooping cough	Bacterium	Mucus of respiratory tract
Cholera	Bacterium	Contaminated water
Ringworm	Fungus	Direct contact with skin of infected person
Schistosomiasis	Helminth	Infected water snails
Malaria	Protozoan	Infected mosquitoes

Table 11.2 *Nobel prizewinning immunologists since 1960*

Year	Prizewinner(s)	Area of research
1960	Macfarlane Burnet, F. Medawar, P.B.	Immunological tolerance
1972	Edelman, G.M. Porter, R.R.	Antibody structure
1977	Yalow, R.	Development of radioimmunoassay
1980	Dausset, J. Snell, G.D.	Transplantation
	Benacerraf, B.	Genetic control of immune response
1984	Kohler, G.J.F. Milstein, C.	Monoclonal antibodies
	Jerne, N.K.	Theories on the control of the immune response

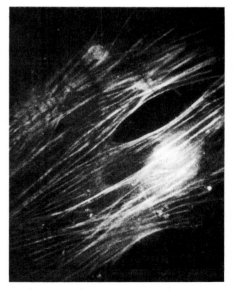

Fig. 11.2 Photomicrograph of mouse fibroblasts stained with a human anti–mouse antibody to a cytoskeletal component (×1300). Courtesy of Dr S. Kumar, Christie Hospital, Manchester, UK.

11.2 Specificity of the immune response

In vertebrates two types of immune defence mechanism work together to combat infection: **non-specific (or innate)** immunity and **specific (or acquired)** immunity. There are several fundamental differences between the two systems. For example, non-specific immunity is shown towards any 'foreign' material and is innate or 'in-born'. Specific immunity is shown towards individual types of organisms or macromolecules and can only be acquired after contact with the organism or macromolecule.

An organism or macromolecule which stimulates a specific immune response is known as an **immunogen**. Non-specific immunity is 'available' as soon as an organism enters the body whereas it may be several days before a specific response becomes effective. On second contact with the same organism, the specific immune system produces a greater response than before, whereas a non-specific response is always of the same magnitude.

Specific immunity is often thought of as more 'important' than non-specific responses but it should be remembered that the non-specific response forms the first arm of the immunological defence mechanism. In addition, there is close interaction between the cells of the two systems. Molecules produced by non-specific cells are able to influence the activity of specific cells and *vice versa*. Moreover, defects in either system result in increased susceptibility to infections.

□ The specific macromolecules and cells involved in specific immunity increase during an immune response and are, in the end, responsible for eliminating the pathogen.

11.3 Non-specific immunity

Non-specific immunity is shown towards any 'non-self' material whether it is a virus, a bacterium or inorganic material such as silica particles or asbestos fibres. This immune system includes several different types of defence mechanism which are summarized in Table 11.3.

Barriers: physical and chemical

The skin is the major barrier to the entry of microorganisms into the body. In addition, lactate in sweat lowers the pH of the skin surface to a level which can only be tolerated by a few types of bacteria. Damage to the skin therefore provides a means of entry for microorganisms: patients who have extensive damage to the skin, perhaps as a result of burns, or who have severe 'weeping' eczema, are prone to infection by this route.

The mucous membranes lining the respiratory, gastrointestinal and genito-urinary tracts separate the 'interior' of the body from the outside. Mucus produced by these membranes is sticky and effectively traps microorganisms and other particulate matter. Food may contain pathogenic microorganisms, particularly if it has not been prepared under hygienic conditions, or has been stored inappropriately. Many of these organisms are killed by the acid conditions of the stomach, the pH of which is about 2.

Proteins

Body fluids contain a variety of proteins which have non-specific bactericidal or anti-viral activities. Amongst these the best characterized are the **complement** proteins and the enzyme **lysozyme**. In addition, the liver of an infected animal synthesizes **acute phase proteins** and virus-infected cells produce **interferons**.

COMPLEMENT is the name given to a group of proteins present in the blood plasma which can be activated to cause *lysis* of bacteria. Some complement

Table 11.3 *Non-specific defence mechanisms*

Mechanism	Examples
Physical barriers	Skin
	Mucous membranes
Chemical barriers	Hydrochloric acid in stomach
	Lactic acid in sweat
Proteins	Lysozyme in secretions, e.g. tears
	Complement in plasma
	Interferon
	Acute phase proteins
Cells	Phagocytes
	Natural killer cells
	Eosinophils
	Basophils

lysis: *literally means 'breaking open'.*

Reference Dawson, M.M. (1987) *Introducing Immunology.* HMSO, London, UK. Biological Sciences series, Continuing Nurse Education programme. Although written for nurses, this open-learning text on the cellular aspects of the immune response is a good starting point for first-year degree students.

proteins which have been activated in this way are able to stimulate phagocytic leukocytes which take up and digest the microorganisms. The complement sequence can be activated directly by bacteria, or indirectly by antibodies, which are the products of the specific immune response, after they have bound to an immunogen. For this reason, details of the activity of complement in both systems will be discussed under specific immunity.

LYSOZYME (Fig. 11.3) is an enzyme which is present in many secretions of the body including tears. Lysozyme hydrolyses the cell walls of some bacteria rendering them susceptible to osmotic lysis or to the lytic action of complement.

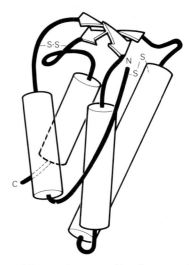

Fig. 11.4 Proposed structure of interferons α and β. The figure shows a predicted tertiary structure for interferon based on amino acid sequences. The four cylinders represent α-helices and the arrows β-strands. From Butler, M. (1987) *Animal Cell Technology: Principles and Products*. Open University Press, Milton Keynes.

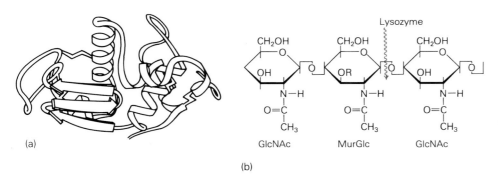

Fig. 11.3 (a) Representation of a lysozyme molecule. Redrawn from Richardson, J.T. (1981). *Advances in Protein Chemistry* **34**, 168–339. (b) The action of lysozyme. The bacterial cell wall has a polysaccharide meshwork of alternating *N*-acetylglucosamine (GlcNAc) and *N*-acetylmuraminate (MurGlc) residues. Lysozyme catalyses the hydrolysis of these residues at the point indicated.

INTERFERONS (IFNs) are proteins secreted by virus-infected cells which protect other cells from virus infection. In non-infected cells, IFNs stimulate the production of proteins that inhibit viral replication. The term 'interferon' refers in fact to several proteins (Table 11.4) with this anti-viral activity, not all of which are produced by virus-infected cells. Two major groups of interferons, designated IFN α and β, form part of the non-specific immune response. A third type, IFN γ is produced by cells of the specific immune system and will be discussed later. IFNs α and β have identical biological activities and are structurally similar (Fig. 11.4). The mechanism of action of IFN is outlined in Fig. 11.5.

□ In addition to their antiviral activity IFNs inhibit the growth of mammalian cells. They have been used to prevent the growth and spread of some types of cancer in humans.

ACUTE-PHASE PROTEINS are synthesized by the liver and appear in the plasma within hours of the onset of an infection. Several of these proteins are not normally found in the blood of healthy individuals; others, such as

Table 11.4 *Classification and properties of interferons (IFNs)*

Interferons	Group	Definition	Structure	Location of genes
IFNα	Classical Type I	Predominant type produced by virus-infected leukocytes	Single polypeptide chain of 165–166 amino acid residues; not glycosylated	IFN gene: chromosome 9 Receptor gene on chromosome 21
IFNβ	Classical Type I	Predominant type produced by virus-infected fibroblasts	Single polypeptide chain of 166 amino acid residues; 29% sequence homology with IFNα; one glycosylation site	IFN gene: chromosome 9 Uses same receptor as IFNα
IFNγ	Immune Type II	Produced by activated lymphocytes	Single polypeptide chain of 166 amino acid residues; no homology with IFNα or β; glycosylated	IFN gene: chromosome 12 Receptor gene on chromosome 6

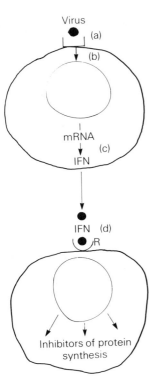

Table 11.5 *The acute-phase proteins*

Protein	Function
C-reactive protein	Binds *Pneumococcal* protein; activates complement
Ceruloplasmin	Increases plasma Cu^{2+} (needed for several enzymes)
Metallothionein	Decreases blood Zn^{2+} and Fe^{2+} (needed for bacterial replication)
Serum amyloid A	Uncertain
Fibrinogen	Blood clotting
α-Globulins, e.g. haptoglobin; α_1-antitrypsin	Various roles
Complement proteins	Lysis of bacteria; stimulation of phagocytosis

complement, have their rate of production increased. Many of the acute-phase proteins are clearly antibacterial, as indicated in Table 11.5, but the function of others, such as serum amyloid A, is unclear.

Cells of the non-specific response

Phagocytic cells form the major cellular defence of the non-specific immune system. Virus-infected cells may also be destroyed non-specifically, by **natural killer** (NK) cells which are not phagocytic.

PHAGOCYTIC CELLS ingest bacteria (Fig. 11.6) and destroy them by one of at least two mechanisms:

Fig. 11.5 Overview of antiviral activity of IFN. A virus infects cell 1 after binding to a receptor (a). The virus switches on the host machinery for the replication of viral DNA (b). The presence of viral nucleic acid also induces the IFN genes (c). IFN is secreted and binds to surface receptors (R) on non-infected cells. (d) Binding of IFN results in the induction of enzymes which interfere with the production of viral protein. One protein inhibits the translation of viral messenger RNA while another stimulates an endonuclease which breaks down viral mRNA. In inducing these two proteins, interferons also inhibit the growth of mammalian cells and for this reason have been used to prevent the growth of cancer cells in humans. Unfortunately, this much publicized 'wonder-drug' has proved only to be of use with a small number of rare tumours.

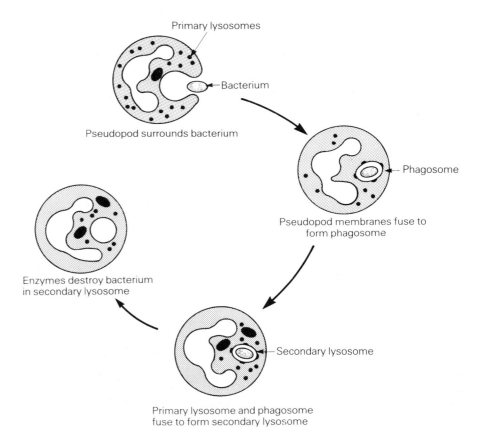

Fig. 11.6 Mechanism of phagocytosis. Redrawn from Dawson, M.M. (1987) *Introducing Immunology*, Biological Sciences series, continuing nurse education programme, Crown copyright.

1. Production of toxic oxygen metabolites. 'Resting' phagocytes obtain their energy by anaerobic glycolysis using glycogen as an energy source. However, the process of phagocytosis is accompanied by a 'burst' of aerobic respiration. This respiratory burst results in the production of antimicrobial oxygen metabolites, such as hydrogen peroxide. In addition, a combination of hydrogen peroxide, chloride ions and the enzyme **myeloperoxidase**, results in the halogenation, and destruction, of bacteria, yeasts and some viruses.

See *Energy in Biological Systems*, Chapter 5

2. The fusion of a primary lysosome with the phagocytic vacuole releases lysosomal enzymes onto the microorganism. The lysosomal enzymes are hydrolytic, and include lysozyme, proteases, phosphatases and nucleases which degrade the ingested parasite.

See Chapter 1

Two types of phagocytic cell are found in mammalian blood: monocytes and polymorphonuclear leukocytes.

Monocytes

Monocytes constitute approximately 5% of the total leukocytes in the blood. They have a characteristic 'horseshoe'-shaped nucleus and abundant cytoplasm (Fig. 11.7). Monocytes are produced in the bone marrow and released into the blood. They circulate in the blood for several weeks, then enter the tissues where they develop into **macrophages** (Fig. 11.7). Macrophages are larger and more actively phagocytic than monocytes. They have a bilobed nucleus and prominent cytoplasmic lysosomes. They are also important in the defence against bacteria that live inside cells, such as *Mycobacterium tuberculosis*, the causative organism of tuberculosis. Macrophages have been shown to kill cultured tumour cells, but whether they are able to destroy cancer cells in the body is uncertain.

Most tissues have some macrophages which act as scavenging cells. However, there are some organs which have a relatively high content of these cells. Together, these organs make up the **mononuclear phagocytic system** (Fig. 11.8), a 'filtering' system in which the macrophages phagocytically remove 'foreign' proteins and particles. The widespread distribution of these tissues ensures that microorganisms entering the body, by whatever route,

□ The size and shape of the nucleus, the relative abundance of cytoplasm, the presence or absence of cytoplasmic granules and the size of the cell are the main criteria used in classifying leukocytes.

□ The mononuclear phagocytic system (MPS) is also known as the reticuloendothelial system (RES). Either term may be found in textbooks although usage of the MPS term is more *recent*.

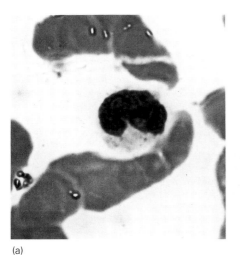
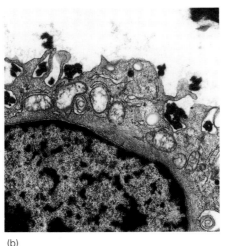

(a) (b)

Fig. 11.7 (a) A monocyte in a blood smear magnified (×630). Note the horseshoe-shaped nucleus and abundant cytoplasm. (b) Electron micrograph of a portion of a macrophage surface showing phagocytosis of extracellular material (×13 600). Courtesy of Dr R. Griffin, Department of Science, Bristol Polytechnic, UK.

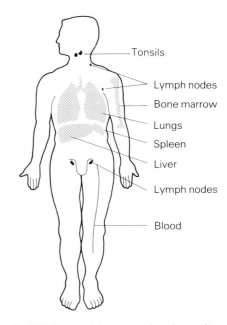

Tonsils
Lymph nodes
Bone marrow
Lungs
Spleen
Liver
Lymph nodes
Blood

Fig. 11.8 Tissues of the mononuclear phagocytic system. Redrawn from Dawson, M.M. (1987) *Introducing Immunology*, Biological Sciences series, continuing nurse education, Crown copyright.

will meet some of these cells. The 'alveolar' macrophages in the lungs take up material which enters via the respiratory tract. Liver macrophages remove material which may have entered from the gut, via the hepatic portal vein, while the spleen filters the blood. The lymph nodes deal with material which, having entered via the skin, trickles with the lymph flow down to a local draining node.

Polymorphonuclear leukocytes

Polymorphonuclear leukocytes (PMNs) are by far the most common leukocyte, comprising 65–70% of the total. They have a characteristic lobed nucleus and granular cytoplasm. There are three different types of PMN in the blood and these are distinguished in a blood smear by their morphology and by characteristic staining patterns (Fig. 11.9). Only one type, the neutrophil, is phagocytic, the remaining two do, however, have other roles in the immune system.

NEUTROPHILS are the most common PMN, accounting for around 60% of blood leucocytes. They are distinguished from other PMNs by the shape of the nucleus with its characteristic three or more lobes. In addition, the cytoplasmic granules do not readily take up stains. Like all blood cells, neutrophils are produced in the bone marrow. Although short-lived, their numbers are maintained by continual replacement from the bone marrow.

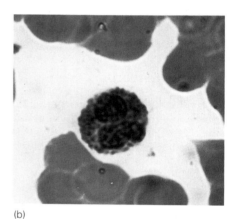
(b)

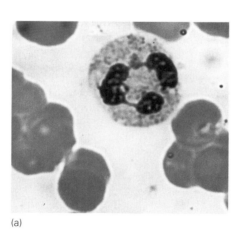
(a)

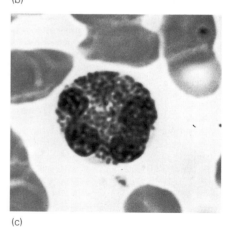
(c)

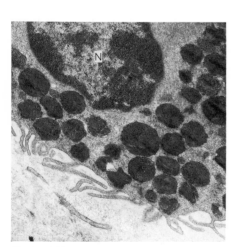

Fig. 11.10 Electron micrograph showing detail of a mast cell. Note the histamine-rich granules. The nucleus is indicated by N. Courtesy of Dr A. Sattar, St Mary's Hospital, Manchester, UK.

Fig. 11.9 Polymorphonuclear leucocytes in blood smear: (a) A neutrophil, (b) a basophil and (c) an eosinophil. (×630 in all cases).

The neutrophils are important in **inflammation**, which is a series of events which takes place at the site of tissue damage. They are the first cells to arrive at sites of tissue damage or injury, and for this reason they are frequently known as the 'inflammatory' cells.

BASOPHILS have a bilobed nucleus and prominent granules which stain blue with Jenner–Giemsa stains. They are produced in the bone marrow and enter the blood. Some basophils may leave the blood and enter solid tissues where they develop into **mast cells** (Fig. 11.10). Mast cells are concentrated in the connective tissue of the respiratory and gastrointestinal tracts and in the skin. Both basophils and mast cells have granules containing chemicals which initiate and promote inflammation. They also contain a *chemotactic factor* for **eosinophils**, the third group of PMNs.

EOSINOPHILS are normally a minor constituent of blood, making up only 1–2% of blood leucocytes in a healthy individual. Their prominent cytoplasmic granules contain several basic proteins including major basic protein (MBP), eosinophil cationic protein (ECP) and eosinophil peroxidase (EPO) which are highly toxic to parasites. The numbers of eosinophils are significantly increased in individuals who have parasitic worm infestations, such as *Toxocara canis*. Eosinophils have the capacity to destroy some parasites and their larvae by releasing their toxic granules.

Increased numbers of eosinophils are also found in individuals suffering from allergies such as hay fever and allergic asthma and this is related to the release of the eosinophil chemotactic factor from basophils. In addition, products of the specific immune response can both increase the number and activity of these cells.

Natural killer cells

Natural killer cells (NK) are large granular lymphocytes (Fig. 11.11) that account for 5–10% of the lymphocyte population in the blood. These cells destroy virus-infected cells in a non-specific manner. They bind to an infected cell and release toxic granules which rapidly lyse the target cell.

Inflammation

The local reaction to tissue damage or injury is known as **inflammation**. A simple scratch on the skin, for example by a splinter of wood, will initiate inflammation. The four classic signs of inflammation quickly become apparent: redness, heat, swelling and pain.

☐ The granules of basophils and mast cells contains histamine which increases the permeability of blood vessels, increases the secretion of mucus by the nasal and bronchial glands, and stimulates the contraction of smooth muscle in the bronchioles and in the gut.

$$HC\!\!=\!\!C - CH_2 - CH_2 - NH_3^+$$

Histamine

☐ *Toxocara canis* is a nematode which is widespread in dogs. The eggs, which are present in the faeces, may be ingested by children following accidental contact in parks and playgrounds. The larvae can migrate to the eye, causing a condition which is difficult to diagnose and may cause blindness.

☐ NK cells also spontanoeusly kill some cultured tumour cells. This cytotoxicity does not depend on previous contact with the tumour. This suggests that NK cells may have a non-specific antitumour effect *in vivo*.

☐ The classic symptoms of inflammation are usually given Latin terminology: *rubor*, *calor*, *tumor* and *dolor*.

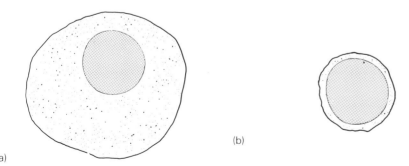

(a)   (b)

Fig. 11.11 The large granular lymphocyte (LGL) (a) drawn here in comparison with a small lymphocyte (b), the most common lymphocyte in blood. Both have prominent rounded nuclei but the LGL has more abundant, granular cytoplasm.

chemotactic factor: *a soluble factor which, at relatively high local concentrations, attracts the movement of cells towards it.*

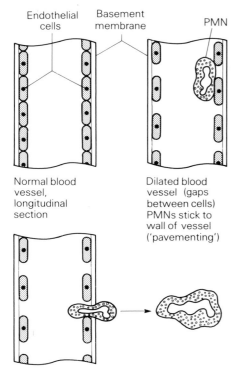

Endothelial cells   Basement membrane   PMN

Normal blood vessel, longitudinal section

Dilated blood vessel (gaps between cells) PMNs stick to wall of vessel ('pavementing')

PMNs migrate between endothelial cells and through the basement membrane into tissue space and migrate to site of inflammation

Fig. 11.12 Entry of neutrophils into a site of inflammation. Redrawn from Dawson, M.M. (1987) *Introducing Immunology*, Biological Sciences series, continuing nurse education, Crown copyright.

The initial damage to the tissue stimulates the release of chemical mediators from mast cells. Histamine (see marginal note) causes the dilation of the blood vessels in the region of damaged tissue, increasing the blood flow, causing reddening and local heating. This results in the 'wheal' at the site of damage. Seconds later, redness may occur further out from the site of damage. This is known as the 'flare' and is of nervous origin. In addition, the blood vessel walls become more permeable to plasma which leaks into the tissue, causing swelling. If large amounts of plasma leak a blister is formed. PMNs may also accumulate at the site, causing more swelling, but this usually occurs only if bacteria are present.

The mechanism by which PMNs move into the area is shown in Fig. 11.12. PMNs cross the blood vessel wall after adhering to the endothelial cells which line the vessel. Dilation of the blood vessel wall produces 'gaps' between the endothelial cells through which the PMNs can squeeze. If bacteria are present at the site of tissue damage, complement (see later) may be activated, resulting in products which attract more PMNs or cause the release of more inflammatory mediators from mast cells. When this occurs, the accumulation of PMNs can be seen as 'pus', which is made up of fluid and dead PMNs.

Later events at the inflammatory site include the accumulation of monocytes and macrophages, particularly if bacteria are present. When macrophages ingest bacteria they secrete a protein, known as **interleukin 1** (IL-1), which also promotes inflammation. Amongst other things, IL-1 stimulates the release of neutrophils from the bone marrow, resulting in greatly increased numbers of these cells in peripheral blood. In addition, IL-1 acts on the brain, inducing a fever. The increased body temperature both inhibits the growth of bacteria and favours the proliferation of cells of the specific immune system. IL-1 also suppresses appetite and induces drowsiness. The latter may be a means of conserving energy. IL-1 stimulates the breakdown of muscle protein, causing an increase in the level of amino acids in the blood. These amino acids are used by the liver to produce the acute phase proteins.

Exercise 2

Organisms A and B gain entry to the body by different routes. Organism A is present in raw eggs and causes severe gastroenteritis. Organism B infects the nervous system and is transferred in the saliva of infected mammals when they bite other animals. List the non-specific mechanisms which organisms A and B must overcome *en route* to infection.

11.4 Specific immunity

Whereas all animals have some form of non-specific immunity, a true *specific* response is found only in the vertebrates. The key features of specific immunity can perhaps best be illustrated with reference to a common childhood disease, such as measles (Fig. 11.13):

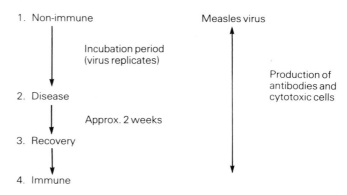

1. Non-immune   Measles virus

   Incubation period
   (virus replicates)

                                        Production of
                                        antibodies and
2. Disease   cytotoxic cells

   Approx. 2 weeks

3. Recovery

4. Immune

Fig. 11.13 Features of specific immunity (see text for details).

1. When a child is initially exposed to the measles virus, non-specific immune mechanisms are insufficient to prevent the replication and spread of the virus throughout the body.
2. There is an incubation period between the initial infection and the production of clinical symptoms (about 2 weeks for measles). During this time the virus is replicating within the cells of the host.
3. During the course of the infection the child produces proteins, called antibodies, which recognize the virus. The antibodies combine with the virus, neutralize it and stimulate the phagocytic cells to remove the virus–antibody complex. Additionally, the immune system produces specialized cells, some of which bind specifically to viral proteins found on the surface of the infected cell, and then kill the cell by the release of toxins which act over a short range. The destruction of the virus-infected cell prevents the virus from replicating, and specific antibodies 'mop-up' released virus. Other cells, closely related to those which develop cytotoxicity, produce proteins which stimulate the activity of macrophages in removing the virus or virus-infected cells.
4. The child is now said to be 'immune' to measles. A state of immunity ensures that when the child is exposed to the measles virus a second time, specific antibodies and *cytotoxic* cells are already present. In addition, more cytotoxic cells and specific antibodies can be produced rapidly. The virus is thus eliminated before symptoms of the disease occur. Subsequent contacts with measles virus will only serve to boost this immunity still further. The state of immunity which has been induced applies only to this virus or very closely related viruses. A child then exposed to a different virus, such as rubella, which causes German measles, must acquire immunity through infection.

The key features of the specific immune response are that the antibodies and cytotoxic cells *are* specific and that, during the first contact, an individual acquires an **immunological memory** for that microorganism. The immunological memory enables a more rapid and heightened immune response the second time around. For these reasons, the majority of people develop measles just once during their lifetime. This life-long immunity is sustained because each exposure to the virus boosts the immune response and the immunological memory. However, almost everyone suffers a head-cold at least once a year and is likely to experience several bouts of influenza during their lifetime. This is not due to a defect in their immune system but to the simple fact that each cold or episode of influenza is the result of exposure to a different strain of virus.

The production of specific antibodies is called **humoral immunity** while the production of specific cytotoxic cells, or of specific cells which release macrophage-activating factors, is known as **cell-mediated immunity** (CMI). Together they permit recovery from infection and endow long-term immunity.

Immunogens, antigens, epitopes and haptens

An **immunogen** is defined as a cell or macromolecule which stimulates a specific immune response. An **'antigen'** is a molecule which combines with a specific antibody but which is not necessarily immunogenic in its own right. Immunogens (Table 11.6) only induce an immune response if they are recognized as 'foreign'. So, for example, the protein bovine serum albumin (M_r 67 000) is immunogenic in sheep but not in cows. In fact, all foreign cells are immunogenic in an appropriate host. Cells may be regarded as complex packages of proteins and glycoproteins each of which is immunogenic. For this reason, cells are strong stimulators of the specific immune response.

☐ The smallest immunogens are macromolecules such as proteins and some polysaccharides. The minimum size for a protein to be immunogenic is frequently quoted as a M_r of approximately 5000. This is based on the finding that insulin (M_r 5200) is the smallest naturally-occurring protein known to be immunogenic. In addition macromolecules have to have internal structural variety, which in proteins is the sequence of amino acids. A synthetic polypeptide made entirely of glycine residues would be a poor immunogen, whereas a natural polypeptide, containing 20 types of amino acid residues would induce a strong immune response. Proteins are, therefore, usually more immunogenic than polysaccharides which contain fewer varieties of residues, although they do have branching. The branched oligosaccharides of glycoproteins are highly antigenic, that is they combine with the products of the immune response but in themselves are not immunogenic, being too small to induce an immune response.

---

cytotoxic: literally, 'cell-killing'. The production of cytotoxic T lymphocytes is characteristic of cell-mediated immunity.

Box 11.2
Immunization

Since Jenner first immunized people against smallpox in 1798, prophylactic immunization has greatly reduced the incidence of many infectious diseases. Smallpox, for example, has been eradicated by an immunization programme sponsored by the World Health Organization.

The aim of immunization is to induce a state of immunity by administering the infectious agent in such a way as to avoid causing the disease itself. This can be achieved by administering:

- a closely related, less harmful strain of the organism, inducing immunity that cross-reacts with the disease-producing strain, for example, cowpox virus for smallpox;
- killed organisms, for example, *Bordetella pertussis* vaccine against whooping cough;
- less virulent (attenuated) strains, for example, polio virus. (Attentuation may be achieved by extensive subculturing of the virus, for example, in cultured mammalian cells.)
- immunogenic extracts such as bacterial toxins which have been treated chemically to render them harmless (toxoids), for example, *Tetanus* toxoid.

New developments in vaccine production include genetically engineered viral proteins able to induce immunity without disease (see also *Molecular Biology and Biotechnology*). One other new development concerns the use of 'anti-idiotype' antibodies. When an animal is immunized with an antigen, it makes an antibody whose unique combining site or idiotype is complementary to the antigen (see figure). An antibody to the idiotype would have a complementary shape to that combining site and would therefore resemble the original antigen. Immunizing people with the anti-idiotype should therefore be as effective as immunizing with the antigen but without the possibility of causing the disease.

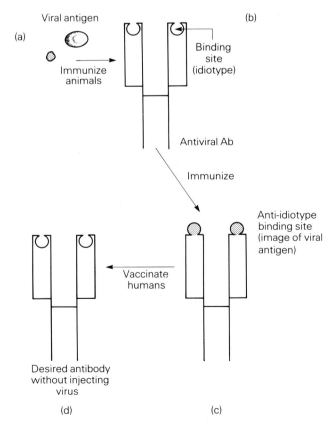

Anti-idiotype vaccines. This rather complex procedure is used to induce antibodies to a virus without injecting the virus itself. An antibody against a virus (b) has a binding site which is complementary to a viral antigen (a). An antibody against this antibody (c) should therefore have a binding site which is an image of the viral antigen. This second antibody can therefore be used as a vaccine (d) to induce antiviral antibodies. It is still experimental, but may prove useful when the virus is particularly harmful, or is difficult to prepare in sufficient quantities for large-scale vaccination.

The term *hapten* is applied to a small chemical group which, while not immunogenic on its own, can elicit the production of a specific immune response when covalently attached to an immunogenic protein. The ability of the immune system to distinguish small molecular groups is often used for the preparation of antibodies which can be used in assays. For example, one sensitive method for determining the level of steroid hormones such as oestrogen, is to use a radioimmunoassay. This requires the use of a specific antibody against the hormone which is itself far too small to be immunogenic. However, if oestrogen is covalently bound to a carrier protein such as bovine serum albumin, and used to immunize a mouse, the animal makes specific anti-oestrogen antibodies. The oestrogen acts as an artificial epitope in this case and is called a hapten. The ability of the higher vertebrates to mount an immune response against a seemingly endless list of haptens, is truly astonishing.

Table 11.6 *List of naturally-occurring immunogens and the predominant response against them*

Immunogen	Type of response which results in elimination of the immunogen
Bacteria	Humoral (except Mycobacteria)
Viruses	Cell-mediated immunity
Fungi	Cell-mediated immunity
Protozoa	Either depending on organism
Parasitic nematodes and helminths	Humoral
Transplanted tissue	Cell-mediated immunity
Proteins	Humoral
Complex polysaccharides	Humoral
Glycoproteins	Humoral
Nucleic acids	Humoral

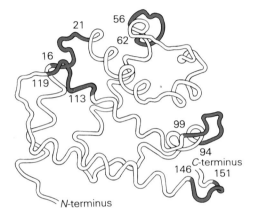

Fig. 11.14 The antigenic determinants of sperm whale myoglobin: the five antigenic determinants are in red. The numbers refer to amino acid residue positions from the amino-terminus. Redrawn from *Biochemistry and Molecular Biology*, Open University third level course. The Open University Press, Milton Keynes.

The immune system responds to an immunogen by the production of specific antibodies and/or cytotoxic cells. However, these specific agents do not recognize the whole of the immunogen but only small regions known as **antigenic determinants**, or **epitopes**. These might be small linear sequences of amino acid residues or branched sequences of carbohydrate, or they might be 'shape' sequences brought about by the folding of a protein molecule (Fig. 11.14). The rest of the protein can be regarded as the 'carrier' for this epitope. A protein such as bovine serum albumin will have several different epitopes on its surface, each of which would stimulate cells with the appropriate receptor.

All immunogens stimulate both humoral and cell-mediated immunity but, depending on the immunogen, one type usually predominates (Table 11.6). For example, when a kidney is grafted to an unrelated donor, an immune response is mounted against the foreign tissue which is subsequently rejected. Although antibodies are made against the cells of the graft, it is the cell-mediated response which triggers rejection.

11.5 The structure and function of antibodies

Antibodies, or immunoglobulins (Ig), are glycoproteins found in the blood plasma, lymph and secretions such as saliva, tears and gastrointestinal fluid. Most antibody molecules are found in the γ-globulin fraction of serum (Fig. 11.15) as can be demonstrated by electrophoresing a small volume of serum before and after immunization.

In humans there are five classes of immunoglobulin: IgM, IgG, IgA, IgE and IgD. These differ in structure, distribution and biological properties (Table

---

Exercise 3

One assay for testosterone levels in infertile men utilizes an antibody to testosterone. Explain how it is possible to raise an antibody against such a small molecule. Why does the production of an antibody against insulin for use in an immunoassay not present similar problems?

Table 11.7 *Comparison of immunoglobulin classes of man*

Antibody class	M_r	Heavy chains	Serum concentration (mg 100 cm^{-3})	Carbohydrate (%)	Complement fixation	Placental transmission
IgM	900 000*	μ	50–200	12	Yes	No
IgG	150 000	γ	800–1600	3	Yes†	Yes
IgA	(160 000)$_2$	α	140–400	8	No‡	No
IgE	200 000	ε	1.7×10^{-3}–4.5×10^{-2}	12	No	No
IgD	185 000	δ	0–40	13	No	No

* IgM consists of five, four-chain units each with an M_r of 180 000. The heavy chains are significantly larger than those of IgG.
† There are four subclasses of IgG and all but one fix complement.
‡ Aggregated IgA can stimulate the alternative pathway.

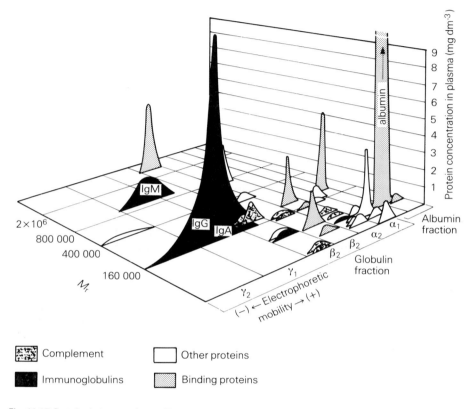

Fig. 11.15 Proteins in human plasma. The three-dimensional graph is obtained by plotting electrophoretic mobility against protein concentration and M_r. The albumin fraction is the fastest moving when electrophoresed at pH 8.6. Redrawn from *Biochemistry and Molecular Biology*, Open University third level course. The Open University Press, Milton Keynes.

11.7). All five classes are found in plasma although IgA is more important as the 'secretory' antibody, which protects the 'external' surfaces of the body. IgG and IgM are the most prevalent antibodies in the blood, and provide the main humoral defence against bacteria. In a primary immune response, when there has been no prior exposure to the immunogen, most antibody is of the IgM class (Fig. 11.16). In a secondary response, following a second or subsequent exposure to the same immunogen, IgG makes up the bulk of the antibody.

IgE, which is normally found in low concentrations in the plasma, is important in immune responses towards multicellular parasites. IgE can trigger the local release of mediators from mast cells in the gut, resulting in inflammation and contraction of the smooth muscle in the gut, helping to

Reference Tonegawa, S. (1985) The molecules of the immune system. *Scientific American*, **253**, 122–31. Scientific American articles are always clearly presented and easy to read; this is no exception.

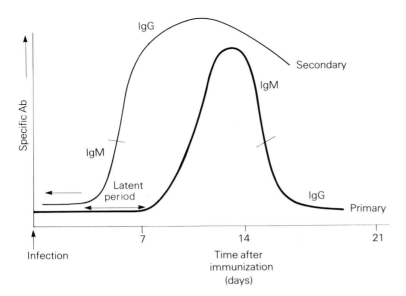

Fig. 11.16 Primary and secondary immune responses. The primary response occurs on first contact with an immunogen. The secondary response occurs in a previously immunized individual. There are differences in the length of the latent period, the amount of antibody produced, the duration of the response and the class of antibody produced. In addition, the antibody in a secondary response has a higher affinity for the epitope.

dislodge intestinal parasites. In addition, an **eosinophil chemotactic factor** attracts eosinophils which attack parasites, particularly if the parasites are coated with antibody.

The structure of antibody molecules is most conveniently discussed with reference first to antibodies of an individual class, in this case IgG.

Structure of IgG

The IgG molecule is a glycoprotein with an M_r of 150 000. It is made up of four polypeptide chains: two identical heavy (H) chains (M_r 50 000) and two identical light (L) chains (M_r 25 000). The chains are held together by disulphide bridges joining L to H and H to H as shown in Fig. 11.17.

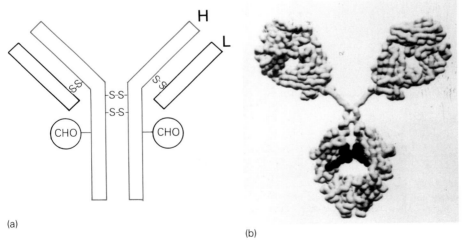

(a)

(b)

Fig. 11.17 (a) Schematic structure of an IgG molecule showing heavy(H) and light(L) chains. The positions of disulphide bonds (–S–S–) and carbohydrate (CHO–) are indicated. (b) Computer-drawn model of an IgG molecule. The darker spheres represent the carbohydrate portions. Courtesy of Dr R.S.H. Pumphrey, St Mary's Hospital, Manchester, UK.

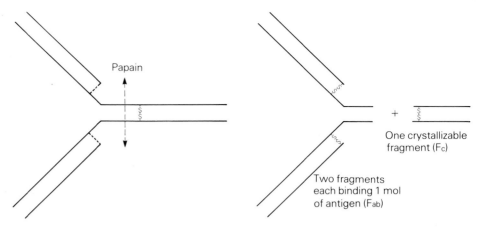

Fig. 11.19 Papain digestion of IgG. Papain cleaves the molecule in 'front' of the disulphide link as shown. This produces fragments which bind antigen Fab and fragments which can be crystallized (Fc).

One crystallizable fragment (Fc)

Two fragments each binding 1 mol of antigen (Fab)

Papain

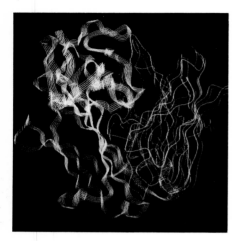

Fig. 11.18 Computer-generated model of the polypeptide backbone of a Fab fragment. Note the extensive β-sheets. Courtesy of Dr C. Freeman, Polygen, University of York, UK.

☐ In humans, IgG is the only immunoglobulin which crosses the placenta. Maternal IgG therefore protects newborn infants against a wide range of bacterial and viral infections. Maternal antibody remains in the baby's circulation for 3–6 months after birth, providing protection during this critical period before the infant's immunoglobulins are produced.

Each IgG molecule has two identical sites for binding to a specific epitope. These sites are situated at the amino-terminal portion of the L and H chains. Each site is on a **Fab (fragment antigen-binding)** region (Fig. 11.18). This name was derived from classic experiments in which the IgG molecule was fragmented with proteolytic enzymes (Fig. 11.19). The carboxy-terminal end of the heavy chains is known as the **fragment crystallizable (Fc)** region. This region determines various biological activities, such as the ability of the molecule to cross the placenta. The **hinge** region of the heavy chain is rich in proline residues (Fig. 11.20). The presence of proline confers flexibility and may enable the two Fab regions to bind to epitopes which are separated at variable distances (Fig. 11.20).

The heavy chain of an antibody molecule determines its class. Thus, all IgG heavy chains are called γ chains. The other classes have μ (IgM), α (IgA), δ (IgD) and ε (IgE) heavy chains respectively. Only two types of light chain are found in immunoglobulins irrespective of class and these are designated κ

(a)

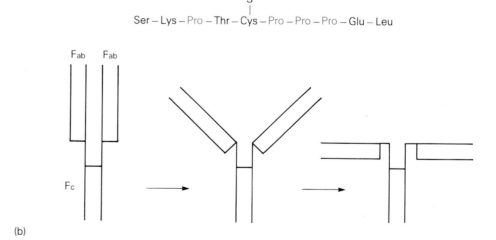

(b)

Fig. 11.20 The immunoglobin 'hinge'. (a) The amino acid sequence at the hinge region, determined for an IgG molecule. Note the four proline residues around the disulphide link. (b) The flexibility at the hinge enables the Fab regions to separate to varying degrees.

Ser – Lys – Pro – Thr – Cys – Pro – Pro – Pro – Glu – Leu

S
|
S

Fab Fab

Fc

Reference Burton, D.R. (1990) Antibody: the flexible adaptor molecule. *Trends in Biochemical Sciences*, **15**, 64–9. This short article uses computer-generated models to discuss the interaction of antibodies with effector cells such as phagocytes.

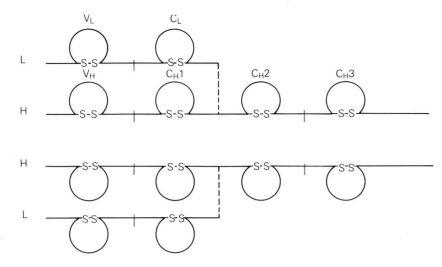

Fig. 11.21 The domains of an IgG molecule. The limits of each domain are indicated by red dashes. Each domain consists of approximately 110 amino acid residues and includes a loop of 60–70 amino acid residues.

and λ chains. A single IgG molecule may contain two κ chains or two λ chains but never one of each. Both the heavy and the light chains contain a regular series of intrachain disulphide bridges which effectively divide the chains into looped domains of approximately 110 amino acid residues (Fig. 11.21). The γ chain has four domains, whereas light chains have two.

Variable and constant regions

Antibodies are secreted by **plasma cells**. All antibody molecules produced by a single plasma cell have identical epitope-specificity. This specificity is conferred by the sequence of amino acid residues in the *Fab* region which is unique to antibodies produced by a single plasma cell. On the other hand, all IgG molecules eliminate immunogens by a very limited number of pathways after binding to it. Therefore, an antibody molecule has to be both different from, and similar to, IgG molecules produced by other plasma cells. Figure 11.22 shows how these opposing requirements are accommodated within the one molecular structure. Each γ chain has a region in which the amino acid sequence is essentially the same in all γ chains from the same animal. This **constant** region occupies approximately 75% of the molecule towards the carboxy-terminus. In contrast, the amino acid residues making up the first quarter of the amino-termini of both heavy and light chains are highly variable in terms of their sequence. However, all antibodies produced by the same plasma cell have an *identical* sequence in this **variable** (V) region. Both κ and λ light chains have a variable and a constant region, each comprising 50% of the polypeptide chain. The antibody combining site is made up of a V region of the γ chain, combined with the V region of the light chain. Within this antibody-combining site are restricted regions which are **hypervariable**, that is, regions where the amino acid residues are even more variable between antibody molecules than elsewhere in the V region. These regions, sometimes known as 'hotspots', are involved in constructing the different types of antibody-combining sites.

Folding of the antibody combining site forms a cleft into which an epitope fits (Fig. 11.23). The better the fit, the greater is the strength of binding or 'affinity' between the two. Assuming that the epitope 'fits' well into the antibody-combining site, other forces strengthen the binding between the epitope and the antibody. These forces include hydrophobic interactions,

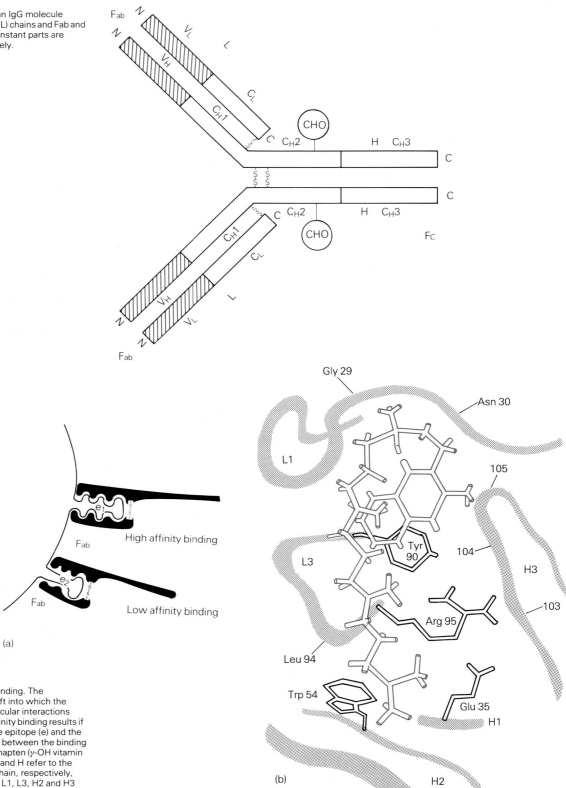

Fig. 11.22 Representation of an IgG molecule showing the heavy(H) and light(L) chains and Fab and Fc regions. The variable and constant parts are indicated by V and C, respectively.

Fig. 11.23 Antigen–antibody binding. The antigen-binding site forms a cleft into which the epitope fits. Non-covalent molecular interactions stabilize the binding. (a) High affinity binding results if there is a good 'fit' between the epitope (e) and the Fab region. Note the difference between the binding of e_1 and e_2. (b) The binding of a hapten (γ-OH vitamin K) to a specific IgG molecule. L and H refer to the regions of the light and heavy chain, respectively, which interact with the hapten. L1, L3, H2 and H3 refer to hypervariable regions of those chains. Redrawn from Zouhair-Atassi, M., Van Oss, C.J. and Absolom, C.J. (1984) *Molecular Immunology*, Marcel Dekker, New York.

Van der Waal's forces, hydrogen bonding and ionic interactions. This non-covalent binding between the antibody-combining site and the epitope can be reversed under the appropriate conditions. For example, the complex can be dissociated by lowering the pH to below pH 2.5.

Other classes of immunoglobulin

All other immunoglobulins conform to the basic four-chain structure of IgG, although they have different heavy chains and different degrees of glycosylation (Fig. 11.17b and 11.24). The other main difference between some classes is the number of these four-chain structures making up the final molecule. For example, IgM is a pentamer made up of five, four-chain structures which are joined together by a protein known as the **J (joining) chain** (Fig. 11.25). IgA exists as a dimer in serum and in secretions. Here too a J-chain joins the structure together. Secretory IgA does, however, differ from the plasma molecule because the former is also associated with a glycoprotein known as the **secretory component** (M_r 70 000). Addition of secretory component appears to confer protection against the 'hostile' environments at some secretory surfaces. For example, IgA is found in the gastrointestinal tract alongside several proteolytic enzymes and the secretory component protects the secretory IgA from proteolysis.

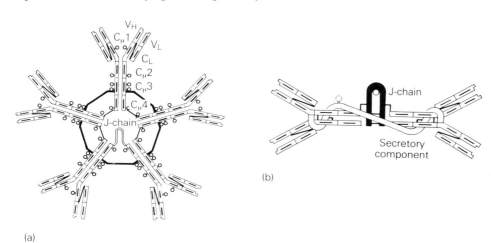

(a)

Fig. 11.25 Structure of IgM and IgA. (a) IgM is a pentamer of the basic four-chain immunoglobulin molecule. See also Table 11.7. The five units are joined to each other via disulphide links, and also via the J-chain as shown. Note the high degree of glycosylation, indicated by Ϙ (b) IgA is a dimer of the four-chain structure, the two units joined through the J-chain. In secretions (but not in blood) this dimer also has a 'secretory component', a protein (M_r, 70 000). This appears to protect the antibody from proteases in secretions. Redrawn from Roitt, I. *et al.* (1989) *Immunology*, 2nd edn, Gower, London.

The role of antibodies

The first, and most important, function of an antibody molecule (Fig. 11.26) is to combine specifically with the epitope it recognizes and then to signal to other components of the immune system that this is a foreign invader to be eliminated. The destruction of the immunogen occurs by one or more of a limited number of pathways outlined below. The class of antibody, and, to a lesser extent, the nature of the immunogen, determines the way in which the immunogen is eliminated from the body.

1. IgG, IgM and IgA are able to agglutinate cells and to precipitate soluble immunogens from solution. These secondary effects take place because

Exercise 4

In the description of antibody–antigen binding, it has been suggested that hydrogen bonding and ionic interactions contribute only in a small way to the strength of binding. Suggest a reason for this assumption. (Clue: look at the pH needed to dissociate them.)

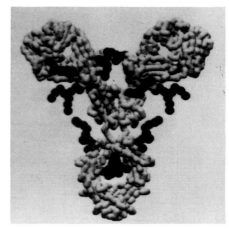

Fig. 11.24 Computer-drawn model of an IgE molecule. The darker spheres represent the carbohydrate portions. Note the extensive glycosylation (darker spheres) compared with an IgG molecule (Fig. 11.17b). Courtesy of Dr R.S.H. Pumphrey, St Mary's Hospital, Manchester, UK.

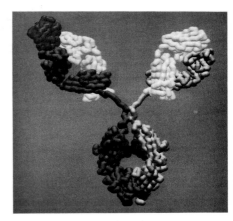

Fig. 11.26 Molecular model showing the binding of an IgG molecule to its antigen (upper left). In this case the antigen is a lysozyme molecule. Courtesy of Dr R.S.H. Pumphrey, St Mary's Hospital, Manchester, UK.

each antibody has at least two combining sites which can cross-link cells or macromolecules (Fig. 11.27). Agglutination of bacteria, or precipitation of possibly harmful proteins, serves to localize an infection and also allows these immunogens to be more easily removed by phagocytes.

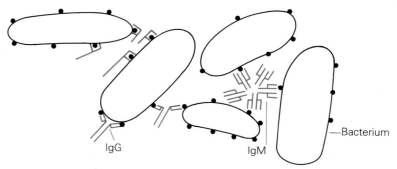

Fig. 11.27 Agglutination by antibodies. Depending on the class, antibodies have two, four or ten Fab sites. Antibodies binding to epitopes on different cells can therefore bind, or agglutinate these cells. IgM, for obvious reasons, is the most efficient at agglutination.

2. The binding of IgG or IgM to immunogens may activate the complement sequence which results in lysis of cellular immunogens. An outline of the complement sequence leading to cell lysis is given in Fig. 11.28.

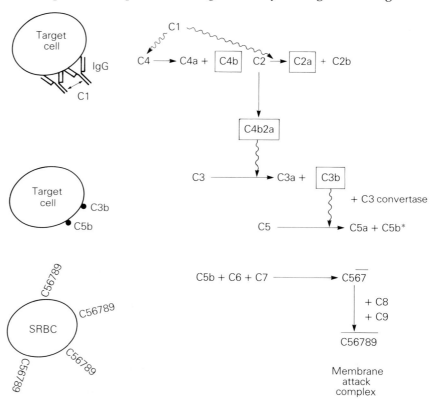

Fig. 11.28 The classical pathway for complement activation. C1 (the first complement component) binds to the F$_c$ regions of two adjacent IgG molecules on a target cell. C1 acquires hydrolytic enzymic activity and cleaves both C4 and C2. The fragments reassemble to form an enzyme, C4b2a which cleaves C3. The C3b fragment binds to the target cell and, together with C4b2a, cleaves C5, C5b also binds to the target cell and the remaining components, C6, 7, 8 and 9 add on, to form a large complex, the **membrane attack complex** (MAC), which 'punches' a hole in the membrane. Because the initial reactions are enzymic, the sequence is an enzyme cascade so that thousands of MACs may result from a single activated C1, resulting in an enhanced response to the infection. An alternative pathway is activated directly by bacterial cell wall components.

Intermediate products of the complement sequence, including the glycoproteins C3b, C3a, C5a, stimulate the uptake of antibody–antigen complexes by neutrophils and macrophages.

3. Neutrophils and macrophages have cell surface receptors for the Fc region of antigen-bound IgG so that this antibody can form a 'bridge', binding the immunogen to the phagocytic cell (Fig. 11.29), resulting in a more efficient phagocytosis.

4. Large granular lymphocytes, possibly the same population as the NK cells, also have Fc receptors for IgG. Although these cells are not phagocytic, binding to an IgG-coated cell triggers the release of toxic granules which destroy the cell. This process is known as **antibody-dependent cellular cytotoxicity (ADCC)**. Eosinophils are able to kill IgG-coated parasite larvae by a related mechanism.

While the binding of antibody to an epitope is determined by the Fab region, the destruction of the immunogen, as well as a number of other biological properties, such as the serum half-life and placental transmission are determined by sites on the Fc region, the structure of which determines the antibody class.

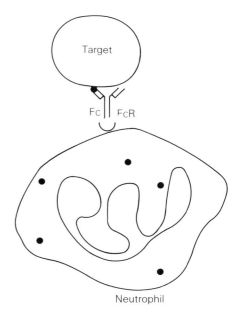

Fig. 11.29 Binding of phagocytic cells to IgG-coated target cells. Both monocytes and neutrophils (shown here) have surface receptors for the Fc region of IgG. Thus, an IgG-coated target cell is bound firmly to a phagocytic cell and phagocytosis is facilitated.

11.6 Cells and tissues of the specific immune response

Humoral and cell-mediated immunity, that is, the production of specific antibodies and cytotoxic cells, are each brought about by **small lymphocytes**. These are indeed the smallest leukocytes, being only 7–9 μm in diameter. They are easily recognized in blood smears by the large and densely-staining nucleus which constitutes most of the cell (Fig. 11.30). A thin rim of cytoplasm, containing few organelles, surrounds the nucleus. Small lymphocytes are also found in large numbers in the **lymphoid tissues** (Fig. 11.31), of which two kinds are recognized:

1. **Primary** lymphoid tissues in which the lymphocytes develop. They include the bone marrow, the thymus, which is present in all vertebrates, and the bursa of Fabricius, a lymphoid organ associated with the hind-gut of birds only.

2. **Secondary** lymphoid tissues in which the lymphocytes fulfil their immune functions. They include the spleen, the lymph nodes and the tonsils. In addition lymphoid aggregates are associated with mucosal membranes and are known collectively as mucosa-associated lymphoid tissue **(MALT)**.

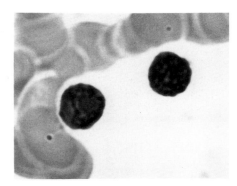

Fig. 11.30 Small lymphocytes in a blood smear (× 750). Note the densely-staining nucleus which occupies most of the cell.

BONE MARROW contains **stem cells**, which are constantly dividing. The daughter cells are initially capable of giving rise to any of the cells of the blood, or to the **megakaryocytes** which give rise to the blood platelets (Fig. 11.32). However, the new cells become increasingly channelled into a developmental pathway, and this pathway is influenced by various glycoprotein growth factors known as **colony stimulating factors (CSFs)**. Some CSFs are produced by cells in the marrow itself while others come from outside. In this way the bone marrow can be directed to produce particular types of blood cells as and when they are needed. The bone marrow exerts most influence on the

---

stem cells: *continually dividing cells, such as those in the bone marrow. Bone marrow stem cells are said to be pluripotent, because their progeny can give rise to any of the cells in the blood.*

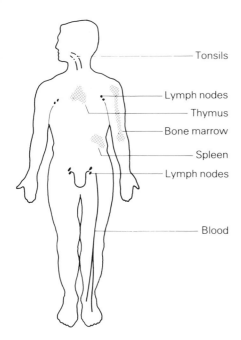

Fig. 11.31 The lymphoid tissues. The thymus and the bone marrow are important for the development of lymphocytes. The remaining lymphoid tissues are concerned with the elimination of immunogens. Redrawn from Dawson, M.M. (1987) *Introducing Immunology*. Biological Sciences series, continuing nurse education, Crown copyright.

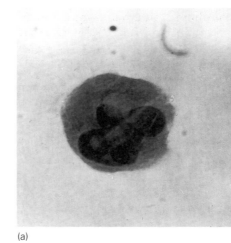
(a)

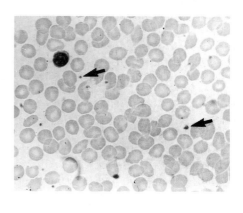

Fig. 11.32 Photomicrographs of (a) a megakaryocyte (×320) and (b) blood platelets (examples arrowed) (×320). Courtesy of Dr L.H. Seal, Department of Biological Sciences, The Manchester Metropolitan University, UK.

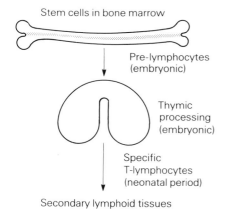

Fig. 11.33 Development of T lymphocytes. The bone marrow and the thymus are important in embryonic life. In the neonatal period mature T lymphocytes leave the thymus for the secondary lymphoid tissues.

production of small lymphocytes during embryonic development. During this period lymphocyte precursors leave the bone marrow and enter other primary lymphoid organs, that is, the thymus and, in birds, the bursa of Fabricius.

THE THYMUS is a primary lymphoid organ found in all vertebrates. In mammals it is a bilobed organ found in the middle of the chest, just above the heart. The thymus is relatively large in babies and reaches its maximum size at puberty. After puberty the thymus slowly shrinks due to a steady loss of cells, a process known as involution. The role of the thymus was established in 1968, following experiments involving the removal of the thymus from progressively younger animals. In mice, removal of the thymus from an adult animal has relatively little effect, but **neonatal thymectomy** results in a severe 'wasting' syndrome resulting in animals which have low weight; are highly susceptible to viral and fungal infection; have low numbers of lymphocytes in the blood; fail to reject skin grafts; have underdeveloped lymph nodes and spleen; and have reduced levels of antibodies in the blood plasma.

Failure to reject skin grafts, and to fight viral infection is indicative of a failure of cell-mediated immunity. It seems, therefore, that the thymus influences the production of lymphocytes which are responsible for this type of specific response. Since these lymphocytes are dependent on the presence of an intact thymus, they are known as **T (thymus-dependent) lymphocytes**. A scheme for the development of T lymphocytes is shown in Fig. 11.33. T lymphocytes leave the thymus during the neonatal period and settle in the secondary lymphoid tissues.

There are at least two distinct populations of T lymphocytes. The first is the precursor of the **cytotoxic T lymphocyte (CTL)** and for this reason is designated T_C. The second type has the potential to produce a range of proteins called **lymphokines**, which are essential for the development of both humoral and cell-mediated immunity. These T cells, which play such a central role in the immune response, are known as **helper** cells or T_H. The loss of T_H cells in mice whose thymus was removed at birth explains the reduced levels of antibodies found in these animals.

neonatal thymectomy: *the removal of the thymus within a few days of birth.*

Reference Playfair, J.H.L. (1987) *Immunology at a glance*, 4th edn, Blackwell Scientific, Oxford, UK. A useful aid to understanding the immune system, relying heavily on diagrams to explain the fundamental concepts.

THE BURSA OF FABRICIUS is an avian lymphoid organ that forms a pouch-like structure connected to the rectum, close to the anus (Fig. 11.34). Like the thymus, the bursa undergoes involution following the onset of sexual maturity. In 1968 it was discovered that the development of the bursa could be severely impaired if eggs containing developing chick embryos were injected with the hormone, testosterone. Birds which have been 'hormonally bursectomized' have severely impaired immune responses as indicated by a high incidence of bacterial infections; low numbers of lymphocytes in the blood (reduced by 30%); negligible levels of antibodies in the blood; reduction in the mass of secondary lymphoid tissue; and a failure in these animals to make antibodies after immunization. However, these birds *were* able to reject skin grafts and combat viral infection. This suggests an impairment of humoral, rather than cell-mediated immunity. The bursa, therefore, influences the precursor lymphocytes derived from the bone marrow, to develop into **B** (**bursa-dependent**) lymphocytes which are responsible for humoral immunity.

Mammals do not have a lymphoid organ similar to the bursa yet they do have the equivalent of B lymphocytes which are responsible for humoral immunity. It seems that, in mammals, the bone marrow, together with the fetal liver, influences the development of B lymphocytes which emerge, preprocessed, and travel directly to the secondary lymphoid tissues (Fig. 11.35).

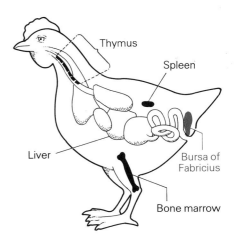

Fig. 11.34 The bursa of Fabricius is a primary lymphoid organ located in the hindgut of birds. It is involved in the development of B lymphocytes. In mammals this role is performed by the bone marrow. Redrawn from Cooper, M.D. and Lawton, A.R. (1974) The development of the immune system, in *Immunology: Readings from Scientific American*, W.H. Freeman, San Francisco.

Role of the secondary lymphoid tissues

Lymphocytes which settle in a particular region of the lymphoid tissue, a lymph node for example, do not remain in that tissue throughout their lifetime. Lymphocytes are constantly recirculating, via the blood and **lymph**, between the different lymphoid tissues (Fig. 11.36). The number of lymphocytes which enter the blood via the subclavian vein in the neck during the course of a day is sufficient to replace the entire circulating pool of lymphocytes several times over. Recirculation of lymphocytes ensures that specific lymphocytes stand a good chance of meeting an appropriate invading immunogen. The lymph nodes and the spleen together form the greatest mass of lymphoid tissue.

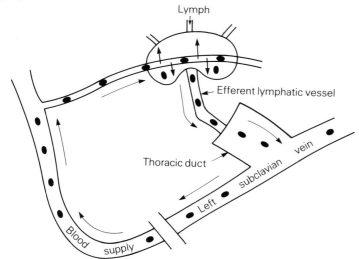

Fig. 11.36 Lymphocyte recirculation. Lymphocytes leave the blood supply in the lymphoid tissues, crossing the walls of specialized blood vessels. Lymphocytes leave the lymph node in the efferent lymph. They re-enter the blood supply at the junction of the major lymphatic vessel, the thoracic duct, and the left subclavian vein in the neck.

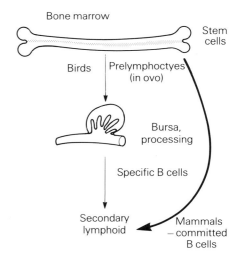

Fig. 11.35 Development of B lymphocytes. In birds, precursors for B cells are processed in the bursa. In mammals this processing takes place within the bone marrow.

lymph: *the fluid, derived from the blood plasma, which drains from the tissues into lymphatic vessels, through lymph nodes, and eventually re-enters the blood.*

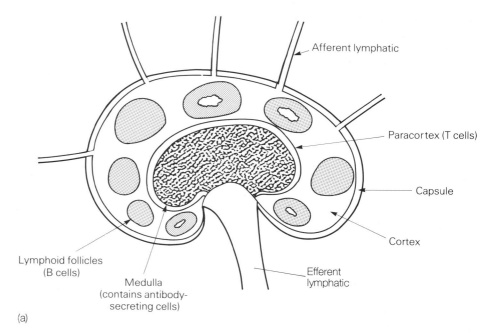

(a)

(b)

Fig. 11.37 The lymph node. (a) The cortex contains B cells in dense follicles; the paracortex contains T lymphocytes; the medulla contains plasma cells. Macrophages are found throughout the node. (b) Photomicrograph of a lymph node (×80). Courtesy of M.J. Hoult, Department of Biological Sciences, The Manchester Metropolitan University, UK.

LYMPH NODES are small kidney-shaped organs which are distributed throughout the body (Fig. 11.37). They 'filter' lymph coming from tissues, and macrophages, found throughout the node, phagocytically remove any immunogens which have entered the tissues. Afferent vessels pierce the capsule of the lymph node. Lymph, draining from the tissues, enters through these vessels and trickles through the body of the node.

The lymph node has a high degree of anatomical organization, with B and T cells being distributed in distinct regions. The cortex contains predominantly B cells in dense lymphoid follicles whereas the paracortex contains predominantly T cells. In this region, too, are specialized venules, where lymphocytes can cross the vessel walls and enter the lymphoid tissue. These lymphocytes can remain in the node or leave via the efferent vessel. Any lymphocytes which are stimulated by an immunogen in the lymph will proliferate and leave via the same route. When B cells are stimulated by the appropriate immunogen, they proliferate and develop into antibody-producing **plasma cells** (Fig. 11.38). Plasma cells do not recirculate but settle in the lymphoid tissues; in the lymph node they are found in the medulla. A specific type of antibody is secreted by individual plasma cells into the efferent lymph which then enters the blood via the thoracic duct.

THE SPLEEN is the largest single mass of lymphoid tissue in the body and is situated in the upper left abdomen, behind the stomach (see Fig. 11.8). The spleen is supplied by a major blood vessel, the splenic artery so that, unlike the lymph nodes, it is bright red in colour.

The spleen has various non-immunological functions such as the removal of old red blood cells. As part of the immune system, it serves to 'filter' the blood, bringing any blood-borne immunogen into contact with macrophages and specific lymphocytes. The spleen is made up of two types of tissue, the red pulp and the white pulp. The white pulp (Fig. 11.39) consists of sheaths of lymphoid tissue surrounding arteriolar branches of the splenic artery. T and B cell areas are found within this **peri-arteriolar sheath (PAS)**. The red pulp contains open 'lakes' of blood or **sinuses** within a fibrous meshwork containing macrophages and antibody-secreting plasma cells. Lymphocytes entering the spleen in the splenic artery may cross the endothelium of blood vessels within the PAS and remain there or travel to the red pulp, gaining access to venules.

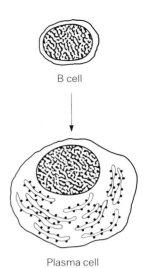

B cell

Plasma cell

Fig. 11.38 After stimulation with an antigen B lymphocytes develop into plasma cells, which have all the morphological features of a protein-secreting cell. Note, for example, the extensive endoplasmic reticulum.

☐ The spleen also acts as a source of stem cells for leukocytes in the embryo and as a reserve of blood in adult life.

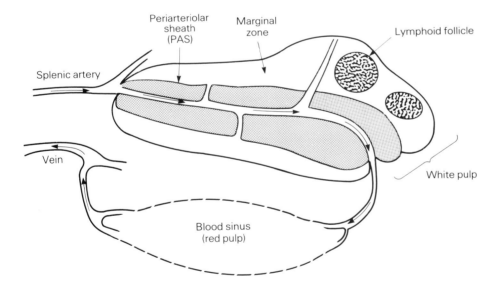

Fig. 11.39 The spleen. The splenic artery branches into smaller arterioles which are surrounded by a sheath of lymphocytes, most of which are T cells. On the periphery of this cylinder is the marginal zone, containing B cells in follicles. The red pulp consists of blood sinuses, containing macrophages and plasma cells, through which the blood is filtered before it enters venules. Redrawn from Roitt, I. (1989) *Essential Immunology*, 6th edn, Blackwell Scientific, Oxford.

11.7 Clonal selection

Each small lymphocyte is capable of binding to a particular epitope. The specificity of binding resides in the protein epitope receptors present in the membranes of the small lymphocytes. The cells acquire this specificity during the period of processing in the primary lymphoid tissues. When a small lymphocyte binds to a specific epitope on an immunogen, a sequence of events is initiated which results in the proliferation and differentiation of that lymphocyte. The result is a *clone* of **effector cells** (Fig. 11.40), all with identical specificity, and which mediate one aspect of the specific immune response.

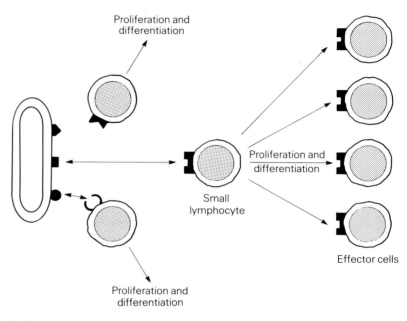

Fig. 11.40 Clonal selection. Each small lymphocyte has receptors for a single type of antigenic determinant. When stimulated in an appropriate manner by this determinant, the small lymphocyte undergoes a period of division to produce a clone of identical cells. These cells differentiate to form effector cells.

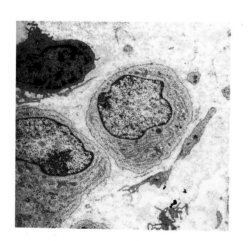

Fig. 11.41 Electronmicrograph of a plasma cell.

T_H CELLS respond to the epitope by proliferating to form a clone of helper cells. These produce lymphokines which are essential for the proliferation of B and T_C cells.

T_C CELLS respond by developing into a clone of cytotoxic T cells with the capacity to kill, specifically, virus-infected cells. In addition, they produce a limited range of lymphokines including interferon γ (IFNγ) which enhances the activity of macrophages.

B LYMPHOCYTES respond by proliferating and developing into plasma cells (Fig. 11.38). Each clone of B lymphocytes secretes an antibody of a single specificity, that is, it is specific to a single epitope. The full pathway from B cell to plasma cell requires several of the lymphokines produced by T_H cells.

MEMORY CELLS are those members of a proliferating clone which do not differentiate into effector cells such as plasma cells (Figs 11.38 and 11.41). Thus an immunized animal has a greater pool of lymphocytes bearing receptors for that same immunogen. This is the reason why a heightened response is produced on second contact with the immunogen.

clone: a population of cells which have all arisen from one cell by repeated cell division.

Immunized mouse

↓

Spleen

↓

Single-cell suspension

Myeloma cells in culture

↓

+ PEG

↓

Hybridomas

↓

Selection and screening

↓

Cloning

A specific immune response in an animal is said to be polyclonal. In terms of antibody production, this means that several B cells are stimulated at the same time, by different epitopes on the immunogen. The antibodies which are produced are heterogeneous with respect to class, affinity and epitope specificity. Monoclonal antibodies are the products of a single clone of plasma cells all derived from a single stimulated B cell. Normal plasma cells soon die in culture but since 1975 (see also *Molecular Biology and Biotechnology*, Chapters 9 and 10) it has been possible to grow plasma cells from an immunized animal in culture providing that they are fused with a cell from an 'immortalized' cell line, such as a plasma cell tumour or myeloma. A single plasma cell, fused with the myeloma to form a **hybridoma**, can be grown indefinitely in culture, forming a clone which secretes the desired antibody into the culture supernatants (see Fig.).

Monoclonal antibodies have many uses. They can be used both to treat and diagnose infectious disease, or, because of their exquisite sensitivity, used in immunoassays to measure extremely low concentrations of antigens. Monoclonal antibodies to tumour antigens have been used to diagnose cancers and hold out the prospect for the treatment of tumours, particularly when combined to a toxic chemical such as ricin. It is hoped that they will 'home in' on the tumour target and thus avoid the damage to normal tissue which occurs with conventional cancer treatments.

Production of monoclonal antibodies. Mice are immunized with the desired immunogen to stimulate the production of antibodies. When tests have established that antibody is being produced, the mice are killed and spleen cells fused to cultured myeloma cells, using polyethylene glycol (PEG) to make the cell membranes fuse. The hybrid myelomas (hybridomas) are cultured, screened, and those cultures producing a desired antibody are cloned and expanded.

Most immunogens have several epitopes, each of which stimulates a specific, small lymphocyte to develop into a clone. For this reason, the normal immune response is said to be **polyclonal**, as it results in the production of many clones of plasma cells with specificities for different regions (that is, epitopes) of the immunogen.

11.8 Antigen-presenting cells

Helper T lymphocytes are not able to respond *directly* to an immunogen. Instead, they respond to immunogen which has been taken up by specialized **antigen-presenting cells** (APCs). These cells then expose epitopes of the immunogen on their cell membranes (Fig. 11.42). A list of known APCs is given in Table 11.8. APCs engulf the immunogen by phagocytosis, and modify it in one of several ways. These modifications have the effect of exposing epitopes of the immunogen. The modifications may include unfolding of protein molecules to reveal hidden determinants, or proteolytic fragmentation of large immunogens. The modified protein, or antigenic

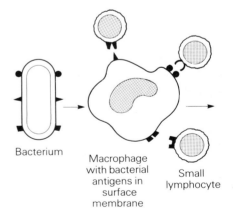

Bacterium

Macrophage with bacterial antigens in surface membrane

Small lymphocyte

Fig. 11.42 Antigen presenting cells take up immunogens, such as bacteria, by phagocytosis. The immunogens are modified, by enzymatic breakdown and antigenic fragments inserted into the APC membrane where they stimulate helper T cells which have the correct receptors.

fragments are then inserted into the cell surface membrane of the APC where it stimulates T_H cells which have the appropriate receptors. When the modified antigen is inserted into the APC cell membrane, it is associated with other 'normal' APC membrane proteins. These 'self' proteins are encoded by genes within by a chromosomal region known as the **major histocompatibility complex (MHC)**. The epitope receptors on T_H cells can only recognize the antigen if it is in association with the MHC proteins. The MHC is important in determining immune responses and is discussed in greater detail in Section 11.9.

As well as presenting the antigen on its cell surface, the APC also produces interleukin 1, which stimulates the T_H cell to proliferate and which also stimulates inflammation.

The lymphokine cascade

Presentation of antigen to T_H cells stimulates a series of reactions known as the **lymphokine cascade** which results in the production of a specific immune response (Fig. 11.43).

A specific T_H lymphocyte responds to antigen on the surface of the APC and to the IL-1 which they produce. The T_H cell produces a lymphokine, known as interleukin 2 (IL-2, Fig. 11.44) which stimulates the T_H to release other lymphokines including IL-3, IL-4, IL-5, IL-6 and IFNγ. B lymphocytes also have specific cell surface receptors and can thus bind to immunogens. These B cells can act as APCs, interacting with the T_H cell and stimulating the release of

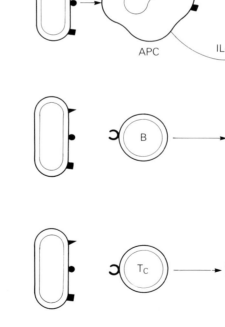

Fig. 11.43 The lymphokine cascade. The APC presents antigen to the helper T cell which responds by producing lymphokines and proliferating. B and T_c cells respond to antigen directly by expressing receptors for IL-2. IL-2 produced by the T_H stimulates proliferation of both B and T_c cells. The B cell, however, requires other lymphokines for complete differentiation into plasma cells. IL-2R, IL-2 receptor; CTL, cytotoxic T lymphocyte.

Fig. 11.44 Crystals of interleukin 2. Photograph courtesy of Dr Chiaki Sano, Technical Dept, Kyushi Factory, Ajinomoto, Japan.

interdigitating cells: relatively large cells with multiple processes found in the paracortex of the lymph node. They act as antigen-presenting cells. The large surface area enables a single interdigitating cell to interact with several lymphocytes simultaneously.

In 1981, doctors in the United States were alerted to an epidemic of a new and fatal syndrome occurring amongst homosexual men. This **acquired immune deficiency syndrome** (AIDS) was characterized by a rare and severe form of pneumonia, and other opportunistic infections. In addition, the same population had an increased incidence of an aggressive form of Kaposi's sarcoma, a previously rare and slow-growing tumour of the skin and blood vessels. All these disorders are indicative of an almost complete failure of the immune system which leads to death, often within months of diagnosis. The modes of transmission of the disease, sexual intercourse, injection of blood products, dirty syringes, pointed to an infectious agent which, by 1983, had been identified as a retrovirus (see also *Molecular Biology and Biotechnology*, Chapters 1 and 3). After several international disagreements on nomenclature, this virus is now called the **human immunodeficiency virus** (HIV) (Fig.).

A retrovirus is an RNA virus which uses reverse transcriptase to make a DNA copy of its genome. This DNA is integrated into the host DNA and used to promote replication of the virus. HIV infects the T helper cell, via a receptor called the CD4 glycoprotein present on these cells. Viral replication leads to lysis and severe depletion of the T cell population. Since these cells are central to the immune response the infected individual becomes susceptible to a wide range of diseases. The presence of Kaposi's sarcoma is thought to be related to latent infection with cytomegalovirus (CMV).

HIV has spread rapidly throughout the world and is endemic in several countries. The large numbers of carriers presents a daunting prospect for the future medical services of many countries since most carriers eventually develop AIDS. At present there is no cure for AIDS, although drugs such as AZT (azidothymidine) can control replication of HIV to some extent. However, this treatment is sometimes associated with severe side-effects. The most hopeful prospects are for the development of a vaccine which will prevent further spread. Even so, by this time millions of people are likely to be infected.

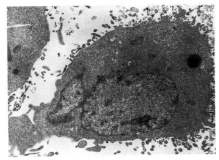

(a)

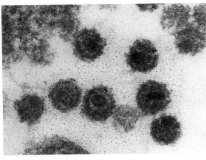

(b)

The human immunodeficiency virus. (a) The virus is emerging from an infected cell. Examples are indicated by arrow. (×3220). (b) At higher magnification (×87 400) the viral particles are clearly visible. Courtesy of Dr Alan Curry, Public Health Laboratory, Manchester, UK.

the lymphokine cascade. In addition, binding of specific epitopes to B cells promotes the production of protein receptors for IL-2. The lymphokines secreted by T_H cells stimulate the proliferation and differentiation of the antigen-stimulated B cells into antibody-secreting plasma cells. This process whereby an antigen-stimulated B lymphocyte develops into a clone of antibody-secreting plasma cells requires IL-2, which stimulates lymphocyte proliferation, and IL-4, IL-5, and IL-6 which promote B-cell proliferation and differentiation. Other lymphokines, such as IFNγ may also influence the class of antibody produced.

Specific T_C cells react to antigen, such as viral proteins on the surface of infected cells, by producing receptors for IL-2. The IL-2 released by the helper cell then promotes proliferation and differentiation of T_C cell into cytotoxic T lymphocytes. IL-4 also stimulates the proliferation of all T cells. The specificity of the immune response resides in the fact that each lymphocyte has specific receptors for a given epitope. Thus, only those lymphocytes with the appropriate receptors will respond to the epitope by producing IL-2 receptors and therefore respond to the proliferative effects of IL-2.

Lymphokines also stimulate non-specific responses. For example, IL-2 and IFNγ both increase the activity of natural killer cells which kill some tumours and virus-infected cells in a non-specific manner. IFNγ activates macrophages, increasing their ability to phagocytose bacteria, kill bacteria and

Exercise 6

Interleukin-2 and interferon-γ have both been tested as possible anticancer agents. The effects of IL-2 are indirect whereas IFNγ can have both direct and indirect antitumour activity. Where are these effects likely to occur?

Reference Smith, K.A. (1990) Interleukin-2. *Scientific American*, **262**(3), 26–33. A clear, simple account of the role of IL-2 in the immune response.

The ability to count the relative proportions of T and B cells in blood is important for the diagnosis of several diseases such as immune deficiency and lymphocytic leukaemias (Table). In addition, the proportion of $T_H : T_C$ cells can be useful in the diagnosis of HIV infection and autoimmune diseases.

Immunofluorescence is used to stain membrane proteins which are found only in one population. For example, all B cells have membrane immunoglobulins. If they are incubated with an anti-immunoglobulin antibody, covalently attached to a fluorescent dye, only the B cells fluoresce when exposed to ultraviolet light. All mature T lymphocytes have a membrane glycoprotein called CD3 (or T3). Incubation of T cells with a mouse monoclonal antibody, anti-CD3, followed by a fluorescent anti-mouse immunoglobulin will enable the T cells to be counted. To distinguish between T_H and T_C cells, use is made of monoclonal antibodies to CD4, a glycoprotein, found on T_H cells and to CD8, a glycoprotein found on T_C cells. The usual ratio of CD4+ : CD8+ cells in health is 2 : 1.

Diseases where an abnormal lymphocyte count forms part of the diagnosis

Disease	Characteristic changes
Glandular fever	Increased numbers of CD8+ T_C cells
AIDS	Decreased numbers of CD4+ T_H cells
Chronic lymphatic leukaemia	Increased B cells
Acute lymphatic leukaemia	Increased T cells
Sex-linked agammaglobulinaemia	B cells absent
Di-George syndrome	T cells absent
Severe combined immune-deficiency	T and B cells absent

tumour cells and also to present antigens. IL-3 stimulates the proliferation of bone marrow stem cells so that they produce more leucocytes, particularly phagocytic neutrophils and monocytes. In addition, both IL-3 and IL-4 stimulate the differentiation of tissue mast cells which are important in inflammation and immune responses to multicellular parasites. IL-5 is also important in anti-parasite immunity because it stimulates the production and differentiation of eosinophils, which are able to kill the larvae of parasitic worms.

Lymphokines are currently of major interest in immunology and the list of lymphokines and activities described is by no means exhaustive.

11.9 The major histocompatibility complex

The major histocompatibility complex (MHC) is a region of genetic material which codes for proteins that regulate immune responses. These proteins are of fundamental importance in determining the ability of an animal to mount an immune response. The MHC contains three groups of genes known as Classes I, II and III. The Class I and II genes code for integral membrane proteins whereas Class III genes code for some serum proteins.

When T lymphocytes respond to an antigen, they also recognize cell surface proteins which are encoded by genes within the MHC. Antigen recognition by T cells is important when APCs are stimulating T_H cells and also for CTLs to destroy virus-infected cells. If the MHC proteins are not also recognized then the immune response does not occur. Therefore, the MHC is said to 'restrict' certain immune responses.

In humans, the MHC is known as the **HLA system**, the genes for which are located on chromosome 6 (Fig. 11.45). The Class I region is encoded by three genetic loci called A, B, and C, while the Class II or D region, is encoded by

□ HLA stands for Human Leukocyte Antigens; these proteins were originally thought to be restricted to leukocytes.

another three loci named DP, DQ and DR. Since there is a pair of chromosome 6 in diploid cells, an individual has two 'A' genes, two 'B' genes, etc. These alleles are **co-dominant** which means that the products of both alleles are expressed in cells. In the human population there are multiple alleles (or alternative forms) of each gene so that the chances of an individual having an identical complement of HLA genes is very small. The different alleles have been assigned numbers so that an individual's HLA 'type' for Class I proteins might be: A3, 10; B21, 27; C3, 5.

CLASS I GENE PRODUCTS are glycoproteins which are found on the surface of all nucleated cells (Fig. 11.46). Each Class I protein is a single polypeptide chain which is associated with a smaller protein called the β_2-microglobulin, which is invariant and is not encoded by the MHC. The proteins specified by different alleles differ in their amino acid composition at the amino-terminal end of the molecules.

Class I proteins restrict the activity of cytotoxic T cells. For example, CTL specific for the measles virus, will only kill cells infected with that virus, if the T cell and the virus-infected cell have the same Class I antigens. What this means is that the T cell must recognize self (MHC) proteins in order to react against the non-self (viral) proteins.

CLASS II GENE PRODUCTS are also glycoproteins. They are found on a restricted range of cells such as APC, B cells and activated (but not resting) T cells. Each Class II protein consists of two dissimilar polypeptide chains called α and β subunits (Fig. 11.46) both of which are encoded within the MHC. Most of the difference between Class II proteins coded by different alleles lies in amino acid sequence of the β chain.

Class II proteins restrict antigen presentation. This means that, when a T_H is presented with antigen on the surface of an APC, the cell must also recognize the Class II molecules on the APC membrane. Again, the T cell has to recognize both 'self' and 'non-self'.

☐ Numbers are assigned to newly discovered HLA antigens at the International Workshops of the HLA Nomenclature committee. Initially the number is prefixed by the letter 'w' (for workshop) until it is established beyond doubt that the number truly represents a new specificity. It is important to know the HLA types of the recipient and donor involved in a transplant since differences, especially in Class I antigens, stimulate rejection of the transplant. Identical twins, being derived from the same fertilized ovum, have an identical HLA type.

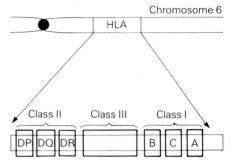

Fig. 11.45 The HLA system. The HLA system is located on chromosome 6. It contains three sets of genes coding for cell surface antigens (Classes I and II) and some plasma proteins (Class III). The letters refer to specific loci (see text).

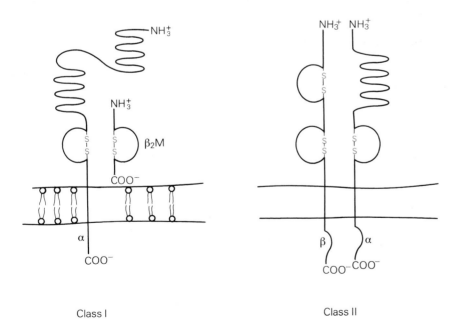

Class I   Class II

Fig. 11.46 Structure of Class I and II molecules. Class I proteins consist of a single polypeptide chain (M_r 44 000) together with a smaller protein, M_r 12 000 called β_2-microglobulin (β_2M), which is not encoded by the MHC. Class II proteins are dimers consisting of an α-chain and a β-chain. In both cases the carbohydrate portion has been omitted for clarity.

In recent years it has been recognized that the possession of certain HLA genes is associated with an increased incidence of particular diseases. The most well-known HLA association to date is between the HLA B27 gene and a disease called **ankylosing spondylitis**, a chronic and progressive crippling disorder of the spine. More than 90% of Caucasians with ankylosing spondylitis have the B27 antigen, an association so strong that HLA typing is used for diagnosing the disease. Although other HLA associations are not as strong as this, many do exist, involving a wide range of diseases including diabetes, rheumatoid arthritis and multiple sclerosis. The mechanism by which a particular HLA type renders an individual increasingly susceptible to an individual disease is not known, but several suggestions have been made. For example, a particular HLA Class II gene may code for a Class II protein which is less effective in stimulating helper T cells during presentation of a pathogen. Alternatively, a pathogen may 'mimic' an HLA antigen so that the immune system fails to recognize that pathogen as foreign. These are by no means the only explanations that exist for HLA disease associations and it is possible that the associations may be the result of several mechanisms.

11.10 Receptors on B and T lymphocytes

The specificity of B and T lymphocytes is determined by the epitope-specific receptors expressed in the membranes of these cells. It is the binding of the epitope to these receptors which ensures that the 'right' lymphocytes are stimulated. The receptors on B lymphocytes are immunoglobulin molecules with identical specificity to that of the secreted antibody. During the development of the B lymphocyte, the class of antibody which is expressed may vary, but the specificity never does. IgD is frequently expressed on B cell surfaces often together with IgM. The role of IgD seems to lie as a cell surface receptor rather than as a soluble immunoglobulin.

The structure of the receptors on T lymphocytes eluded immunologists for many years, this having finally been achieved through the powerful techniques of recombinant DNA technology. The T-cell receptor, called T_i, is a protein made up of two non-identical polypeptide chains termed α and β respectively. The amino-terminal ends of these chains show the variability associated with a specific antigen-binding structure. The T cell receptor is always associated with another protein, called CD3, which does not vary and is found on all mature T lymphocytes (Fig. 11.47). The T_i–CD3 complex is called the **T-cell receptor complex**. The binding of a specific epitope to the T-cell receptor causes an activation signal to be transmitted across the cell membrane. The CD3 glycoprotein appears to be involved in this signal transmission.

See *Molecular Biology and Biotechnology*, Chapter 9

□ The structures of many molecules in the immune system (Table 11.9), bear certain resemblances to the structures of immunoglobulins, particularly in their domain structure. For this reason, these molecules are grouped together as the 'immunoglobulin super-gene family' and are all thought to have evolved from a common ancestral gene.

Table 11.9 *Some members of the immunoglobulin supergene family*

Gene product	Distribution	Function
Immunoglobulin chains	B lymphocyte membranes	Antigen receptor
	Plasma cells	Soluble antibody
T-cell receptor chains	All mature T cells	Antigen receptor
Class I antigens of MHC	All nucleated cells	Control cytotoxic T cell responses
Class II antigens of MHC	Antigen-presenting cells	Control antigen presentation
	B cells	
	Activated T cells	
CD4 protein	Helper T cells	Binding to APC
CD8 protein	Cytotoxic T cells	Binding to virus-infected cell

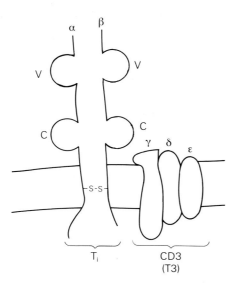

Fig. 11.47 The T-cell receptor complex. The specific antigen receptor, T_i, on T cells consists of two polypeptide chains, α and β (M_r 55 000 and 45 000, respectively), joined by a disulphide bond. Each chain has a variable and a constant region as shown. Each T_i is associated with another protein, CD3, which does not vary. The CD3 is made up of at least three polypeptide chains (γ, δ and ε).

11.11 Diversity of the immune response

Higher vertebrates seem to have the potential to make specific immune responses to all possible epitopes. The production of an immune response to a given epitope is determined by the possession of lymphocytes with the 'correct' receptor molecules. It therefore follows that each animal is capable of producing a tremendous diversity of receptor molecules, and this ability would pose considerable logistical problems, if each receptor polypeptide chain were encoded by a different gene. The generation of the diversity of receptor molecules will first be discussed in relation to B lymphocytes where the immunoglobulin receptors have the same specificity as the antibody which the B cell has the potential to secrete.

Immunoglobulin diversity

Each immunoglobulin heavy or light chain is the product of several genes which are discontinuous in the germ-line cells. During the development of a B cell within the bone marrow, these genes are rearranged by a process which involves DNA splicing, deletion and re-joining. This process is illustrated in Fig. 11.48 for heavy chain genes, but a similar process also occurs with the genes coding for light chains. A B-cell precursor has several hundred genes each encoding a different V region sequence, and a stack of genes, each encoding a constant region for a particular heavy chain class. In between, there is a set of **Joining (J)** genes and a set of **Diversity (D)** genes. During the development of a B lymphocyte, a gene rearrangement brings together a single D gene next to a single J gene. This DJ arrangement is then joined to one V region gene. All intervening DNA is deleted from that cell. It is the VDJ sequence which determines the specificity of the antibody molecule which is expressed in the membrane or will eventually be secreted. In addition to the diversity which is generated by the number of permutations of V, D and J, it is known that mutations can also occur in the DNA of the VDJ arrangement after

☐ The Joining (J) genes involved in generating antibody diversity are not to be confused with the Joining (J) protein which links IgA and IgM molecules together (p. 303).

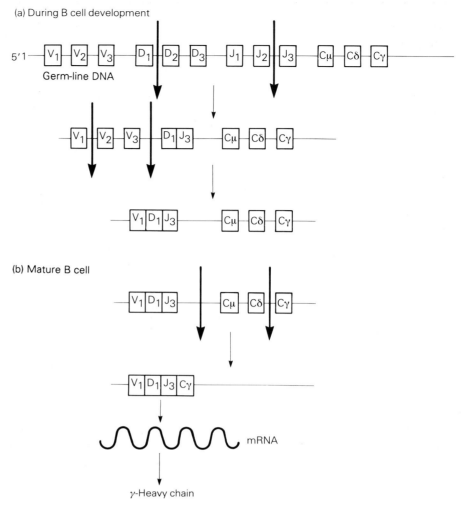

(a) During B cell development

5'1 — V_1 — V_2 — V_3 — D_1 — D_2 — D_3 — J_1 — J_2 — J_3 — $C\mu$ — $C\delta$ — $C\gamma$ —

Germ-line DNA

— V_1 — V_2 — V_3 — D_1J_3 — $C\mu$ — $C\delta$ — $C\gamma$ —

— $V_1D_1J_3$ — $C\mu$ — $C\delta$ — $C\gamma$ —

(b) Mature B cell

— $V_1D_1J_3$ — $C\mu$ — $C\delta$ — $C\gamma$ —

— $V_1 D_1 J_3 C\gamma$ —

mRNA

γ-Heavy chain

Fig. 11.48 Rearrangements in the immunoglobulin heavy chain genes. (a) The DNA in the earliest lymphocytes is not rearranged. These cells have several hundred genes each coding for a different variable sequence of a heavy chain (three only are shown, for simplicity). D, J and C represent Diversity, Joining, and Constant region genes. During the development of the B cell, a D gene is joined to a J gene by random splicing of the intervening sequence, and rejoining. In a similar fashion, one of the V region genes is joined to the DJ rearrangement. (b) In a mature B cell, the specificity of the antibody is determined by this rearranged sequence. Later splicing and joining to a single constant region gene may restrict the cell to production of a single immunoglobulin class.

it has been produced. These mutations tend to increase the affinity of the antibody for the epitope, without altering its specificity.

Thus, a mature B cell is capable of producing antibody of a single specificity, although several classes of immunoglobulin may be produced. Later in development, the VDJ may be permanently rearranged next to a gene specifying a particular constant region gene so that the cell will only produce antibody of, for example, the IgG class. Precisely when, and indeed if, this last permanent change occurs is not known. At all times, initiation of the immune response is triggered by binding of epitope to those B cells expressing specific receptors.

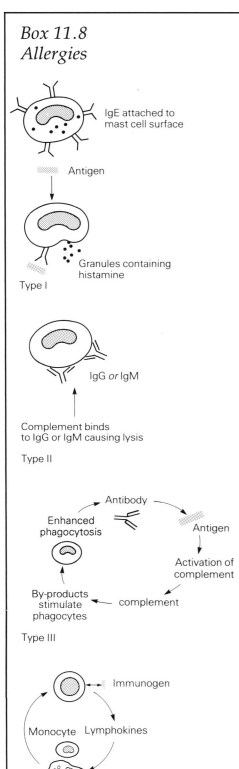

IgE attached to mast cell surface

Antigen

Granules containing histamine

Type I

IgG *or* IgM

Complement binds to IgG or IgM causing lysis

Type II

Antibody

Enhanced phagocytosis

Antigen

Activation of complement

By-products stimulate phagocytes

complement

Type III

Immunogen

Monocyte   Lymphokines

Macrophage

Type IV

Immunological allergy, or more correctly, hypersensitivity, refers to those reactions when, by responding normally to an immunogen, the immune response itself causes tissue damage. Hypersensitivity is classified into four types according to the underlying reaction (Fig.).

Type I, Anaphylactic: individuals produce IgE rather than IgG in response to the immunogen. IgE fixes to the surface of tissue mast cells and blood basophils. Further contact with the immunogen can stimulate the release of the mast cell granules, containing mediators which dilate blood vessels causing reddening of skin and contract smooth muscle causing diarrhoea and breathing difficulties. The systemic release of these mediators can sometimes cause death due principally to contraction of the muscles of respiration. Examples of Type I allergies include hay fever and asthma, caused by immunological reactions to pollen, animal dandruff, faeces of house mites, and food allergies, commonly to shellfish, mushrooms, strawberries, eggs.

Type II, Cytotoxic: IgG or IgM antibodies combine with the immunogen and activate complement, causing lysis of the cell. This is a normal reaction but can result in damage if, for example, the immunogens are cells of a mismatched blood transfusion. Lysis of the mismatched red blood cells by naturally occurring antibody causes a massive release of haemoglobin which can overload the kidneys and cause kidney failure.

Type III, Complex-mediated: complexes of antigen/antibody stimulate the activation of complement, causing inflammation wherever the complexes are deposited. Precipitation of complexes in the lung can be caused by allergic reactions to fungal spores. The inflammation occurring over a period of years can cause fibrosis of the lung tissue (Farmer's lung). Soluble complexes may circulate and tend to be deposited in the kidney, the joints, and the walls of blood vessels, causing glomerulonephritis, arthritis and arteritis. Glomerulonephritis occurring after bacterial infections unrelated to the kidney is thought to be complex-mediated.

Type IV, Delayed: brought about by the action of T lymphocytes and seen, for example, in an individual who is sensitized to nickel salts formed when cheap nickel plated jewellery is in contact with the skin. T lymphocytes arrive in the area and release lymphokines which stimulate a slow, steady build-up of monocytes and lymphocytes at the site. This results in swelling, pain and irritation at the site.

Most allergic reactions are best treated by avoidance of the immunogen, but if this is impossible, drugs can be of some use. For example, antihistamine cream or tablets can relieve some of the discomfort in Type I reactions, and asthmatics can inhale the drug sodium cromoglycate (Intal) which prevents mast cell degranulation. Corticosteroids can relieve inflammatory reactions in Type III and Type IV.

Allergies. Redrawn from Dawson, M.M. (1987) *Introducing Immunology*, Biological Sciences Series, continuing nurse education, Crown copyright.

Reference Roitt, I., Brostoff, J. and Male, D. (eds) (1989) *Immunology*, Gower Medical, London, UK. For those who want to learn more immunology, this book is excellent. It includes several chapters on the clinical aspects of the immune response.

Diversity of the T-cell receptor

It is now known that both polypeptide chains which make up the T-cell receptor, T_i, are also the products of V, D, J and C genes which have undergone gene rearrangements during the development of the T cell in the thymus. Thus, each mature T lymphocyte expresses receptors which are specific for a single epitope. Binding of the epitope to these receptors triggers the onset of the T-cell response.

11.12 Overview

Immune responses in higher vertebrates involve complex systems in which the interaction of specific and non-specific cells and molecules protect the body from infectious agents. Non-specific immune mechanisms form the first arm of the defence against microbial pathogens until specific immunity is developed. Products of the specific immune system act principally by stimulating non-specific effector cells. These interactions can be summarized as follows:

Macrophages act as antigen-presenting cells for specific helper T lymphocytes (T_H). These macrophages produce IL-1 which stimulates the activation of the T_H cells. Antigen-stimulated T cells produce IFNγ which increases the activity of macrophages and natural killer (NK) cells. In addition, T_H cells produce IL-2 which increases the activity of NK cells, and IL-3, which increases the number of phagocytic cells in the blood and the number and activity of mast cells in the tissues. Helper T cells also produce IL-4 and IL-5, which increase the number and activity of mast cells and eosinophils respectively. Both of these activities may be important in immunity to multicellular parasites. Immunoglobulins IgG and IgM both activate complement, and thereby stimulate phagocytosis. IgE may cause the release of mast-cell mediators, increasing inflammation and attracting phagocytes.

The efficiency of this system is shown by the disastrous effects of immunodeficiency diseases, which can be congenital, or can be acquired by infection with the human immunodeficiency virus (HIV).

1

Type of organism	Diseases (e.g.)	Source of pathogen
Virus	German measles	Inhalation
	AIDS	Body fluids
	Chicken pox	Inhalation, contact
	Mumps	Inhalation
Bacteria	Diphtheria	Inhalation
	Tetanus	Puncture
	Typhoid	Water
Fungus	Thrush (candidiasis)	Sexual contact
	Athlete's foot	Direct contact
	Aspergillosis	Inhalation
Protozoan	Dysentery	Food, water
	Toxoplasmosis	Cat faeces, raw meat
Helminth	Tapeworm	Contaminated food
	Ascariasis	Contaminated food

2. Organism A: Lysozyme in saliva; HCl in the stomach; normal bacterial flora in the gut which compete for nutrients. Organism B: Macrophages in muscle and connective tissue (following bite). Interferon produced by infected cells.

3. Small molecules such as testosterone can be linked covalently to a carrier protein such as bovine serum albumin (BSA). The immune system recognizes the testosterone as an epitope and produces antibodies against it. Insulin does not present this problem since it is a protein and is immunogenic. Thus, the administration of pig insulin to diabetics may stimulate production of anti-pig insulin.

4. In order to dissociate antigen–antibody complexes, it is necessary to lower the pH to 2.5. If ionic interactions made a greater contribution to the binding, minor changes of pH would result in dissociation.

5. Patients with agammaglobulinaemia lack the capacity to make antibodies and are therefore prone to bacterial infections. Antisera are available for a wide range of bacterial infections. Pooled immunoglobulins from a range of healthy individuals will contain antibodies to most of the common pathogenic bacteria. However, because of the dangers of transferring diseases such as hepatitis and AIDS, it is important to screen for the viruses which cause these diseases.

6. Interferon γ has direct effects on the replication of the cancer cell since it induces the production of enzymes which interfere with nucleic acid and protein synthesis. Indirectly, IFNγ, like IL-2 increases the activity of NK cells.

7. The diagram should look something like this:

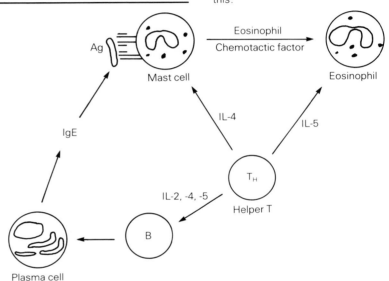

IL-4 and IL-5 increase the number and activity of mast cells and eosinophils respectively. T$_H$ 'help' B cells to produce IgE which attaches to mast cells. Contact with antigen (Ag) cause release of a factor which attract eosinophils.

FILL IN THE BLANKS

1. The small lymphocyte is the central cell in the _____ immune response. These cells are found in the _____ and the lymphoid tissues. There are two types of lymphoid tissues. The _____ lymphoid tissue is concerned with the development of lymphocytes. Examples of such tissues are the _____ and the _____ _____ _____ . The _____ lymphoid tissue deals with the immunogens. The biggest single mass of this lymphoid tissue is the _____ .

There are two types of small lymphocyte: the B and the T cells. The role of the B lymphocyte is to make _____ . This type of immunity is known as _____ . There are at least two types of T lymphocyte. The _____ T cell is central to the immune response because it produces _____ such as _____ _____ , which stimulates lymphocyte proliferation, and _____ _____ which is known to stimulate macrophages. The second type of T cell can develop into _____ _____ _____ which are important in immunity to _____ .

Choose from: antibodies, blood, bursa of Fabricius, cytotoxic T cells, helper, humoral, interferon γ, interleukin-2, lymphokines, primary, secondary, specific, spleen, thymus, viruses.

MULTIPLE-CHOICE QUESTIONS

2. State, in each case, the immunoglobulin class(es) (A, D, E, G, M) which:
A. are most efficient at agglutinating cells
B. stimulates antibody-dependent cellular cytotoxicity (ADCC)
C. activates complement
D. cross the placenta
E. attach to mast cells

SHORT-ANSWER QUESTION

3. List three activities of interleukin-1 (IL-1) in promoting the acute phase response to infection.

4. A sample of IgG was treated with 2-mercaptoethanol and applied to a column of Sephadex G-100 for gel filtration. Draw a diagram to show what the elution profile might look like and explain what you draw.

5. When antibodies are used as tools in biochemistry and histology to locate specific molecules, they may be labelled with fluorescein or with a molecule of enzyme ("enzyme-linked"). Explain how each of these techniques works and what their relative advantages and disadvantages are.

6. Why has the use of monoclonal antibodies to treat human diseases, such as cancer, proved problematic?

ESSAY QUESTION
7. Write an essay outlining the importance of macrophages in immune responses.

12

Differentiation and development

Objectives

After this chapter you should be able to:

☐ describe the stages of development common to a variety of species;

☐ explain the central role of cell differentiation in development;

☐ discuss the role of differential gene activity as a major mechanism by which cells express their differentiated characters;

☐ understand that the zygote has the ability to develop into a number of cell types, but that individual cells formed as the embryo develops usually lose this ability, particularly in animals;

☐ discuss the importance of cell–cell communication in the establishment of the patterns of cells which form the tissues and organs of the embryo.

12.1 Introduction

Developmental biology is concerned with the morphological, physiological and biochemical changes which occur during the life of all multicellular (and some unicellular) organisms. It is a well-established science, which has relied historically on the physical manipulation of developing organisms (experimental embryology) to formulate general theories as to how various cellular processes involved might be controlled.

Biochemists have focussed their attention on the study of development only recently. By combining classical techniques in experimental embryology with the use of powerful biochemical and molecular biological tools, significant contributions have been made to the study of development.

Central to development is the process of **differentiation**, whereby individual cells or groups of cells undergo ultrastructural and metabolic changes which distinguish them functionally from other cells in the developing organism. In virtually all multicellular organisms the cells of the adult are descended from a single cell, the fertilized egg or **zygote**.

In human beings there are about 200 types of differentiated cells and study of the ultrastructural and metabolic changes which are involved in their formation has led to some understanding of development. Such studies provide many important details, but development is much more than cell differentiation alone. In particular, **cell division** (an adult human contains many millions of cells), **cell movement** (various stages of development involve co-ordinated movements of cells to new locations), and **pattern formation** (groups of cells of two or more types establish, maintain and refine particular spatial relationships) have to be considered.

The instructions for development are contained in the zygote, specifically in the DNA. The process of activation of these instructions, **gene expression**,

☐ The seventeenth century physician Harvey was first to realize that all organisms are derived from a single egg. The title page of his book on the generation of animals carried the statement *Ex ovo omnia*, all things come from the egg.

See *Biological Molecules*, Chapter 8

Reference Alberts, B., Bray, D., Lewis, J. et al. (1989) *Molecular Biology of the Cell*, 2nd edn, Garland, New York, USA. A comprehensive student text, emphasizing the central role of cell biology in biology. It includes several excellent chapters on development.

Reference Gilbert, S.F. (1991) *Developmental Biology*, 3rd edn, Sinauer, Sunderland, MA, USA. An up-to-date text on animal development: good source of recent references and very good diagrams.

is an important set of biochemical events throughout development, as is the sending of signals to appropriate cellular destinations, by intra- and inter-cellular communications.

Box 12.1 *The cellular society*	In human societies it has proved beneficial for individual members of the society to choose or be allocated different tasks. For example, farmers, builders, teachers, perform functions essential to the society as a whole. Similarly, an individual is made up of a vast number of cells, which are organized into groups performing specific tasks for the benefit of the individual as a whole. Just as individuals gradually acquire the knowledge, skills and experience to become farmers, builders and teachers, cells in the developing individual gradually become specialized for particular tasks. Cell differentiation is the process of development of these specialized functions.

12.2 Stages of development in animals

Developmental stages are each characterized by a set of cellular events. Some of these stages are common to all multicellular organisms whereas others are peculiar to one or more groups. Animal development shows three major phases:

1. **cleavage**, where the zygote undergoes a series of rapid cell divisions without any change in the overall mass of the embryo;
2. **gastrulation**, which involves a co-ordinated series of cell movements resulting in the formation of a multilayered structure surrounding a rudimentary gut cavity;
3. *organogenesis*, where cells differentiate and establish the patterns which constitute the tissues and organs of the embryo.

The cellular events which characterize each stage overlap. For example, general cellular patterns are established early during organogenesis but refinements of pattern and many cell divisions are involved in the subsequent growth of the early embryo through to adulthood. These stages will now be considered in more detail.

Fertilization, cleavage and formation of the blastula

Fertilization takes place as a result of the fusion of the male and female gametes to form the zygote. Development is then activated. The zygote begins the rapid series of cell divisions characteristic of cleavage (Fig. 12.1). This process is particularly easy to study in amphibian species, such as *Xenopus laevis* (Fig. 12.2). Raw materials and energy for this process are derived from the yolk of the egg. In particular, large amounts of protein, RNA and membrane material are already available. **DNA synthesis** is thus the only major biosynthetic process that must take place, and the extremely rapid rate of cell division is made possible by the large number of replication origins present in the DNA.

The first three divisions of the zygote take place along mutually perpendicular axes to give eight cells. Already some asymmetry can be observed. The four cells at the lower end of the embryo (corresponding to the **vegetal pole** of the egg) contain more yolk and are larger than the other four cells (corresponding to the **animal pole**). Cleavage continues for many more cycles. Initially the cell divisions are synchronized, that is, all the cells divide together, but eventually (after about 12 cycles in *Xenopus*) this synchrony is lost. The cells formed by this process are called **blastomeres**. In the centre of

organogenesis: the development of tissues and organs.
blastomere, blastcoel, blastula: *all derived from the Greek* blastos *meaning a sprout.*

Reference Graham, C.F. and Wareing, P.F. (eds) (1984) *Developmental Control in Plants and Animals*, 2nd edn, Blackwell Scientific, Oxford, UK. A series of papers which give in-depth accounts of particular topics. Gives the experimental background to current thinking on the control of development.

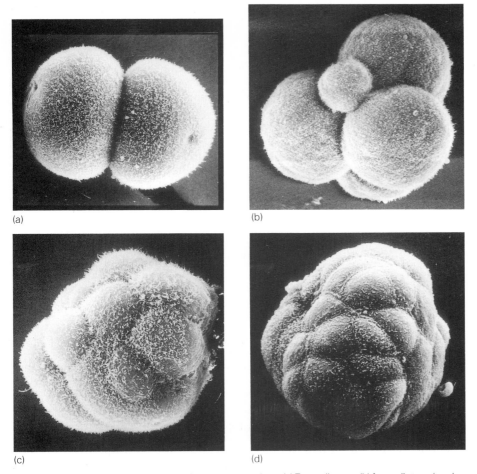

Fig. 12.1 Scanning electron micrographs of early mouse embryo. (a) Two-cell stage, (b) four-cell stage (a polar body, a small cell with little cytoplasm, produced during the formation of ova is also visible), (c) eight-16 cell stage and (d) blastocyst. Reprinted from Calarco, P. and Epstein, C.J. (1973) Cell surface changes during preimplantation development in the mouse, *Developmental Biology* **32**, 208–13, and Alberts, B. *et al.* (1989) *Molecular Biology of the Cell*, 2nd edn, Garland, New York, p. 895.

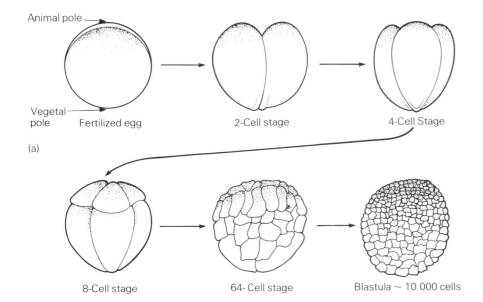

Fig. 12.2 (a) Stages in the formation of the blastula in *Xenopus*. (b) Cross-section of the blastula showing the individual blastomeres and the blastocoel. Redrawn from Alberts, B. *et al.* (1989) *Molecular Biology of the Cell*, 2nd edn, Garland, New York, p. 880.

Reference Slack, J.M.W. (1983) *From Egg to Embryo: Determinative Events in Early Development*, Cambridge University Press, Cambridge, UK. A comprehensive account of the events leading to cell commitment.

Reference Elinson, R.P. (1989) Microtubules and specification of the dorsoventral axis in frog embryos. *BioEssays*, **11**, 124–7. Fascinating short essay on the role of microtubules in frog embryos.

Mammals undergo an extra developmental stage prior to gastrulation during which the rudiments of the placenta and other extra-embryonic tissues are formed.

See Section 4.4

the embryo the space between the cells enlarges to form a fluid-filled cavity called the **blastocoel**. The accumulation of fluid is a result of the action of a Na^+-pump which actively transports Na^+ into the extracellular space. Consequent differences in osmotic pressure allow water to fill the developing cavity. The contents of the blastocoel are prevented from leaking around the gaps between cells by **tight junctions**, and thereafter the contents are maintained by transport of materials from and to the surrounding cells. The formation of cavities surrounded by **epithelial cells** which selectively transport material is a common feature of development.

Even at this early stage in development, cells can communicate with one another and communications between groups of cells are mediated by **gap junctions**. These structures allow the movement of small molecules, such as nutrients, and chemical signals from cell to cell. The biochemical characterization of these signals is not well advanced. It is likely, however, that gap junctions are important in inter-cellular communication at this stage, since a circulatory system has not yet developed. In some species at least, the movement of ions and small molecules via gap junctions between yolk cells and the other cells of the embryo is found only at stages prior to the point at which the blood begins to circulate.

By the end of the series of cleavage divisions and associated events the embryo consists of a hollow ball of cells, the **blastula**.

Gastrulation: the division of the embryo into layers

At the blastula stage cells remain largely undifferentiated (at least in ultra-structural terms) and it is only at the next stage, the formation of the *gastrula*, that cells begin to differentiate. The major feature of gastrulation is a co-ordinated series of **cell migrations**, the result of which is to separate the embryo into layers.

The process begins close to the vegetal pole of the blastula, by a few cells which change shape, broadening their inward-facing surfaces. This causes the outer sheet of epithelial cells to bend inwards. Cells then attach to the inner surface of the blastula and migrate further, pulling more cells behind them. The opening through which cells migrate remains narrow, just like the neck of a bottle (Fig. 12.3). The inner surface is gradually coated with migrating cells until the blastocoel disappears; at the same time a new cavity is formed, which eventually forms the gut of the developing organism. The area between this new cavity and the outer surface of the embryo is several cells thick and gradually becomes differentiated into the three **primary germ layers**: the **endoderm** (inner layer); the **mesoderm** (middle layer) and the **ectoderm** (outer layer).

The endoderm gives rise to the digestive tract, the mesoderm to the various support tissues (skeleton, muscles, connective tissue), blood, and the organs of the urinogenital system, and the ectoderm to the epidermal and nervous systems. The remainder of the interior of the embryo is filled by residual yolk cells, the **yolk plug**.

The changes in cell shape, adherence and movement, together with metabolic changes which occur during differentiation are collectively responsible for *morphogenesis*, the process by which the shape and form of the embryo is established. Morphogenesis begins during gastrulation and continues through subsequent stages. In recent years some of the biochemical events responsible for the co-ordinated cellular migrations during morphogenesis have been characterized. Changes in cell shape involve reorganization of the **cytoskeleton**. A variety of signals initiate these changes during development. For example, components of the **extracellular matrix** interact with **membrane receptors** which then activate intracellular enzymes (such as

See Chapters 6 and 7

gastrula: little stomach; from the Greek gaster meaning belly.
morphogenesis: derived from the Greek morphos meaning shape. It is the process which defines the shape or form of the embryo.

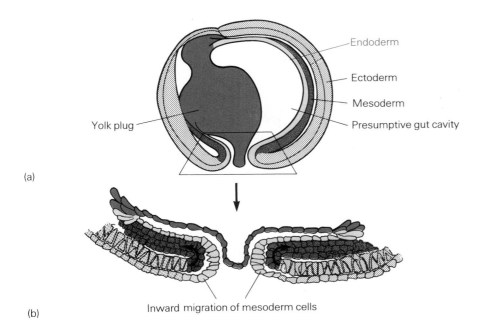

Fig. 12.3 The formation of the gastrula in *Xenopus*: (a) the formation of the three-layered structure; (b) the migration of mesodermal cells through the narrow neck.

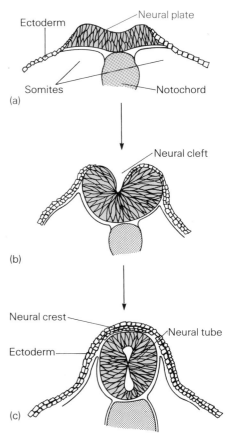

Fig. 12.4 The induction of the neural tube in *Xenopus* begins with the rolling up of the neural plate under the action of the underlying notochord (a). The neural cleft forms (b) and as the neural tube is formed the neural crest closes above it (c).

protein kinases) which are involved in the control of cytoskeletal dynamics. Also, cells which undergo co-ordinated movements are usually joined by **adhering junctions**, such as belt or spot desmosomes. Groups of cells undergoing such movements are attached to one another physically, but, in addition, these junctions can transmit signals from cell to cell. **Belt desmosomes** are often associated with bundles of **actin filaments** and these could initiate contractile movements. **Spot desmosomes** are linked to **intermediate filaments**, the most rigid components of the cytoskeleton which must, therefore, undergo reorganization during any cell movements or changes in cell shape.

Organogenesis: specifying the patterns of organs and tissues

The three primary germ layers formed during gastrulation will eventually gives rise to the tissues and organs of the developing organism (organogenesis). Normally in the adult a particular area of an organ is composed of cells originating from more than one of the primary germ layers. For example, a piece of intestine consists of epithelia (endodermal), blood vessels, smooth muscle and connective tissue (mesodermal), and nerves (ectodermal). The formation of such structures illustrates another general feature of development. Cells with very different developmental histories may migrate to adjacent locations where they can interact and differentiate to give particular cellular patterns. Such interactions are examples of **embryonic induction**.

An instructive example of induction takes place soon after gastrulation, namely the interaction of mesodermal and ectodermal tissue to initiate the formation of the nervous system. A specialized group of mesodermal cells known collectively as the **notochord**, interacts with adjacent ectodermal cells directly above it. The ectodermal cells elongate, thicken and roll up to form the **neural tube** (Fig. 12.4). The tube grows and differentiates to form the neural tissues of the **central nervous system**. As the tube closes a few cells

□ Tissues and organs formed by induction include the ear, lens of the eye, mammary gland and epidermal layers.

break loose and form a layer between the tube and the newly closed ectoderm above. This layer of cells is the **neural crest**. Virtually all the cells of the **peripheral nervous system** develop from the neural crest.

In vertebrates the notochord, together with other mesodermal cells, develops into the **vertebral column**. In 'lower' chordates, which do not have vertebrae, this development does not take place, and the notochord remains as a primitive structure. Thus, as well as having a developmental relationship, the notochord and the vertebral column are linked in evolutionary terms. This is just one example of a common feature of **evolution**. A structure which is present in a rudimentary form in one species often evolves into a more advanced structure in another species. Studies of such evolutionary events can give important clues to the mechanisms of evolution of species.

The notochord forms the central axis of the embryo, thus dividing the embryo into left and right sides. After the neural tube has formed the blocks of mesoderm on either side of the notochord begin to divide into segments, or *somites*. This process takes place in a sequential manner from head to tail, resulting in the repetitive segmented pattern which characterizes the vertebrate body plan. During somite formation the embryo elongates substantially, as a result of cell division, cell migration and rearrangements of cell interactions.

Each somite (or segment) subsequently differentiates into a segment of the organism forming the rudiments of the tissues and organs characteristic of the future adult. Development as far as the formation of somites is broadly similar in all vertebrates, but beyond this point the different groups (fish, amphibians, birds, mammals, etc.) show divergent developmental pathways. Some segments will go to form internal organs, whereas others initiate the formation of **limb buds**, which will eventually form the arms (or wings) and legs of the adult organism. The end result of all these processes is the formation of an embryo which is a miniature form of the adult. All the major tissues and structures are present, albeit in a rudimentary form.

Further growth and development

Although differentiation and morphogenesis continue during subsequent development, the most important feature of the remaining stages is the growth of the miniature form to give an adult. This involves many cell divisions, but not all parts of the embryo divide at the same rate. Further refinements of the organizational patterns of cells in tissues occur, leading to an organism which grows with constantly changing minor features. An obvious example in humans is that at birth the head of the baby is proportionately larger in relation to the rest of the body, than in the adult.

Following birth further changes occur, such as those to the genitalia and other tissues that accompany sexual maturation. Also, development does not cease once adulthood is reached. There is a continuing need for the renewal and replacement of cells throughout life. This requirement varies from one cell type to another. Some cells (for example blood cells) need to be replaced regularly, whereas others, usually metabolically inactive cells such as the elongated fibre cells of the lens of the vertebrate eye, are laid down early in development and remain throughout life.

Mechanisms of cell renewal vary (Fig. 12.5). Some cells are replaced simply by division of already differentiated cells, accompanied by limited cell movements to replace dead cells and maintain the organizational pattern of the tissue. This type of renewal is shown by many endothelial cells such as those of the liver which have the capacity to **regenerate**, or those cells which are required to reform blood vessels after wounding. Other cell types are renewed by **stem cells** which remain undifferentiated in the adult. These

☐ Evolution need not lead to more complex structures or species. Evolution can involve degradation of a structure that is no longer required, such as the tail in the evolution of earlier primates to human beings. Similarly, parasitic flatworms are examples of organisms which have evolved from more complex free-living ancestors.

somites: refers to the body of the organism (from Greek soma), *excluding the germ cells.*

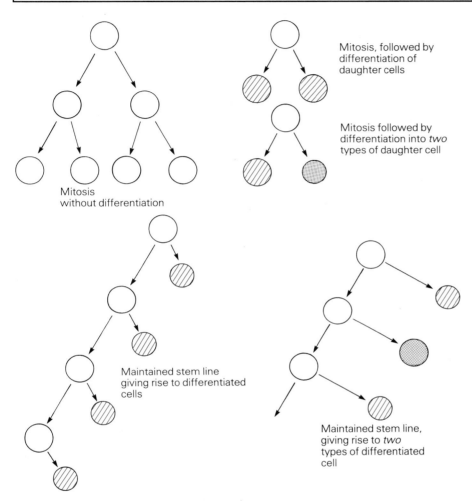

Mitosis, followed by differentiation of daughter cells

Mitosis followed by differentiation into *two* types of daughter cell

Mitosis without differentiation

Maintained stem line giving rise to differentiated cells

Maintained stem line, giving rise to *two* types of differentiated cell

Fig. 12.5 Determination in differentiating cells. Some examples of possible sequences of mitotic and differentiation events in cells during development.

divide to maintain the stem cell population, and some of the progeny then undergo **terminal differentiation**, losing any further capacity to divide in the process.

Although undifferentiated (at least ultrastructurally), stem cells are said to be **determined**, that is they are committed to a future differentiation pathway. That pathway may lead to a single type of differentiated cell, in which case the stem cell is said to be **unipotent**, or to a small number of different but related cell types, in which case it is said to be **pluripotent**. Some of the cells which replace the epidermal tissues of the skin are unipotent; cells of the bone marrow which replace blood cells are pluripotent.

Totipotency is found to a much greater extent in plant cells than in animal cells and is considered in more detail in Section 12.5.

☐ Tissues and organs renewed by stem cells include epidermal tissues, sweat glands, blood cells and skeletal muscle.

12.3 Development in plants

There are many differences between the structural organization of plants and animals, and so it is not surprising that plant development has unique features. At the cellular level the presence of a cell wall in plants imposes some developmental constraints. Plant cells are unable to form the complex direct cell interactions found in animals nor are they able to move and change shape as rapidly as animal cells. As a result the processes of cell migration, inter-action, and reorganization that are such a feature of animal development are replaced by a strictly regulated series of orientated cell divisions and expansions. Environmental cues exert a more obvious influence on plant development than they do in animals. Simple observations show the effects of the environment on plant development. The effect of moving the position of, or administering feed or water to a houseplant demonstrates how the availability of light, nutrients and gravity can affect development significantly.

Despite these differences many of the cellular events in plants are similar to those in animals. In particular cell division, growth and differentiation are the crucial events. Some details of these processes in relation to stages in plant development are now considered.

Stages in plant development

As in animals, asymmetry is established very early. The zygote divides asymmetrically and this leads by a complex series of events to the production of a **seed** (Fig. 12.6). In flowering plants the seed contains the embryo proper derived from the original diploid zygote, and a rich nutritive support tissue or **endosperm**, which is derived from a triploid endosperm nucleus formed during the process of fertilization. The endosperm provides many of the nutrients required for further development. Development is often arrested at the seed stage, until environmental conditions suitable for further development prevail.

Development past the seed stage is centred around specialized regions called **meristems**, which are first formed during seed production. Cell division is confined almost entirely to these specialized regions and they are responsible for the development of the variety of plant structures. **Apical meristems** are found at the tips of growing roots and shoots, and from these will develop new root tissue on the one hand and new stem and leaf tissue on the other. Figure 12.7 shows the features of typical **root** and **shoot** apical meristems. **Lateral meristems** (cambium) are found around the circumference of the plant stem and are responsible for adding to its girth as the plant grows.

During growth the shoot apical meristem produces a series of repeating structures, **nodes**. Each node comprises a branch in the stem, a leaf and a bud. Nodes are connected together by **internodes**, the intervening stem. Each **module** (node plus internode) in a particular plant initially has a similar structure. However, the local environmental conditions encountered influence the development significantly. In dim light plants produce an abundance of dark green leaves, whereas in a very bright environment, the foliage is lighter in colour and less extensive. The shaping of the development of individual modules is brought about primarily by controlling the direction and extent of cell division, growth and differentiation. Plant cells do not migrate in the manner of animal cells and therefore the choice of direction for particular developmental events is crucial. The arrangement of cytoskeletal elements, both microtubules and actin filaments, define the plane along which an individual cell will divide. This direction is established prior to **mitosis** and is maintained during division itself.

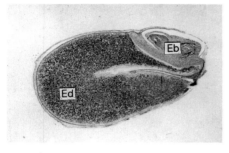

Fig. 12.6 Photomicrograph of longitudinal section through seed of wheat. Embryo (Eb) and endosperm (Ed) are indicated. Courtesy M.J. Hoult, Department of Biological Sciences, The Manchester Metropolitan University, UK.

☐ Fertilization in flowering plants entails a double fusion event which is unique to flowering plants. The pollen grain produces two haploid male nuclei. One of these fuses with the haploid egg to give a diploid zygote which develops to form the embryo. The other male gamete fuses with the two polar nuclei of the embryo sac to produce a triploid primary endosperm nucleus. Repeated mitotic divisions of the endosperm nucleus eventually produces the endosperm tissue of the seed, which is an energy reserve for the developing embryo.

Reference Steeves, T.A. and Sussex, I.M. (1989) *Patterns in plant development*, 2nd edn, Cambridge University Press, Cambridge, UK. A splendid book giving very a good introduction to plant development.

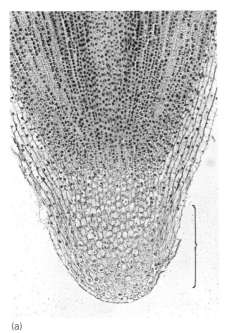

(a)

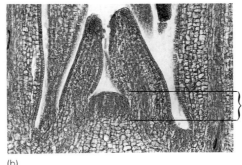

(b)

Fig. 12.7 Photomicrographs of typical apical meristems of flowering plants. (a) LS of root tip (×42) and (b) LS of shoot tip (×40). In both cases the regions of rapid mitotic division are indicated by a bracket.

12.4 The organization and control of development

To gain a deeper understanding of development it is necessary to consider ideas on how the whole process is organized and controlled. The ultimate aim must be an understanding of human development, but many experiments are not possible in humans and so other organisms have to be used.

Mammals in general are difficult to use in developmental studies, as most development takes place *in utero* and thus cannot be observed easily. A further difficulty is that although the study of abnormalities can give many insights into normal development, most mammalian developmental abnormalities are fatal at an early stage. Experimental studies have therefore concentrated on a few key species. Each has one or more features which makes it particularly suitable for developmental studies. Some of the most notable examples are:

- Amphibians such as *X. laevis* (Fig. 12.8(a)). This organism has large eggs which can be manipulated easily, for example in microinjection experiments. Amphibians also have the advantage of clear membranes through which the many developmental stages that occur in the aquatic larval form can be viewed.
- The fruit fly *Drosophila melanogaster* (Fig. 12.8(b)). A large number of mutations, including an increasing number which affect development, have been characterized both genetically and biochemically in this organism. The fly is a relatively simple organism and many abnormalities, some of them quite bizarre, are tolerated. As a result, adult flies or reasonably advanced embryos with gross developmental abnormalities may be studied.
- The nematode *Caenorhabditis elegans* (Fig. 12.8(c)). This very simple organism shows many of the developmental features of more complex species. The major attraction of *C. elegans* for developmental biologists is

Reference Schierenberg, E. (1989) Cytoplasmic determination and distribution of developmental potential in the embryo of *Caenorhabditis elegans*. *BioEssays*, **10**, 99–104. A short essay emphasizing the importance of this nematode to development studies.

Fig. 12.8 Organisms commonly used in developmental studies. (a) The African clawed toad, *Xenopus laevis* (b) the fruit fly, *Drosophila melanogaster*, (c) the nematode *Caenorhabditis elegans*, (d) the albino mouse, *Mus musculus* and (e) common wall cress, *Arabidopsis thaliana*. (a) and (d) courtesy Ms J. Gilpin, Department of Biological Sciences, The Manchester Metropolitan University, (b) Department of Biochemistry and Molecular Biology, University of Leeds, (c) reprinted from Sulston, J.E. and Horvitz, H.R. (1977). Post-embryonic cell lineages of the nematode *Caenorhabditis elegans*, Dev. Biol. **56**, 110–56. (e) Courtesy of the Smith Collection.

that its development can be studied cell by cell. The complete developmental history of every cell, or **cellular fate map**, is known, and varies little, or not at all, from one individual to another. Such knowledge is extremely useful in studies where the development of the nematode is manipulated experimentally. Any deviation from the normal cellular pattern, however small, can be detected.

- The mouse *Mus musculus* (Fig. 12.8(d)) is the mammal used most in developmental work. It is the first choice when features peculiar to mammalian development are being investigated. Mice have a short gestation period which facilitates studies. Many of the techniques now used in the **in vitro fertilization** of human embryos or in the cloning of embryos from domestic farm animals were developed as a result of initial experimentation on mouse embryos.

- The common wall cress, *Arabidopsis thaliana* (Fig. 12.8(e)) is gaining in popularity as a species suitable for the study of plant development. Plant development is less well understood than animal development, in part because the organisms used for genetic analysis have been those with commercial importance, for example, maize and potato. Both of these examples, in common with many plants, have long life-cycles and large genomes which makes genetic analysis difficult. *A. thaliana* has a small genome in comparison to other plants, and this, coupled with the fact that it can be grown indoors in large numbers in a few weeks, is leading to its adoption as a 'model plant' for molecular genetic analysis.

Many developmental events are similar in species which are distant in evolutionary terms. As a result, knowledge gained from studies on one species can often be used as a starting point for studies on an unrelated species. Also, a few developmental determinants have been identified biochemically and some of these are highly conserved over a long evolutionary time-scale, suggesting that similar mechanisms for controlling development operate in diverse species.

12.5 Totipotency, gene activity and differentiation

Stem cells have limited developmental potential although they can be unipotent or pluripotent. This illustrates a general feature of normal development. The zygote is totipotent but as development proceeds individual cells are produced which have a capacity to differentiate into an increasingly limited variety of cell types. The cellular and biochemical changes that accompany this sequential loss of potential have been investigated to try to discover how this change in potency occurs. An important clue to this comes from experiments which demonstrate that some fully differentiated cells can demonstrate totipotency when removed from their normal environment. In plants this was first demonstrated by Steward and co-workers who took secondary phloem cells from a mature plant and established them in culture. It was found that a single cloned cell in the culture could be induced to grow and develop into a fully mature plant (Fig. 12.9). Similar experiments in animals proved more difficult, but Gurdon and co-workers, using *Xenopus*, produced similar results (Section 1.3).

From these and other experiments the idea that all cells contain a full genetic complement of DNA, and that **variable gene activity** is the basic biochemical mechanism by which differentiated cells manifest their individuality, was established. Although it is thought that most cells have a

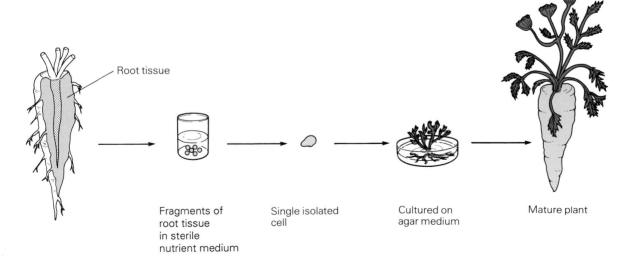

Root tissue

Fragments of root tissue in sterile nutrient medium

Single isolated cell

Cultured on agar medium

Mature plant

Fig. 12.9 Plant totipotency demonstrated in the carrot. A single cell was isolated from fragments of root tissue and after culture grew into a fully mature plant. Redrawn from Becker, W.M. (1986) *The World of The Cell*, 1st edn. Benjamin Cummings, Menslo Park, CA, p. 592.

See *Molecular Biology and Biotechnology,* Chapter 10

Fig. 12.10 Scanning electron micrograph showing characteristic layered structure of human lens fibre cells. The cells are enucleated and are filled with lens-specific proteins, the *crystallins*. From Kessel, R.G. and Kardon R.H. (1979) *Tissues and Organs: A Text-Atlas of Scanning Electron Microscopy*, Freeman, San Francisco, p. 100.

See *Molecular Biology and Biotechnology*, Chapter 2–5

See Chapter 10

full developmental *potential*, an environment in which this can be realized has been demonstrated in only a few cases, such as those described above. There are exceptions to this general rule. Germ cells, of course, have only half as much DNA as normal somatic cells and certain specialized cells, such as red blood cells (in some species) and the lens fibre cells of vertebrates, lose their DNA in the final stages of differentiation (Fig. 12.10).

Variable gene activity is believed to be the central mechanism in the expression of differentiated character, but other mechanisms make a contribution in certain specialized cells. For example, in antibody-producing cells, **gene rearrangements** occur before transcription (see below) is activated, and this contributes to the diversity of antibodies that can be produced (Section 11.11). Another possibility which has been suggested is that of selective gene amplification in cells where large amounts of a particular gene product is required. Evidence for this has not been found. Indeed, in situations where it might be thought likely, for example the **globin genes** in red blood precursor cells, they have been shown to be few in number. This number does not vary from cells which are synthesizing large amounts of haemoglobin to cells which do not produce haemoglobin. Some examples of amplified genes are known, such as those which code for the ribosomal RNAs (Section 3.9), but this amplification is not a part of specialized cell differentiation.

Control of gene activity

The expression of genes is a complex process involving many steps, from initial gene transcription to the synthesis and maintenance of the active protein product. These steps include **transcription**, **RNA processing**, **RNA transport**, **selection of mRNA** for **translation**, and **mRNA degradation**. Control of the process of cell differentiation could be exercized at any or all of these levels (Fig. 12.11).

Cells contain a large number of different proteins and this can be demonstrated using **two-dimensional gel electrophoresis**. More than 1000 different proteins can be separated from a single cell type in this way (Fig. 12.12). Comparing the proteins resolved from different cell types shows certain distinctive features. Each type of cell contains large amounts of a small number of proteins, termed **luxury proteins**. These are characteristic of that particular cell type. For example, actin and myosin are prevalent in muscle cells. Luxury proteins characteristic of one cell type may be found in other cell types, but usually only in vastly reduced amounts. In addition, all cells contain many common proteins, termed **housekeeping proteins**. Their name gives a clue to their function. Housekeeping proteins include the enzymes of the central metabolic pathways or the vital structural components. The amounts of housekeeping proteins vary from one cell type to another, but these variations are on a much smaller scale than those seen with luxury proteins.

It should be realized that large numbers of proteins which are present in only very small amounts in differentiated cells are not detectable following

Fig. 12.12 Two-dimensional electrophoretic separation of the total proteins extracted from bovine endothelial cells. Courtesy Dr D.C. West, Christie Hospital, Manchester, UK.

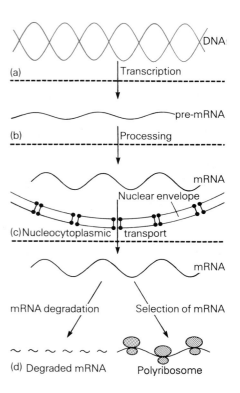

Fig. 12.11 Gene activity may be controlled at the level of (a) transcription, (b) processing, (c) nucleocytoplasmic transport and (d) selection of mRNA for translation or degradation.

electrophoresis. This is highlighted when the results of studies on the numbers and types of mRNAs are compared with the protein detected after electrophoresis. Upwards of 10 000 different types of mRNA can be found attached to **polysomes**, where presumably they are being translated into proteins.

Estimates of mRNA populations are based on studies of the kinetics of association of a population of mRNAs with a complementary DNA (**cDNA**) mixture, produced using the enzyme **reverse transcriptase**. The sequences of cDNA in such a mixture will mirror exactly the sequences in the mRNA population from which it was derived. For example, if there was a large amount of one type of mRNA there will be an equally large amount of that type of cDNA in the mixture. Measurements of the rate at which **mRNA–cDNA hybrids** are formed can be used to give broad estimates both of the *number* of different mRNA sequences present and their relative *concentrations* (Table 12.1, and see Box 12.4). The composition of other cellular RNA fractions such as **non-polysomal mRNA** and **nuclear pre-mRNA** can be studied in the same way. Generally a non-polysomal mRNA fraction is much

---

Box 12.4
Finding a partner at a children's party

A simple way to sort out partners at a children's party is to ask all the boys to stand at one end of the room and all the girls at the other end. On a signal the boys and girls are asked to find partners. In a molecular hybridization reaction mRNAs from a differentiated cell similarly find their partner cDNAs. Imagine at the children's party that there is a rule that boys and girls must find a partner whose name begins with the same letter as their own. There will be large numbers of Joes, Jims, Jeans, Jackies etc. who will quickly form partners (these correspond to the mRNAs/cDNAs for **luxury** proteins), but Quentins, Zoes etc. will be present in much fewer numbers and will take much longer to find their partners (these correspond to the mRNAs/cDNAs for **housekeeping** proteins). Thus the longer the time taken to find a partner, the less common is that type of mRNA.

Table 12.1 *Populations of mRNA from chick oviduct and liver. In each case classes of mRNA are grouped into three fractions: I, highly abundant; II, moderately abundant; III, rare. From Axel, R. et al. (1976) Analysis of the complexity and diversity of mRNA from chicken liver and oviduct. Cell 7, 247–54.*

Fraction	No. of different mRNAs	Copies per cell	% of total mRNAs
Oviduct			
I	1	100 000	50
II	8	3750	15
III	14 000	5	35
Liver			
I	1	32 800	16
II	106	750	40
III	11 600	7	44

more complex, containing approximately ten times as many different mRNAs as the corresponding polysomal fraction. Nuclear pre-mRNA populations are more complex still.

These experiments provide evidence for the idea that in differentiated cells a large number of genes are expressed as transcribed pre-mRNA, but that only a small proportion of these RNAs (probably less than 5%) are actively **translated** into protein at any one time. This demonstrates the importance of control of gene expression at the **post-transcriptional** level in *qualitative* terms.

In *quantitative* terms the picture is quite different. When large amounts of a few luxury proteins are required in one type of differentiated cell the most important control is at the level of **gene transcription**. This has been demon-strated in a number of systems. An example is the action of steroid hormones (Section 8.8), where specific receptor molecules have been characterized and shown to interact with the **chromatin** in the **nucleus** where they affect gene transcription. The large changes in gene expression in such systems make them relatively easy to study, but subtle changes affecting the *qualitative* pattern of gene expression are much more difficult to characterize. It is possible that these changes may hold important clues to the *initial* events which trigger cells to proceed along particular differentiation pathways. What is clear is that a sophisticated intracellular production and distribution system has evolved in cells which can respond to vastly differing requirements in response to appropriate developmental stimuli.

Box 12.5
The protein supply business

Individual cells in the embryo initially have the ability to produce all types of protein. They synthesize small amounts of a very large number of different types of blueprints (pre-mRNAs) from which the individual proteins can be made. Some of each type of blueprint is stored in the factory (the nucleus). Types for which there is likely to be a demand are processed into mRNAs and transported to a distribution centre in the cytosol. As required, selections of mRNAs from the distribution centre are translated into proteins. As the cells differentiate they respond and supply more of the proteins for which there is greatest demand. Small changes in demand can be met by using mRNAs from the distribution centre or the factory store. If demand for a particular protein becomes high a signal is sent to the factory to increase production. To keep the system efficient, the amounts of a particular mRNAs stored in the factory and distribution centre are kept to a minimum.

12.6 Determination and differentiation

Stem cells become determined even though they have not undergone any large-scale ultrastructural or biochemical changes. Subtle changes have no doubt occurred but these are extremely difficult to detect. Thus experimental work on cell determination has concentrated on embryo manipulation experiments. If a group of cells is transplanted to a new location in the embryo and subsequently develop as if they were still in their *original* location then they are said to be **determined**. If they develop according to their *new* position then they are **not determined**, and can respond to alternative developmental signals. In theory, at least, determination is quite different from differentiation, where the cell has undergone some recognizable ultrastructural and/or metabolic change. In practice it is difficult to identify the point at which such changes become recognizable and so the boundary between the two states is somewhat blurred.

Some of the most useful studies of determination have been carried out using the fruit fly *Drosophila*. In *Drosophila* the fertilized egg first develops into a **larval form**, and then undergoes further extensive developmental changes (*pupation* and *metamorphosis*) before the adult form emerges (Fig. 12.13). The larval form is quite different in structure from that of the adult, in that the larval cells which are going to form the adult structures remain undifferentiated. Instead they are arranged in 19 separate groups, termed ***imaginal discs***, each of which is determined to differentiate into a particular adult structure (Fig. 12.14). Differentiation of the adult form from the imaginal disc cells is triggered during metamorphosis by the action of a steroid hormone, **ecdysone**. After the adult has formed the level of the hormone declines, but some imaginal disc cells remain and these proliferate throughout the life of the fly.

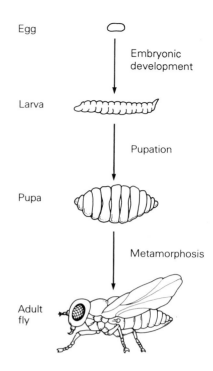

Fig. 12.13 A synopsis of the major stages in the development of *Drosophila melanogaster*.

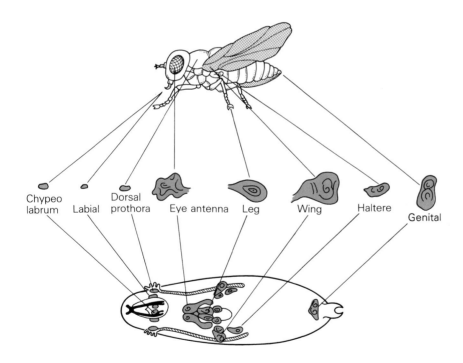

Fig. 12.14 Schematic diagram of the imaginal discs of *Drosophila* and the corresponding adult structures. Redrawn from Fristom *et al.* (1969) in Hanley (ed.) *Problems in Biology: RNA in Development*, University of Utah Press, Salt Lake City.

pupation: *this stage of development, intermediate between the larval and imago, is usually relatively passive.*

metamorphosis: *a change of shape or form during development. For example caterpillar to butterfly, tadpole to frog.*

imaginal: *from the Latin* imago *meaning a likeness which is imagined rather than real.*

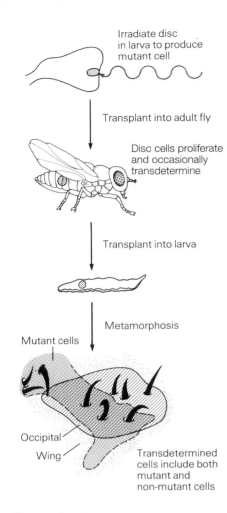

Irradiate disc
in larva to produce
mutant cell

Transplant into adult fly

Disc cells proliferate
and occasionally
transdetermine

Transplant into larva

Metamorphosis

Mutant cells

Occipital

Wing

Transdetermined
cells include both
mutant and
non-mutant cells

Fig. 12.15 Experiment to show that a transdetermined tissue (in this case occipital to wing) can arise from more than one cell. The transdetermined region (red) contains cells from both non-mutant and mutant (grey) tissue. Occipital–wing diagram redrawn from Gehring, W. (1967) Clonal analysis of determination dynamics in cultures of imaginal discs in *Drosophila melanogaster*, *Developmental Biology* **16**, 438–56.

Experimentally the state of determination of a group of imaginal disc cells in the adult can be tested at any time by transplanting it into a larva which is about to undergo metamorphosis. This process can be repeated many times so that a clone of cells can be followed through many generations. Normally the cells differentiate into the expected structure, but, in some cases, cells **transdetermine** to form a structure typical of that from a different imaginal disc. These transdeterminations occur as a result of a change affecting a local group of cells and not as a result of a clone arising from a single mutated cell. This has been demonstrated by experiments in which some of the cells in an imaginal disc in an adult were marked (for example by irradiation) so that they were distinguishable from normal cells (Fig. 12.15). A sample of the disc containing both marked and unmarked cells was transplanted into another larva, as before. In some cases trandetermined structures originated from *both* marked and unmarked cells, suggesting that transdetermination occurs as a result of a mutation in a **control gene** which is involved in specifying a particular developmental signal. This signal is presumably spread to a local group of cells either *directly* by the altered protein product of this gene or *indirectly* as a result of its activity.

In recent years a number of control gene mutants, including some which correspond to transdetermination events, have been characterized in *Drosophila*. Some of these mutants merit a more detailed description since they provide clues to mechanisms of development in many species, including humans.

Genetic studies of *Drosophila* development

Three major events characterize the development of the *Drosophila* embryo: the establishment of polarity in each of the three axes; the dividing of the embryo into the segments of the larva and the specification of the parts of the adult fly. Early in development the polarity of the embryo is established along three mutually perpendicular axes (Fig. 12.16): the **proximo-distal** (from head to tail); the **dorso-ventral** (from back to front) and the **anterio-posterior** (from side to side).

The polarity of the embryo along its three axes is controlled by a group of related genes. An example is *dorsal*, one of the genes controlling polarity along the dorso-ventral axis. Some mutations in this gene result in the production of a 'dorsalized' embryo, totally lacking ventral components. Less severe *dorsal*

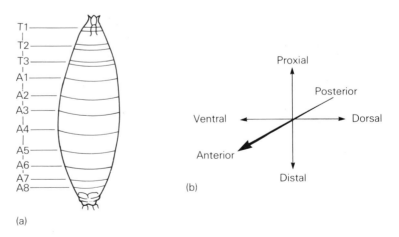

Fig. 12.16 Development of polarity in (a) *Drosophila* (T, thoracic and A abdominal segments) embryo. (b) Shows the orientation of the three axes of polarity in the early *Drosophila* embryo. (a) Redrawn from Nüsslein-Volhard, C. and Wieschaus, E. (1980) Mutations affecting segment number and polarity in *Drosophila*. *Nature*, **287**, 795–801.

Reference Sang, J. (1984) *Genetics and Development*, Longman, London, UK. A broad, useful introduction to developmental genetics.

mutants allow some ventral structures to develop, particularly those nearest to the dorsal end of the axis. This suggests the presence of a gradient of some developmental signal. Interestingly, some of the mutants can be 'rescued' by the injection of cytoplasm from non-mutant *Drosophila* eggs into the embryo. Recently this activity has been shown to reside in the maternal **poly A$^+$ RNA**, at least in some cases. However, most of the biochemical details of such 'rescues' remain unresolved.

Mutations in *dorsal* and the other genes whose products are already present in the egg are known as **maternal-effect mutations**. Maternal effects are thought to be responsible for specifying the early patterns of development in a wide range of species. For example, *parthenogenetic* mouse embryos (unfertilized but containing a full diploid complement of maternal genes) can undergo significant development before they degenerate and die.

Considerable progress has been made in the genetic and biochemical analysis of the genes which control the division of the *Drosophila* embryo into segments, the **segmentation genes**, and those which are involved in specifying the parts of the adult, the **homeotic genes**. Mutations in segmentation genes lead to an altered segment pattern, such as the deletion of alternate segments (pair-rule mutations), or the loss of a group of adjacent segments, i.e. gap mutations (Figs 12.17 and 12.18). *Homeotic mutations* produce quite bizarre effects. One part of the embryo is replaced by a structure typical of a different part, such as homeotic legs in place of antennae (Fig. 12.19).

Many homeotic and segmentation genes have been cloned using **genetic engineering** techniques. These clones have been compared and used to probe the *Drosophila* genome for related sequences. As a result a number of segmentation, homeotic and other as yet uncharacterized genes have been shown to contain a region of about 180 base pairs of DNA, known as the **homeobox**. Using the genetic code this sequence can be translated into a corresponding protein segment of 60 amino acid residues, the **homeo domain**. This protein domain is highly conserved in all the sequences studied and contains a high proportion of basic residues (Fig. 12.20). Homeobox sequences are found not only in *Drosophila* but in all species which have a segmented body pattern, from insects to humans. This suggests an important developmental role for homeo domain-containing proteins, but what mechanisms are used?

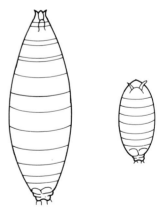

Fig. 12.18 The hunchback mutation in *Drosophila*. This shows an altered segment pattern in the embryo. Labial head segments and thoracic segments T1–3 are missing, and abdominal segments A7 and A8 are fused (see Fig. 12.16(a)). Redrawn from Hulskamp, M. *et al.* (1989) Posterior segmentation of the *Drosophila* embryo in the absence of a maternal posterior organiser gene. *Nature*, **338**, 629–34.

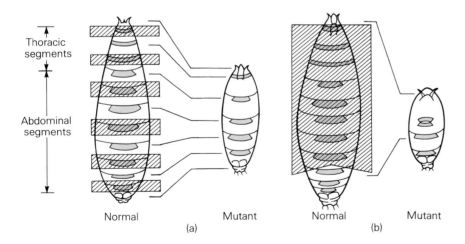

Fig. 12.17 Examples of two types of segmentation mutants in *Drosophila*. Grey shaded areas show segments of the embryo which are lost as a result of the mutation: (a) pair-rule mutant; (b) gap mutant. Redrawn from Nusslein-Volhard, C. and Wieschaus, E. (1980) Mutations affecting segment number and polarity in *Drosophila*, *Nature*, **287**, 795–801.

parthenogenetic: *by virgin birth. It is derived from Greek words* **parthenos** *(virgin) and* genesis.
homeotic mutation: *when one part of the body acquires the characteristics of another part or somite.*

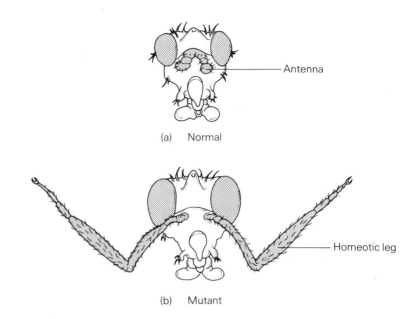

(a) Normal

(b) Mutant

Antenna

Homeotic leg

Fig. 12.19 The effect of a homeotic mutation in the *antennapedia* locus in *Drosophila*. The diagram shows the head with the antennae (red) in the (a) normal and (b) the homeotic leg (also red) in the mutant. Redrawn from Becker, W.M. (1986) *The World of The Cell*, 1st edn. Benjamin Cummings, Menlo Park, Ca, p. 643.

	1																			20
Antennapedia	Arg	Lys	Arg	Gly	Arg	Gln	Thr	Tyr	Thr	Arg	Tyr	Gln	Thr	Leu	Glu	Leu	Glu	Lys	Glu	Phe
Ultrabithorax	Arg	Arg	Arg	Gly	Arg	Gln	Thr	Tyr	Thr	Arg	Tyr	Gln	Thr	Leu	Glu	Leu	Glu	Lys	Glu	Phe
Frog MM3	Arg	Lys	Arg	Gly	Arg	Gln	Thr	Tyr	Thr	Arg	Tyr	Gln	Thr	Leu	Glu	Leu	Glu	Lys	Glu	Phe
Mouse MO-10	Ser	Lys	Arg	Gly	Arg	Thr	Ala	Tyr	Thr	Arg	Pro	Gln	Leu	Val	Glu	Leu	Glu	Lys	Glu	Phe

	21																			40
Antennapedia	His	Phe	Asn	Arg	Tyr	Leu	Thr	Arg	Arg	Arg	Arg	Ile	Glu	Ile	Ala	His	Ala	Leu	Cys	Leu
Ultrabithorax	His	Thr	Asn	His	Tyr	Leu	Thr	Arg	Arg	Arg	Arg	Ile	Glu	Met	Ala	Tyr	Ala	Leu	Cys	Leu
Frog MM3	His	Phe	Asn	Arg	Tyr	Leu	Thr	Arg	Arg	Arg	Arg	Ile	Glu	Ile	Ala	His	Val	Leu	Cys	Leu
Mouse MO-10	His	Phe	Asn	Arg	Tyr	Leu	Met	Arg	Pro	Arg	Arg	Val	Glu	Met	Ala	Asn	Leu	Leu	Asn	Leu

	41																			60
Antennapedia	Thr	Glu	Arg	Gln	Ile	Lys	Ile	Trp	Phe	Gln	Asn	Arg	Arg	Met	Lys	Trp	Lys	Lys	Glu	Asn
Ultrabithorax	Thr	Glu	Arg	Gln	Ile	Lys	Ile	Trp	Phe	Gln	Asn	Arg	Arg	Met	Lys	Leu	Lys	Lys	Glu	Ile
Frog MM3	Thr	Glu	Arg	Gln	Ile	Lys	Ile	Trp	Phe	Gln	Asn	Arg	Arg	Met	Lys	Trp	Lys	Lys	Glu	Asn
Mouse MO-10	Thr	Glu	Arg	Gln	Ile	Lys	Ile	Trp	Phe	Gln	Asn	Arg	Arg	Met	Lys	Tyr	Lys	Lys	Asp	Gln

Fig. 12.20 The sequence of amino acid residues in the homeo domain of proteins corresponding to homeobox sequences. The *antennapedia* gene from *Drosophila* is taken as the 'standard'. Other sequences are from a second *Drosophila* gene, *ultrabithorax*, and genes from the mouse and frog. Sequences which are common with the *antennapedia* sequence are shown in red.

An important clue comes from studies on a group of yeast genes, the **MAT genes**, which show some homology to homeobox sequences. Each MAT gene codes for a protein which regulates all the other genes involved in the control of differentiation of the yeast into one of the two mating types or into a sporulating form. MAT proteins act by binding to specific DNA sequences upstream (that is, towards the 5′ end) from these genes. The amino acid sequence of the MAT proteins suggests that the homeo domain proteins, with their similar sequences, may be **DNA-binding proteins**. These are therefore likely to act in a manner similar to the MAT gene products, that is as master gene regulators.

One might think that the molecular mechanisms of development are close to being solved, but it should be realized that the recent rapid advances in the

Reference De Robertis, E.M., Oliver, G. and Wright, C.V.E. (1990) Homeobox genes and the vertebrate body plan. *Scientific American*, **263**(1), 26–32. A short, well illustrated account of the group of genes which determine body shape.

characterization of homeoboxes and homeo domains represent only part of a complex story. For example, the homeo domain is only one domain of the structure of these proteins. Little is known about the other domains of homeo domain-containing proteins. Other unrelated developmental control proteins will, no doubt, be discovered. In addition, differentiation often involves related changes in a local *group* of cells, suggesting that inter- as well as intracellular communication is involved.

12.7 Positional information and the formation of pattern

During development groups of cells organize themselves into particular patterns, which are modified and refined as development proceeds. A local group of cells which are at first alike and then become organized into a pattern of different cell types is said to form an **embryonic field**. To co-ordinate such pattern formation, cells in a local group in the embryo must communicate with each other. Mechanisms of communication within an embryonic field may be direct, via cell junctions and other cell–cell and cell–extracellular matrix interactions. Alternatively, communications may be indirect via chemical signals which diffuse through the extracellular fluid to specific target cell receptors, in a manner similar to the action of some hormones.

One way in which it is envisaged that such communication takes place is by varying the amount of a chemical signal from point to point, for example in a **concentration gradient**. A specialized type of chemical signal which specifies the development of a particular cellular pattern in an embryonic field is called a *morphogen*. Although the existence of morphogens has been predicted for a considerable time it is only recently that progress in their biochemical identification has been made.

The central idea of the morphogen theory is that it provides positional information allowing the cell to 'sense' its position in relation to other cells in the embryo. Cells are envisaged as being able to respond to a series of environmental cues, such as the concentration of a morphogen. If the level of the morphogen reaches a certain threshold then the cells become determined and differentiate in one way. If that threshold is not reached then the cells follow a different differentiation pathway. Of course this is a gross over-simplification. In reality, individual cells detect and respond to a variety of signals, whose concentration gradients will be established in different directions. Thus the choice of differentiation pathway will be a result of a combined response to a variety of spatial cues.

Some of the most successful theoretical and experimental work on the characterization of the effects of gradients of morphogens has been based on studies of chick limb bud development. Experimental manipulation of the tissue along the anterio-posterior (from thumb to forefinger) axis of the chick limb bud can affect the pattern of digits formed. This provides a convenient experimental system. A small region of cells known as the **polarizing region**, found at the posterior margin of the limb, orchestrates the pattern of digits along the anterio-posterior axis (Fig. 12.21). If cells from this region are transplanted from one limb bud to either the anterior edge (Fig. 12.21(b)) or the tip of another (Fig. 12.21(c)) then digit duplication occurs. The extra digits are formed almost exclusively from cells originating from the limb bud into which the transplant was introduced and *not* from the transplant itself.

The pattern of extra digits formed can be explained in terms of the production of a morphogen by polarizing region cells. Cells along the anterio-posterior axis respond to variations in the concentration of this morphogen by specifying the pattern corresponding to particular digits. An important recent

☐ Many insect and amphibian species can regenerate limbs after amputation. The structure of these limbs can be explained in terms of the positional values of the cells which regenerate the new limbs.

---

morphogen is a substance involved in defining the form of the embryo.

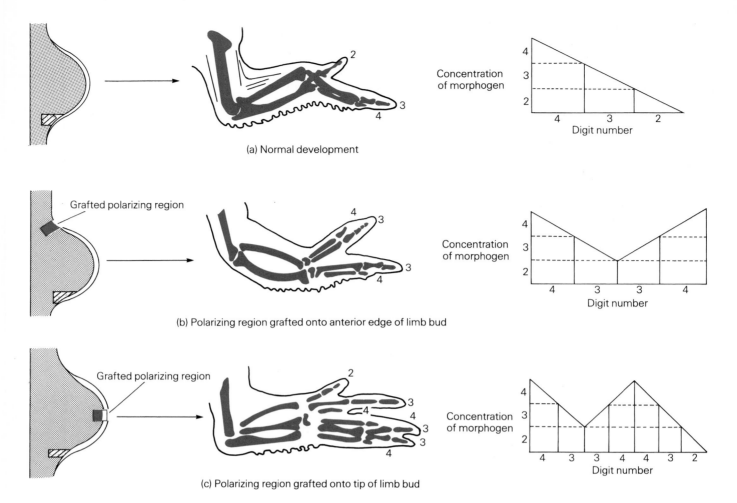

(a) Normal development

Grafted polarizing region

(b) Polarizing region grafted onto anterior edge of limb bud

Grafted polarizing region

(c) Polarizing region grafted onto tip of limb bud

Fig. 12.21 Grafting of polarizing region of the chick limb bud produces digit duplication along the anterior–posterior axis. The graphs demonstrate how a concentration gradient of a morphogen produced by polarizing region cells could account for the pattern of digits.

Fig. 12.22 Molecular model of retinoic acid (retinoate) which is believed to be a morphogen. Courtesy Dr E.E. Eliopoulos, Department of Biochemistry and Molecular Biology, University of Leeds, UK.

Box 12.6
French, British and American flags

The French, British and American flags all have three colours: red, white and blue. The distribution of these three colours is quite different in each, producing characteristic patterns. If the three colours each correspond to types of differentiated cells then it is easy to see how a simple pattern (the French flag) or complex ones (British and American flags) can be produced using the same basic units.

Reference Brickell, P.M. and Tickle, C. (1989) Morphogens in chick limb development. *BioEssays*, **11**, 145–9. Excellent account of the role of retinoic acid in the development of chicken limbs.

Reference Green, J.B.A. (1990) Retinoic acid: the morphogen of the main body axis? *BioEssays*, **9**, 437–9. A short overview of the role of retinoic acid in development.

advance in this system is the accumulation of evidence that the morphogen involved is retinoate (retinoic acid), the acid corresponding to retinol or vitamin A (Fig. 12.22). If confirmed this would be the first morphogen to be identified biochemically. Retinoate is small enough to pass between cells via gap junctions and this may give a valuable clue as to how such signals are transmitted.

Morphogens have been proposed as being involved in module development in the apical meristems of plants, but as yet these putative plant morphogens have not been characterized biochemically. A series of compounds, known collectively as **plant growth regulators**, do however have profound effects on plant development and their biochemistry is at least partially understood. Plant growth regulators include **auxins**, **gibberellins**, **cytokinins**, abscisic acid (ABA), **oligosaccharins** and **ethylene** (Section 8.9). These are all small molecules capable of penetrating the cell wall. Their effects are quite complex but an example illustrates the way in which they can affect development. Gibberellic acid and ethylene have opposing effects on young pea shoots. Ethylene-treated cells develop a longitudinal orientation of microtubules within the cell resulting in the growth of short fat shoots. In contrast, gibberellic acid-treated cells develop a transverse microtubular orientation leading to the growth of long thin shoots.

Much work needs to be done on the molecular mechanisms involved in the mode of action of morphogens and growth regulators, but at this stage there is no reason to suppose that the mechanisms will differ greatly from those that have been elucidated for other regulatory substances such as hormones.

12.8 Cell lineage studies

Useful information can be gained by following the fates of groups of cells which form identifiable structures during development. Several examples of such studies have been mentioned in previous sections. Ultimately it is desirable to study the development of all the individual cells of the organism. In vertebrates, cell lineage studies are difficult because development is complex and significant variations occur between individuals. In certain invertebrates, such as nematodes, molluscs and annelids, development is much more ordered, varying little or not at all between individuals.

The complete cellular fate map for the nematode *C. elegans* is known. This map specifies the historical origin of each of the about 1000 somatic and approximately 2000 germ cells of the adult (Fig. 12.23). The body plan of *C. elegans* has a considerable similarity to those of vertebrates, in that it is bilaterally symmetrical and the various structures differentiate from the three primary germ layers. Tissues are composed of cells of different lineages even in this relatively simple organism and in only a few cases are structures derived from a single clone of cells. In contrast to many vertebrates, most of the cells develop autonomously and are not dependent on interactions with other cells to promote differentiation. The evidence for this comes from **laser microsurgery** experiments, where focussed lasers are used to destroy individual cells. In most cases structures derived from the destroyed cell are missing in the adult, but surrounding structures are unaffected. In vertebrates such detailed lineage studies are not yet possible, but in contrast to *C. elegans* **embryonic induction** by specific cell–cell interactions is common.

A useful technique for the study of cell lineages in higher vertebrates, such as the mouse, is the production of chimeras. A *chimera* is an embryo which develops from aggregates of genetically different groups of cells. Different combinations of cells isolated from more than one embryo at an early stage of development can develop into a chimeric adult. Marking cells in some way,

---

Exercise 3

An extract from eye tissue promotes lens cell differentiation in an explant culture of head ectoderm, whereas extracts from other tissues do not. This suggests that the eye extract contains a morphogen. Two experiments were carried out: (a) the extract was treated with trypsin and then added to the culture; (b) the extract was dialysed extensively before it was added to the culture. In neither case did the treatment affect the ability of the extract to promote differentiation. What do you conclude about the nature of the morphogen?

Exercise 4

Give examples of species to which the following techniques for the study of development are particularly suited. Give reasons for your choice. (a) Laser ablation of individual cells. (b) Genetic analysis of developmental mutants. (c) Transgenosis.

---

chimera: a word used a lot in mythology to describe an animal made from parts of different animals.

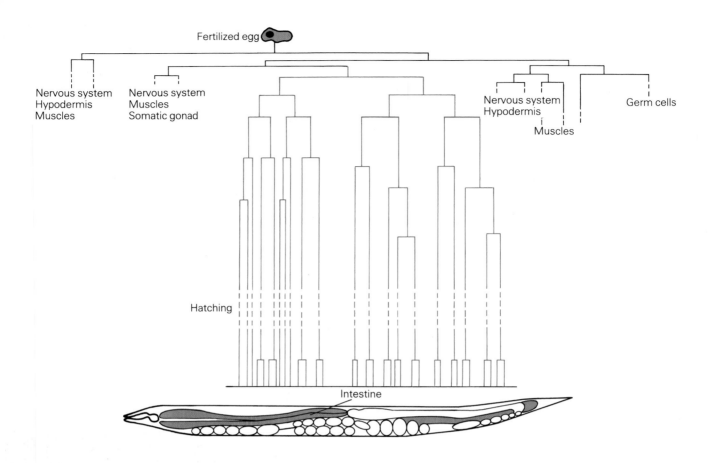

Fertilized egg

Nervous system
Hypodermis
Muscles

Nervous system
Muscles
Somatic gonad

Nervous system
Hypodermis

Muscles

Germ cells

Hatching

Intestine

Fig. 12.23 Part of the cell lineage of *C. elegans* showing the development of the cells that form the intestine (red). Redrawn from Alberts, B., Bray, D., Lewis, J. *et al.* (1989). *Molecular Biology of the Cell*, 2nd edn. Garland, New York, p. 904.

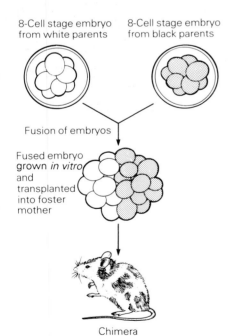

8-Cell stage embryo from white parents

8-Cell stage embryo from black parents

Fusion of embryos

Fused embryo grown *in vitro* and transplanted into foster mother

Chimera

Fig. 12.24 The production of chimeric mice. The chimera contains cells from four parents (two black, two white).

for example cells from a black strain of mouse combined with cells from a white strain, allows subsequent lineages to be followed (Fig. 12.24). Cells in the very early mammalian embryo are totipotent. Studies with chimeras have shown that it is the *position* of the cell in the embryo rather than its genetic composition which determines its subsequent lineage.

Cell lineages can also be studied by the introduction of some foreign genetic material by **microinjection** at an early stage of development. The foreign DNA can become incorporated into the host and the site of incorporation controlled experimentally. Thus the expression of the foreign DNA can be confined to particular cell types, and to particular stages of development. This process is known as **transgeneosis** (producing transgenic organisms) and has a variety of applications.

12.9 Cell differentiation and development in the nervous system

The structure of the human brain and, indeed, the brain of other species, is more complex than any of the other organs of the body. An understanding of how such a complex structure is formed is, therefore, the ultimate challenge for developmental biologists. The patterns of connections between

> *Box 12.7*
> *The brain as a parallel processing computer*
>
> One of the features of the brain is that it processes information in a parallel fashion, or deals with many bits simultaneously. Computers also process large amounts of information, but until recently only one bit at a time. An understanding of how the connections in the brain are made and how the brain processes information perhaps could be used to design a much more powerful computer. In fact this is already happening. Computers that use parallel processing, albeit in a rudimentary form when compared to the brain, are being developed. Such computers are very useful to biologists. They can be used to test models of brain function and development, to see which most closely model reality.

neurons must be precise and an understanding of how these patterns are formed will provide clues as to how the brain can store and process so much information.

The brain and the rest of the nervous system consist primarily of neurons, which transmit the electrical impulses both within individual cells and from cell to cell via **synapses**. In addition there are a number of types of helper or **glial cells**. Glial cells perform a variety of functions from forming the **myelin sheath** around certain neurons, to lining the cavities of the brain, guiding growing nerve axons, and performing protective functions in a manner similar to macrophages.

See Chapter 9

During the development of the nervous system neurons differentiate and produce the long appendages called ***axons*** and ***dendrites*** (Figs 9.1, 9.12 and 12.25) along which the electrical impulses are transmitted. The directions taken by these growing appendages have to be carefully controlled, so that on reaching their destination (for example, a synapse or neuromuscular junction) the appropriate connections can be made.

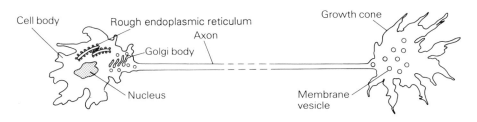

Fig. 12.26 Schematic diagram of a developing neuron.

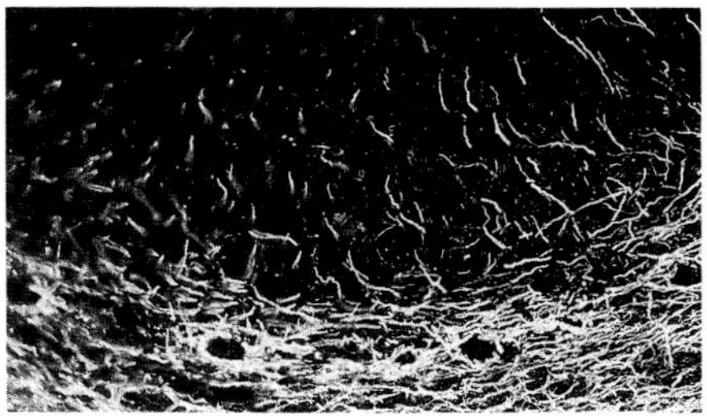

Fig. 12.25 Photomicrograph of nervous tissue (×1000). Courtesy M.J. Hoult, Department of Biological Sciences, The Manchester Metropolitan University, UK.

Intracellular transport in the developing neuron

The developing neuron has an elongated structure (Fig. 12.26). This poses particular problems for transport of materials within the cell, particularly between the **cell body**, where synthesis of new cellular material takes place, and the **growth cone**, the point where axon growth occurs (similar but smaller structures are found at the extremities of the dendrites). During axonal growth the position of the cell body remains fixed while the axon extends and migrates through the substratum. The internal structure of the axon shows a highly organized array of cytoskeletal components, mainly microtubules and neurofilaments. The growth cone has high concentrations of actin filaments (Fig. 12.27).

The role of these components in neuron development is shown by the action of chemicals on isolated neurons in cell culture. In the presence of **cytochalasin B** (which prevents the polymerization of actin into filaments) the growth cone ceases movement, but the overall structure of the neuron

See Chapter 6

axon: *the main impulse-carrying appendage of the nerve cell; derived from the Greek word for axis.*
dendrite: *the highly branched smaller appendages forming connections to other cells. Dendrite is derived from the Greek* dendron *meaning tree.*

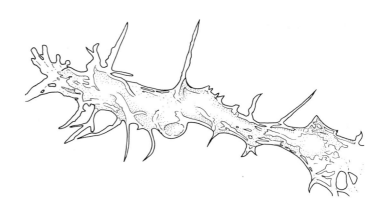

Fig. 12.27 Axon-like extension of a neuroblastoma cell. Actin is concentrated in the growth cone-like regions. Redrawn from Osborn, M. *et al.* (1978) Stereo immunofluorescence microscopy: I. Three dimensional arrangement of microfilaments, microtubules and tonofilaments, *Cell,* **14,** 477–88.

remains intact. If **colchicine** (which disrupts microtubule organization) is added the growth cone retracts back into the cell body. Cytoskeletal components comprise about 50% of the total protein of the brain, a figure which emphasizes their importance in neuron development.

Guidance of the growing axon

When the growth cone moves forward, extending its attached axon, it has to pass across a variety of surfaces, either those of other cells or components of the extracellular matrix. How is this movement guided? In cell culture only the cell body and growth cone are attached to the **substratum**, the axon is not attached. Interaction between growth cone surface components and the substratum can guide the direction of growth. Coating the surface of cell culture dishes with a variety of substances in a particular pattern can induce cultured neurons to follow prescribed paths (Fig. 12.28). In general, axons will extend along paths which allow the greatest degree of interaction between the neuron and the substratum.

In vivo, similar contact guidance phenomena seem to occur. Cell–extracellular matrix interactions are involved in the guidance of pioneer growth cones, although the details of these interactions have still to be determined. In addition, pioneer neurons serve as guides for further neurons, and as a result nerve axons group together in bundles. Recently significant advances in the biochemical identification of the cell surface molecules involved in this cell–cell **adhesion** have been made. One of the best characterized of these is **N-CAM (neuronal cell adhesion molecule)**, a large glycoprotein. If N-CAM binding sites are blocked with specific antibodies then the tendency of neurons to group tightly together is greatly reduced. Cell-specific cell adhesion molecules are now thought to be a mechanism for the establishment and maintenance of cell–cell contact in a variety of cell types.

Chemical signals present in the extracellular fluid are also involved in guiding growth cones towards their destination. In the presence of **nerve growth factor (NGF)** certain neurons will grow in the direction of the source of this factor. Indeed, some neurons require a constant supply of NGF for their survival and without it they degenerate and die.

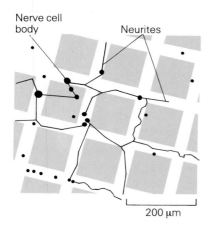

Fig. 12.28 The movement of growth cones on the surface of a culture dish. The dish is coated with a symmetrical pattern of polyornithine and palladium. The growth cones adhere strongly to the polyornithine and stay off the palladium. Redrawn from Letourneau, P.C. (1975) Cell-to-substratum adhesion and guidance of axonal elongation, *Developmental Biology,* **44,** 92–101.

Nerve cell body

Neurites

200 μm

Reference Levi-Montalcini, R. and Calissano, P. (1979) The nerve growth factor, *Scientific Amercian,* **240**(6), 68–77. The story of her Nobel prizewinning work on NGF by Levi-Montalcini.

A group of climbers at the foot of a rockface that they wish to scale are in a similar position to a group of developing nerve cells. Both require to reach specific targets by finding the easiest route over the surface. A pioneer climber (growth cone) starts up the face (extracellular matrix), feeling for the best holds (adhesion sites). He is attached to the foot of the rockface (cell body) by a rope (axon) which extends as he proceeds. He provides extra anchorages (CAMs) as he goes for those who will follow. All the time he is checking on his target (monitoring the level of NGF) and adjusting his direction towards its source. Other climbers follow the same route, using the extra anchorages provided.

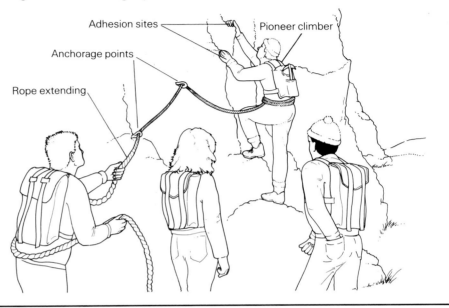

When a growing axon reaches its final destination it undergoes a series of changes to form a permanent junction (for example, synapse or neuromuscular) with another cell. In particular, the membrane of the growth cone changes from being a structure involved in feeling its way along the substratum to one with a main function of **neurosecretion** (Section 9.4). The details of this transformation remain obscure.

Connecting up the system

This section considers the development of the nervous system in relation to one or a small number of axons, but in the human brain, for example, there are about 10^{10} axons and up to 10^{13} separate connections. How are the correct connections made and how are the numbers of axons and target cells co-ordinated? Patterns of nerve connections in vertebrates are most easily

Following an accident which causes an amputation, a surgeon can sew back the amputated part using microsurgical techniques. As well as reconnecting blood vessels to provide nutrients it is vital that nerve fibres are carefully joined, as any muscle that loses its connection to the central nervous system will degenerate. Not all nerve axons in a bundle will be reconnected exactly as they were before the accident, and so some changes in nerve signals will result. This explains why amputees have to relearn how to use the amputated part (see also Box 10.2).

References Davies, J.A. and Cook, G.M.W. (1991) Growth cone inhibition – an important mechanism in neural development? *BioEssays*, **13**, 11–15. Tosney, K.W. (1991) Cells and cell-interactions that guide motor axons in the developing chick embryo. *BioEssays*, **13**, 17–23. Strittmatter, S.M. and Fishman, M.C. (1991) The neural growth cone as a specialized transduction system. *BioEssays*, **13**, 127–34. These three essays give an up-to-date view of the motor neuron growth cone.

studied using motor neurons. Motor axons grow out from the neural tube during development and make connections with the peripheral muscles at neuromuscular junctions. Initially there is a large overproduction of neurons and so several can be attached to each muscle fibre. In a competitive process, the details of which are poorly understood, most of these connections are eliminated as development proceeds, with the consequent degeneration of the corresponding neurons (Fig. 12.29). Such overproduction and **programmed cell death** may seem wasteful but ensures that each muscle fibre does become connected to a motor neuron. If a muscle fibre fails to become innervated by a certain stage in its development then it too degenerates and dies. It is paradoxical that programmed death of some neurons should be an important process in the growth and development of others, but this seems to be a common phenomenon in a variety of developing tissues.

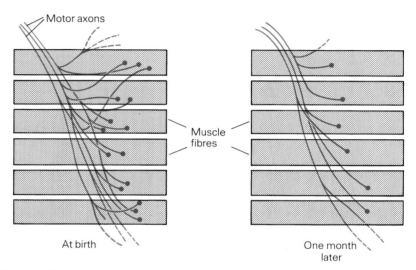

Fig. 12.29 Simplified schematic diagram to show the elimination of surplus neuromuscular junction in mammalian skeletal muscle after birth. Redrawn from Alberts, B. *et al.* (1989) *Molecular Biology of the Cell*, 2nd edn. Garland, New York, p. 1126.

12.10 Overview

Developmental biology is concerned with the growth of the fertilized egg and the differentiation of the cells produced to form the tissues and organs of the adult. Development progresses through a number of defined stages many of which are common to a variety of species.

Stages in animal development are characterized by particular cellular processes. During cleavage a series of rapid cell divisions occurs; gastrulation is accompanied by large-scale cell migration; and during organogenesis groups of cells organize themselves into patterns.

In general, all differentiated cells have the same genetic complement, and specialized characteristics are a result of variations in gene activity. Gene activity can be controlled at a number of levels, although in quantitative terms the most important is gene transcription. Cell differentiation is preceded by determination, when cells become committed before ultrastructural and metabolic changes become recognizable.

Studies of *Drosophila* mutants have characterized many developmental control genes. A number of these contain a highly conserved region of DNA, the homeobox. The corresponding protein segment, the homeodomain, may form part of the structure of a family of DNA-binding proteins, which act as master gene regulators. Homeobox regions have been found in the genes from a variety of species, including humans.

Cells form specific patterns as a result of interpreting positional information in relation to surrounding cells. Experimental manipulations can be used to produce cellular fate maps, charting the lineages of individual cells.

A study of the development of the nervous system, involving the most complex of developmental events, illustrates the challenges to be faced.

Answers to Exercises

1. 1.35%. The remainder is accounted for by introns and control elements of genes active in that cell type, genes which are not transcribed in that cell type, pseudogenes, repeated DNA.
2. Red blood cell, haemoglobin; lens fibre cell, crystallins; mammary gland cell, caesin, lactalbumin; skeletal muscle cell, actin, myosin, troponin; thyroid gland cell, calcitonin; fibroblast, collagen; gastric zymogen cell, pepsinogen; B lymphocyte, immunoglobulins.
3. 1. It is not a protein. 2. It is too large to pass through the dialysis membrane. Note that these conclusions are not absolute. For example, it could be a protein which is not degraded by trypsin. Similarly it could be a small metabolite which is bound to something else in the extract which prevents it from being removed by dialysis.
4. Possible answers are: (a) *C. elegans*. Cell lineage is known and is virtually invariant. (b) *Drosophila*. Many mutants non-lethal. (c) Domestic farm animals. Can be used as biological factories for the production of a particular protein.

QUESTIONS

FILL IN THE BLANKS

1. Following _____ the egg begins a series of rapid cell divisions, a process called _____ . Soon after division starts in _____ some asymmetry is seen. Cells at the _____ pole are larger than those at the _____ pole. The individual cells formed are called _____ and the cavity in the centre is the _____ . Fluid is contained within the _____ by the formation of _____ junctions between the surrounding cells.

The hollow ball of cells formed by the _____ divisions, the _____ is largely _____ . During gastrulation considerable cell _____ take place, forming the three _____ _____ _____ of the embryo.

The organs and tissues of the embryo are formed during _____ . Individual organs and tissues contain cells which originate from more than one germ layer. For example, a limb will contain both _____ and _____ tissues. The central nervous system develops from the _____ _____ and the peripheral nervous system from the _____ _____ . Interaction of cells during development to form defined structures is called _____ _____ .

Choose from: animal, blastocoel (2 occurrences), blastomeres, blastula, cleavage (2 occurrences), ectodermal, embryonic induction, fertilization, mesodermal, migrations, neural crest, neural tube, organogenesis, primary germ layers, right, undifferentiated, vegetal, *Xenopus*.

MULTIPLE-CHOICE QUESTIONS

2.   Indicate which of the following are examples of homeobox-containing genes in *Drosophila*:
A. maternal-effect genes
B. segmentation genes
C. polarity genes
D. homeotic genes
E. morphogen genes
F. housekeeping genes

3.   Indicate which of the following are characteristic features of morphogens:
A. they are molecules of low M_r
B. they specify the pattern of cells in an embryonic field
C. they bind to specific receptors on the surface of target cells
D. they require to be synthesized continuously throughout development, otherwise the cells on which they act will degenerate and die
E. they are present in small amounts
F. they are molecules of high M_r

4.   Which of the following is true of stem cells?
A. they are undifferentiated
B. they are not determined
C. they remain undifferentiated in the adult, but are determined
D. they are totipotent
E. their population is maintained throughout the life of the organism by cell division
F. they are reponsible for cell regeneration in the liver

5.   The growth cone of a developing neuron has several functions. Which of the following are included?
A. secretion of nerve growth factor (NGF)
B. attachment to components of the extracellular matrix
C. introduction of new membrane into the growing axon
D. actin filament-based movement
E. synthesis of membranes and proteins
F. formation of a permanent junction with a target cell

6.   State which of the following are true:
A. both endosperm and plant embryo are triploid
B. endosperm is a diploid tissue
C. endosperm is a triploid tissue
D. embryo is diploid while endosperm is triploid
E. both endosperm and plant embryo are diploid
F. the embryo is diploid while endosperm is haploid

SHORT-ANSWER QUESTION

7.   Give examples of biochemical and cellular processes involved in cell migration.

Answers
to questions

Chapter 1

1. The cytoskeleton consists of microfilaments, intermediate filaments and microtubules. The first consists largely of the protein actin, the last tubulin. Intermediate filaments are composed of a variety of proteins which are cell-specific. Microtubules form a scaffold in the cell. They are organized by microtubule organizing centres (MOCs). Centrosomes consist of two centrioles and pericentriolar material (PCM). The spindle is formed during cell division by a reorganization of the microtubules.

2. A. lysosome
    B. lysosome
    C. centrosome
    D. microfilaments

3. ATP synthesis — mitochondria
    protein synthesis — ribosomes
    genome — nucleus
    glycosylation — Golgi apparatus
    hydrolytic enzymes — lysosomes
    H_2-production — hydrogenosomes

4. A. False
    B. False
    C. False
    D. True

5. coated pit, coated vesicle, endosome, primary lysosome, secondary lysosome, residual body.

6. Increase in $[H^+]$
    $$= 10^{-5}-10^{-7}\,mol\,dm^{-3}$$
    $$= 9.9 \times 10^{-5}\,mol\,dm^{-3}$$
    $$= 9.9 \times 10^{-5}\,mol/10^{15}\,\mu m^3$$

    Volume of lysosome
    $$= \frac{4}{3}\,\pi r^3 = \frac{4}{3}\pi 0.5^3 = 0.524\,\mu m^3$$

Increase in mol H^+
$$= \frac{0.524 \times 9.9 \times 10^{-15}\,mol}{10^{15}}$$

Increase in H^+
$$= 5.18 \times 10^{-21}\,mol\,H^+$$

Increase in H^+
$$= 5.18 \times 10^{-21}\,g\,H^+$$

Number of molecules ATP
$$= \frac{5.18 \times 10^{-21} \times 6.02 \times 10^{23}}{2}$$
$$= 1559$$

7. (a)

Step	Volume (cm^3)	Total protein (mg)	Markerase activity (U)	Specific activity (U mg^{-1})	Purification (fold)
Homogenate	120	1800	5.4×10^{-3}	3×10^{-6}	1
Differential centrifugation	20	180	5.1×10^{-3}	2.83×10^{-5}	9.4
Rate-zonal centrifugation	6	40	4.9×10^{-3}	1.23×10^{-4}	41.0

(b) 90.7%

8. ^{14}C-Leu activated as tRNA-^{14}C-Leu. Used for protein synthesis on rough endoplasmic reticulum (RER). ^{14}C-Leu incorporated into putative lysosomal enzymes. Transported: RER → Golgi apparatus → coated vesicle → primary lysosome. Cellular and biochemical details of each stage should be given.

Chapter 2

1. Bacteria are cells <u>without</u> a differentiated nucleus. Viruses are <u>not</u> cellular. Bacterial plasma membranes do not contain <u>sterols</u> but possess <u>branched</u> fatty acyl chains. Only Gram-negative bacteria contain <u>lipids</u> in their cell walls.

In archaebacteria, the initiation complex for protein synthesis contains <u>Met-tRNA</u>, while in eubacteria, it contains fMet-tRNA, and, in eukaryotes <u>Met-tRNA</u>.

The protein envelope surrounding the viral genome is termed <u>capsid</u>; this is formed by a number of <u>protein</u> <u>subunits</u>, the capsomers.

The genome of tobacco mosaic virus contains single-stranded <u>RNA</u>, ϕX174 contains <u>single-stranded</u> DNA while poxvirus contains <u>double-stranded</u> <u>DNA</u>.

2. A.

3. D.

4. D.

5. B.

6. (a) and (b): see main text.

7. The volume ratio will be 1000 : 3, and the percentage total volume 0.3%.

8. Useful for catalysing DNA synthesis under conditions where nucleases or other enzymes are inactive. In particular, this enzyme is not heat denatured in the process of separation (melting) of the newly synthesized strands of DNA at 95°C. This allows the amplification and detection of a single copy of a target DNA sequence in as little as $10\,\mu$g of genomic DNA.

9. Since viruses are obligatory parasites, they must presumably follow strictly the host cell rules concerning replication, transcription and translation.

10. Perform an infection run in the presence of an endonuclease.

11. 6×10^7 (g/mol)/
6×10^{23} (particles/mol)
$= 10^{-16}$ g/particle.

Chapter 3

1. <u>DNA</u> is the genetic material in organisms. In bacterial cells it is found in the <u>nucleoid</u> associated with basic proteins. <u>Supercoiling</u> of the bacterial DNA shortens the overall length of the molecule, allowing the DNA to be packed into the cell. In eukaryotic organisms DNA is associated with basic proteins called <u>histones</u>. These proteins are enriched in the amino acid residues <u>lysine</u> and <u>alginine</u>. The fundamental structural unit of chromatin is the <u>nucleosome</u>. These consist of about two and a half <u>turns</u> of DNA wrapped around an <u>octomer</u> of <u>histone</u> proteins. The <u>H1</u> protein is found outside the <u>core</u> attached to <u>linker</u> DNA. Further <u>folding</u> <u>(coiling)</u> of the chromatin produces greater packing of the DNA ultimately forming the condensed <u>chromosome</u> present at <u>mutaphase</u> <u>(mitosis)</u>. Chromatin is attached to the <u>lamina</u> of the nuclear <u>skeleton</u>, which in turn is attached to the <u>inner</u> <u>membrane</u> of the nuclear <u>envelope</u>.

2. nucleoid – bacterial cells
nucleus – chromatin
H1 histone – linker DNA
annulus – porosome
nucleolar – organizing region – nucleolus

3. A. T
B. T
C. F
D. F
E. F

4. (a) Chromosome is 1.1×10^6 nm long
length increased by $750 \times 0.34 \times 2$ nm s^{-1}
therefore time $= 1.1 \times 10^{-6}/$
$750 \times 0.34 \times 2$ s
$=$ about 36 min.
(b) Cell division begins before DNA replication is complete. Since the length of the cell cycle is about half the time needed for DNA replication, *E. coli* cells growing rapidly contain about two DNA molecules.

5. (a) 21 mm $= 21 \times 10^6$ nm
therefore no. of bp $= 21 \times 10^6/$
$0.34 = 6.18 \times 10^7$
(b) Length of mitotic DNA $= 21 \times 10^3/$
$1.2 \times 10^4 = 1.75\ \mu$m.

Chapter 4

1. Biological membranes contain protein and lipid molecules which are associated <u>non-covalently</u>. The weight ratio of these components varies according to the cellular locality and <u>function</u> of membrane.

A characteristic feature of the lipids found in biological membranes is that they are <u>amphipathic</u>. Examples of these lipids are glycerophospholipids such as <u>phosphatidyl</u> <u>ethanolamine</u> and sphingolipids such as <u>sphingomyelin</u>.

Proteins found in biological membranes generally have specific functions. These functions include membrane <u>transport</u>, hormone <u>reception</u> and <u>transmembrane</u> signalling. Some proteins are deeply embedded in the lipid bilayer and are referred to as <u>integral</u> proteins. Extraction of these proteins requires the use of <u>detergents</u>.

Both membrane proteins and membrane lipids may diffuse rapidly in the <u>plane</u> of the membrane. Important in maintaining the fluid nature of biological membranes is the <u>length</u> and degree of <u>unsaturation</u> of the constituent fatty acyl chains. In mammals the presence of <u>cholesterol</u> is also important in modifying membrane fluidity.

2. A. T
B. F
C. F
D. F

3. A. F
B. F
C. T
D. T

4. A. F
B. F
C. T
D. F

5. A. T
B. F
C. F
D. T

6. Description of sodium dodecyl sulphate (SDS) polyacrylamide gel electrophoresis, where all membrane proteins (both integral and peripheral) are solubilized in a solution of SDS and the solubilized components separated by electrophoresis.

7.   Examples include: membrane transport, transmembrane signalling, energy transduction, enzymic functions, cell movement, endocytosis, exocytosis, phagocytosis.

8.   Examples of common properties:

- Contain lipid and protein in a non-covalent association. Both consist of a continuous double layer of polar lipid molecules in which a number of membrane proteins are embedded.
- Both membranes maintain an electrical potential with the outside kept positively charged with respect to the inside which is negatively charged.

Examples of dissimilar properties:

- The lipid component of the plasma membrane contains a much higher proportion of cholesterol than the inner mitochondrial membrane.
- The inner mitochondrial membrane lacks carbohydrate, whereas the plasma membrane contains carbo-hydrate as glycoproteins and glyco-lipids. The proportion of protein with respect to lipid is generally much greater in the inner mitochondrial membrane than in the plasma membrane.
- The maintenance of membrane potential in the two membranes is by a different mechanism.

9.   Because the transport of each substance is mediated by a separate carrier the final ratio depends *only* on the magnitude of the Na^+ gradient. Therefore, the ratio of internal to external concentration for glucose and each of the four amino acids is maintained at 20 to 1.

Chapter 5

1.   The mitochondrion has two membranes: the outer membrane, which is permeable to small ions, and the inner membrane, which is impermeable to small ions. The internal aqueous phase is termed the matrix and contains the enzymes of the TCA cycle. The inner membrane contains the electron transport chain, the ATP synthetase and several transporters . Submitochondrial particles have a transverse membrane orientation.

The chloroplast is a plastid found in plants. The function of the chloroplast is to assimilate carbon dioxide into carbohydrate in a reaction process called photosynthesis . This consists of two sets of reactions. The initial light reactions use the energy of light to transfer hydrogen atoms from water to $NADP^+$, and to generate ATP. The enzymes of this set of reactions are embedded within the thylakoid membrane.

2.   A. F
     B. F
     C. F
     D. T
     E. T

3.   A. T
     B. T
     C. F
     D. F
     E. T

4.   See Fig. 5.3b and text p. 120.
5.   See text p. 132.

Chapter 6

1.   The six different kinds of actin in cells do not differ very much in amino acid composition. Thus, they are highly-conserved molecules with important roles in the cell. All are globular proteins 5.5 nm in diameter with an M_r of 42 000. α-actins are found in muscle tissues, while β- and γ-actins occur in non-muscle cells. N-γ-methyl-histidine occurs in all actins and may be used as a marker in its structural identification.

Group 2 proteins control actin polymerization e.g. gelsolin is involved in amoeboid movement. Capping proteins stabilize polymerized filaments and allow fairly permanent extensions of cytoplasm e.g. microvilli . Group 3 proteins cross-link actin into stress fibres for cell migration, into microvilli and into stiff actin rod associated with fertilization of an ovum by a spermatozoon.

Spectrin cross-links actin to form the rigid erythrocyte cytoskeleton. Group 4 proteins anchor actin filaments to membranes. Without anchorage, stress fibres would not function to allow cells to change shape and particularly to spread. Signals received at the plasmalemma often stimulate changes in cell shape via these anchoring molecules.

The proteins making up microfilaments and microtubules are the highly conserved molecules actin and tubulin . Intermediate filaments are coded for by multiple gene families and are composed of two double α-helices coiled about one another into a cable with a left-handed twist. The amino and carboxy termini are randomly coiled with variable amino acid sequences as are three regions linking four blocks of the rod domain. There are more than 20 kinds of cytokeratin intermediate filaments. Type I are acidic while type II are basic proteins. Neurofilaments exist as a triplet of proteins in neurons . Cells that are stimulated to divide rapidly in culture, re-express vimentin .

2.   A. T
     B. F
     C. T
     D. F
     E. F
     F. F

3.   A. T
     B. T
     C. F
     D. T
     E. T
     F. F

4.   A. T
     B. T
     C. F
     D. T
     E. F
     F. F

5.   TEM uses thin sections, giving little idea of the three-dimensional structure. (a) The electron beam used in high-voltage electron microscopy has enough energy to penetrate whole thin cells to give a three-dimensional image. (b) Minimal tissue processing gives minimal distortion of the cytoskeleton. A surface view of an etched surface gives a three-dimensional view.

6.   An antibody is an immunoglobulin with complementary structure to that of its antigen. Thus, it binds only to its antigen and not to any other molecules. If a marker molecule is first bound to the antibody, the marker then shows up the position of the antibody bound to antigen, either on gels or in cells or tissues. Markers include fluorescent dyes, colloidal gold and enzymes such as peroxidase or alkaline phosphatase.

7. See Table 6.1.

8. Stress fibres are F-actin filaments cross-linked into thick bands. They act as skeletal struts giving support to a cell.

9. Intermediate filament content is related to the tissue origin of a cell. It is an important means of determining the tissue origin of tumour cells, as characteristic tissue structure has often disappeared. This helps decide what the tumour is and what treatment to give the patient. Different tumour types respond differently to radio- and chemotherapy.

10. Polymerization at (+) ends of polar microtubules makes the spindle lengthen between the poles. Depolymerization at (−) ends of kinetochore microtubules shortens them. The effect of these is to push and pull (respectively) chromatids apart and towards the poles.

11.
desmin – muscle
vimentin – mesenchymal cells
glial fibrillary acidic protein – glial cells
cytokeratins – epithelial cells
neurofilament proteins – neurons

Chapter 7

1. Focal contacts, or adhesion plaques, are points of close contact between the cell and the extracellular matrix . Immunofluorescent studies have shown that cytoskeletal proteins, such as actin, talin and vinculin , together with fibronectin and heparan sulphate proteoglycans co-localize to these sites. More recently, certain integrins have also been localized to these structures suggesting that they form a transmembrane link between the matrix and the cytoskeleton.
Fibronectin is an extracellular matrix protein which stimulates cell adhesion and migration . It binds to several integrins, through an RGD sequence, and can also bind fibrin , heparin and denatured collagen. Its ability to bind to the cytoskeleton through a transmembrane receptor, and to self associate forming long fibrils, suggests that it plays an important role in the cellular organization of the extracellular matrix .

Integrins are a family of integral membrane receptors, which bind a number of structural proteins via an internal RGD sequence. They are composed of two large subunits termed the α- and β-subunits. In some cases the α-subunit is itself a disulphide-linked heterodimer. Both subunits are transmembrane proteins which co-operate in attaching to the cytoskeleton through talin . This interaction with the cytoskeleton is regulated by phosphorylation of the cytoplasmic domain of the β-subunit.

2. A. T
    B. F
    C. F
    D. F
    E. T
    F. T

3. A. T
    B. T
    C. F
    D. T
    E. T
    F. F

4. A. T
    B. T
    C. F
    D. T
    E. T
    F. F

5. A. F
    B. F
    C. T
    D. T
    E. T
    F. F

6. A. F
    B. T
    C. T
    D. F
    E. F
    F. F

7. A. T
    B. F
    C. T
    D. F
    E. T
    F. F

8. Proteoglycans, collagens, elastin, structural proteins and hyaluronan. The types and proportions of these basic components show great variations between the different tissues.

9. Figure should show: (1) A multimeric protein (i.e. at least two chains); (2) cell-binding site; (3) at least one binding site for a matrix component, and (4) a site of self-association needed for fibril formation.

10. Initially by amino-terminal association followed, by interchain disulphide exchange. Interchain covalent bonds may also be formed by the action of transglutaminase.

11. Fig. 7.28.

12. The presence of an RGD sequence in the ligand.

13. Phosphorylation of a tyrosine residue in the cytoplasmic tail of the β-subunit appears to be the most common control mechanism. Phosphorylation disrupts/prevents the formation of the talin–integrin complex.

14. Glanzmann's thrombosthenia (lack of GpIIb/IIIa) and leukocyte adhesion deficiency (lack of β-subunit).

15. Should include: (1) the structure of the large aggregating proteoglycan–hyaluronan complexes of load bearing tissues; (2) the water-binding properties; (3) the small non-aggregating proteoglycans and integral membrane forms; (4) their non water-binding functions, i.e. regulation of collagen deposition, cell-surface regulation of bound enzymes and growth factors and the heparan sulphate proteoglycans and basement membrane filtration.

16. Adhesion protein: (1) is composed of several disulphide-bonded protein chains; (2) has binding affinity for cell-receptor and at least one extracellular matrix component; (3) has domain structure shared with other proteins; and (4) self-aggregation, binding to cell surface via cytoskeleton-linked receptor and presence of multiple matrix binding domains implicates the protein in matrix aggregation and structuring.

17. Basic properties of integrins are: the specificity for RGD sequence (YIGSR) laminin integrin may be discussed; the α/β-chain transmembrane subunits, and the ability to interact with the actin cytoskelton via talin. Structural and immunological homologies are outlined in the text.

18. Evidence that cytoskeleton elements and adhesion/structural proteins; (1) adhesion plaques – co-localization of cytoskeleton elements and adhesion/structural proteins; (2) surface receptors that bind to both cytoskeleton and adhesion proteins; (3) matrix organization ability of structural proteins – the 'matrisome', and (4) role of cytoskeleton. The effect of the extracellular matrix on cellular phenotype and its possible relationship to cell shape are discussed in the last section of the text.

Chapter 8

1. Hormones which breakdown phosphatidylinositol 4,5 bisphosphate yield inositol 1,4,5-triphosphate and diacylglycerol as second messengers. Release of intracellular Ca^{2+} from the endoplasmic reticulum is caused by inositol 1,4,5-trisphosphate, while diacylglycerol causes an activation of protein kinase C. Protein kinase C has to be bound to the plasmalemma to be active. Ca^{2+} binds to a protein called calmodulin which can then activate several types of protein kinase. Two examples of protein kinases are myosin light chain kinase and phosphorylase kinase.

2. C is false; the others are true.

3.

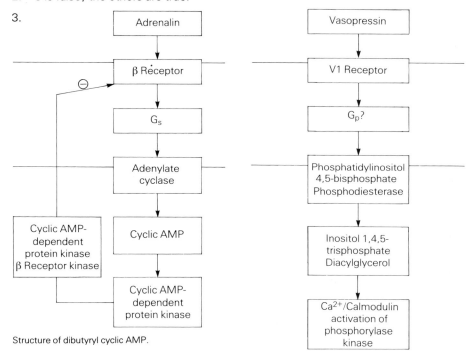

Structure of dibutyryl cyclic AMP.

4. (i) The hormone should increase the concentration of cyclic AMP in the tissue; (ii) the rise in cyclic AMP concentration should precede the biological effect of the hormone, (iii) when the hormone is removed the cyclic AMP concentration should fall back to basal with a similar time-course to the fall in the biological response.

5. Dibutyryl cyclic AMP crosses the lipid bilayer of membranes more readily than cyclic AMP because it is less polar than cyclic AMP. Dibutyryl cyclic AMP should mimic the action of the hormone. It would do this better than cyclic AMP because it would penetrate the cell membrane more easily, and it would be broken down much more slowly than cyclic AMP.

6. An inhibitor of cyclic AMP phosphodiesterase should potentiate the action of the hormone. The rise in cyclic AMP content induced by the hormone would be greater in the presence than in the absence of the phosphodiesterase inhibitor.

Name	Structure	Subtypes	Activators	Method of activation	Substrate specificity
Cyclic AMP-dependent protein kinase	R_2C_2	I and II	cyclic AMP	Dissociation of R_2	Serine/threonine—broad
Cyclic GMP-dependent protein kinase	P_2 (dimer)	?	cyclic GMP	Conformational change	Serine/threonine—broad
Protein kinase C	Single polypeptide chain	Several	Ca^{2+}, phosphatidyl serine, diacylglycerol	Conformational change, binding to cell membrane	Serine/threonine—broad

8. (a) The insulin receptor has two α and two β subunits. (b) The insulin receptor is a tyrosine kinase. It can phosphorylate tyrosine side-chains in proteins. (c) Increases in phosphorylation could be caused by activation of a serine/threonine kinase or inhibition of a serine/threonine phosphate phosphatase by insulin. Decreases in phosphorylation could be caused by inhibition of a serine/threonine kinase or activation of a serine/threonine phosphate phosphatase.

9. Oestrogen will switch on genes normally activated by glucocorticoids.

10. The mass of plant cells (callus) would differentiate and produce roots. Shoot development would be inhibited.

11. Answer should contain the following points:

Steroid hormones
Site of action	Intracellular (nucleus)
Mechanism of action	Control of mRNA and hence protein synthesis
Speed of action	Several hours

Hormones which activate adenylate cyclase
Site of action	Initially cell membrane then intracellular
Mechanism of action	Protein phosphorylation
Speed of action	Seconds to minutes

Chapter 9

1. Although all cells exhibit a resting potential, only excitable cells can elicit action potentials. Polarized membranes have an electric potential negative inside; depolarization occurs most commonly through an entry of Na^+. Depolarization of the pre-synaptic ending produces a release of neurotransmitter. The main neurotransmitter in parasympathetic neurons is acetylcholine, while the sympathetic system uses catecholamines.

All receptors are intrinsic membrane proteins. Some of them have been purified and reconstituted into artificial lipid bilayers.

Sound waves are transmitted to the endolymph that fills the inner ear. Vibrations of this fluid produce depolarization of the ciliated cell.

In retinal rod outer segments, the disc membrane contains rhodopsin, a protein possessing 11-cis-retinal as a prosthetic group.

2. A

3. D

4. B

5. D

6. D

7. (i) C; (ii) A; (iii) E; (iv) D; (v) B.

8. $-62\,mV$

9. See main text, 'The ion channels'

10. They are all quaternary ammonium salts

11. As an agonist of cholinergic systems

12. $\approx 200\,kJ$

Chapter 10

1. Muscle is a specialized tissue designed to produce movement in animals. The basic unit in skeletal and cardiac muscle is the fibre. Alternating light and dark bands give rise to striations, the distance between two Z-lines constituting one sarcomere. The contractile unit is built up out of arrays of thick and thin filaments. These are largely constructed from the proteins myosin and actin. The filaments interact via cross-bridges and use the energy source, ATP to drive the process.

The major thick filament protein myosin is made up of six subunits. The use of proteolytic enzymes to dissect the molecule leads to subfragments such as HMM and LMM. ATPase activity is located in the S-1 region of the molecule. Repeating sequences are found in the heavy chain polypeptide which adopts a coiled-coil conformation.

Thin filaments are constructed of the globular protein actin. Regulatory proteins include tropomysin and troponins. Ca^{2+} are released by the sarcoplasmic reticulum in response to the arrival of a nerve impulse. A cyclic series of interactions between thick and thin filaments, known as the Lymn–Taylor cycle, drives shortening of the sarcomere.

2. A. T F. F
   B. F G. F
   C. T H. F
   D. T I. F
   E. T J. F

3. α-Helices in heavy chain polypeptides; coiling of the two α-helices further in the coiled-coil; helical arrangements of the myosin cross-

bridges; helical arrangement of the actin monomers.

4. They are both thick filament proteins.

5. In striated muscle the myosins pack in a parallel fashion except in the bare-zone; in smooth muscle the packing is antiparallel (see Figs 10.18 and 10.20).

6. In Type I anaerobic fibres, because myoglobin provides a reservoir of oxygen for use by mitochondria.

7. False. ATP would cause S-1 to dissociate from actin (see Fig. 10.35).

8. The thick filaments of muscle are composed of mycin. The thin filaments of muscle are composed of actin. Different kinds of actin and other intermediate filaments are described in Chapter 6 (Cytoskeleton) and Chapter 7 (Extra cellular matrix).

Chapter 11

1. The small lymphocyte is the central cell in the specific immune response. These cells are found in the blood and the lymphoid tissues. There are two types of lymphoid tissues. The primary lymphoid tissue is concerned with the development of lymphocytes. Examples of such tissues are the thymus and the bursa of Fabricius . The secondary lymphoid tissue deals with the immunogens. The biggest single mass of this lymphoid tissue is the spleen .

There are two types of small lymphocyte: the B and T cells. The role of the B lymphocyte is to make antibodies . This type of immunity is known as humoral . There are at least two types of T lymphocyte. The helper T cell is central to the immune response because it produces lymphokines such as interleukin 2 , which stimulates lymphocyte proliferation, and interferon γ which is known to stimulate macrophages. The second type of T cell can develop into cytotoxic T cells which are important in immunity to viruses .

2. A. IgM
   B. IgG
   C. IgM and IgG
   D. IgG
   E. IgE

3. IL-1 induces the synthesis of acute phase proteins, is responsible for fever

and the increase of free amino acids in the plasma.

4. You would expect to see an elution profile with two peaks of approximately equal areas. Treatment of IgG with 2-mercaptoenthanol splits the disulphide bridges releasing heavy (M_r 50,000) and light M_r (25,000) chains.

5. Fluorescein-labelled antibodies are visible under ultra-violet light. Enzyme-labelled antibodies are made visible by adding the substrate of the enzyme which is converted into a product visible under a normal light microscope or following Western blotting on to a membrane. Because each enzyme molecule can convert several thousand substrate molecules into coloured product in a few minutes, enzyme-linked systems provide *amplification* of the signal.

6. The aim might be to develop monoclonal antibodies against specific proteins that appear on the surface of tumour cells but not on normal cells. Some sort of toxic molecule would be coupled to the antibody molecule, the hope being that it would seek out and destroy tumour cells. So far this has not been realised. At present, monoclonal antibodies are of rodent origin and if injected more than once the human immune system will treat them as foreign proteins and raise antibodies against them. In trials, the toxic molecules attached to the antibodies have reacted with other things before they have arrived at the proposed site of action.

7. The essay should include reference to macrophages as phagocytic cells, as antigen presenting cells and as being influenced by products of specific immunity such as antibodies and interferon γ.

Chapter 12

1. Following fertilization the egg begins a series of rapid cell divisions, a process called cleavage . Soon after division starts in Xenopus some asymmetry is seen. Cells at the vegetal pole are larger than those at the animal pole. The individual cells formed are called blastomeres and the cavity in the centre is the blastocoel . Fluid is contained within the blastocoel by the formation of tight junctions between the surrounding cells.

The hollow ball of cells formed by the cleavage divisions, the blastula , is largely

undifferentiated . During gastrulation considerable cell migrations take place, forming the three primary germ layers of the embryo.

The organs and tissues of the embryo are formed during organogenesis . Individual organs and tissues contain cells which originate from more than one germ layer. For example, a limb will contain both mesodermal and ectodermal tissues. The central nervous system develops from the neural tube , and the peripheral nervous system from the neural crest . Interaction of cells during development to form defined structures is called embryonic induction .

2. B, D

3. B, E

4. A, C, E

5. B, C, D, F

6. C, D

7. Contraction of actin filaments using ATP; assembly/disassembly of microtubules; formation/breakage of cell junctions and adhesion sites; cell–extracellular matrix interactions; guidance using chemical signals such as NGF.

Glossary

A

Actin: *is a protein and a major constituent of microfilaments.*

Adenovirus: *a type of virus, named from the Greek* aden, *gland.*

Adrenalin (epinephrine): *meaning roughly 'on top of the kidney'. Refers to the location of the adrenal glands (above each kidney – Latin* renes, *kidney) from which adrenalin was first extracted.*

Agnatha: *a class of fish-like, primitive, jawless vertebrates. It comprises the cyclostomes (lamprey and hagfish) and their extinct relatives, the ostracoderms.*

Amphipathic: *from the Greek for 'both feelings'. Molecules containing two regions or domains, one is lipophilic and one is hydrophilic, hence their peculiar physical properties.*

Appressed: *closely pressed together but not united.*

Archaebacteria: *a group of bacteria formed with the Greek root* archae, *ancient. (Bacterium is Greek for 'little staff'.) (See also* Eubacteria.)

Astrocyte: *a specialized supporting cell with branching processes, found in the central nervous system. From the Greek* astros, *star, and* kytos, *cell.*

Autophosphorylation: *ability of a protein with kinase activity to phosphorylate itself.*

Autotrophic: *a term applied to those organisms which can manufacture their own organic constituents from inorganic material. Autotrophic organisms can be phototrophic or chemotrophic. (See also* Heterotrophic.)

Axon: *the main impulse-carrying appendage of the nerve cell; derived from the Greek word for axis.*

B

Bacteriophage: *a virus which infects bacteria, therefore 'bacteria eater'; compare macrophage, phagocyte.*

Band proteins: *the proteins stained, usually in a gel, following electrophoresis.*

Basal lamina: *a network or matrix of extracellular connective material secreted by some types. It is composed of proteoglycans and other glycoproteins, reinforced by collagen fibres, and serves as an attachment site for sheets of epithelial cells.*

Basement membrane: *a sheet of extracellular matrix seen in light microscopy underlying epithelial and endothelial cells and surrounding adipocytes (fat storage cells), Schwann cells and muscle cells. Usually separates these cells from the mesenchymal or connective tissue cells.*

Blastomere, blastocoel, blastula: *structures associated with the development of an animal embryo, all derived from the Greek* blastos *meaning a sprout.*

C

Capsid: *protein coat of a virus, from the Latin* capsa, *box.*

Carcinoma: *tumour of epithelial origin (malignant).*

Catabolism: *degradative reactions, from the Greek* kata, *down, and* bolism, *to throw.*

Catecholamines: *adrenalin and noradrenalin are jointly called catecholamines, since they contain both an amino and a catechol (1,2-diphenol) group. Catecholamines are important mediators in the nervous (sympathetic) and hormonal systems.*

Cell: *basic structural unit of organisms, derived from the Latin for small room.*

Centriole: *a short hollow cylinder of microtubules. From the Greek* kentron, *centre.*

Chaperonins: *protective proteins. From the French* chape, *a hood or covering protector.*

Chemotactic factor: *a soluble factor which, at relatively high local concentrations, attracts the movement of cells towards it.*

Chimera: *a word used a lot in mythology to describe an animal made from parts of different animals.*

Chondrocyte: *cartilage-forming cell.*

Cisternae: *expanded, flattened spaces within the cytoplasm surrounded by a membrane. From the Latin* cisterna, *water reservoir.*

Clone: *a population of cells which have all arisen from one cell by repeated cell division.*

Connective tissue: *tissues which provide structural support for other tissues and organs throughout the body, i.e. cartilage, tendon, bone, blood vessel walls. Adipose tissue is also included in some definitions.*

Cytokinesis: *the division of the whole cell after nuclear division. From the Greek* kinesis, *movement, and* kytos, *cell.*

Cytotoxic: *literally 'cell-killing'. For example the production of cytotoxic T lymphocytes is characteristic of cell-mediated immunity.*

D

Dendrite: *the highly branched smaller appendages forming connections between nerve cells. Dendrite is derived from the Greek* dendron *meaning tree.*

Desmin: *an intermediate filament protein that binds microfilaments together. From the Greek* desmo, *bond ligament.*

E

Embryogenesis: *the development of the embryo from the zygote.*

Endergonic reactions: *chemical reactions which yield energy. (See also* Exergonic reactions.)

Endoplasmic reticulum: *enclosed membrane system within the cytoplasm, from the Greek* endon, *within, plasma, form, and the Latin* rete, *net (as used by gladiators).*

Endosymbiont: *symbiosis (Greek* symbios, *a companion) is when two species live together to their mutual benefit. An endosymbiont lives within its companion.*

Epimerization: *inversion of configuration about one chiral centre, i.e. conformational change from one isomer to another.*

Erythrocyte ghost: *purified erythrocyte plasma membrane obtained by breaking open the intact red cell and washing away the cytoplasm.*

Eubacteria: *formed with the Greek root* eu, *good (or true). (Bacterium is Greek for 'little staff'.) (See also* Archaebacteria.)

Eukaryotes: *organisms with a nucleus. From the Greek* eu, *good, well and* karyon, *nut. See also* prokaryotes.

Exergonic reactions: *chemical reactions which yield energy (See also* Endergonic reactions.)

F

Fluorescent chromophore: *a molecule which emits visible light when irradiated with ultraviolet. Examples include fluorescein, which emits green light, and rhodamine, which emits red light.*

G

Gastrula: *the stage in development when animal cells begin to differentiate; little stomach; from the Greek* gaster *meaning belly.*

Gelsolin: *a protein which polymerizes soluble G-actin into insoluble F-actin gel. From the Latin* gelare, *to congeal, and* solutus, *dissolved.*

H

Heterotrimer: *protein constructed by the binding together of three different polypeptide chains.*

Heterotrophic: *a term applied to those organisms which require a supply of organic material from which they can make most of their organic constituents. (See also Autotrophic.)*

Hijacked: *slang term meaning to steal shipment of goods. Said to have originated during prohibition in the USA when bootleggers fell prey to other criminals. When a hold-up took place, the command was 'Stick'em up high, Jack!'*

Histones: *basic proteins, rich in arginine and lysine, found in the nuclei of eukaryotic cells.*

Homeotic mutation: *when one part of the body acquires the characteristics of another part or somite.*

Homogenate: *of the same composition throughout. From the Greek homos, same; genos, kind. Despite the term, homogenates are not of consistent composition throughout, and contain connective tissue fibres, unbroken cells, and organelles at various stages of disruption.*

Hyaluronan: *this term is increasingly applied to hyaluronate (hyaluronic acid) to emphasize the macromolecular nature of the molecule.*

I

Imaginal: *from the Latin imago meaning a likeness which is imagined rather than real.*

Immunohistochemistry: *the study of antigens, antibodies and their interactions in tissues. From the Latin immunes, free from, and the Greek histos, tissue, and chemeia, alchemy.*

Interdigitating cells: *relatively large cells with multiple processes found in the paracortex of the lymph node. They act as antigen-presenting cells. The large surface area enables a single interdigitating cell to interact with several lymphocytes simultaneously.*

L

Lamina densa: *microscopically dense region of the glomerular basement membrane. (See also Lamina rara.)*

Lamina rara: *microscopically light region of the glomerular basement membrane close to epithelial layer. (See also Lamina densa.)*

Low-density lipoprotein: *a complex of about 1500 esterified cholesterol molecules encapsulated by a lipid bilayer containing several copies of a single large protein.*

Lumen: *cavity or tubular part of an organ or vessel.*

Lymph: *the fluid, derived from the blood plasma, which drains from the tissues into lymphatic vessels, through lymph nodes, and eventually re-enters the blood.*

Lyse: *to break open a cell, releasing its contents. Originally derived from haemolysis: when red blood cells are placed in water (lower osmotic pressure than cell contents) they burst.*

Lysis: *literally means 'breaking open'.*

M

Mesenchymal: *relating to connective tissue, composed of stellate cells embedded, with fibres, in a gelatinous ground substance. From the Greek mesos, middle, and egchyma, infusion.*

Metachronal rhythm: *the pattern set up by rows of beating cilia. From the Greek meta, change of position, and chronos, time.*

Metamorphosis: *a change of shape or form during development. For example, caterpillar to butterfly, tadpole to frog. (See also Morphogenesis.)*

Mitochondrion: *organelle involved in electron transport and oxidative phosphorylation. Introduced by Benda in 1898, this term originates from the Greek for 'thread-like' granules and was used to describe their appearance during spermatogenesis.*

Morphogenesis: *derived from the Greek morphos meaning shape. It is the process which defines the shape or form of the embryo. A morphogen is a substance involved in defining the form of the embryo. (See also Metamorphosis.)*

Mosaic: *a group of plant diseases, of viral origin, characterized by mottling of the foliage.*

Mutation: *a change in the DNA. This can be as simple as a single base change, loss or addition, or as complicated as the loss of a large section of DNA or even a whole chromosome. A mutation is usually recognized by the resulting change in the characteristics of a cell or organism. In fact a change may have no effect at all (silent mutation) at one extreme or may be fatal at the other.*

Myopathy: *abnormality in biological function.*

Myotube: *the primitive muscle-forming cells (myoblasts) elongate to become multinucleated myocytes. The elongated tubes are termed myotubes.*

N

Negative feedback: *the ability of a product of a particular stimulus to inhibit its own further formation.*

Neonatal thymectomy: *the removal of the thymus within a few days of birth.*

Nuclear envelope: *consists of the complex of two nuclear membranes, the perinuclear space and the nuclear pore complexes.*

Nucleoplasm: *the ground substance of the nucleus. The term is derived from Latin and Greek words meaning a nut and form.*

O

Organogenesis: *the development of tissues and organs.*

Osteosarcoma: *tumour of bone cells.*

P

Paramagnetism: *a property of unpaired electrons, whereby they oscillate between low- and high-energy states when placed in a magnetic field.*

Parietal endoderm cells: *primitive epithelial cells in the embryonic endoderm (inner layer).*

Parthenogenetic: *by virgin birth. It is derived from the Greek words parthenos (virgin) and genesis.*

Pathogen: *an organism which causes disease.*

Pathogenesis: *from the Greek for 'the origin' of a disease.*

Potentiation: *the response to the two agents together is greater than the sum of the responses to the two agents acting independently.*

Prion: *protein particles which are associated with a number of diseases of the central nervous system; from proteinaceous infectious material.*

Prokaryotes: *organisms without a nucleus. From the Greek pro, before and karyon, nut. (See also Eukaryotes.)*

Proteases: *enzymes that break proteins down into peptides and amino acids by catalysing the cleavage of peptide bonds.*

Proteoglycans: *glycoproteins found in the extracellular matrix which are rich in carbohydrate, the amount of carbohydrate often exceeding 90–95%.*

Protoplasm: *the material of the cell. It is composed of nucleoplasm, the material of nucleus; and cytoplasm, the rest of the cell. From the Greek protos, first; kytos, hollow vessel; plasma, form; and the Latin nucis, nut.*

Protoctista: *from the Greek protos, first; ktistos, to establish. The term protista has been used since the nineteenth century to denote single-celled organisms; however, the kingdom Protoctista includes some multicellular types, such as certain algae.*

Pupation: *a stage of development, between the larval and imago stages.*

R

Reversions: *a reversion is a back mutation.*

S

Sarco-: *derived from the Greek sarkos meaning flesh. Sarcoplasm, sarcolemma, sarcoplasmic reticulum and sarcosol are all applied to muscle.*

Somites: *refers to the body of the organism (from the Greek soma), excluding the germ cells.*

Stem cells: *continually dividing cells, such as those in the bone marrow. Bone marrow stem cells are said to be pluripotent, because their progeny can give rise to any of the cells in the blood.*

Synapse: *the narrow gap separating two excitable cells. From the Greek synapsis meaning conjunction or union.*

Syncytium: *a multinucleated cell. The term is derived from a Greek word meaning together.*

Synergism: *when two or more components act together to produce an effect greater than the sum of the individual effects. From the Greek synergetikos, to co-operate.*

T

Talin: *a link protein between integrins and the actin cytoskeleton. Present in adhesion plaques.*

Thylakoid: *portion of internal structure of a chloroplast. From the Greek thulakos, empty pouch.*

Tonofilaments: *a class of cytoskeletal filaments also known as intermediate filaments or stress fibres. These filaments, with a diameter of 8–10 nm, are built from various non-motile structural proteins, among them keratins, vimentin and desmin.*

Topography: *the detailed description of a particular area.*

Transducer: *element which carries a signal from one system to another.*

Transduction: *transfer of energy or a signal (information) from one system to another.*

Transferrin: *a plasma glycoprotein of M_r 80 000 which binds iron (Fe^{3+}). There are two binding sites for Fe^{3+} per molecule of transferrin. Transferrin is used to transport iron in the blood because free Fe^{3+} is toxic and in any case precipitates as $Fe(OH)_3$.*

Transformation: *the genetic modification of a bacterial cell as a result of the entry of extraneous DNA into that cell. It occurs naturally and is used experimentally in genetic engineering. The word transformation is also used to describe the changes that occur in cultured cells after treatment with tumour viruses or treatment with carcinogens.*

V

Vacuole: *a large, membrane-limited, fluid-filled cavity within the cell; from the Latin vacuus, empty space. Also called the tonoplast; from the Greek tonos, stretch, and plastos, formed.*

Very late antigens: *group of adhesion receptors found on stimulated lymphocytes that appear at a late period of stimulation.*

Vesicle: *any small, fluid-filled, membrane-limited space within the cytoplasm. From the Latin vescula, meaning a small bladder.*

Vimentin: *an intermediate filament protein. From the Latin vimineus, made of strong pliable strands.*

Virus: *nucleoprotein particles which are all obligate parasites. From the Latin for 'slimy liquid, poison'.*

X

Xenobiotic: *foreign to a living organism, usually applied to compounds such as drugs and pesticides.*

Z

Zygote: *cell resulting from the fusion of oocyte and sperm which can develop to form an organism. From the Greek zygon, a yoke.*

Index

Page references to Tables are in *italic* and those to Figures are in **bold**. References to Boxes and Side-notes are indicated by (B) and (S) after the page numbers respectively.

Ion channels 238–42
Ionophores 110–11 (B)
Isocitrate dehydrogenase 221
Isotropic 260

J (jointing) chain 303
J (jointing) genes 318

Karyotyping 76, 77 (B)
Keratan sulphate 184 (B)
Kinetochore 159
Kinetosomes 35

Lactate dehydrogenase 277
Lamina densa 195
Lamina rara 194
Laminin 172, 188, 193, 194–5, 197 (B)
Larva 337
Laser microsurgery 343
Lateral meristems 330
Lattice theory 108
Leiotonin 273
Leucoplasts 16
Leukocyte adhesion deficiency 198
Leukocytes 292–3
Light meromyosin (LMM) 146 (B), 265–6
Light microscopy 2–5
Light reactions 120, 127
Limb buds 328
Link protein 181
Lipid A 50
Lipids, membrane 92–3
Lipoprotein, murein 51
Liposomes 93
Lithium and mental illness 220 (B)
Local chemical mediators 207
Low-density lipoprotein 105–6
Low-density lipoprotein receptors 106 (B)
Lumen 129
Luxury proteins 334, 335 (B)
Lymn–Taylor scheme 278–9
Lymph 307
Lymph nodes 308–9
Lymphoid tissues 304
    primary 304
    secondary 305, 307–11
Lymphokine cascade 312–14
Lymphokines 306
Lysosomal enzymes 26–9
Lysosomes 11, 29 (B), 26–30
Lysozyme 288, 289
Lysyl oxidase 177

M-line 260, 269
M-protein 269
Macrophages 291
Magnetosomes 57 (B)
Major histocompatibility complex (MHC) 312, 314–16
Malleus 250
Mannose 6-phosphate residue 28
Mast cell 293
MAT genes 340
Maternal-effect mutations 339
Matrisome 172
Matrix 13, 121, 127, 180 (B), 185
Megakaryocytes 305
Membrane
    asymmetry 99–100
    basement 175
    cell 10, 12–13, 58, 91
    chemical components 92–4
    chloroplast 128
    compartments 12–33
    components, organization of 94–100
    as a dynamic entity 104–7
    fluidity 98–9
    freeze-fracture 94, 95
    functions 92
    inner mitochondrial 125–6
    outer mitochondrial 51, 125
    plasma 44–5
    post-synaptic 240
    viral 63–4
Membrane attack complex 305
Membrane lipids 92–3
Membrane proteins 22, 94, 96 (B), 98
Membrane receptors 197–201, 326
Membrane transport 107–15
    facilitated and active transport 109–14
    mechanisms 114–15
    passive diffusion 107–9
Memory cells 310
Meristems 330
Mesenchymal tissues 155
Mesoderm 326
Mesosomes 45
Metachronal rhythm 159, 162
Metamorphosis 337
N-gamma-methylhistidine 144
Microbodies 12, 32 (B), 30–3
Microfibrils 174 (B)
Microfilament accessory proteins 147

Microfilaments 33, 139, 143–51
Microscopy 2–8
    electron microscopy 5–7
    light microscopy, modern 2–5
        conventional 3–4
        dark-field 4
        fluorescence 4–5
        phase-contrast 4
Microsurgery 347 (B)
Microtubule organizing centres (MTOCs) 34, 158, 159
Microtubule-associated proteins (MAPs) 158
Microtubules 34, 139, 156–64
Middle lamella 174 (B)
Miniband 85
Mitochondria 9, 12–13, 120–7, 260
    biogenesis of 133–6
    development in meristematic cells 134 (B)
    energy-transduction in 118, 119–20
    evolutionary origin of 136–7
    isolation and purification 123–4
    ultrastructure 121–3
Mitochondrial enzymes, compartmentation 124–7
Mitochondrial membrane transporters 125–6
Mitochondrial myopathies 123 (B)
Mitoplasts 124
Mitosis 84, 162 (B), 163
Mitotic chromosomes 85
Mitotic division 87
Mitotic spindle 158, 159
Mollicutes 47
Monoamine oxidase 246, 247
Monoclonal antibodies 311 (B)
Monocytes 291–2
Monogalactosyldiacylglycerol 128, 129
Mononuclear phagocytic system (MPS) 291
Morphogen 326, 341
Morphogenesis 326
Mosaic disease in plants 74 (B)
Motor end plate 280
mRNA 334, 335
mRNA–cDNA hybrids 335
Mucosa-associated lymphoid tissue (MALT) 305
Multicellular organisms 1
Murein 49
Murein lipoprotein 51
Mus musculus 332

Muscarinic receptors 247
Muscle
    contraction 257, 275–80, 280–3
    fibres 258–62
    growth 261 (B)
    heat production 276 (B)
    hormonal effects on 282 (B)
    structural proteins 264–75
    structure, maintenance of 274–5
    transformation to meat 279 (B)
Muscle genes 267 (B)
Muscular dystrophy 274 (B)
Mutations 75
Mycoplasms 47
Myelin sheath 345
Myeloperoxidase 291
Myoblasts 262
Myosin 139, 141, 261, 270
    cleavage sites 147
    filaments 268–9
    heavy chains 264, 265–6
    light chains 264, 265–6
    in non-muscle cells 145–6
    substructure 265
    type 267 (B)
Myotubes 195

Na$^+$-K$^+$ATPase 111–12
Natural killer cells (NKC) 290, 293
Nebulin 270, 271
Negative-staining 121
Neonatal thymectomy 306
Nerve growth factor (NGF) 346
Nerve impulse 234
Nerve-muscle interface, disorders of 281 (B)
Nerves 234
Nervous system, cell differentiation and development 344–8
Neural crest 328
Neural tube 327
Neurofilament proteins 139, 154
Neuron 345–6
Neuronal cell adhesion molecule (N-CAM) 346
Neurosecretion 347
Neurotransmitters 207, 242, 244–8
Neutrophils 292–3
Nicotinic acetylcholine receptors 247, 248–50
Nidogen 172
Nodule 330